Christmas 1998

For David –

As this book demonstrates, Confucian filial piety is alive and well in China!

Filially yours,

Richard

DENG XIAOPING

DENG XIAOPING: MY FATHER

DENG MAOMAO

BasicBooks
A Division of HarperCollins*Publishers*

Published by BasicBooks, A Division of HarperCollins Publishers, Inc.

Translated from the Chinese first edition, published in 1993 by Zhong Yang Wen Xiang Publishing House, Beijing. The translation was done by Lin Xiangming, Lu Min, Li Tongnian, Ma Hairong, Jin Shaoqing, Wang Yuling, Zhang Zhijun, Chen Mingming, Chen Naiching, and Cong Jun.

Designed by Joan Greenfield

LIBRARY OF CONGRESS CATALOGING-IN-PUBLICATION DATA

Mao-mao, 1950–
[Wo ti fu ch' in Teng Hsiao-p'ing. English]
Deng Xiaoping: my father /Deng Maomao.
p. cm.
Includes index.
ISBN 0–465–01625–1
1. Teng, Hsiao-p'ing, 1904– 2. Heads of state—China—Biography.
I. Title.
DS778.T39M3613 1995
951.05'8'092—dc20 94–37429
[B] CIP

95 96 97 98 ◆/RRD 9 8 7 6 5 4 3 2 1

THIS BOOK IS DEDICATED TO MY FATHER

Father and his comrades-in-arms are of a generation closely linked to the destiny of the whole century. They have written and created history and devoted their whole lives to the motherland and the people.

THIS BOOK IS ALSO WRITTEN FOR OUR CHILDREN

As the younger generation, you love your grandfather's generation. I hope you will know and understand your grandfather's generation through this book. I hope that you, like them, can make brilliant achievements for the Chinese nation.

CONTENTS

Written from the perspective of a daughter about a father she obviously loves, but, more important, admires and respects, this volume presents information that will be mined and evaluated by historians and biographers for years to come. It tells an important part of the story of modern China and one of its most dynamic figures, chronicling the first forty-five years in the life of Deng Xiaoping, now ninety. The anticipated second volume will cover Deng's second forty-five years, encompassing events that are more familiar to Westerners: the disasters of the Great Leap Forward and the Cultural Revolution; the launching of economic reform and the opening policy in the late 1970s; June 4, 1989; and China's rapid emergence as a rising world power.

Deng is the contemporary world's last major leader to have lived through both world wars, played a major leadership role in a protracted national revolution, come near the summit of power throughout much of the Cold War, and survived to guide his nation into the post–Cold War era. This generation in general, and Deng Xiaoping in particular, had a capacity to see broad vistas while overseeing the thousand and one small things that lead to the transformation of national societies and the world system.

In telling the story of her father, Mao Mao (born Deng Rong and known to many in China and abroad as Xiao Rong) provides additional perspective on the development of modern China itself: its revolution, the forces tearing at its fabric and impelling its people forward in the twentieth century, and the terrible price that has been paid by both individuals and society for the change that has occurred. Through the lens of this volume we see how China's Long March generation, and to some extent that generation's successors, view the external world.

This book speaks to the author's countrymen, particularly her generation that has known China only under Communism. She is asking them not to forget what her father and his generation have contributed to China. In the process of pointing to that generation's struggle, Mao Mao seems implicitly to argue that today's youth in China ought not surrender to materialism, to a life devoid of the aspiration to serve the nation.

Of no small interest is the fact that Mao Mao provides glimpses of life in the family of China's supreme leader, revealing emotions of love, sorrow, resentment, and uncertainty that one rarely finds in contemporary historical accounts of PRC leaders. Discussions of the day of Deng's 1989 retirement, his first and second wives, and the background of his current spouse (Mao Mao's mother), Zhuo Lin, are particularly revealing. In addition to being able to relate vignettes from daily life in the Deng household, Mao Mao has had extensive access to archives and interview sources in the People's Republic and has traveled within and outside the country to acquire additional documentary and anecdotal information.

This volume provides a wealth of new information that will help successive generations of historians and biographers address some of the key questions about one of the giant figures of the twentieth century: What are the chief goals and aspirations that Deng Xiaoping brought to his exercise of power, and how did he acquire these commitments? How did Deng rise to power, and what were the tactics he so deftly employed? How much of his rise is attributable to skill, and where did luck intercede? And how has that rise to power shaped his political style and the way in which he subsequently exercised power?

Almost every chapter in this book suggests that Deng Xiaoping has had a simple agenda for China since at least his work-study days in France during the early 1920s: to contribute to making it a strong, wealthy, and unified country. My reading of this volume suggests that Deng was less committed to Communism as a sacred ideology than to the idea of an organized political party that held out the prospect of achieving wealth, power, and the capacity for self-determination for the Chinese nation. Deng was (and, I believe, is) more a practical Leninist and national patriot than an ideologically committed utopian Marxist. His experiences with warlords in his native province of Sichuan (and elsewhere in China) and his struggles with colonial and semicolonial occupiers were the anvil on which his character and goals were forged in the first half of this century. In this book the force of Chinese nationalism is evident in its full power and depth.

In terms of Deng Xiaoping's developing leadership style, this volume describes the role Zhou Enlai played from their earliest days together. From his first organizing among fellow work-study students in France, Deng worked with Zhou and manifested a style of listening and delegating; a practical approach to addressing specific problems; a capacity to identify methodically the steps necessary for the completion of a large task; the ability to give simple, clear orders; and the willingness to follow orders when required. Deng learned the skills of administration and management from Zhou Enlai, whom he characterized as something akin to an elder brother.

From Mao Zedong (and his longtime military colleague, Marshal Liu Bocheng) in the 1930s and 1940s, Deng acquired and honed his ability to think strategically. Deng Xiaoping's leadership style, therefore, represents a combination of Zhou's administrative sense and knowledge of the West, Mao Zedong's strategic breadth, and an instinct for survival and swiftly administered discipline that his mentors displayed in great abundance.

This volume also points to perhaps the key lesson that Deng learned before 1949, and which was reinforced by his experiences after 1949. That lesson was that leftist, ideologically driven attempts to achieve rapid progress lead to disaster. Almost every debacle in which the rising young Deng Xiaoping was involved—from ill-advised uprisings in the late 1920s and 1930s to the Great Leap Forward—had its origin in the excessive zeal of the political left, often in league with misguided foreign military and political advisers. Indeed, Deng is clear that he was mistaken when he acquiesced or played a leadership role in such endeavors. By the time Deng Xiaoping had assumed supreme power in the late 1970s, he was a leader who had had his fill of leftist dogmatism and foreign advice devoid of grounding in China's reality. He was interested in positive, tangible results. He had all along wanted China to become strong and rich, and when he assumed power in the late 1970s he found himself at the helm of a nation that was still poor, weak, and isolated to a considerable extent.

Beyond what this book reveals about Deng's early experiences and how they shaped his subsequent policy and leadership behavior, considerable new detail is provided about his professional associations and activities, family life, specific activities and battles in the war against Japan and in the civil war periods of the late 1920s, 1930s, and late 1940s, and material that enriches our understanding of Communist party history. Of particular interest is the account of the murky events relating to the military setbacks of the 7th Corps in 1931 and Deng's "consign[ment] to limbo" by what he considered to be the Soviet-influenced Central Committee of the Chinese Communist Party in Shanghai.

For instance, one finds accounts of Deng Xiaoping's early childhood that portray a basically stable, supportive family and a father who encouraged his son's aspirations. This situation contrasts with the explosive tensions that existed between Mao Zedong and his father. One cannot help but wonder whether their later divergent styles and behavior can, in part, be accounted for by these contrasting childhood experiences. Whereas Mao emphasized struggle and contradiction, Deng placed greater emphasis on consensus and organizational solidarity; Mao seemed to dwell on past injustices and problems, while Deng seems to look to the future and not linger on past missteps and difficulties.

Finally, by sheer weight of detail, this volume reveals why Deng has proved so durable in, and central to, Chinese politics. The tentacles of his influence and connections extend into each of the three indispensable, traditional bases of political power in the PRC: the Party apparatus, the military, and the geographic regions of China's vast hinterland.

Biography is often also autobiography. In the process of providing a wealth of new material about her father, Mao Mao simultaneously provides insight into herself and the small group of children and their senior-cadre parents with whom she grew up in the high-walled surroundings of the leadership compound of Zhongnanhai in Beijing. She is therefore able to present a perspective on the leaders and personalities of this small community that few others could provide. At the same time, she has spent time in America and has traveled extensively; she knows many of the concerns and questions of interest abroad.

Here, then, is the first half of the story of an exceedingly intelligent, pragmatic, and tough individual who has sought to make China strong and unified, who by the late 1970s had gained the power to move China in that direction, and who by the early 1990s had presided over profound change in his nation. This volume recounts how he developed from a young adolescent in Sichuan to a politically mature revolutionary leader by 1949. It is fitting to close this foreword with the words of Deng Xiaoping himself, in 1981:

> I am a son of the Chinese people. I deeply love my motherland and her people. We Chinese have created a glorious and ancient civilization. But we have also undergone manifold sufferings and hardships and waged unremitting struggles, paying dearly for them.

David M. Lampton
President, National Committee on
United States–China Relations
New York City, September 1994

PREFACE

Late at night on January 25, 1950, I was born in the city of Chongqing, Sichuan Province, in southwestern China. My first response to the world was a soft cry, followed by a sound sleep. Little did I know who I was or that my country had just undergone an earthshakingly historic revolution. After losing eight million troops, the Kuomintang (KMT) had been overthrown, and the new People's Republic of China had been established on the vast Chinese land.

The Chinese revolution was a proud, soul-stirring movement for the Chinese people. By contrast, my birth was fairly insignificant. My mother's first glance at this little newborn with thin, brownish black hair inspired her to nickname me "Mao Mao"—literally, "hairy," a loving reference to my fuzziness. I was the fourth child born to the family. I have two elder sisters and one elder brother. One and a half years after my birth, my younger brother was born. Our big family then consisted of my parents, their five children, and my grandmother, who is my father's stepmother and comes from his hometown of Guang'an in the countryside of Sichuan Province.

The spring had not yet come. There was a damp chill in the night air. As a newborn baby, I knew nothing about the extraordinary life journey on which I had embarked. It would turn out to be filled with great extremes of happiness and frustration.

I was born and grew up in a special environment. I had firsthand experience of history in the making. Many historical persons surrounded my life, and many historic events unfolded around my family and me. My relatively short life has borne a rich abundance of stories and experiences. The more I have come to know, the more deeply I have felt driven to record these events and experiences and my environment. What I know may be limited and shallow, but what I have to share should not be lost or forgotten.

Most especially I mean what I know of my father. He was originally called Deng Xixian or Deng Bing. Later he changed his name to Deng Xiaoping. At the age of sixteen, he left China and crossed the wide oceans to travel to the West in search of a way to realize his ideals.

By eighteen, he was an ardent Communist devoted to saving his country and his people. In his sixty-year revolutionary career, he did underground work and served as a military commander, an important government official, and a key leader of the Communist Party of China. One section of the lengthy annals of Chinese history will be engraved with his name.

He has said that he cannot write an autobiography, nor does he want a biography written about him. As his daughter, however, I would be ashamed to face history if I did not recount what I know. If I achieve nothing else in my lifetime but this long-cherished aspiration, I will have a very special sense of satisfaction.

In this book, I have written about the first half of the life of a man who represents a generation of heroic, larger-than-life leaders. What I can recount covers only a short period of time in history, but that period is an important link in the magnificent chain of Chinese history, which has lasted for thousands of years.

Although what I have recorded concerns only the past, I believe that we can learn lessons by reflecting on the past and by confidently approaching the future as our forefathers did. If my book leaves some impression on future generations, I will be satisfied.

To My English-Speaking Readers:

China is an ancient civilization with a culture stretching back over 5,000 years. Our most recent 150-year history has been fraught with hardships. Amid this torrent of history a number of names are familiar to you: Mao Zedong, Zhou Enlai, and Chiang Kai-shek. Today, many more people also think of Deng Xiaoping.

I am the daughter of Deng Xiaoping. In writing about my father I do not intend to give a complete appraisal. Rather, I want to tell you the story of Deng Xiaoping and the period of Chinese history that coincides with his life.

Deng Xiaoping has traveled a ninety-year course. His life has been both difficult and extraordinary. I hope my book will enable you to better understand this man and this faraway Oriental land. And perhaps it will no longer seem as strange to you.

I give you my best wishes and hope you will enjoy this book of mine.

MAPS

ROUTE OF MARCH OF THE 7TH CORPS OF THE RED ARMY TOWARD THE CENTRAL REVOLUTIONARY BASE AREA

(October 1930–July 1931)

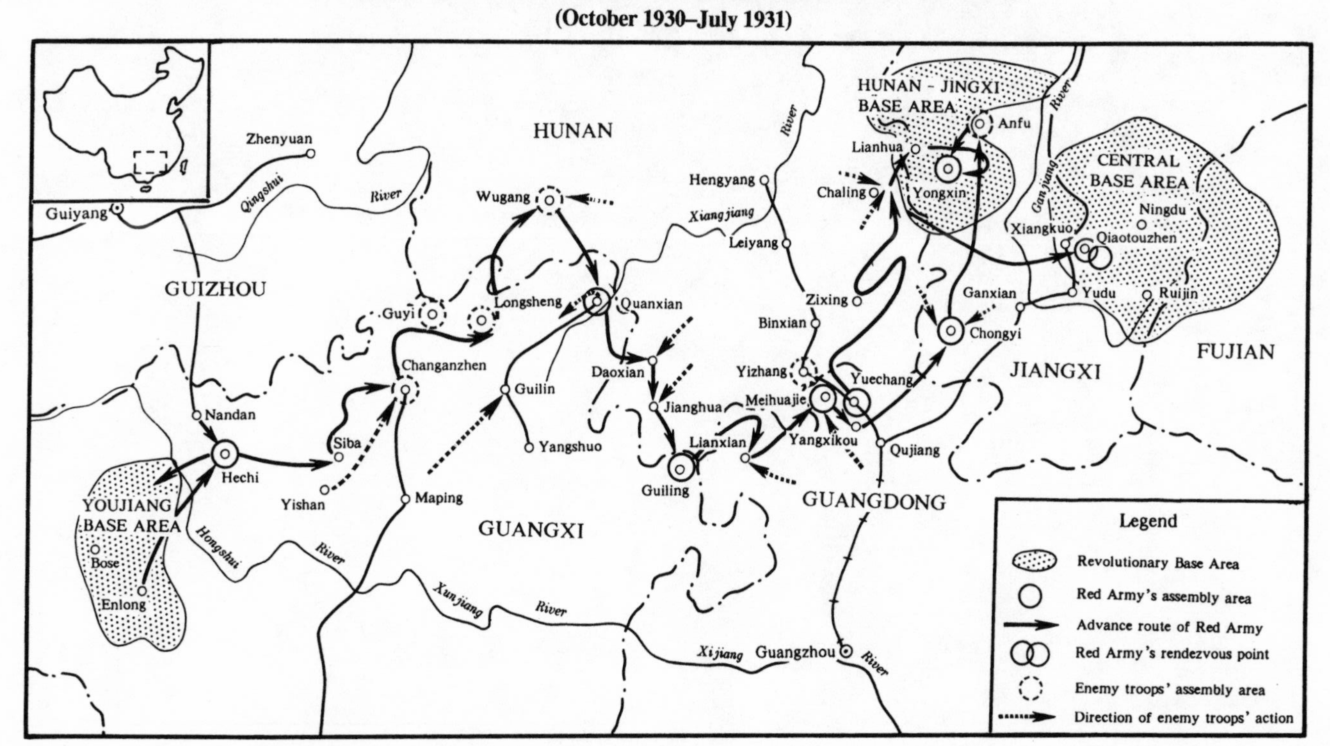

ROUTE OF LONG MARCH OF THE FIRST FRONT ARMY OF THE CHINESE WORKERS' AND PEASANTS' RED ARMY

(October 1934–October 1936)

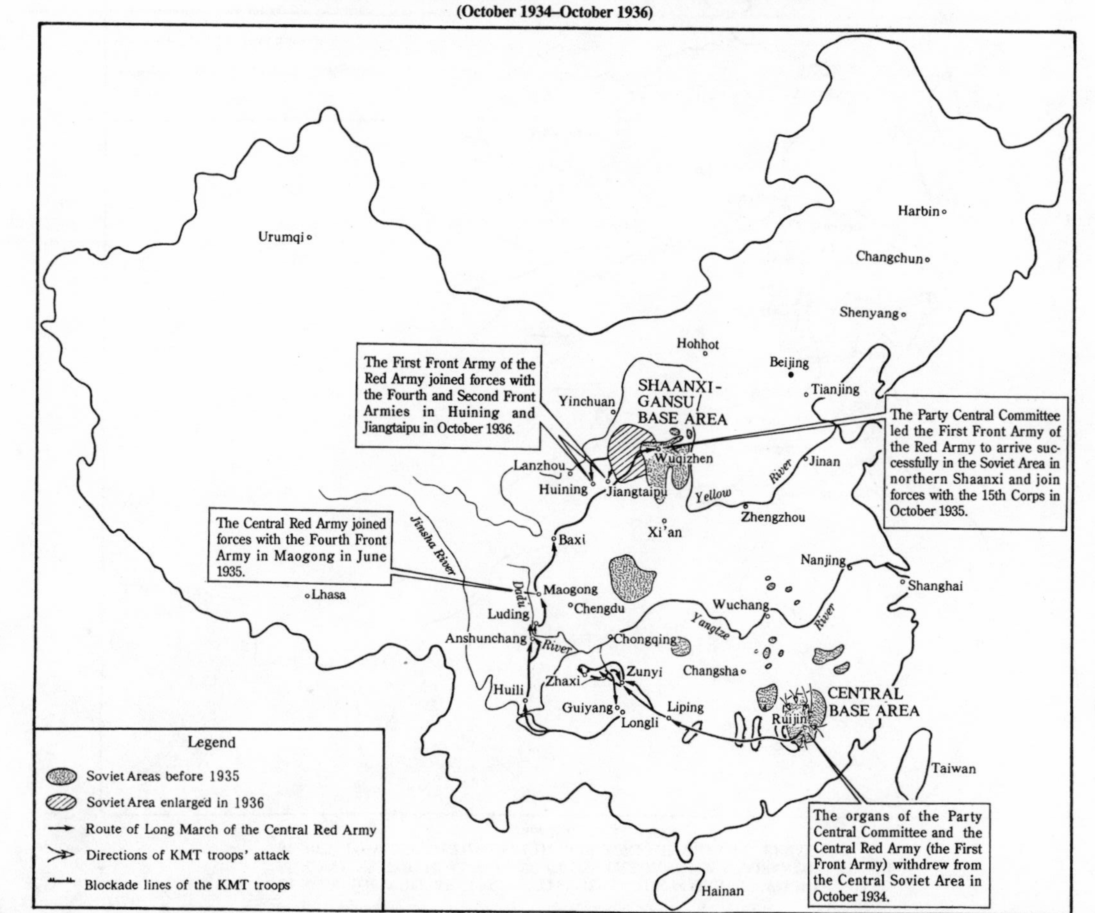

DENG XIAOPING'S MILITARY ACTIVITIES IN THE MARCH TOWARD THE DABIE MOUNTAINS, THE HUAI-HAI AND CROSS-THE-YANGTZE CAMPAIGNS, AND THE MARCH TOWARD SOUTHWEST CHINA DURING THE WAR OF LIBERATION

(June 1947–October 1950)

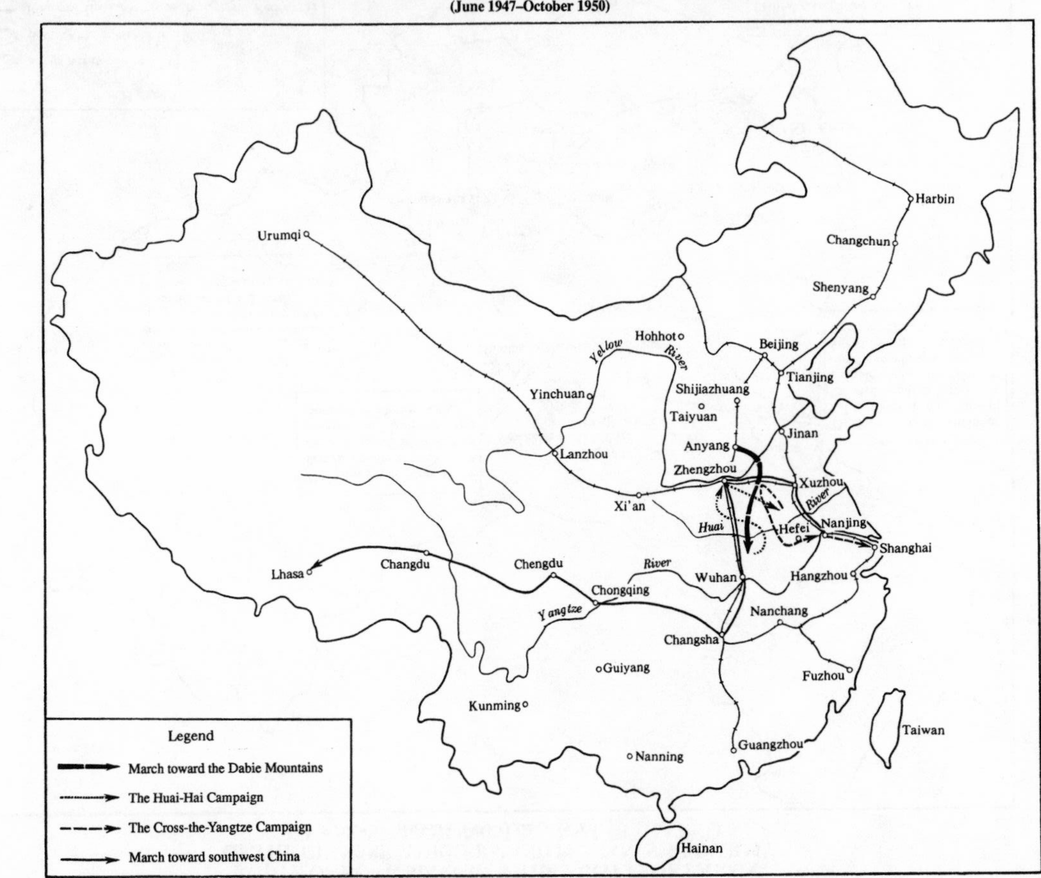

MILITARY GEOGRAPHY OF THE SHANXI-HEBEI-SHANDONG-HENAN LIBERATED AREA

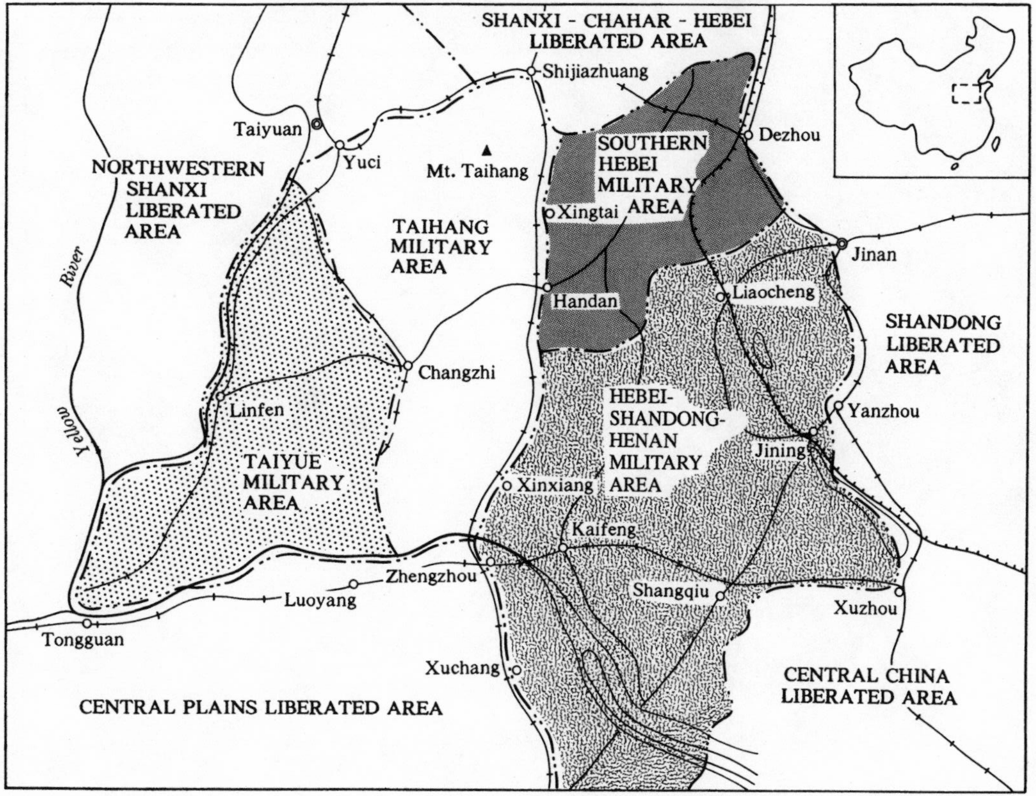

DENG XIAOPING

CHAPTER 1

THE DAY OF RETIREMENT

It was November 9, 1989.

Early in the morning it was still dark. Drizzle dampened the late autumn land of Beijing.

Father rose and had breakfast at his usual hour. Then, as was his custom, he sat down to read newspapers, books, and documents. His youngest grandson had stayed home from kindergarten with a cold. I took him to see Grandpa.

"Is it still raining?" Father asked me. "It's turning to snow," I replied.

Father stood up abruptly, opened the window, and walked outdoors.

The air was cold and damp. As a light north wind blew, wet snowflakes, mingled with icy raindrops, drifted down. Watching the sleet fall, Father said with feeling, "Beijing needs the snow badly. This wet weather is good news."

Perhaps it was because of the "greenhouse effect" that both fall and winter came so late that year. Although it was already November in Beijing, the weather was mild. The snow that day was not very heavy, but it was the first snow of the year.

At around nine o'clock, the director of Father's office, Wang Ruilin, came in and gave him an update on the Central Committee plenary session that was being held. One of the issues addressed by the session was the issue of Father's retirement. Mr. Wang told Father that after the participants read some relevant documents and held discussions, they had come to know his determination to retire and the significance of his retirement. There had been emotional speeches and a considerable amount of debate. A vote was scheduled for that afternoon, with the news to be announced that evening. Father was very pleased with Wang Ruilin's briefing, saying, "We are finally close to home."

It was time for lunch. The whole family sat together around the two round dining tables. Discussions were focused on Father's retirement. One of my elder sisters proposed a family celebration. My elder brother offered a bottle of wine of a famous brand. Mother said that if she felt up to it, she'd like to go and watch the participants take a picture with Father that afternoon.

Father expressed his desire to lead the simple life of a common man after retirement, to walk freely in the street and pay visits here and there. The eldest granddaughter, Mian Mian, said with a smile, "Grandpa is really an idealist."

At about three o'clock that afternoon, the Fifth Plenary Session of the Thirteenth Central Committee of the CPC (Communist Party of China) voted to accept Father's resignation from the post of chairman of the Military Commission of the Central Committee.

At about four o'clock, Father drove to the Great Hall of the People to have a photo taken with the participants in the plenary session just concluded.

As Father entered the reception area, senior comrades of the Central Committee went out of the meeting hall. One comrade after another approached him to shake hands. Jiang Zemin, the newly elected chairman of the Military Commission, stepped forward and grasped Father's hand. Jiang proposed taking a group photo of Party leaders with Father. As Father posed with Jiang Zemin, Yang Shangkun, Li Peng, Yao Yilin, Qiao Shi, Song Ping, Li Ruihuan, Wang Zhen, Bo Yibo, Wan Li, Song Renqiong, and Hu Qiaomu, journalists rushed forward with cameras flashing to record the historic moment.

The group comprised China's Party and state leaders. Some were aging, with white hair. Others were younger, with hair still black. Together they stood, side by side.

When Father and the group entered the banquet hall of the Great Hall of the People, there was a roar of applause. Father walked past the members of the Central Commission for Discipline Inspection, the Central Advisory Commission, and finally, the Central Committee.

Standing at the microphone, Father said with a broad smile, "I wish to express my thanks for your understanding and support. The plenary session has accepted my request to retire. My sincere thanks to the plenary session and to all comrades." Then a photo was taken of Father together with all participants and observers of the plenary session. As Father was about to leave the Great Hall of the People, Jiang Zemin went with him to the door. With a fervent handshake, Jiang pledged to dedicate his life to the cause.

Although night had fallen, our house was brightly lit. The whole family had been busy that afternoon preparing for the celebration of Father's retirement. When it was time for dinner, four grandchildren ran to invite Grandpa to the table. They gave him a congratulatory card they had made for him. On the card were four beautiful butterflies, representing the four grandchildren. The card's message was: "Wish Grandpa to be as young as we are, for ever and ever." One by one they gave Grandpa a big kiss. The youngest, three-year-old Xiao Di, planted a very wet kiss on Grandpa's cheek, leaving quite a trail of saliva. All family members laughed heartily.

In the dining room, our family's cook, Yang, who had worked in our house for thirty years, prepared a sumptuous feast. On the sky-blue wall was hung a bright red poster: "1922—1989—Eternity." Looking at the poster, Father smiled.

As the dozen family members happily gathered around Father and raised their glasses of red wine in a joyful toast, my heart was filled with deep feelings. A lifetime of over eighty years and a revolutionary career of more than sixty years would certainly be full of hardships for anyone. It is time to rest and relax!

Father had been ready to retire for many years. When he was reinstated in his position and took charge of the work of the Party and the state, he began to look for successors. In the early 1980s, overriding different views, he took the lead in resigning from some of his posts.

We supported his retirement because we wanted him to be in good health and enjoy a long life. He insisted on his retirement, taking into consideration the future of the state and the interests of the Party. At last, his wish had come true. This brought him great peace of mind and made his family happy. On the following day, November 10, 1989, the *People's Daily* published Father's letter requesting that his retirement be approved, as well as the resolution adopted at the Fifth Plenary Session of the Thirteenth Central Committee of the CPC.

In his letter to the Central Committee, Father wrote:

> As early as 1980, I proposed reforming the leadership system of the Party and the state and abolishing the lifelong tenure of key leadership positions. In recent years, many veteran comrades have quit their leadership positions in the Central Committee. In 1987, prior to the Thirteenth National Congress, I expressed my desire to retire so as to sincerely demonstrate my resolve to abolish the lifelong tenure of leading posts. At that time, after repeated deliberations regarding my opinion as well as those from within the Party, the Central Committee agreed to my resignation from the membership of the Political Bureau and its Standing Committee, from the Central Committee, and from the chairmanship and membership of the Central Advisory Commission. The Central Committee decided that I would continue to serve as chairman of both the Party and state military commissions. Since then, when consulted on major issues by the collective leadership of the Central Committee, I have respected and supported the views of the majority within that body. I have, however, insisted on assuming no responsibility for day-to-day matters, and I have been looking forward to replacing the older generation with the younger generation as soon as possible so that I can realize my wish to resign completely from leadership positions.
>
> The Fourth Plenary Session of the Thirteenth Central Committee elected a leading nucleus headed by Comrade Jiang Zemin, which is now working effectively. After careful consideration, I wish to resign from my current post while I am still in good health so as to realize my long-held aspiration. This will be conducive to the interests of the Party, the state, and the armed forces. I sincerely hope the Central Committee will approve my request.
>
> As a veteran Party member and a senior citizen who has fought for decades for the cause of communism, national independence, unity, construction, and reform, my life belongs to the Party and the state. After I retire, I shall continue to be loyal to the cause of the Party and the state. The achievements our Party, state, and armed forces have made are the result of the efforts made by several generations. Our cause of reform and opening to the outside world has begun quite recently. The tasks before us are arduous, and the road ahead will be tortuous. But I am convinced that we will be able to overcome difficulties and carry forward from generation to generation the cause our predecessors pioneered. Since the Chinese people have the ability to stand up, they will surely have the ability to stand ever more proudly among the nations of the world.[1]

The resolution of the Fifth Plenary Session read as follows:

> Comrade Deng Xiaoping is an acknowledged leader of the people of all nationalities in China who enjoys high prestige, and he has made tremendous contributions during all the historical periods of the Chinese revolution and national construction under the leadership of the Party.

> The plenary session highly evaluates Comrade Deng Xiaoping's outstanding contributions to our Party and the country. Revolutionary practice during several decades shows that Comrade Deng Xiaoping is an outstanding Marxist, a staunch Communist, a brilliant proletarian revolutionary, a statesman and military strategist, and a long-tested leader of our Party and the country. The series of viewpoints and theories he put forth according to the principle of combining Marxism-Leninism with China's practice constitute an important component of Mao Zedong Thought. They have inherited and developed Mao Zedong Thought under new historical conditions and represent valuable ideological wealth for the Chinese Communist Party and the Chinese people.[2]

The Party and the Chinese people highly evaluated Father's contributions because he dedicated his life to and worked tirelessly for his motherland, the Party, and the people. From then on, Father was indeed retired, and he was able to rest and relax. He said his retirement should be a genuine one. We sincerely hope that he lives happily and peacefully in his later years and enjoys a healthy and long life.

Father has another aspiration, which has yet to be fulfilled. He wishes to set foot on Hong Kong after China resumes sovereignty over it in 1997. He said that even if he has to visit Hong Kong in a wheelchair, he will; even if he stands on Hong Kong soil for only one minute, he will be satisfied. By then he will be ninety-three years old. Our family believes that he will be able to visit Hong Kong and we will do all we can to help him fulfill this wish.

Although Father is retired, people still care for him a great deal. Both Chinese people and foreigners are very concerned about his health. People are interested in the merits and demerits in his long career. Many commentaries and research papers have been written, both at home and abroad, about his political achievements and thought. His rich and complex experiences fascinate people. Biographies have been written and are being written by authors in Germany, Hungary, and Hong Kong and by some Chinese writers.

Father is an introvert and a man of few words. He is unpretentious and prefers to keep his experiences to himself. Even we his family know little of his past. As a result, many people know his present but not his past, the outer Deng Xiaoping but not the inner man. There are many misconceptions about his life.

Father's life has been an extraordinary and remarkable one. I am not qualified to write his biography. However, I want to record what I know about him to set the record straight and to add something more real and valuable to the already rich literature on him.

All things on earth have their origins, and every story has its beginning. To write about Deng Xiaoping, I must begin with his hometown—Guang'an County in Sichuan Province.

AN AFFECTION FOR SICHUAN

Known as "the land of abundance," Sichuan belonged to the states of Ba and Shu in ancient times.

Sichuan has a very long history of civilization. Two million years ago, the ancestors of mankind lived and multiplied there.[1] Later, the two small states of Ba and Shu were established in the eastern and central-western parts of present-day Sichuan. In the late Spring and Autumn Period, the emperor of the Qin Dynasty marched his army south to attack the state of Ba first and conquered the state of Shu subsequently. Both states were annexed by the Qin Dynasty in 316 B.C. Not long after that, the Qin Dynasty established the two prefectures of Ba and Shu in the area near present-day Chongqing and in the Chengdu area, respectively. Since then, the states of Ba and Shu have belonged to a unified China.

The name of Sichuan was given in the Song Dynasty. It was officially named Sichuan Province in the Qing Dynasty.

There are plentiful resources in Sichuan. Because the weather is mild and humid and the four seasons are distinctive, the region is well suited to agriculture. The Sichuan Basin has been known as a granary since ancient times; many military strategists had their garrison troops claim wasteland and produce grains there. The land of Ba and Shu abounded in rice, silk and linen, fruits, tea, and well salt. Spinning and weaving, the production of well salt and chinaware, and metallurgy had been fairly well developed industries since the Song Dynasty.

Many prominent figures have brought glory to Sichuan. More than 2,000 years ago, in the Han Dynasty, there emerged the highly gifted master of poetry and prose Sima Xiangru. Numerous men of letters and other celebrities played an important role in the historical arena of Ba and Shu. Among them were Li Bai and Du Fu, leading poets of the Tang Dynasty, Liu Bei and Zhuge Kongming, key figures of the Period of the Three Kingdoms, and Li Bing and his son, great masters of water conservancy of the Period of the Warring States. The inhabitants of the states of Ba and Shu were good at farming, silkworm breeding, mineral smelting, and fine silk weaving. They were known for their ability to endure hardships and hard work and for their diligence and honesty.

As a native of Sichuan, how could you not love this land and be proud of it!

All Chinese know Sichuan, and the name of Sichuan is known even to some foreigners. However, our hometown, Guang'an County, is a part of Sichuan that is rarely known to people outside the well-known province. All members of our family have called themselves Sichuanese; we are natives of Guang'an only to other Sichuanese.

Guang'an is over 200 kilometers to the east of Chengdu, the capital of Sichuan Province, and 100 kilometers to the north of Chongqing, a city of strategic importance along the Yangtze River. Guang'an falls under the jurisdiction of the Nanchong Prefecture. It has no railway line, so communications rely mainly on overland vehicles and boats. Guang'an, a hilly land, is located at the edge of the Chengdu Plains. Its soil is not impoverished, but neither is it a prosperous land. Fortunately, the mighty Qujiang River flows down across the whole county.

Guang'an was part of Liangzhou in ancient times. Its inhabitants were Congs who created the pre-Ba culture together with some other tribal aboriginals.

In the late Spring and Autumn Period, the Bas who had lived along the middle reaches of the Hanshui River moved into Sichuan. The Bas mingled with the aboriginals living in east Sichuan and established the state of Ba, a slavery state based on an alliance of tribes. This state was conquered by the Qin Dynasty in 316 B.C.[2]

In the Period of the Warring States, Guang'an already belonged to the state of Ba. When the Qing Dynasty conquered Ba, it established a county in present-day Guang'an. In the Song Dynasty, it was given the name of Guang'an.[3]

Accordingly, it should be said that Guang'an was named in the Song Dynasty and the inhabitants there are the descendants of two merged tribes—the Congs living in ancient Liangzhou and the Bas living in the Hanshui River area. Probably because the Bas moved from the Hanshui River area to Sichuan, some people say that our ancestors were natives of Hubei Province.

Today, Sichuan has a population of 100,000,000, the largest among all provincial-level administrative divisions of the country. As part of the most populous province, Guang'an County does not lag behind in this regard. According to the records, in the years of Emperor Xuanzong's reign during the Tang Dynasty, what was then Qujiang County already had a population of over 18,000. The population of Guang'an increased abruptly in the Qing Dynasty and reached more than 131,000 by the years of Emperor Xianfeng's reign. Now the small county of Guang'an has a population of more than a million.

To the outsider, Guang'an is ordinary and has nothing worthy of high praise. But the natives of Guang'an are especially proud of their hometown.

According to the annals of Guang'an County, written toward the end of the Qing Dynasty, there was no wasteland or fallow land there; the soil was fertile. Farmers were worried about neither flood nor drought. There were mulberry, hemp, elm, and jujube trees, and the inhabitants kept oxen, horses, chickens, and pigs; the vegetables and fruits grown there included onions, chives, melons, gourds, and potatoes. Guang'an abounded in special resources, including bamboo and cypress in the wooded mountains, wild fowl and animals in the wilderness, fish and insects in the ponds and on the mountain slopes, both water and land plants, precious medicines and stone needles for acupuncture in the caves, as well as plentiful animal fur and horns. Guang'an rice and maize were especially fragrant and nourishing, and they were known as the food of gold and jade. The silk, bright, clean, and transparent, was of superior quality. The fine

Cong cloth woven there was recorded in the literature of the Han Dynasty and was very expensive.

The educational level in Guang'an is not very low. As early as the years of Emperor Jing's reign during the Han Dynasty, 100 B.C., schools were established in the Ba Prefecture. A school was set up in Guang'an in the third year of Emperor Ping's reign during the Han Dynasty (A.D. 3). In the nearly 2,000 years since, schools have been run in Guang'an. In the year of the founding of the Republic of China, a primary school already existed and a middle school was established. Compared with some areas, the educational level in Guang'an was certainly low, but not as low as that of many other areas of China at that time.

In the ordinary course of events, the natives of Guang'an, with such good natural conditions, should have lived contentedly on farming. However, they were plagued by numerous internal troubles and external invasions.

The first trouble was war. In the Sui and Tang dynasties, military commanders went on expeditions; late in the Song Dynasty, the north and the south were at war; late in the Ming Dynasty, roving bandits invaded; in the Qing Dynasty, natives of Yunnan came to pillage. As a consequence of the chaos of war, the natives of Guang'an had hardly lived in peace since ancient times.

The second trouble was natural disaster. The terrain of Guang'an is high, with the river far below, so drought has been rampant. According to the records, in the years of severe drought, there were some 50 kilometers of barren land, a scene of utter desolation. Victims of disaster fled, and beggars could be seen all along the way.

The third trouble was starvation. Because of the frequent disasters, grain prices rose abruptly, causing panic among the inhabitants. They had to eat roots of grass and barks, and the area was thus sparsely populated. Lamentably, the natives of Guang'an lived in a rich place but were reduced to famine victims.

The fourth trouble was epidemic disease. Disease was prevalent on a small scale every three years and on a large scale every five years. Infectious disease spreads from one family member to another, from one village to another. In the years of Emperor Tongzhi's reign during the Qing Dynasty, 5,000 people died of ordinary dysentery!

Poor communications kept Guang'an closed to the outside, and the natural and man-made calamities impeded its development. In the more than 2,000 years before the founding of New China, notwithstanding the passage of time and the changes of dynasties, the industrious and honest natives of Guang'an could never shake off the yoke of fate and always lived in poverty and backwardness.

CHAPTER 3

A TRIP TO MY HOMETOWN

Father did not return to his hometown and did not allow us to do so either, on the grounds that our arrival there would disturb a lot of people as well as the local government.

Therefore, it was not until 1989 that I and my second aunt, Deng Xianfu, "returned" to Guang'an County. Actually, I had never been there, nor had I lived there. But as Guang'an is the hometown of my ancestors, I use the word *return* nevertheless.

We got up early one morning in October and drove from Chengdu, the capital of Sichuan Province, to Guang'an. We traveled quickly, stopping in Suining first and then in Nanchong. When we arrived at the area of Guang'an, it was late at night. We had to stop at the county guest house for the night and continue our journey to my hometown the next day.

I tossed about in bed, unable to fall asleep, perhaps because of the damp bedding, perhaps because of excitement. When day began to break, I got up and went outdoors.

In the autumn, the weather in the south is a bit cool. The air was fresh and humid. Half of the hill was shrouded in morning mist. The surrounding slopes were all dark green, and green leaves were wet with sparkling dew. Unlike the dry, fresh, cool, and bright morning air in Beijing, the mist here was dim, and so was the sunshine.

The guest house we stayed in was located halfway up to the hill. Below it was the building of the Guang'an County Party Committee. It was quite a unique building. Upon inquiry, I learned that it had been the residence of the well-known warlord of Sichuan Province, Yang Sen.

The residence was built on hilly terrain. The gate was on the lowest level. It was faced with four huge sago cycas trees inside. Up the steps were several courtyards, around which were tall, tile-roofed houses. They were now used as offices and probably had been the living rooms of Yang Sen and his family. Further up the hill was the back garden of Yang Sen. On both sides of the stone steps were flowers and trees that looked indistinct in the mist. On top of the hill was a cave, called Han Xu, where Yang Sen probably sat in meditation. Although Guang'an was countrified, its warlord absolutely was not. Even a tennis court was built in his residence. It was said that Yang Sen enticed a coach from Shanghai to teach him to play tennis by offering him a handsome salary. Possibly it was Yang Sen who sponsored the sport of tennis in Sichuan!

The most interesting part of Yang Sen's residence were the four sago cycas trees inside the gate. These trees were not from Guang'an; Yang Sen had them shipped from as far away as Guangdong Province. It was said that although they produced many branches and leaves after being transplanted in Guang'an, they never blossomed. But in 1978, after Father was reinstated in his post for the second time, these trees blossomed. The leaves were full of magnificent golden flowers. The elders of my hometown were surprised to see the blooms and took special pictures to send to Deng Xiaoping in Beijing. Of course, we did not know whether they had really never blossomed before, but the episode showed that the people of our hometown respected and loved Father by and large.

After breakfast, we visited several sites nearby in Guang'an County town. One was Guang'an Primary School, a small two-story building with gray brick walls, wooden doors, windows, and banisters, and gray tile roofing. It was very, very old. Only two rooms upstairs and two rooms downstairs were now used by the county land bureau. It seemed likely that these remaining rooms would soon be pulled down.

We took pictures in front of the building because my aunt and I knew that this was the school where Father had studied more than seventy years before. Although the school would not have been so old and shabby then, it could not have been very magnificent. I could imagine how those children with cotton-padded gowns and skullcaps and books under their arms ran up and down the stairs. Father entered this school around the age of ten, and after completing the higher primary school courses, he was admitted after an examination to the Guang'an County Middle School. He studied there for a short time and then entered the preparatory school for studying in France on a work-study program in Chongqing. Therefore, the primary school was the sole regular school he had attended in his hometown. I imagined he must have been impressed by the school because it was where he had spent his whole childhood.

One could know at a glance that Guang'an was an old county town. It had many modern buildings, but it had many more of the old-fashioned houses typical of the south. Along the street were small, mostly two-story houses of brick-wood structure with all-gray walls painted white. Every second floor had a balcony surrounded by a row of wooden banisters. It seemed that Guang'an produced stones. The steps on the street and house foundations were all made of square stones. We also saw some stone troughs by the roadside, probably for washing clothes or vegetables.

On a commercial street were people bustling about and an endless array of beautiful articles for everyday use. Modern commodity economy had already played its role in this small mountain town. Walking on the street were both country folks carrying bamboo baskets on shoulder poles or on their backs and youths wearing quite fashionable clothes. Their multicolored clothes, their hair style (one popular in the big cities), and the color TV sets and audio equipment gave a sense of modernity to this remote mountain area.

Cars and carts pulled by oxen, color TV sets and bamboo baskets—all characterize present-day China. It is impossible to shake off poverty and backwardness overnight in a country so populous and with so poor an economic base. We have, however, started to take the first steps and are striding forward toward abundance and prosperity.

Just look at these peasants. They had glossy, dark green seedlings in their fields and heavy baskets of rice. They were barefooted no more and wore Western-style clothes. What a great change!

The Sichuanese are fond of eating. On the stands and shelves were fresh fish and meat as well as round, bulging Sichuanese sausages. A great variety of steaming hot dainty snacks were being cooked in the pots of street peddlers. The fresh and tender cabbage, radishes, pea seedlings, and broad beans sold here were vegetables that city dwellers found hard to buy.

I got to love this place in such a short time, probably because it is my hometown.

Our hometown is called Xiexing Township, which is ten kilometers from Guang'an County town. After a tour of the county town, we hurried right to the north. On our way, we first saw the Qujiang River, a tributary of the Yangtze River, in its upper reaches, that has its source in the mountainous area of northeastern Sichuan. The Qujiang River flows down from the northeast and joins with the Jialing River from the west and then into the Yangtze River. We could see the water of the Qujiang River in the valleys. Although the Qujiang River is not so imposing and broad as the Yangtze River, it surges as mightily as the latter. Wherever there is water, there is life. For thousands of years the Qujiang River overflowed, causing flood, and was also too dry to mitigate drought, but it has become the water of life and happiness for the people of my hometown today.

Our car left the highway and entered the countryside. Here we could see the true appearance of my hometown, a typically hilly area that is not part of the Chengdu Plains. A boundless stretch of the plains is indeed very beautiful, but the rise and fall of hilly land has a greater charm and romance. At that moment, the mist dissipated and it cleared up. Under the autumn sunshine, everything looked bright and warm, and our minds suddenly became clear and open.

In northern China, winter comes in October; leaves begin to fall and grass turns yellow. But here it was still a green and luxuriant world. There were rows upon rows of gardens. Rice had been brought in, and the tall rice straws growing in the fields would be plowed into the soil to fertilize it for the next year.

My aunt said that in her childhood rice straws were cut completely and used in kitchen fires because of the lack of firewood in Guang'an. But today most people use coal for cooking and leave rice straws in the fields to use as fertilizer. Well-fertilized soil looks black; the following year the sprouts will be strong, the rice seeds will be plump, and peasants will have a good harvest.

Between the farmlands we could see detached houses for one or several households. Although the peasants there could not afford to build houses of two or more stories, like those in prosperous areas, their houses were not the old-fashioned thatched cottages anymore. Most of the commodious green-tiled and gray-walled farmhouses and clumps of bamboo were set against each other.

The bamboos were really wonderful. They were as tall as the trees, and their tips wept like phoenix tails. Ordinary farmland and farmhouses were full of romantic charms because of the bamboos. Under the dark shade of the bamboo grove, people must have felt very cool and pleasant, and endless numbers of moving stories must have been told. The bamboos were just like the souls of the peasant families.

Vegetables were grown everywhere, beside the farmland, by the roadside, on the slopes, by the ponds, and in front of and behind the houses. Their leaves were dark green, indicating the use of adequate amounts of fertilizer and water. When we lived in Beijing, people of our hometown often brought us different kinds of vegetables. We only know that the vegetables grown in Sichuan are far more delicious than those in northern China. Only on this trip to my hometown did I get to know that vegetables are grown especially well in Sichuan. My aunt told me that the Sichuanese know how to grow vegetables. Furthermore, they value the land very much. They grow vegetables even on tiny plots and in a very meticulous way, just like embroidery. It is no wonder that vegetables are so cheap in Sichuan. When the natives of Chengdu go to Beijing, they always complain of the expensive vegetables in Beijing.

The soil of Guang'an is not as fertile as that of the Chengdu Plains. The soil of the slopes is worse still, and not even grass grows luxuriantly there, as we were told by local people. But we saw green trees all over the hill. Upon inquiry, we learned that they were orange and tangerine trees. These trees on the hills and mounds formed a forest. Beneath the orange trees, people usually grew vegetables to present beautiful scenery. My aunt told me that these trees had been planted over the past ten years. What could peasants do when even grass did not grow on the loess? The local government told them to buy some gunpowder at a price of 0.8 *yuan* to blow a hole in the stony hills and plant a bundle of tangerine trees. One hole after another was blown, and one bundle of trees after another was planted. The hills were covered with orange and tangerine trees that grew, formed a forest, and bore fruits. The hills became green, and the peasants became rich because they grew these trees. All the villagers praised the Communist Party.

In my childhood I heard from Grandma that in Guang'an there was a hill where there was a malus toringo. I did find that tree this time. The hill has slight slopes, and the top of it is flat. On the hilltop stands the malus toringo, alone. Its trunk is tall and straight, and its crown is huge. It looks like an opened umbrella or an unfolded flag at a great distance. Surprisingly, this tree always stands on the summit, and there is only one tree on each hilltop. Nobody knows whether it is natural or man-made, but it is unique. It dwarfs all other trees and towers aloft. When those who have been away from home for a long time see the tree, they feel that they are back home. The natives of Guang'an never forget it when they are away from their hometown.

The proudly towering malus toringo is like the god of mountains and rivers.

CHAPTER 4

THIS IS OUR OLD HOME

Here we are. This was Father's birthplace and our old home.

In front of the gate, a cement road had been constructed. On both sides of the road, flowers and grass as well as banana trees had been planted. Of course, our fellow villagers did all this with good intentions. Originally, the flat ground and road in front of the gate were all muddy and there were no flowers, grass, or banana trees.

Like other farmhouses, this house had limed walls, wooden doors, and tiled roofing. The principal rooms, all in a row, were slightly taller than the several wing-rooms on both sides. In the center of the rooms on the left, central, and right sides was flat ground. In those years there must have been many chickens and ducks playing in front and at the back of the house. I heard Grandma say that her family raised several geese not to eat but for guarding the house. Geese are loud and peck people violently but are not as fierce as dogs, so the villagers used them to protect their houses.

A wooden horizontal board was hung above the gate of the central principal room, on which was inscribed "The Old Home of Comrade Deng Xiaoping." As I stepped into the room, I immediately felt cool. Some of Father's pictures were on display in several rooms. They were taken in different periods—when he studied in France, when he was in the Eighth Route Army and in the War of Liberation, and after liberation. Oh, there was a picture of the whole family, including me, that was taken in the 1950s!

A portrait of Father by an unknown painter was hung in a room. In it, Father is standing with a smile and holding a cigarette between his fingers. Blossoming flowers are before him, and a high mountain surrounded by floating clouds is in the background. The portrait was not painted very skillfully but has a local flavor. On either side of the portrait are a pair of antithetical phrases: "A great many families are pleased with correct policies and good relations between peoples," and "Many households are happy about the prosperity of the country and the people." This pair of phrases were certainly not written by a master, but they express the popular opinion.

In the room on the left were a bed, a cabinet, and a tea table. They were all very old articles. The bed was very unique. It was made of quality wood with a wide bedstead. It had railings around and a ceiling above, all of which had pattern carvings and must have been very nice earlier. If the bed were new and fitted with silk and satin curtains, it would be very magnificent. But it was old and worn-out, no longer elegant. The villagers said this was the bed, used by my grandparents, on which Father had been born. But my aunt and the people from the county told me that all the household articles of the Deng family had been distributed to the villagers during land reform and that these old articles of Father's family were later returned. As there were many beds of that type in the locality and nobody knew which bed was the right one, one of them had to be selected.

I thought it did not matter whether that bed was the right one. It would be enough if I could get a general picture of the old time from the bed, the cabinet, and the tea table.

Father has never paid attention to unimportant matters. He has never concerned himself with his old home in his hometown and has been quite unwilling to have it turned into any kind of exhibition home. When my grandmother and other family members left the old home after liberation, this house was distributed to local poor peasants during land reform. Perhaps some ten families lived there. They did not move out until 1987 and 1988, when they had more money and could afford to build new houses. Then the house was changed into a special exhibition home, which has been open to visitors.

When I was back in our old home, a brother of Father's mother was still alive. Although he was not a member of the Deng family, he had lived in the house for many, many years. He and his own family had moved out only several years before, when they built a new house. I visited his new house, which, to my surprise, was a small, two-story building. All his neighbors were living in such houses. With such houses, who would want to live in that small old house! Naturally, it was vacated. Otherwise, Father would never have allowed others to ask the villagers to move out of his old home.

Walking out of the small back door, I saw right away a lush, exuberant growth of jade green bamboos at the back of the old house. When several bamboos are planted, they grow quickly into groves. The slender bamboos looked like the beautiful fingers of young girls, while the big ones could be compared with the arms of healthy lads. Bamboo leaves grow while falling and fall while growing. You could find yellow leaves all over the ground, whereas those on the branches were still dark green. These bamboo trees had grown there for perhaps 100 years already, but they were still full of vitality and seemed younger than me. Indeed, at that time I wanted to sit on a small bamboo stool under the small bamboo grove with a bamboo fan in hand, quietly listening to the rustling of bamboo leaves, smelling the delicate fragrance of bamboo branches, and looking at the sun through the densely growing branches and leaves.

This was both my home and not my home. I was totally unfamiliar with the place, but I had a deep feeling for it. I had never been there before, but I felt it was a familiar place—because it was Father's birthplace, and the place where my ancestors had lived.

In front of the gate, there was a long antithetical couplet. The golden characters were inscribed on both sides of the gate. The couplet, written in the autumn of 1983 by Ma Shitu, a famous man of letters of Sichuan, read:

When the mansion was about to fall down, an outstanding person was born at this nice place;
This great man helped the distressed and succored those in peril, and all the people put their hands on their foreheads in jubilation.
He made vigorous efforts to turn the tide, and this place abounds in treasures;
He worked to harness rivers, afforest mountains, promote industry, and support agriculture, and praises will be heaped on Guang'an for years to come.

Local people told us that not many people from other places came to Guang'an and very few came to Xiexing Township. But all those who came here paid visits to the old home. Foreign visitors also came here from thousands of kilometers away to do research or just to have a look.

We have a big family in Beijing, but we have almost no relatives in our hometown. As I said, a brother of Father's mother, named Dan Yixing, lived there. He and Father were the same age and were schoolmates. They were close to each other during their childhood. He was honest, tolerant, and kindhearted, but also stupid and weak and lacked competence and ability for life. When he was young, he was an opium addict. He sold all his properties and even nearly his children. His wife left him with the children. Then he came to live in his sister's home, the home of our Deng family. After liberation, Mother sent him some money every month for his living expenses. When he received the money, he often spent it all for a meal with some of his friends.

After the start of the Cultural Revolution, we did not know his whereabouts. After the Cultural Revolution was over, we were surprised to hear that he was still alive! We did not know how he survived the ten-year disaster. Later on, to look after him, the county government made him a member of the committee of the county Political Consultative Conference and granted him living expenses every month. We also remitted some money to him as before, for use as pocket money. When I went to his home to see him, he was living with his wife, sons, and grandsons. He was then eighty-five years old. He was thin, dim-sighted, and deaf, with a white beard. To our surprise, he could still recognize my second aunt but knew nothing about me.

Upon seeing him, I wanted to laugh, mainly because I remembered a story about him. Some years before, he had suddenly insisted on going to Beijing. Someone asked him, "How do you go to Beijing?" "I shall go by train," he answered. "Do you have enough money for going by train?" To the other's surprise, he replied, "I am the uncle of the 'monarch,' do I need to buy the ticket?" He was just such a pedant. He and Father were dear to each other as uncle and nephew, but the difference between them was so great! He died in 1990, the year after we returned to our old home. Soon his wife died, too. Since then, no relatives of ours have lived in our hometown.

While staying in our hometown, we went at last to have a look at the tomb of my grandfather and the grave of my grandmother. I wondered very much why these graves were not dug out during the Cultural Revolution.

The tomb of my grandfather was located in a field not far away from our house. In front of the tomb stood a stone tablet on which were carved the characters "The Tomb of Deng Shaochang." The stone tablet was erected in 1937. The grave of my grandmother was located on a hill where several grandmothers of elder generations of the Deng family had been buried. My grandmother, my father's own mother, was buried there. Her family name was Dan, and her name was unknown. She died when Father was twenty-two years old. Her tombstone was erected in the name of her sons, including Father. Actually, Father was then in France, far away from home, and might have known nothing at all about the erecting of the tombstone.

During that trip to our old home, I mistook the grave of my grandmother for our ancestral grave. Afterwards, I learned that the ancestral grave of the Deng family was located elsewhere, a long boat ride away. The ancestors of our family might have all been buried there. In old society, only the males of the family were allowed to go to the ancestral grave to offer sacrifices to the ancestor.

I had no time, and there was no need, to go to the ancestral grave. To find the tracks of my ancestors, however, it was necessary to find the family tree—which I did.

CHAPTER 5

TRACING FAMILY ROOTS

Father hardly told us anything about his family. When he left his hometown he was only fifteen, so perhaps he knew little about his family and clan. Only Grandma sometimes told us some stories about our hometown.

I know that my grandfather's name is Deng Shaochang, also known as Deng Wenming. People generally called him Deng Wenming. My grandmother's family name is Dan, but her name is unknown. Among our ancestors, one was chosen and admitted to Hanlin Yuan (the Imperial Academy) and was thus known as Deng Hanlin (that is, a member of the Imperial Academy). Maybe he was the greatest, most noted man in Deng's clan.

Many people who have researched Father's life have studied the history of our family and clan. Some say that our family migrated from Hubei; others maintain that members of Deng's clan were the Hakkas of Guangdong Province; and still others believe that Deng Xiaoping's family name is not Deng but Kan, and that his name is Kan Zegao. As opinions vary, no consensus can be reached. Even my uncle has said that when he was a child, he learned that Deng's family had moved from Hubei Province to Sichuan.

There is no doubt that Deng Xiaoping's family name is Deng. It is a pure misunderstanding to say that his family name is Kan. An old gentleman named Li Huang, who studied in France, mentioned in his memoirs that Deng Xiaoping's name was not Deng Xiaoping but Kan Zegao. His words misled quite a few researchers. For some time, many people really believed what he had said. I once said to Father jokingly, "Do you know that someone has changed your ancestors' name?"

Father changed his own name several times in his life, but he has never changed his family name.

Our family is not handed down in a direct line from the ancestor but is a collateral branch. So we do not know if there is a genealogy of the Deng family. In feudal society, orthodox ideas were absolute. In a clan, only names of lineal descendants from the ancestor could be recorded in a dignified way in the genealogy. In other words, only family members of the eldest son, the eldest grandson, the eldest great-grandson, the eldest great-great-grandson, could be recorded in the book. Our genealogy was kept by a lineal descendant of Deng's ancestor.

Nobody knows when the genealogy began to be compiled or who wrote it. According to the notes, the genealogy started from the Ming Dynasty; there are no facts to trace it back to earlier times. When the genealogy was being compiled, the text was based on the inscriptions on the memorial tomb tablets of our ancestors, and therefore, "the genealogy was based on facts and caused no objections." The inscriptions on the memorial tomb tablets of our ancestors are recorded at the end of the genealogy, perhaps just to prove its accuracy and genuineness!

Deng's Family Genealogy starts from the Ming Dynasty and ends in the early years of the Republic of China. The first ancestor was Deng Hexuan, who was from Luling County, Ji'an Prefecture, Jiangxi Province. In the thirteenth year under the reign of Hongwu (that is, the year 1380, or the thirteenth year after Emperor Zhu Yuanzhang established the Ming Dynasty), he went to Sichuan with his family as a supernumerary military officer of the ministry of war and assumed office in Guang'an. From then on, the Deng family records were kept in Guang'an, Sichuan. As to the details of the ancestors of Deng's family in Jiangxi Province before the Ming Dynasty, probably only Deng Hexuan was familiar with them, but he did not tell his descendants. Therefore, the details were lost. Among our ancient ancestors, there were no accounts of prominent figures or heroic deeds to be passed down to later generations. If they had been descendants of Confucius or Mencius, our ancestors would not have forgotten to record all the historical facts about the clan.

A clan's genealogy is compiled only by its lineal descendants, and naturally, they would write most about the bright side of the family's history. Besides, after the disastrous ten-year Cultural Revolution, who knows whether the genealogy is true or not. Therefore, I tried to get a set of the annals of Guang'an County so as to prove the genealogy's genuineness.

A paragraph on "Deng's family, Yaoping Village, Wangxi Township," in the "Clan Records" in volume 2 of book 2 of *The Annals of Guang'an Prefecture* reads: "Deng's ancestors were from Luling County, Jiangxi Province. During the reign of Emperor Hongwu of the Ming Dynasty, Deng Hexuan was recommended to be a supernumerary military officer of the ministry of war and was sent to Sichuan. His family went with him and settled down in Yaoping Village, Guang'an Prefecture. Their ancestors' tombs and temple were all in Yaoping."

Deng's Family Genealogy is consistent with the family origin mentioned in *The Annals of Guang'an Prefecture*. It seems that the genealogy might be genuine. Using both sources, I want to trace the footsteps of Deng's clan in Guang'an Prefecture over the course of the past 500 years.

A supernumerary military officer of the ministry of war at that time was a junior officer equivalent to a present-day section chief. But in ancient times, there were not as many grades for officials as there are now. So the junior military officer might have had a little more power than the section chief.

In *Deng's Family Genealogy*, Deng Hexuan was the first generation; during the Ming Dynasty, there were altogether nine generations. The compiler of the genealogy certainly sang the praises of his ancestors: among our ancestors there were several "officials." Some were true officials, but most of them were not really officials at all. According to the *Annals*, only two of our ancestors passed the imperial examinations at the national level. One was the eighth ancestor, Deng Shilian, and the other was his brother, Deng Shichang.

Borrowing from these sources, I would like to tell a few anecdotes about my ancestors. Since they are hard to verify, the reader might just take them as legends or stories. Our second-generation ancestor in the Ming Dynasty was Deng Hexuan's son, whose name was Deng Xian, also known as Deng Meizhuang. It was said that he was famous in Sichuan for his writing skills. When King Xian of Sichuan heard of his talents, he invited our ancestor several times to serve as an official in the court, but Deng Xian never accepted the post. His deeds are recorded in *The Annals of Guang'an Prefecture*. He was a virtuous man in Deng's clan.

Our eighth-generation ancestor in the Ming Dynasty was Deng Shilian, also known as Deng Renlin. He passed the imperial examinations at the national level during the reign of Emperor Chongzhen. He was quite talented and cherished high aspirations. He had a good memory and could recite books by Confucius and other ancient philosophers after skimming them once. He served for some time as the magistrate of Haiyang County, Guangdong Province, and then as deputy minister of civil personnel. In the late Ming Dynasty, he went with the king of Guangxi Province to Yunnan Province and Burma to serve as the minister of civil personnel and later to be promoted to the post of president of the Imperial Academy. In the autumn of the eighteenth year under the reign of Emperor Shunzhi of the Qing Dynasty, after being entrapped by the Burmese, he and forty-one other high-ranking officials were killed. In the forty-seventh year under the reign of Emperor Qianlong, the title of "a noble man with integrity" was bestowed on him. He was a martyr in Deng's clan.

Deng Shilian had a cousin named Deng Shichang, also known as Deng Longmen. Having passed the imperial examinations at the national level during the reign of Emperor Wanli of the Ming Dynasty, he was first appointed an official in the ministry of revenue in Nanjing and later was promoted to the post of head of Chuzhou Prefecture, Zhejiang Province. The soil there was infertile, and the peasants were poor. So he organized water conservancy projects among the peasants, who derived much benefit from them. Because of his merits, he was further promoted to the post of deputy chief prosecutor in charge of Yongzhou and Hengzhou prefectures in Hunan and Guangdong provinces. His rivals were jealous of him and managed to impeach him. So he had to leave his post and go back home. He was regarded as a man of virtue in Deng's clan.

Before beginning my research, I only knew that a member of the Imperial Academy was from the Deng family. Now I know that there were also loyal officials and martyrs in the Ming Dynasty among our ancestors!

From the Qing Dynasty to the present time covers ten generations of the Deng family. The reign of Emperor Qianlong was the prime period of the Qing Dynasty. At that time, the Deng family in the remote southwestern part of China enjoyed a period of prosperity, and a member of the family had become a member of the Imperial Academy, bringing honor to his ancestors. It seems that when something reaches its prime, it declines. The Deng family, after enjoying prosperity, gradually declined. Not only were the scholarly traditions not passed down to the younger generation but the farmlands were also sold piece by piece. My relatives told me that finally the family became so poor that they had to sell the house where the academy member had once lived. This showed how the family had declined.

Our eighth-generation ancestor, Deng Shilian, had a son named Deng Fang, who was the last generation of the Deng family in the Ming Dynasty—our ninth-generation ancestor. Toward the end of the Ming Dynasty, he went with his wife and two sons to Guangdong, where his father held an official post. According to *Deng's Family Genealogy*, when the family reached the San'yi River in Gaoyao County, Guangdong Province, they encountered pirates. Unfortunately, all the family members were drowned, except for Deng Fang's two sons. Some of the pirates showed mercy on these two and, instead of killing them, threw them onshore. The elder, Deng Sizu, was seven, and his brother, Deng Shaozu, was four.

Deng Sizu, also known as Deng Shengqi, was the first-generation ancestor of the Deng family in the Qing Dynasty. After his parents and all the family servants were drowned, he had no money or property. Sizu and his younger brother Shaozu had to beg. When they reached Wujia Village, they met a landlord who took a liking to them. After learning about the misfortunes that had befallen the two brothers, he kept them in his house. He gave them food and taught them how to read and write.

When Sizu grew up, the landlord married his daughter to him. Not long afterwards, a son was born to Sizu, named Deng Lin, in Guangdong. During a national examination in Guangdong, Deng Sizu met with Li Xiangen, whose father was a friend of Deng Sizu's grandfather, Deng Shilian. At that time, Li Xiangen happened to be an envoy on an educational inspection tour. Li told Sizu how his grandfather had been killed and then issued an order to send Sizu to his hometown. Sizu must have been very sad to learn about all this. He took his wife, son, and younger brother back to Sichuan in the tenth year under the reign of Emperor Kangxi (A.D. 1671). They had stayed in Guangdong for twenty-eight years; now Sizu and Shaozu were at last returning to their hometown to inherit the family property. It was said that Sizu was very kindhearted and enjoyed high prestige in his village. That reputation might have had something to do with the misfortunes he suffered in his childhood. This story may be the most touching one in Deng's family history.

Deng Sizu had two sons: one was Deng Lin, who was born in Guangdong and went back to Sichuan with his family when he was three, and the other was Deng Yan, who was born in Guang'an, Sichuan Province, after the family's return. Deng Lin had six sons, and Deng Yan four. From then on, the Deng family in Guang'an was divided into two major branches.

According to the *Annals*, Deng Lin, also known as Deng Shishan, could write classical prose when he was a child. When he grew up, he studied classics, history, and economics. In the thirteenth year under the reign of Emperor Yongzheng (A.D. 1735), he was appointed dean of students of Zhongjiang County. Having great learning for an educator, he could give very good instruction. His eldest son, Jianlin, and third son, Liangzhi, both became successful candidates in the provincial examinations, and his sixth son, Shimin, was a successful candidate in the national examinations and later became a Hanlin (member of the Imperial Academy).

According to the genealogy, Deng Yan, also known as Deng Yinghua, was not a good student, so he had to engage in farming. Loving philanthropic work, Deng Yan was able to inherit his forefathers' properties. He treated Deng Lin's sons as his own, and when he discovered that his nephew Deng Shimin studied very hard, he gave him sixty *mu* (one mu equals 0.0667 hectares) of farmland as an allowance for his studies. Deng Yan died at the age of eighty-one. Though he achieved nothing in learning, he managed his family affairs well. It is possible to deduce that if he gave his nephew sixty mu of land, he must have owned at least several times more than that. Though his property could not be compared with that of the rich landlord in northern China, his family was quite influential in that area. And he helped a member of the Deng family, Deng Hanlin, to become a member of the Imperial Academy.

Deng Hanlin's name was Deng Shimin, also known as Deng Xunzhai; his assumed name was Deng Mengyan. According to the *Annals*, Deng Shimin was amiable and modest. In the tenth year under the reign of Emperor Yongzheng (A.D. 1732), he passed the provincial examinations, and in the first year under the reign of Emperor Qianlong (A.D. 1736), he passed the national examinations. Afterwards, he became a Hanlin and was appointed editor at the Imperial Academy.

"Hanlin" was an official title in ancient times. In the Tang Dynasty, the scholars in the Imperial Academy were responsible for writing confidential documents. During the Ming and Qing dynasties, the Imperial Academy kept scholars of great ability, and during the imperial examinations some were chosen to be officials in the academy.

Although, as editor, Deng Shimin was only a junior official in the Imperial Academy, his appointment was a great honor to the ancestors of the Deng family in Guang'an. Later, Deng Shimin was promoted to the post of lecturer of the Imperial Academy. Then he served as the imperial commissioner of the area south of the lower reaches of the Yangtze River, senior lecturer of the Imperial Academy, and deputy secretary-general of the court, and finally he was further promoted to head of the Imperial Court in the tenth year under the reign of Emperor Qianlong (A.D. 1745).

The Imperial Court was the central judicial institution in ancient times and was responsible for examining criminal cases and making verdicts. The head of the court was called "Qing." The Imperial Court was equivalent to the present-day supreme court, and the post that Deng Shimin took up was equivalent to being president of the supreme court.

After Deng Shimin's father Deng Lin died, Shimin presented a memorial to the emperor and asked for permission to go back to his hometown and look after his mother. After his return to Guang'an, he reedited *The Annals of Guang'an Prefecture*.

In the twenty-ninth year under the reign of Emperor Qianlong (A.D. 1764), Shimin went to the court again and resumed his former position. The *Annals* say that when Shimin was head of the Imperial Court, he often made painstaking efforts to redress mishandled cases. If he could not have a mishandled case set right, he would present a memorial to the emperor for justice. Being upright and outspoken and tolerating no evil, he was held in awe. Later on, because he was getting old, Shimin was permitted by the emperor to retire and return to his hometown, and the honorary title of senior counselor was conferred on him by imperial mandate. He passed away at the age of sixty-six in his hometown. Shimin had sons but no grandchildren, so he had no descendants.

If the information in the *Annals* is true, Deng Shimin was a man of great knowledge and merits. Members of the Deng family of his generation left a brilliant chapter in Deng's genealogy during the 500 years of the Ming and Qing dynasties.

It is a difficult and quite painstaking task for a man, a family, or a cause to make progress or to be successful, whereas decline after success can happen in a flash. The success of a cause depends on opportunity, on geographical advantage, and especially on the support of the people. Deng's ancestors were officials and scholars and owned much property. Nobody knows how the family gradually declined. The land was sold, and the family members became poor. They could neither enter schools nor engage in trade.

The national fortunes were as bad as the Deng family financial situation. China had been so big and powerful and strong! But when the foreigners with powder and guns came, even Empress Dowager Ci Xi was driven out of the capital. As a result, the Qing Dynasty began to pay indemnities. The Chinese had to pay an indemnity regardless of the outcome of a war. The silver, together with the blood, sweat, and tears of the Chinese, were taken away by the foreigners.

The indemnities were paid and the territory was ceded, but the foreigners, instead of leaving China, became all the more arrogant. They bullied the Chinese, and the Chinese also bullied the Chinese. Senior officials bullied junior ones, and the rich bullied the poor. Up to the generation of my grandfather, the country had fallen into the hands of foreigners and the wars continued. Exhausted, the people had no food or clothing. The natural and man-made calamities had caused a once strong and powerful country to fall into ruin. Of the 400,000,000 Chinese, about 320,000,000 people lived in hunger and poverty.

The 2,000-year feudal culture had left China in a shambles of national betrayal and humiliation, poverty, and backwardness. Its people had no means of livelihood, and the country was riddled with gaping wounds.

Maybe heaven would first put a man in a state of destruction and then let him survive. Only after a country, a family, or a man suffers from setbacks can they possibly come to realize the truth. Only after such realizations can a nation be reformed and rejuvenated.

THE TRAGIC HISTORY OF A DECLINING NATION

CHAPTER 6

People of my grandfather's age had no luck. They were born in troubled times, grew up in troubled times, and died in troubled times. They did not spend even a single day enjoying times of peace.

My grandfather was probably born in 1886, the twelfth year under the reign of Emperor Guangxu. At that time, Guangxu, only fifteen years old, was a puppet emperor, with Empress Dowager Ci Xi holding court from behind a screen. Since 1840, under the rule of the increasingly corrupt and incompetent Qing government, China had experienced difficulty and turmoil.

The Opium War, launched by Britain against China, broke out in 1840. In the end, Britain was the victor, and China was forced to sign the Treaty of Nanking. The British aggressors forcibly occupied Hong Kong, extorted the huge indemnity sum of twenty-one million silver dollars, and forced the Qing government to accept a number of unfair provisions. This was the first unequal treaty in the modern history of China that brought on national betrayal and humiliation.

Such a great and proud country of the East, such an illustrious government, could not withstand a single blow and was bullied by others at will. Who would not grab benefits so easily obtained? As a result, other Western countries followed closely in Britain's footsteps and came to China.

In 1844, with the same gunboat policy, the American aggressors forced the Qing government to sign the Sino-American Treaty of Wangxia. In the same year France also lost no time in sending warships to blackmail China and easily signed with China the Sino-French Treaty of Whampoa. In 1849 the Portuguese colonialists did not even bother to sign a leasing agreement. They simply drove Chinese officials away and occupied Macao by force.

After Britain, the United States, and France gained many benefits in China, czarist Russia, Italy, and Portugal did not want to lag behind. They swarmed into China, demanding a share. The Qing government was so "magnanimous" that it agreed to all their demands "without discrimination." In this way, China's door was opened wider to Western capitalism. China was gradually reduced to the status of a semicolonial society.

During the nine years from the First Opium War to the time when Portugal occupied Macao by force, China's territory was seized forcibly and its natural sovereign rights over judicial affairs, customs, and territorial waters were infringed on one by one. Foreign colonialists willfully bullied, humiliated, and exploited the Chinese people, and Western missionaries rode roughshod over them as well. Sectors of China's economy were soon controlled by foreign aggressors. Instead of promoting the development of China's economy, the capitalist invasion dealt a heavy blow to the country's already quite backward urban and rural handicraft industry, destroyed the natural economy of self-sufficiency on which the Chinese people relied for existence, aggravated the country's poverty, and made the lives of the Chinese people even more miserable.

As the foreign aggressors violated China's rights by force, the Qing government yielded in humiliation, and the Chinese people moaned, China's modern history began.

The Chinese nation had glorious traditions, and its people were not reconciled to humiliation. Filled with indignation, some insightful Chinese denounced the treasonous acts and planned to save China through reforms. Patriotic Chinese officers and men fought heroically against foreign aggressors and burned opium at Humen. In defiance of brute force, the broad masses of the Chinese people rose against the aggressors, "showing the punishment by Heaven and expressing public indignation."[1] Dedicated to a just cause, some people raised the standard of revolt and fought everywhere against the invaders. Their great influence across the nation shook the reactionary rule of the Qing government.

The just, patriotic struggle of these people continued wave upon wave. Though their struggle was heroic and brave, they were confronted with the frenzied Qing government and the heavily armed Western powers. They had only the will to save the country but not the slightest power, so their struggles were defeated and their revolts put down again and again.

The Qing government not only strangled the people's patriotic resistance movement but continued to be obsequious toward the imperialist powers and to betray China shamelessly. Lenin penetratingly pointed out that war is an inevitable outcome of capitalism. The avarice of imperialism and colonialism is always insatiable. In the 1850s the capitalist economy further expanded in the West, so the bourgeoisie was even more eager to conduct overseas expansion and dominate markets.

In 1856, with Britain taking the lead, the four aggressor nations of Britain, France, the United States, and Russia formed an aggressive bloc against China and provoked the Second Opium War. The allied forces of Britain and France easily seized Guangdong Province. They committed all kinds of atrocities against the Chinese people, such as burning, killing, and looting. The fatuous and decadent Qing government hit upon a good idea—not to fight and not to resist but to console. It signed treaties as it had done before!

In 1858 the Qing government signed the capitulationist and traitorous Treaty of Tientsin with Britain, France, the United States, and Russia. This treaty provided the Western aggressors with many more colonial rights in China. It stipulated that foreign ships (not just commercial ships) could sail at more Chinese ports, and the Qing government was compelled not only to pay an indemnity of four million taels of silver to Britain and two million taels of silver to France but to invite the British to China to handle customs, tariffs, and so on. The treaty also shamelessly stipulated that the "opium trade is legal"! Karl Marx indignantly denounced this treaty as "one of out-and-out swindle."[2]

In 1860 the allied forces of Britain and France forcibly seized Zhoushan, Dalian, Yantai, and Tianjin and finally captured the heart of the Chinese nation—Beijing, the capital of China. The heroic patriotic officers and men shed blood and gave their lives to resisting foreign aggression, whereas the ignoble officials of the Qing government fled helter-skelter. The ancient city of Beijing was subjected to barbarous plunder, and the magnificent Yuanmingyuan Garden was burnt to ashes by the aggressors.

Shame upon shame. China signed again with Britain and France the Convention of Peking and was forced to cede territory (Kowloon) and pay indemnity (eight million taels of silver). As Britain and France gained benefits, the United States and Russia naturally did not look on unconcerned. They also acquired equal rights.

The Second Opium War ended. The aggressive wars launched by the capitalist powers against the backward nations and countries were designed to turn the latter into their colonies or semicolonies, commodity markets, sources of raw materials, and places for investment. Therefore, in all the treaties imposed upon China by Western aggressors after the First Opium War, besides the provisions concerning the ceding of territories, the payment of indemnities, and a series of other provisions infringing on China's sovereignty, another important article was added, namely, the so-called opening of trading ports. Chinese ports were to be opened not only for "trading." So long as they were trading ports, the signatory states could send resident consular officials there and dispatch warships to China's inland rivers for "protection." Another aggressive act of semicolonialism was the appointment of foreigners to the official posts in Chinese customhouses and tax offices in these trading ports.

In 1840, in accordance with the Treaty of Nanking, the five ports of Guangzhou, Fuzhou, Xiamen, Ningbo, and Shanghai were opened. In 1858, in accordance with the Treaty of Tientsin, the ten trading ports of Niuzhuang (later changed to Yingkou), Dengzhou (later changed to Yantai), Tainan, Danshui, Chaozhou (later changed to Shantou), Qiongzhou, Hankou, Jiujiang, Nanjing, and Zhenjiang were opened. In 1860, in accordance with the Treaty of Peking, Tianjin was opened.

In 1876, in accordance with the Treaty of Yantai, Yichang of Hubei Province, Wuhu of Anhui Province, Wenzhou of Zhejiang Province, and Beihai of Guangxi Province were opened. The treaty allowed foreign officials to go to Yunnan for investigation and to Tibet via Gansu, Qinghai, and Sichuan provinces.

By 1887, when the Sino-French War ended, the southeastern coastal provinces, the Yangtze River Basin, and southwestern China had all been subjected to the encroachment of foreign powers. But the foreign imperialists and colonialist forces were not satisfied. Their ultimate aim was to carve up China, occupy it, and reduce it to colonial status.

The czarist Russia empire north of China had long coveted China's territory. Unequal treaties, such as the Sino-Russian Treaty of Aigun signed in 1858, the Sino-Russian Treaty of Peking signed in 1860, and the Sino-Russian Ili Treaty signed in 1881, allowed Russia to seize over one million square kilometers of territory of northeastern China—in the northern part of the Heilongjiang River and the eastern part of the Wusuli River—and over half a million square kilometers of territory in northwestern China to the east and south of the Balkhash Lake. Russia occupied a total area of one and a half million square kilometers. In 1895, under the pretext of leasing, czarist Russia seized Lüshun and Dalian, two important ports in eastern Liaoning.

In Japan, China's neighbor to the east, the forces of militarism were rapidly swelling in a vain attempt to contend for hegemony in the world. In 1894 Japan outrageously launched an aggressive war against China and Korea, the Sino-Japanese War of 1894–95. The capitulationist policy of the Chinese government led to the Sino-Japanese Treaty of Shimonoseki, which stipulated that China pay 200 million taels of silver in reparations for Japan's military expenditures and cede to Japan the Liaodong Peninsula, the Taiwan Island and its subsidiary islands, and the Penghu Islands. Thus, China again lost vast territories in the northeast and on the East China Sea.

As Russia and Japan obtained benefits from China, other countries also stepped up their efforts to carve up China. From 1895 to 1900, China experienced the disaster of extermination; with a 4,000-year history of civilization, the nation was facing the danger of subjugation. Let's remember how the subservient Qing government had sold out China!

The Qing government agreed that:

the whole of northeastern China was to be placed under the control of czarist Russia;

Shandong Province was to be the German sphere of influence;

the Guangzhou Bay and its surrounding waters were to be leased to France for ninety-nine years;

Weihaiwei and its surrounding waters, as well as the whole of Kowloon Peninsula, were all to be leased to Britain;

the three provinces of Yunnan, Guangxi, and Guangdong were to be the French spheres of influence;

the Yangtze River Basin was to be the British sphere of influence;

Fujian was to be identified as the Japanese sphere of influence; and

the United States hurriedly came and also got its share.

All Chinese rights were sold out and all Chinese territories were carved up. Empress Dowager Ci Xi and Li Hongzhang were the guilty persons condemned by the nation, and they have been nailed to the cross of humiliation forever.

This is Chinese history from 1840 to 1900. Could a man of conscience read on after reading all this? How could a man of any patriotism not heave a sigh of strong indignation?

I am not a student of history and have never dared to write about it. I want to inform people about the journey of the Chinese nation. If you do not know how the Chinese nation traversed its path in the past, you will not be able to see the road ahead for it in the future.

CHAPTER 7

THE PEOPLE'S HEROIC REVOLT

While the Qing government betrayed the country and surrendered in utter disgrace and without shame, the courageous Chinese people revolted in defiance of brutal oppression.

As the feudal rulers of the Qing government became increasingly corrupt and the invasion by foreign capitalism more ferocious, the national and class contradictions in China grew more acute and the social crises thus intensified.

Mao Zedong once said, "The Chinese never submit to tyrannical rule but invariably use revolutionary means to overthrow or change it."[1]

The people rose in resistance. In different localities, they rose in rebellion on various scales and through different means. They rebelled against the cruel rule of the Qing government and against the barbaric aggression of foreign powers. They suffered dismissal from office, exile, suppression, and even strangulation for their resistance. But never did they cease their fight.

In 1839, Lin Zexu, a patriot, and the indignant people of Guangzhou took an action that shook the world. They burned 1,188,127 kilograms of opium of the British and American aggressors and eradicated the humiliation of the Chinese nation. The burning of the opium at Humen marked the beginning of the people's resistance against colonialism in the modern history of China.

In May 1841 the people of Sanyuanli, Guangzhou, armed with swords, killed and wounded many of the British troops who had been killing, burning, and committing other atrocities against the Chinese people. The drumbeats thundered on the battlefields. Chased everywhere by the local inhabitants, the defeated British troops fled helter-skelter.

The anti-British struggle waged by the people of Sanyuanli greatly boosted the morale of the Chinese people. From then on, their struggle against internal and external reactionary forces intensified. In the ten years before 1850, more than 100 large-scale peasant uprisings broke out. In January 1851 the Jintian Uprising in Guangxi triggered a widespread peasants' revolutionary storm. The heroes of the Taiping Heavenly Kingdom, led by Hong Xiuquan, directed their attack at the Qing Dynasty from the very beginning. Their goal was to overthrow the corrupt regime and establish a new peasant government. They fought both the feudal rulers and foreign aggressors, thus winning the support of millions of people. Their tremendous struggle lasted fourteen years and exerted great influence in eighteen provinces. The whole nation was stirred by their courage.

Under the influence of the Jintian Uprising, many areas witnessed people's uprisings, such as the Small Dagger Society in Shanghai, the Heaven and Earth Society in Guangdong, Guangxi, and Hunan, the Bianqian Society in Jiangxi, the Nian Army in the north, and the uprisings by the people of all nationalities in Yunnan and Guizhou. The peasant revolutionary movement surged forward, undermined the rule of the Qing Dynasty, and quickened the pace of the bourgeois democratic revolution. All these heroic peasants' revolutionary struggles were eventually suppressed, but the peasant heroes nurtured the Chinese land with their blood. From the blood-soaked soil grew the flowers that were as red as blood.

The Taiping Heavenly Kingdom failed in 1864, but people's struggles did not cease. The remainders of the Taiping Army continued to fight. In southwestern China, the Miao people in Guizhou and the Yi and Hui people in Yunnan joined the fight, as did the Hui people in Gansu in northwestern China.

When foreign enemies were invading our country, bringing humiliation to China, the national contradictions became prominent and the Chinese people started a new round of struggle against them.

In the late nineteenth century, world capitalism began its transition to imperialism, and the imperialists' scramble for colonies reached a climax. They were no longer content with their established interests in China, and they stretched out their tentacles for its territories. Their insatiable plunder naturally met with the fierce resistance of the people of all nationalities of China.

In 1876, with the support of the people, a senior general, Zuo Zongtang, recovered the lost territories of Xinjiang despite British obstruction and consequently thwarted the Russo-British plot to partition the Tianshan Mountains.

In 1883 the French invading troops launched attacks on the Chinese army, and the Sino-French War broke out. Within half a year, Vietnam was formally reduced to the status of a French colony. The arrogant French army soon began to attack Taiwan with its Far East Fleet but retreated in panic when it unexpectedly encountered fierce counterattack by the Chinese troops. The French army did not resign itself to defeat but initiated an even larger sea battle at Mawei. Following the surrender of the Qing government, the French army intensified its aggression against China.

When the Qing government surrendered, the people were infuriated. Under the intense pressure of public opinion, in both the court and the commonalty, the Qing government reluctantly ordered its troops to be ready to fight the French army. The Chinese army would not accept humiliation. When the French army launched an assault at the Zhennan Pass, the Chinese army was determined to fight a last-ditch battle there. At the fogbound Zhennan Pass, a fort in southwestern China, on March 23, 1885, the Chinese army fought with enormous courage. The French aggressors suffered crushing defeat and fled.

Strange things do happen in this world. The incredible result of the Sino-French War was that France was "victorious while defeated," and China was "defeated while victorious"! Although the Sino-French War ended with China's acceptance of another unequal treaty—the Treaty of Tientsin—the great victory at the Zhennan Pass helped spread the great influence of the Chinese army.

In 1894 the Japanese militarists invaded China. Under attacks by the Japanese army, the Chinese government was compelled to declare war. Because the Qing government repeatedly made concessions, the Sino-Japanese War ended with the Treaty of Shimonoseki, which stipulated that China cede territories, pay indemnity, and open up trading ports. In the Sea Battle of 1894, the Chinese navy fought bravely against the Japanese navy. More than 250 officers and men on the Chinese warship *Zhiyuan* died a heroic death. Their heroic dedication to their motherland and the national hero Deng Shichang will be remembered by the Chinese people forever.

The raging flames of the Chinese people's struggles against the Qing government and imperialism could never be put out. In 1894 the indignant people on the Kowloon Peninsula put up resistance against the British troops. In 1895 tens of thousands of the Hui people in Gansu rebelled, in protest against exorbitant taxes and levies. In June 1897 the people of Shandong opposed the German aggressors' acquisition of land for the construction of railways, and some twenty Chinese people were killed. The indignant people rose in armed struggle, forcing Germany to stop the railway project for some time. In 1898 the Chinese army and the inhabitants of the Liaodong Peninsula rose against the invasion of czarist Russia, forcing the Russian army to retreat. In the same year the people of Guangzhou rose against the French aggressors, who were trying to survey the boundaries of ceded territories; the workers and shopkeepers of Shanghai went on strike to halt the expansion of the French concessions, suffering scores of casualties; the Yulin Uprising in Guangxi against the Qing army broke out; tens of thousands of people participated in uprisings in northern Jiangsu, northern Anhui, and Henan; and in November, the unyielding people of Guangzhou Bay revolted and killed French soldiers.

In 1900, the Yihetuan Movement broke out and swept across the country. Holding high the anti-imperialist and patriotic banner, the Yihetuan heroes wanted to eliminate foreigners and their religions and to topple the Qing government. This mammoth peasants' revolutionary storm originated in Shandong, spread rapidly to every corner of China, and culminated in the second nationwide revolutionary upsurge in the modern history of China.

The Qing Dynasty was like a rotten and dilapidated palace that was about to crumble. China resembled a long-suppressed volcano that was on the point of erupting all over a bleeding land. Humiliation, agony, bitterness, and indignation had been pent up for too long. A tremendous revolutionary change was maturing and about to happen.

CHAPTER 8

DR. SUN YAT-SEN'S ENDEAVOR

There was once a man of whom the Chinese people cherish a fond memory and the people in other countries speak highly, and whose portrait still stands at the heart of the capital of China. He was Dr. Sun Yat-sen, the great forerunner of the Chinese revolution.

Born in Xiangshan (now Zhongshan) County, Guangdong Province, Sun Yat-sen was formerly a medical practitioner. Because he accepted progressive ideas, he was determined to oppose the Qing rulers and called for political reform.

In the late nineteenth century, national capitalism developed steadily and the national bourgeoisie came into being. From the very beginning of its development, however, Chinese national capitalism experienced the dual oppression of domestic feudalism and foreign capitalist aggressors. To secure their own development, the Chinese national bourgeoisie attempted to oppose feudalism by way of reform, but their moderate approach failed. The constitutional reform and modernization drive, endorsed by many people with lofty ideals, including Kang Youwei, Liang Qichao, Tan Sitong, and Yan Fu, was stifled by the Qing government. In the Coup of 1898, the reformists were removed from office, sent into exile, put in jail, or executed. Before dying a martyr's death, Tan Sitong cried bitterly, I had the will to fight the enemy, but no power to reverse the desperate situation.

The failure of the reformist cause enlightened a group of revolutionaries. The Eight-Power Allied Forces—imperialist invading troops sent by Britain, France, Germany, Italy, Japan, Russia, the United States, and Austria—launched a savage war of aggression against China. They frenziedly plundered China's treasures and massacred Chinese people in cold blood. The atrocities committed by the imperialists infuriated the entire Chinese nation and awoke the people of insight devoted to a democratic revolution.

A group of revolutionary pioneers started stirring up public opinion in favor of the revolution. Zhang Binglin, Zou Rong, Chen Tianhua, and Qiu Jin took the lead in advocating revolution, calling upon the people to sharpen the sword, raise the flag, and exert all efforts, to march to the battlefields. They called for uprisings so as to establish an independent, democratic, prosperous, and powerful bourgeois republic!

Their words were bold and radical, and their deeds brave and heroic. Zhang Binglin was put in jail and then forced into exile. Zou Rong died in prison at the age of only twenty-one. Chen Tianhua took his own life in grief and indignation at the age of only thirty. Qiu Jin died a martyr's death at the age of twenty-eight. Although these revolutionaries failed to attain their goal, they certainly helped push the revolution forward. The widespread dissemination of democratic ideas accelerated the establishment of revolutionary organizations.

On August 20, 1905, a group of revolutionaries led by Sun Yat-sen founded Tong Meng Hui (Chinese Revolutionary League), the first bourgeois political party in China. Sun Yat-sen proposed to "accomplish the task of the three major revolutions—national, political and social revolutions—at one stroke." The Chinese Revolutionary League established a revolutionary program calling for "the expulsion of the Manchus, restoration of China, founding of a republic, and equalization of land ownership." In addition, Sun Yat-sen put forward the Three People's Principles, i.e., the Principle of Nationalism, the Principle of Democracy, and the Principle of People's Livelihood.

While opposing the feudal royalists and the pro-Confucian anti-legalists and constitutionalists, the bourgeois revolutionaries staged a series of armed uprisings, displaying a militant resolution to carry out the revolution even at the cost of their lives. All the uprisings failed, but the revolutionaries did not lose heart and continued to fight on the blood-stained battlefields.

By that time in China, the rising wind foreboded the coming storm. The wind, like a hurricane, fanned the flames of fury of the Chinese nation. The whole nation was thrown into a great chaos!

In 1909 more than 130 mass revolts took place all over the country. The following year the number rose sharply to more than 290 and the unrest of rice-seizing in Changsha spread rapidly to engulf the whole country. In Shandong there was a powerful struggle against exorbitant taxes and levies. In 1911 the mighty movement to protect China's railway rights shook many provinces.

It was a time when revolutionary struggles could be triggered at any moment and supported by the masses. On October 10, 1911, the Wuchang Uprising broke out, shocking the country and the whole world. The victory of the Wuchang Uprising revitalized the bourgeois democratic revolution. As the shots were fired in the uprising, revolutionaries across the country immediately responded. All of China was quickly engulfed by the flames of armed revolution.

On New Year's Day 1912, Sun Yat-sen was sworn in as the provisional president in Nanjing, and he declared the founding of the Republic of China. On February 12 of the same year, the Qing Dynasty, which had ruled China for 267 years, ended with the abdication of the last emperor. The feudal monarchy, which had lasted more than 2,000 years, was eventually overthrown in the Revolution of 1911.

However, disaster and distress haunted the Chinese nation. When feudal emperors were overthrown, feudal forces lingered on. The old emperor was dethroned, but someone else wanted to step up to the throne. Before the foreign aggressors left, the warlords in various parts of China started to set up their separate regimes and carve up the country. Actually, people of all descriptions and motivations participated in the Revolution of 1911 and wanted a share of the fruits of its victory.

In less than half a year after the revolution—Sun Yat-sen served as the provisional president for just over a month—the fruits of the revolution, namely the Republic of China, were wrested away by the arch usurper of state power, Yuan Shikai, who became the provisional president. Externally, he continued the Qing government's policies of selling out China and signed with Japan the humiliating Twenty-one Demands. Internally, he retaliated against the people and had Song Jiaoren, a revolutionary, assassinated. Yuan Shikai did not stop there; he went on to restore monarchy in open defiance of world condemnation. His leadership lasted, however, for only eighty-three days.

On March 12, 1925, the great forerunner of the democratic revolution, Dr. Sun Yat-sen, died in Beijing. The Revolution of 1911 led by him went down in history. But the program he advocated for a bourgeois republic did not materialize. Dr. Sun Yat-sen devoted all his life to the national revolution and the democratic revolution and finally proposed the Three Major Policies—alliance with Russia, cooperation with the Communist Party, and assistance to the peasants and workers. He was the father of China's bourgeois democratic revolution.

After the feudal monarchy ended, China entered a new era. It was to be a semicolonial and semifeudal era, a turbulent time plagued by separatist warlords' regimes, incessant civil wars, foreign aggressions, and the utter misery of the people.

CHAPTER 9

MY GRANDFATHER

From the late Qing Dynasty to the whole period of the Republic of China, the turmoil and chaos of war occurred frequently. There was no time for people to live in peace.

Perhaps it was because of the natural and man-made disasters that our family began to decline from one generation to the next, just as the saying goes, "Even the country is on the verge of subjugation, is there any family to sustain?"

It was said that the father of my grandfather was very poor. He owned little land. But he lived a simple life and was very diligent. He knew how to spin and weave. He saved on food and clothing every day, knew no sufferings, and tried to accumulate wealth for his family bit by bit. He often went to the country fair to sell his cotton yarn and cloth. When he went, he took only some broad beans with him, and when hungry, he just put a few beans into his mouth and washed them down with cold water. Gradually, he began to save some money and bought a piece of land. When my grandfather was born, the family had about a dozen mu of land.

The father of my grandfather, my great-grandfather, perhaps could not be counted as a craftsman at all. At most, he was a peasant-craftsman.

My grandfather was born in 1886. It was said that until my grandfather was born, each generation had had one son only. In old feudal society, women were looked down upon. In fact, my grandfather had several sisters, but they amounted to little, so my grandfather was also the only child of the family.

My grandfather was called Deng Shaochang and also known as Wen Ming. People always called him Deng Wenming. We never saw our grandfather, and Father never mentioned him. We got to know a little about him from Grandma. So our generation has little impression of him. We only knew that his name was very funny, especially when pronounced in the Sichuan accent.

It was only recently that I have got to know something about him from some relatives. My grandfather studied for a while at school when he was young, but he was far from being an intellectual. Because the family had a little land, there was no need for him to till the land. Because he hired some farmhands, he can be regarded as a landlord in terms of class status. But as the amount of land was not much, he was a small landlord at best.

In the words of my uncle Deng Ken, my grandfather was a typical man of the old society, and his thought and style of life all belonged to the old days. However, he was discontented with the old society. He even said: "This society is really shocking. It needs a revolution."

The Revolution of 1911 broke out. As early as the beginning of the twentieth century, northeastern Sichuan, with Chongqing as the center (where Guang'an County is situated), had been influenced by reformist ideas and drawn to the notion of a bourgeois democratic revolution. Zou Rong, an activist of the bourgeois democratic revolution, was a native of Baxian County in the Chongqing area.

In 1906 the Tong Meng Hui founded by Dr. Sun Yat-sen established a branch in Chongqing, further promoting the revolutionary struggle in Sichuan. From 1907 the Tong Meng Hui staged several large-scale armed uprisings in Sichuan. On the eve of the outbreak of the Revolution of 1911, on September 25, 1911, Wu Yuzhang and other members of the Tong Meng Hui led an uprising in Rongxian County, Sichuan Province, and declared independence. In November the Tong Meng Hui led uprisings in Changshou and Fuling in the Chongqing area. On November 21 the Tong Meng Hui in Guang'an led an army to seize Guang'an and set up the Northern Sichuan Military Government of China. On November 22 the Military Government of Chongqing was established by the Tong Meng Hui, marking the collapse of the feudal autocratic rule of the Qing Dynasty there.

In Sichuan as a whole, especially at a time when the bourgeois democratic revolution was very active and the revolutionary movement was surging in eastern Sichuan, my grandfather, who was about twenty-five at the time and full of youthful vigor, rose to support the revolution and took part in the local armed rebellion. The aim of the rebellion was to end the rule by the Manchus and revitalize the rule by the Hans. In the revolutionary army in Guang'an, he served as a kind of low-ranking commander similar to a platoon leader. The revolutionary army set up two barracks opposite the county town of Guang'an—Dazhai and Xiaozhai—with 100 to 200 troops stationed there. Society was in great turmoil. Therefore, it was entirely of their own accord that people joined the revolutionary army. When the Revolution of 1911 broke out, Father was only seven years old. When my grandfather was in the army barracks, Father went to see him and stayed there for two days and nights. Although Father was very small, the revolutionary flames certainly left a deep impression on his mind. He can still remember this incident.

My grandfather was perhaps not a good businessman and knew little about making a fortune. However, he was loyal to friends and participated in the handling of some important occasions, thus earning himself a reputation.

In Sichuan there was a secret society called Pao Ge Hui, or Ge Lao Hui (the Society of Brothers). The organization participated in the campaign against Western churches, the movement to protect the railways, and the Revolution of 1911. It played an important role in the modern history of Sichuan. My grandfather once served as *San Yie*, meaning the number-three person in the society in Xiexing Township. Because he was also probably in charge of the society's routine affairs, he was also called "steward." Later on he was promoted to the top position.

In about the third year of the Republic of China, my grandfather served as the commissioner of the Security Agency (also called the Home Guards

Bureau) of Guang'an County. The post was commissioned by the county magistrate. When the county magistrate lost his post, my grandfather was also dismissed. After that, he served as the head of our township.

It was said that when he was the chief of the Home Guards Bureau, he led his men to suppress bandits headed by a certain person called Zheng in the Huaying Mountains, thus causing a feud. Later, Zheng accepted the government's amnesty and was appointed division commander. A division commander was much higher than a bureau chief, so my grandfather had to go into hiding in Chongqing, where he lived for eight consecutive years. There he made quite a number of friends, who told him the news that students were being sent to study in France on work-study programs. Thus, he had his son in the countryside come to Chongqing and sent him to study in France, thereby enabling him to take an extraordinary road in life.

Even after my grandfather became the head of the family, he devoted little energy to its affairs, perhaps because he was very concerned about the external world and social affairs. When he served as the chief of the Home Guards Bureau, he earned some money and increased the wealth of the family. At the time, the family owned more than twenty mu of land. Even though my grandfather had one or two hired hands to work for him, the family was not very well off. Sometimes they even lived a hard life. To support his son in his studies and bear other expenses, my grandfather had to sell some of the land. Though he had a lot of old ideas and habits, on the whole he was fairly enlightened. Not only did he send his son to study in France, but when he learned that his son was engaging in revolutionary activities abroad, he had no objections. When his son could not support himself in his revolutionary activities and wrote asking for money, he sold land and grain to raise money to help his son. He kept in a big box all the revolutionary books and magazines his sons sent back and did not burn them until the Kuomintang (KMT) searched closely. My grandfather was such a typical person, holding both old and new ideas. He died in 1936.

The eldest of his four sons left home as a teenager and never returned. The second son left home for school and later joined the revolutionary ranks. He did not return either. The third son also wanted to go and see the world. But this time my grandfather insisted that he stay home; perhaps he wanted the third son to inherit what was left of the family property. However, the third son turned a deaf ear to his proposal and fled. My grandfather pursued him. For days on end, he walked. He used to pass blood in his stool. His anger and fatigue aggravated his disease, and later he died on the way. He was then barely fifty years old. At the sudden bad news, the family fell into deep grief. They had to buy a piece of land to bury him at a place not far from home.

My grandfather had married four times. He married his first wife, named Zhang, when he was about thirteen. She died only two years later, leaving no children.

His second wife, my grandmother, was named Dan. The Dan family was a prosperous one in Wangxi Township, Guang'an County. During the Qing Dynasty, some of the Dan family members served as county magistrates in Tongcheng of Hubei Province, Jiading of Jiangsu Province, and Weiyuan of Gansu Province. When this girl of the Dan family was married to Deng Wenming, her family was more prosperous than the Deng family. My grandfather married her about 1901. The following year, their first daughter was born. She was Deng Xianlie. My grandfather was then sixteen years old. In 1904 their first son was born. He was Deng Xiansheng, my father. In 1910 the second son (my second uncle) came into this world, called Deng Xianxiu (changed to Deng Ken later). The third son was Deng Xianzhi, and his name was later changed to Deng Shuping.

My grandmother did not know how to read and write, but she was very capable and good at reasoning. When there were family disputes in the area, people asked her to mediate. She also knew how to raise silkworms and reel silk. She sold the silk to earn additional income for the family. All her sisters were very capable, while her brothers were good for nothing. I mentioned earlier in this book my uncle-in-law Dan Yixing, who was the same age as my father and the younger brother of my grandmother. He was a useless person who could do nothing. As my grandfather spent little time at home, all the family affairs had to be handled by my grandmother. My father respected her very much. He once said that it was his mother who kept the family going. It was said that my grandmother loved her eldest son very much. When he left home, she was very worried when she did not hear any news from him. Some people said that she died because she missed her son too much. Perhaps they were right. This traditional woman had to handle the family affairs and, at the same time, missed her son. Hard work and grief led to her early death. She died in 1926, never seeing again the son whom she missed so much.

The third wife of my grandfather was Xiao by surname. She gave birth to the fourth son of the Deng family before she died. Then my grandfather had another wife, Xia Bogen, the Grandma who is still living with us and whom we love so much. She is over ninety but still very healthy. Her life was ordinary and, at the same time, not so ordinary. Her father was a boat tracker on the Jialing River, a true poor laborer, without an inch of land. She had an elder brother, but he died when he was a child. Her mother was so grieved that she left for the other world soon afterwards. Grandma, the only daughter, lived with her father. When she was a teenager, she was married to a "middleman" (something like a notary public) and gave birth to a daughter. But unfortunately, her husband died soon afterwards. She was married to my grandfather, taking her daughter along. With my grandfather, she gave birth to three daughters. The first one was my second aunt, Deng Xianfu; the second one was Deng Xianrong, who died at about age ten; and the third one was Deng Xianqun. When my grandfather died, she was barely one year old.

The death of my grandfather was a big loss to Grandma. As she was a remarried widow and had no sons, she had no position in the family at all. However, she was loved by her fellow villagers because she was clever, capable, reasonable, and ready to help others. She was good at weaving cloth and knew how to till the land. She was especially good at cooking. My third uncle, who was the head of the family, did not in fact care too much about the family. The whole burden fell on Grandma. Like my grandmother Dan, she was the prop of our family that kept it going.

So, my grandfather had seven children (not counting those who died early) in all: Deng Xianlie (female), Deng Xiansheng, Deng Xianxiu, Deng Xianzhi, Deng Xianfu (female), Deng Xianqing, and Deng Xianqun (female).

My first aunt, Deng Xianlie, was two years older than my father. She was married to a landlord named Tang, who was much richer than the Deng family. She is still alive.

My second uncle, Deng Ken (Deng Xianxiu), joined the Communist Party in 1937. He was the only intellectual in the Deng family. He worked as a journalist and then as a cultural worker. After liberation, he served as a deputy mayor of Chongqing, deputy mayor of Wuhan, and vice-governor of Hubei Province. He has retired and now is living in Wuhan. In our childhood, we felt that he was most unlike his eldest brother because he was tall and handsome. But now, when he sits by my father, we feel that he is like his elder brother after all, though one is tall and the other shorter.

My third uncle, Deng Shuping (Deng Xianzhi), was a small landlord before liberation. He did nothing else in particular and smoked opium. After liberation, Father sent him to a rehabilitation center, where he received some kind of revolutionary education. Later on, he did some work in the Liuzhi Prefecture of Guizhou Province. During the Cultural Revolution he was persecuted to death because of his class status as a landlord and his brother's loss of power.

My second aunt, Deng Xianfu, studied at a middle school before liberation and had connections with the local underground Party organization at that time. After Sichuan was liberated, she was sent to study at the Southwest Military and Political School run by what was then the Southwestern Bureau of the CPC Central Committee. All along she did confidential office work. She has now retired, retaining only the post of a member of the Sichuan Provincial Committee of the Chinese People's Political Consultative Conference. Because she worked in Sichuan for a long time, is always ready to help others, and devotes all her energy to public good, people in Sichuan lovingly call her "Sister Deng" or "Deng Niang Niang" (Aunt Deng). Her husband, Zhang Zhongren, served as the director of the Archive Bureau of Sichuan Province. He is a humble, honest, industrious, and dependable person. My grandmother loves him the most.

My fourth uncle, Deng Xianqing, lost his mother when he was a child, and it was my grandmother who raised him. From an early age, he has not been physically fit and just does what he can in Sichuan.

My youngest aunt, Deng Xianqun, is the youngest of the older generation of our family. Grandma has described her as a "wild cat" ever since she was a child. She climbed trees to search for birds' nests and went down to the river to fish. She was up to all kinds of mischief. For all this, she was beat the most. After liberation, she came to Beijing with us. She finished her studies at the Experimental Middle School and entered the Harbin Academy of Military Engineering. Except during the Cultural Revolution, she has always worked in the army, and now she is the director of the Mass Work Department of the General Political Department of the Chinese People's Liberation Army. She is one of the few women major generals in the army. She is both our senior and our good friend because of her age. She has a lively temperament. We played together when we were children, making no distinctions according to family status. Her husband is Li Qianming, deputy commander of the Second Artillery Corps of the Chinese People's Liberation Army. She and her husband were schoolmates at the Harbin Academy and are now leading cadres in the army. As her husband is a major general, we do not call him uncle but, jokingly, "Great General Li."

Grandma likes my second uncle most among her sons-in-law, but her favorite daughter is my youngest aunt. The two aunts all have close relations with us. Their children live in our family, and Grandma looks after them. They serve as "toys" of our family as well.

My grandfather was the only son in his generation, but there were many more children in our generation and the next. The Deng family, which had only one son for three generations, is most populous now.

CHAPTER 10

FATHER'S CHILDHOOD

It was the year 1904. On July 12 in the Chinese lunar calendar—August 22 in the Western calendar—Father was born in Baifang Village, Xiexing Township, Guang'an County, Sichuan Province.

His father, Deng Wenming, was then eighteen years old, and his mother, Dan, was twenty. It was the happiest occasion for the Dengs to have a baby boy. However happy an event it was in the family, it was nothing in Guang'an, Sichuan, and throughout the country. Father was one of tens of thousands of babies born that year. Because of his fame now, some relatives and villagers spread a story about a good omen at the time of his birth. Those were nothing but fabrications.

At that time Baifang Village was poor and remote. Certainly no one had anything like a camera. So we have no clue as to what Father looked like at his birth. However, when my younger brother's son Xiao Di was born, every family member said he resembled his grandfather. I jokingly said that if someone was going to make a film on Deng Xiaoping's childhood, Xiao Di was qualified to play his grandfather as a baby. What does my nephew Xiao Di look like? A round face, wide forehead, light eyebrows, white skin, small eyes, plus a round nose tip that is typical of our family and has been handed down from our ancestors.

At the age of five, Father went to Si Shu, the old-style Chinese private school. It was located in the "imperial academy," the old house of the former academy member Deng. The teacher of Si Shu thought that my father's name, Xiansheng (meaning sage), was not a good one (maybe he thought it a little disrespectful to Sage Confucius), so he changed my father's name to Xixian (meaning, hoping to be a wise man). Father used this name for twenty years.

At the age of six, Father went to the lower primary school in Xiexing Township. To be educated in Si Shu, he only read *Sanzijing*, a three-character-phrase textbook for children, and *Baijiaxing*, a book of Chinese clan names. At the lower primary school, only *Sishu* (*The Four Books*, namely, *The Great Learning*, *The Doctrine of the Mean*, *The Analects of Confucius*, and *Mencius*), and *Wujing* (*The Five Classics*, namely, *The Books of Songs*, *The Book of History*, *The Book of Changes*, *The Book of Rites*, and *The Spring and Autumn Annals*) were taught. The teaching method then was mainly reciting. True, it was not a correct method to recite without comprehending the meaning. However, what one recited in childhood was remembered forever. Moreover, as far as Chinese was concerned, the more one learned by heart, the greater impact it would have on one's cultural traits. Nowadays, teaching emphasizes comprehension, not cramming. But I believe that because they do little reciting or reading of the classics, a lot of children are still at a very low level culturally when they graduate from middle schools and even colleges. It is beyond my purposes here to say which is the correct teaching method.

In 1915, when he was eleven years old, Father passed the examination and entered the primary school of Guang'an County, which was the only one there. The school was situated on a hill in Guang'an County town. When my second aunt and I returned to our hometown, we visited the school. There were only a few old rooms left. On the second floor was a wooden corridor with wooden banisters. The classroom was rather small, for some twenty pupils only. Because it was the sole higher primary school in the county, it was hard to gain entrance. I presumed that Father studied hard when he was young, so he could go to this school.

The school enrolled only one or two classes of students at a time. It offered very limited courses, and almost no courses in the natural sciences. In addition to the works of Confucius and Mencius, Chinese courses included such articles from *The Selected Works of the Best Classics* as the proses composed by Liu Zongyuan and Han Yu. The teaching method was the same as that in the lower primary school, that is, recitation. The pupils differed in ages; some were around ten, and some over twenty. All those pupils whose homes were not in the county town were boarders. Father was one of them and went home once a week. My aunt told me that there was no highway whatsoever at that time. When she went to school, she had to take a ferry across the river, climb hills, and walk on the stone road.

At the time, Guang'an County had a population of between 200,000 and 300,000, but the whole county had only one higher primary school and one junior middle school.

In 1918, at the age of fourteen, Father finished his studies in the higher primary school and was admitted to Guang'an County Middle School.

Many great men distinguish themselves and show their talents in their childhood and youth. But I consider Father's childhood to have been very ordinary. Some legends about my father told by some relatives and fellow villagers are mostly unreliable. But one thing is certain: in his childhood my father was a bright good boy dearly loved by his parents as well as a good hard-working pupil at school.

Father left Guang'an County Middle School not long after being admitted to it. My grandfather was then in Chongqing. When he heard that a preparatory school for studying in France on a work-study program would be established, he sent a message back home, asking his son to come to Chongqing to study in the school.

CHAPTER 11

THE MOVEMENT TO STUDY IN FRANCE ON WORK-STUDY PROGRAMS

The movement to study in France on a work-study program was a component of the new cultural movement that emerged during the late nineteenth century and the early twentieth century.

After the First Opium War broke out in 1840, China was gradually reduced to the status of a semifeudal and semicolonial country with a corrupt government, a chaotic social order, and a backward economy. While the Chinese people were waging unyielding struggles against imperialist aggression and oppression by feudal emperors, some people with lofty ideals began to explore various ways of saving the nation and the people. Under the influence of reform ideas and the "Westernization movement," many people, especially the intellectuals, believed that they had to learn from the Western countries to save the country. Before the Revolution of 1911, some ardent youths went abroad group by group to study foreign languages, politics, law, and military science. Mao Zedong said, "Chinese who then sought progress would read any book containing the new knowledge from the West. The number of students sent to Japan, Britain, the United States, France and Germany was amazing. At home, the imperial examinations were abolished and modern schools sprang up like bamboo shoots after a spring rain; every effort was made to learn from the West."[1]

Some young Chinese revolutionaries who were drawn into the old democratic revolution not only studied and explored foreign ideas for saving China and its people but also carried out and initiated revolutionary causes in foreign countries far away from the hinterland of feudal rule. Zou Rong, Qiu Jin, Chen Tianhua, Huang Xing, and other renowned revolutionary pioneers all studied in Japan, and the great revolutionary forerunner Dr. Sun Yat-sen also took a foreign land as a revolutionary base when he formed in Japan the first national bourgeois revolutionary party, Tong Meng Hui.

The movement to go abroad for advanced studies prevailed during the period before the Revolution of 1911. The number of Chinese students studying in the United States exceeded 800, and some 500 students went to Europe each year. The largest number of Chinese students studying abroad—over 20,000—went to Japan.

After the Revolution of 1911, autocratic feudal rule in China was overthrown and all the imperialist powers were busy fighting the First World War. During this period, national capitalism developed further in China, and as a result, Chinese educators and advocates for studying abroad radically changed their thinking. Some thought that France was a country where the bourgeois revolution had been thoroughly carried out and where most of the new scientific ideas had also originated. They therefore believed that Chinese students would do better to study in France than in Japan. Besides, living costs in France were lower than in other European countries. So France was a suitable country for Chinese students who had to cross the ocean to study, especially those who studied at their own expense.

In the year following the Revolution of 1911, in April 1912, Cai Yuanpei,[2] Wu Yuzhang,[3] Li Shizeng,[4] and others in Beijing sponsored the establishment of the Society of Part-Work and Part-Study in France, as well as a preparatory school for students going to France. By June 1916 it had organized two groups of students, totaling eighty, and sent them to France for further study.

After the Revolution of 1911 failed, Cai Yuanpei and Wu Yuzhang had to go into exile abroad. In June 1915, with their support, the Chinese workers in France set up the Work-Study Society, which was designed to "carry on study through hard work."

The First World War broke out in August 1914. One million French people died or were wounded in the war. As a result, there was an acute shortage of labor on the French home front. The French government hurriedly sent officials to China to hire cheap Chinese workers. In this way, during the First World War the number of Chinese workers in France grew to over 100,000. The wages of these Chinese workers in France were very low (the daily pay was only five to ten francs, less than one-third of that earned by French workers), and they had little chance to receive education. In view of this, in June 1916 Cai Yuanpei and others sponsored the establishment of the Sino-Franco Education Association (SFEA).

In the same year, in China, Yuan Shikai's scheme of restoring the autocratic monarchy shamefully failed, so many revolutionaries who had gone into exile abroad could come back to China. In 1917 the Sino-Franco Education Association and the Association for Studying in France on a Work-Study Program were established in Beijing.

The Association for Studying in France on a Work-Study Program set up three preparatory schools in Beijing, Baoding, and another city for students preparing to go to France. Very soon the movement to study in France on a work-study program spread to all parts of the country. The SFEA opened or started various preparatory schools and courses for France-bound students in Shanghai, Chengdu, Chongqing, Changsha, Guangzhou, Jinan, Tianjin, Wuhan, and other cities.

In November 1918 the First World War came to an end.

After 1919 the movement to study in France developed rapidly and reached a new climax after the May 4th Movement of 1919. Between 1919 and 1920 the two associations organized seventeen groups of young students, totaling over 1,600—most from Hunan and Sichuan—to go to France for further study. The movement to study in France on a work-study program was unprecedented.

This movement spread across the country with a powerful momentum mainly owing to the direct influence of the international situation and the situation in China itself at that time. After the Revolution of 1911, the bourgeois revolutionaries led by Dr. Sun Yat-sen were forced to give up the fruits of their victory under pincer attacks by powerful Chinese feudal forces and imperialist forces. When the Qing emperor gave up the throne in 1912, the fruits of the Chinese bourgeois revolution were first seized by Yuan Shikai, the arch usurper of state power, and later were grabbed by the Northern Warlord government represented by Li Yuanhong and Duan Qirui. The "Second Revolution" and the "War of Upholding the Provisional Constitution," both led by Dr. Sun Yat-sen, ended in failure.

China entered a miserable period of separatist warlord regimes, both big and small, and internecine warfare among them. The peasants became impoverished, and the workers lost their jobs. Natural calamities also occurred. The people's lives became more and more miserable, and the rural and urban economies languished. The chaotic domestic situation compelled some young intellectuals to seek ways to save the nation and the people from abroad. The failure of the old bourgeois democratic revolution also prompted some people with courage and consciousness to explore new truth.

During this time, two major events shook China like thunder and lightning. The first was the outbreak of the October Revolution in Russia in 1917. The victory of the October Revolution greatly encouraged the Chinese revolutionary intellectuals, spurred the development of the new Chinese cultural movement, and promoted the quick dissemination of Marxism in China. Vast numbers of progressive Chinese intellectuals yearned for and absorbed the new thinking and the new culture. Li Dazhao, Chen Duxiu, and others spared no efforts in actively propagating, advocating, and disseminating the great news of the October Revolution and Marxism. These efforts undoubtedly drew the attention of many Chinese revolutionaries and gave them new hope.

The second event was the outbreak of the May 4th Movement in 1919. After the First World War ended, in January 1919 in Paris, the victorious nations held a "peace" meeting that was actually held for the purpose of "dividing the spoils." As one of the victorious nations, China put forward a claim for the return of its sovereignty over Shandong Province and the abolishment of the unequal treaty, the Twenty-one Demands. Britain, France, the United States, Italy, and Japan not only turned down China's just claims but also transferred to Japan all the rights and interests seized by Germany in Shandong Province. The Northern Warlord government signed this humiliating "peace treaty." The whole nation immediately made an indignant outcry. On May 4, over 3,000 students from Beijing University and other institutions of higher learning gathered in Tian'anmen Square and held a demonstration. In spite of being suppressed by the army and police and foreign police, the students gave a good beating to the Northern Warlord government officials and set fire to the residence of Cao Rulin, the vice foreign minister of the Northern Warlord government. The Chinese people immediately responded to the patriotic actions of the Beijing students. Before long, students, workers, and businessmen all went on strike, and the whole nation joined in the protest.

The May 4th Movement was a great anti-imperialist and antifeudal revolutionary movement and marked the beginning of the new democratic revolution. A large number of revolutionary intellectuals across the country were tempered in the May 4th Movement, an involvement that laid the foundation for their continued quest for the truth and their further revolutionary activities.

The dissemination of new ideas and the outbreak of the May 4th Movement further propelled the movement to study in France, which became widespread throughout the country. Hunan was one of the provinces where the movement proceeded very smoothly. In April 1918 a twenty-five-year-old man named Mao Zedong and a twenty-three-year-old man named Cai Hesen set up in Changsha, Hunan Province, a revolutionary organization—Xin Min Society (New People's Society)—and very soon it started the work-study program in France. Between 1919 and 1920, of the 346 students from Hunan Province who studied in France, 18 were members of the New People's Society. At home, they were the core members who organized the movement to study in France, and in France they were the backbone of progressive study groups and were involved in establishing communist organizations in Europe. Among them were Cai Hesen, Xiang Jingyu, Li Weihan, Li Fuchun, Zhang Kundi, and Cai Chang, as well as Xu Teli, a famous educator, and Ge Jianhao, mother of Cai Hesen.

In Sichuan the movement to study in France was sponsored and led by Wu Yuzhang himself. In 1918 the schools that prepared students for study in France were set up in Chengdu and Chongqing, respectively. By the end of 1920, 378 students, including Chen Yi, Nie Rongzhen, Deng Xixian (Deng Xiaoping's name at the time), Jiang Zemin (Jiang Keming), and Zhou Weizhen, had gone to France for further study.

By the end of 1920, many progressive youths had gathered in France. Wang Ruofei and his uncle, Huang Qisheng, a venerable gentleman, left Guizhou for France in December 1919. Chen Yannian and Chen Qiaonian (sons of Chen Duxiu), Li Weinong, and others from Anhui left for France in 1920. In September 1920 Zhao Shiyan from Sichuan left for France. In November 1920 Zhou Enlai, an organizer and leader of the Society of Awakening in Tianjin, went to France. These young Chinese students did not travel such a long way to seek Western civilization or to learn some professional skills. Cherishing patriotism and the ideal of saving the nation, they broke through the yoke of feudalism and went to the outside world to acquire knowledge and seek truth so that one day, when they came back to their motherland, they could save the country and serve it.

The twenty-two-year-old Zhou Enlai wrote in his poem:

Bear in mind your spirit,
your determination,
your courage,
Make progress joyfully,
and all depends on your brave struggle.
Go abroad,
Cross the East Sea, South China Sea, Red Sea and Mediterranean Sea,
With surging waves
rolling on in the vast oceans,
You are bound for the coast of France, the hometown of freedom.
On that land
Take up the tools,
be wet with sweat in labor,
make your brilliant achievements.
Develop your ability and maintain your innocence.
When you are back someday,
Unfold the banner of freedom
and sing the song of independence.
Strive for women's rights,
Seek equality
and conduct experiments in society.
To negate old doctrines
depends entirely on the ideal in your mind.

CHAPTER 12

A LONG JOURNEY IS STARTED BY TAKING THE FIRST STEP

Wu Yuzhang initiated a preparatory school for students to study in France in Chengdu in Sichuan in the spring of 1918, and the first group of students was sent to France in June 1919.

Chongqing, an important city in eastern Sichuan and a trading port area, would certainly not lag behind Chengdu in culture and education. Wang Yunsong, chairman of the Chamber of Commerce of Chongqing, and Wen Shaohe, director of the Chongqing Educational Bureau, raised tens of thousands of silver dollars from celebrities in the city to set up a school in Chongqing to prepare students for study in France.

My grandfather, who presumably had heard the news, sent word from Chongqing to my Father in Guang'an asking him to come to Chongqing to attend the school. In the second half of 1918, Father arrived in Chongqing, together with Uncle Deng Shaosheng, a distant relative, and Hu Mingde, a fellow villager. Both Deng Shaosheng and Hu Mingde (also known as Hu Lun) were students at Guang'an County Middle School.

The preparatory school enrolled both government-funded students (or those financed by government loans) and self-supporting students. Deng Shaosheng qualified for the first category, whereas Deng Xixian and Hu Lun fell under the second category. The travel expenses for the trip to France were 300 silver dollars, of which over 100 were provided by the school board of directors; the rest had to be raised by the students themselves. Father's travel expenses were paid, of course, by Grandfather. Deng Shaosheng had to borrow from others, and Hu Lun was subsidized by his friends.

In early September 1919 the Chongqing Preparatory School for Students to Study in France on a Work-Study Program was formally opened. Wang Yunsong was the chairman of the board of directors of the school. Under him were the schoolmaster, the dean of studies, and the administrative staff. The school was located in the Confucian Temple in Chongqing.

Students of the school were either middle school graduates or youths who had acquired the same educational level. The school enrolled over 100 students and divided them into two classes. All middle school graduates studied in the advanced class, the rest in the elementary class. The courses offered included French, algebra, geometry, physics, Chinese, and elementary knowledge of industry. French was the main subject. Two teachers taught the French course. Wang Meibo, an interpreter working in the French Consulate in Chongqing, taught the advanced class, and a man named Zhang, who had studied in France, taught the other class. Jiang Zemin, a schoolmate of Father's, recalled that the school had poor teaching facilities and was loosely organized.

Students went to school only to attend classes and left right after class.

There were no dormitories or playing fields. The purpose was to teach students a little French and impart some knowledge of industrial technology so as to prepare them for their study in France. Father once said that the school was at that time the best one in Chongqing, so it was not easy to be admitted.

When Father went to this school, he was only fifteen. Jiang Zemin recalled: "Deng Xiaoping entered this preparatory school a little bit later. He looked quite smart and was always full of energy. He did not talk much and always studied very hard."

Not long after the first students were enrolled, something happened that they would never forget for the rest of their lives. Chongqing was a passage to southwestern China, a hub of communications, by both land and water, and the largest industrial and commercial city on the upper reaches of the Yangtze River. After China signed the special supplement to the Treaty of Yantai with Britain in 1890 and the Treaty of Shimonoseki with Japan in 1895, Chongqing formally became a trading port to foreign countries. France, the United States, Japan, and other countries set up their consulates in Chongqing. Chongqing was in the spheres of influence of the British and French imperialists. From then on, Britain and France shared all rights in Sichuan. Foreign commercial ships sailed in frequently, and the warships of the imperialists ran amok on the Yangtze River. Britain, France, Germany, Japan, and the United States took turns controlling Chongqing customs. Worse still, they occupied the piers, set up barracks, and occupied the concessions by force. As it was very hard to go to Chongqing, the feudal forces there were stronger, the economy was more backward, and the class oppression was more severe compared with conditions in other coastal cities.

After the Revolution of 1911, Chongqing became the focus of the internecine warfare among the warlords of Sichuan. The gigantic forces of the separatist warlords in Sichuan and the wars among them year after year brought untold sufferings to the people in Chongqing.

The May 4th Movement broke out in Beijing on May 4, 1919. Owing to Sichuan's poor communications, the news did not reach Chongqing until mid-May. The students, youth, and people from all walks of life in Chongqing were burning with patriotism and responded immediately. They not only denounced the traitors in various forms but also boycotted Japanese goods and opposed all transactions with Japanese businessmen.

In November of that year, Zheng Xianshu, commissioner of the Chongqing Police Bureau, misappropriated over 4,000 silver dollars of public funds to purchase over 80 cartons of cheap Japanese goods from a Japanese firm, then sold them openly in the name of the Police Bureau, triggering an immediate wave of protest by the patriotic students. The morning of November 17, over 1,000 students from Chuangdong Normal School, Chongqing Associated Middle School, the Preparatory School for Students to Study in France on a Work-Study Program, and other schools held a protest rally in front of the Police Bureau, strongly demanding that Zheng Xianshu hand over the Japanese goods. Zheng was too scared to come out. The students surrounded the bureau day and night. Their patriotic action won the support of the city dwellers, who sent them food. This support in turn heightened the students' morale. The next morning, Zheng Xianshu was forced to agree to hand over the Japanese goods. In the clash between the students and Zheng's armed escort, two students were wounded. The indignant students fought with the escort and disarmed him. Zheng escaped through a window. That afternoon, the students burned the Japanese goods that Zheng handed over at Chaotianmen in Chongqing. The students finally won victory in the struggle, and the Sichuan authorities had to dismiss Zheng Xianshu from his post.

Jiang Zemin recalled:

> To boycott Japanese goods and oppose traitors, the students of our preparatory school went together to present a petition to the headquarters of the Chongqing Garrison, where we fought for two days and one night and won some initial results. After returning to school, we threw all our daily necessities with Japanese trademarks, such as tooth powder and basins, on the ground and had them burnt. To show our determination not to use inferior Japanese goods, we even tore those clothes made of Japanese cloth. Our minds were influenced by the patriotic trend of thought—the pulse of the times. I was deeply impressed by the high patriotism of the young students and people from all walks of life.

Father, who was then fifteen years old, took part in the movement, together with all his schoolmates. He once recalled that his patriotic ideas of national salvation were enhanced after participating in this movement. However, the kind of national salvation they talked about at that time was nothing but the idea of saving the country by means of industry. In his still naive mind, he was anxious only to go to France to study so that he could learn something useful for his country.

At the time, Father had only some patriotic and progressive ideas; he had not yet acquired the distinct outlook on life and the world that he had later. However, his May 4th Movement experience had a great influence on him and was of considerable significance in the development of his outlook and his revolutionary practice.

After completing a year of study, the students of the preparatory school had their graduation ceremony at the Chongqing Chamber of Commerce on July 19, 1920. Present at the ceremony were the French Consul and the French businessmen and missionaries in Chongqing, as well as the schoolmasters of other schools. Over eighty students passed the school examinations, the oral examinations given by the French Consulate in Chongqing, and the physical checkup. Deng Xixian was one of them, and the youngest.

Wu Yuzhang was generally acclaimed as the initiator of the movement in Sichuan to study in France, whereas Wang Yunsong was generally believed to be the hero of the movement in Chongqing. Wang Yunsong called himself De Xun. Having witnessed the grand occasion of the students from Chengdu going to France via Chongqing, he immediately prepared to establish the Chongqing Branch of the Association of Studying in France on a Work-Study Program. Besides serving in various positions associated with the school, he took personal charge of everything. He helped to establish the branch association and the preparatory school, raised funds, applied for visas, and finally saw the students off. His students were so impressed by his enthusiasm that they had not forgotten him even decades later.

One day in 1949, after Chongqing was liberated, the headquarters of the Southwestern Military Area sent several people to Wang Yunsong's home. Not knowing whether this visit augured well, he did not dare to meet with them. A jeep came the next day and brought him to the headquarters. To his surprise, he was invited to dine with Deng Xiaoping, political commissar of the Southwestern Military Area. Overjoyed, he told anyone he met after coming home that, "Xiaoping is pretty good. Only now I come to know that the Communists do not forget their old acquaintances." When the Second National Political Consultative Conference was held in 1950, Wang Yunsong was invited to attend the conference in Beijing as an observer. After coming home, he told people that he attended the banquet given in the Huairen Hall at Zhongnanhai. Chairman Mao was the host of the first table, and Comrade Xiaoping was at the second one. Wang Yunsong was also at the second table. Xiaoping and Chen Yi had a division of labor: Comrade Xiaoping played host to Wang Yunsong and Chen Yi sent him back to the guest house by car after dinner.

Wang Yunsong had served in the Qing government for a while as a fourth-grade district magistrate. He had been an official of the Qing government, but he was reform-minded. He did not mean to set up the school to train Communists but to train industrialists who would save the country by means of industry.

The venerable gentleman Wang loved his country and also loved the Communist Party after liberation. He donated all the precious historical relics he had collected to the state. He had a pair of ancient porcelain vases that he loved dearly. He had the three characters "Dong Fang Hong" (the East is red) carved on the box made of phoebe nanmu that the vases were in and presented them to Chairman Mao as a birthday present. According to the rules, leaders of the Chinese Communist Party do not celebrate their birthdays, nor do they accept birthday presents. A comrade, who worked in the United Front Work Department of Chongqing at that time, said when Comrade Xiaoping learned about that, he told the department it had to "understand Wang Yunsong." The United Front Work Department made an exception and accepted the present.

That same comrade from the department said that Comrade Xiaoping mentioned later that Wang Yunsong had fostered for us two vice-premiers (referring to Comrades Xiaoping and Nie Rongzhen).

At 3:00 P.M. on August 27, 1920, eighty-three students from Chongqing

Preparatory School lined up and were ready to leave. They boarded the steamer, a French commercial ship named the *Jiqing*, which was heading for Wanxian County. On August 28, the *Jiqing* weighed anchor and sailed slowly toward Yichang. We can imagine that many people went to see them off on the river bank and that quite a few of them were in tears. With great expectations, they were sending off their sons and brothers. They must have felt that in front of them was an endless road to the remotest world.

Those eighty-three students may have been reluctant to bid farewell to their dear family members and their hometowns. Yet they must also have been eager to face the future and to see a foreign land. They must have been quite excited and longing for new things.

CHAPTER 13

MAKING A LONG AND DIFFICULT JOURNEY

It is harder to travel in Sichuan than to try to reach the blue sky.

Sichuan is located in the interior of southwestern China. It borders on the loess plateau in the north, the Qinghai-Tibet Plateau in the west, the Yunnan-Guizhou Plateau in the south, and some barren mountains and hilly land in the east. This flat green land finds itself among countless mountains and valleys, looking quite like a piece of shining green jade inlaid in a fine wood carving.

Sichuan is indeed a precious land richly endowed by nature. However, it has kept itself closed off from communication with the outside since ancient times. Mountains like Qinling, Dabashan, Minshan, Daxueshan (Snow Mountain), Daliangshan, and Daloushan surround the Sichuan Basin. It took people in ancient times several months to enter Sichuan by horse or cart.

In modern times, someone who traveled in Sichuan once said that every day people there have to look for roads. He took a ship, finding no way ahead or behind when it sailed along the Three Gorges of the Yangtze River. In his eyes, all water in Sichuan is dangerous water and every mountain is a high mountain. As the saying goes, "Sichuan is already in good order when other parts of the country are still chaotic. Sichuan is in disorder when order has been restored across the country." Such a characteristic has been favorable to Sichuan's political and economic independence. Yet it has made Sichuan even more closed. When people from other provinces get to Sichuan, sometimes they feel like they are in a kettle.

The roads out of Sichuan are nothing but dangerous mountain paths. Fortunately, the surging Yangtze River flows down to the east, going through many twists and turns and providing a lifeline for people to get in and out of Sichuan since ancient times.

If it is hard to enter Sichuan, it is also difficult to get out of Sichuan. Even traversing the Yangtze River, one has to experience numerous precipitous mountains and dangerous rapids. For ages, countless travelers have lost their lives in the torrents and become food for the fish.

Nevertheless, the people of Sichuan aspire to higher living standards and new life. They know that a wider world awaits them outside of Sichuan. Therefore, when they have sailed down the river with a favorable wind, they must have felt excited. The hazy mountain scenery on the way must have made them feel especially carefree and joyful.

The eighty-three students from Chongqing Preparatory School for Students to Study in France on a Work-Study Program had a smooth journey sailing eastward on the *Jiqing*. The Chongqing government did not send anyone to escort them. The students had organized themselves and helped each other. They were divided into four groups, each consisting of about twenty students. Each group had a leader.

After eight days of travel through Yichang, Hankou, and Jiujiang, they finally reached Shanghai, safe and sound. It was a rather short journey, but their first one. We can imagine how excited they must have been in those eight days. They could enjoy the mountain scenery, then the plains, all of which passed by them in a twinkling of an eye. The scenery was similar to that of Sichuan, yet so different.

Shanghai was the commercial and trading center in eastern China. It was also an important port city through which China engaged in trade with other countries. All the students organized by the Sino-Franco Education Association and the Association for Studying in France on a Work-Study Program first gathered in Shanghai and embarked for France from there. The Shanghai SFEA was located at 247 Xiafei Road in the French Concession in Shanghai. Since only one ocean liner went from Shanghai to France each month, the Shanghai SFEA put the students up at the Mingli Hotel and helped them book tickets and get their visas from the French Consulate. A week later, at 11:00 A.M. on September 11, 1920, the eighty-three students from Sichuan boarded the French ocean liner *André-Lebom*. It rained cats and dogs that day. But the young students were so eager to see the new world as soon as possible that the heavy downpour was nothing to them.

After boarding the ship, the students checked their luggage. As the siren wailed, the *André-Lebom* weighed anchor and left Huangpu Harbor in Shanghai. Before long, the ship passed by Wusongkou and disappeared into the vast ocean.

The French ocean liner sailed between Europe, Asia, and America. It was over 166 meters long, 20 meters wide, over 33 meters high, and weighed tens of thousands of tons. It had three classes of cabins, each with a capacity of several hundred people. The top deck was a game ground where the passengers could do physical exercises. The cargo holds were on both ends, each with a large capacity. There were two hoists on the ship for loading and unloading cargo.

There were altogether ninety Chinese students on the ship, of whom eighty-four[1] were from Chongqing Preparatory School and the rest from Zhejiang Province. Of the eighty-four students from Chongqing, forty-six had obtained loans to finance their studies and thirty-eight were paying their own way. Jiang Zemin and Deng Shaosheng had loans, whereas Deng Xixian (Xiaoping) and Hu Lun paid their own expenses.

Jiang Zemin recalled that a first-class ticket on the ship cost 800 silver dollars, a second-class ticket 500 silver dollars, and a third-class ticket 300 silver dollars. The Chinese students paid 100 silver dollars for fourth-class (classless) tickets. The ship actually had no fourth-class cabin. Such a cabin was set up especially for the poor Chinese students. It was in essence a cargo hold located at the bottom deck of the ship, and it was poorly lit. Cargo was everywhere, and there were no facilities except some bunk beds for the students to sleep on. The unventilated air was hot and stuffy. Worse still, there were a lot of mosquitoes and other bugs. Many students bought deck chairs so as to kill time and get some sleep on the deck. When the sea was calm, they could drink in the ocean scenery. When it became stormy, huge waves hit the ship and the turbulence made them sick.

By that time, the students' excitement had probably cooled down to a great extent. From the ship, one could see nothing but the ocean and the sky on the horizon. The ship was all alone. There were no mountains, trees, or land in sight. It was a bleak picture that made the students feel lonely. Some students tried to while the time away by reading the books they had brought with them.

One of the students was Feng Xuezong from Baxian County, Sichuan Province. In his letters to his kinsfolk and friends, he gave a detailed account of the voyage through which we can get some idea of the trip and the students' feelings.

> On the 14th, we arrived in Hong Kong where we stayed for one day. With hills behind and the sea in the front and with green trees as nice shades, Hong Kong is crowded with merchants and tourists. The streets are wide and houses neatly lined up. Though the businessmen are Chinese, all the rights to govern the place are in the hands of the British. After acquiring the place, the British formulated many rules and regulations to control the Chinese. It has recently become the most prosperous and most important commercial city along the coast.
>
> The ship reached Saigon on the 18th. It is a flat area. Since the French captured the city, it has been well managed with harbors built along the coast and buildings constructed along the banks. However, there is one sad thing. That is the Vietnamese people, who have lost their own country. Since their country became a colony, they have no choice but to yield to the rule by the foreigners. People who have better education have to answer the calls of the French, thus becoming apathetic.
>
> People who have little or no education have to serve the French either by engaging in agriculture or by pulling rickshaws. They are lashed if they are indolent. It is sad enough to see them in such a misery. Yet the French enforced all kinds of evil rules so that [the Vietnamese] would never have the chance to recover their sovereignty. For instance, people have to study French if they want to study and pay tax if they want to wear shoes. The French plans not only to eliminate the Vietnamese language but also the race. Where is justice? Where is humanism? With disheveled hair and bare feet, the Vietnamese people led such a life throughout the year. Wouldn't that be a problem?
>
> Being situated in a strategically important place of Europe and Asia, Saigon is densely populated with people from all continents. There are about sixty to seventy thousand Chinese residing there. As one rigorous way to restrict the people from outside who are going to Saigon, the French requires each adult to pay dozens of dollars every year as poll tax. Since we Chinese have been known as "sick men," every country places strict check upon us.
>
> After we reached Saigon, I saw my companions pass through all kinds of check-ups before they lined up and walked to the police station for registration. Otherwise, they could not go ashore. You see, we Chinese are like would-be slaves to foreigners.

Three days later, our ship left for Singapore on the 21st.

It took us three days to reach Singapore. The streets and houses there are almost as neat as those of Saigon, but the city is larger and cleaner than Saigon. There are hundreds of thousands of Chinese residents and the Chinese businessmen are managing pretty well. Many families are quite well off. Yet most of the people are still laborers.

We left Singapore on the 25th. On the following day, the sea was terribly turbulent. Huge waves tossed the ship up and down, left and right. My companions were like sick people who could neither stand straight nor had any appetite. It lasted for three long days. We longed to see the shore as if people expected the clouds in a bad draught. Day in, day out, and we were all desperate until the ship finally reached Columbus.

It was the 30th when we reached the British colony, Columbus, where it was hard to anchor because of the strong winds and waves. Fortunately, there was a harbor where our ship could be berthed. Yet, we were not able to see the place because it was the British colony which required British visa to go ashore and since we only had the French visa, we were not allowed to do so.

By October 17, while the ship sailed on the Arabian Sea, it was close to the mouth of the Red Sea. A lot of dangerous explosives were placed under the water during the war in Europe and they were not taken out after the war. All the ships passing by the area have to be prepared for contingencies. We learned how to use the lifesaving appliances that day. Still we felt quite unsafe.

It was the 8th when the ship reached Djibouti. It was located at the mouth of the Red Sea and belonged to Africa geographically, but was the French colony. Being a desert, it was extremely hot. Even grass would not grow. There were very few inhabitants. The reason why the French have not abandoned it is that it could serve as a rest place because all ships have to pass by it. For the same reason, the French have put in a lot of efforts and funds. The local people are all black with black faces and black bodies. Even their teeth are black. Most of them do not wear anything on the upper part of the body.

A piece of cloth is wrapped on the lower part of the body and quite resembles the Chinese skirts. There is nothing special there except some fruits, camel wool, and some ornaments.

We reached the Red Sea on the 10th. The air was dry and the sun was hot. After days of watching the reflections of the sun and water, we found that the blue waves looked red, thus the name of Red Sea was given, I guess. That day marked the ninth anniversary of the founding of the Republic of China. All the Chinese students on the ship gathered in the auditorium in the afternoon, each holding a national flag. After we bowed three times to the national flag, the national anthem was played. We also celebrated it by telling stories and putting on new plays. Everyone was happy and foreigners who watched the show also applauded for us. The sound of clapping was like thunder. That was something very interesting that happened during the voyage.

We reached the mouth of the Suez Canal on the 13th.

Several hours later, we continued with our voyage. We entered the mouth at dusk. Trees lined up the banks of the canal. Lights sparkled on the flowing water.

> We were so excited that we did not want to sleep. In the morning of the next day, I had a good look of the canal which is over 30 meters wide, enough for two ships to pass side by side. Before I realized, the ship reached Port Said at the north end. Since we could not go ashore, we did not see anything. Before we left the canal, we saw bronze statue erected at the bank, which looked quite impressive. He turned out to be Lesseps, the man who had the canal cut. At 5:00 o'clock in the afternoon, we entered the Mediterranean Sea.
>
> We passed by the Italian Peninsula on the 17th.
>
> Although most of it is hilly, the Italian already constructed many railway lines which made transportation convenient. Beautiful and magnificent cities could be seen one after another. Finally, in the ocean a huge island with thick smoke over it came to sight. My companions told me that it was an active volcano which look like that throughout the year.
>
> After breakfast on the 19th, we saw at a distance many masts and lamp stands that were drawing close to us.[2]
>
> Before we realized, we were already in Marseilles, a city which is located in the south of France.

Feng Xuezong's account provides us with relatively complete information about their voyage, and we are lucky that it has been preserved.

Jiang Zemin, another student on the ship, also wrote about the voyage:

> We were hit by a big storm in the Indian Ocean. The wind swept up the sea water, creating waves as big as little hills. The ocean liner with a tonnage of 40,000 was now pushed to the top of the waves and now placed at the bottom of the waves. It became dark in the daytime and our huge ship floated in the vast ocean like a boat which was as small as a piece of tree leaf. That was really terrifying. We could not take in any food. Furthermore, we vomit everything in our stomach. We suffered for three days and three nights. Fortunately, we all came through.
>
> The other side of the coin was that it was eye-opening for us. The ocean liner stopped at every big harbor for two or three days to load and unload cargoes. Those rich passengers would go ashore shopping and eat at restaurants whereas we poor students would go ashore for sightseeing and visiting museums, scenic spots and historical sites. Though in many cities there were tall buildings and well-dressed people, there were also beggars and people who were shabbily dressed. At some harbors, I saw poor children swim around the ships, and they were begging for money. Some passengers would throw coins into the water and the children would dive into the sea to pick them up. The passengers were amused by that, whereas the children lived on it, which was so sad to see. It made me realize that though the people of the world lived under the same sky, they led such drastically different lives. Injustice was everywhere, indeed. Of course, at that time I did not know that it was caused by capitalism and colonialism.
>
> The beautiful memory I had of the voyage was the residual flames of the volcano we saw in the Mediterranean Sea. At night, the flames that resembled the colorful fireworks shot into the dark blue sky, leaving inverted reflections on the water. That was such a wonderful night view. Though from time to time we were told to put on life buoy in case the ship hit the mines left from the war, our ship passed the

area safely. After nearly 40 days of voyage, we finally landed on the French soil in Marseilles in mid-October.[3]

Since this was their first trip abroad, each student must have had deep thoughts about and unforgettable impressions of the foreign scenery and of the experience in the vast ocean, where the sky looked so high.

In 1974, when the Cultural Revolution proceeded like a raging fire, Jiang Qing and her like created the *Fengqing* Ship Incident. Taking advantage of the return of the China-made *Fengqing* ship from its voyage abroad, the Gang of Four gave enormous publicity to it. Making use of the issue of whether to buy or make our own ships, they criticized the tendency to "worship everything foreign" as "national betrayal," directing their criticism at then Premier Zhou Enlai and other leaders of the CPC Central Committee. After Father was forced out of his post at the beginning of the Cultural Revolution in 1966, he just resumed his position and served as the first vice-premier to take charge of the State Council. He looked down on Jiang Qing and her like for what they did and waged a tit-for-tat struggle against them. Father mentioned the incident later more than once. He said, "They bragged so much about a ship with a tonnage of 10,000 tons. I told them that there was no need to brag. As early as 1920, when I went to France, the ship I took weighed several times more than that!" We can see that his first voyage made a deep impression on Father.

A venerable gentleman named Li Huang who had also studied in France recalled that when he went to Marseilles to meet the students from that ship, he met Deng Xiaoping, who was probably the head of those students. I asked Father about it. He said with a smile, "I was the youngest among more than eighty students, and I did not even have a say." So it must be someone else whom Li Huang met at that time, not my father.

CHAPTER 14

FROM SCHOOL TO FACTORY DESPITE HARDSHIPS TO PURSUE STUDIES

Situated at the outlet of the Rhone River by the Mediterranean, Marseilles was an important port as well as an industrial and commercial city in southern France.

On October 19, 1920, the ocean liner *André-Lebom* sailed into the port of Marseilles. After a journey of thirty-nine days and more than 15,000 kilometers, the Chinese students on board had finally arrived at the western part of the European continent and set their feet on the soil of France, a country they had long yearned for.

The *Petit Marseillais* reported on October 20 that 100 Chinese youths had arrived in Marseilles. Aged between fifteen and twenty-five, dressed in Western- or American-style suits and wearing bowler hats and pointed shoes, they seemed polite and cultivated. The Sino-Franco Education Association had sent some people from Paris to Marseilles to meet the newcomers. Mr. Liu, chief of the student division of the Sino-Franco Education Association, made a welcoming speech. The young Chinese students had arrived in France after a very long journey, and their happiness was reflected in their utterances and manners.[1]

A student called Feng Xuezong recalled that the students got off the liner with their luggage and left Marseilles the same day, heading straight for Paris by bus.[2] Sixteen hours later, they arrived.

Jiang Zemin recalled that "we arrived in Paris the next day and were greeted by many of the students who were studying there on a work-study program. Among them was Comrade Nie Rongzhen who had come to France a year before. Meeting on a foreign land, we felt very happy and close to each other."[3]

Nie Rongzhen, a native of Jiangjin, Sichuan Province, later became a marshal of the People's Republic of China. He came under the influence of progressive ideas during his early years and participated in the May 4th Movement of 1919 when he was studying in a middle school. During the summer vacation of 1919, because he was eager to change the existing state of affairs, he raised 300 silver dollars and went to Chongqing with a dozen fellow students. There, with the help of Wang Yunsong, the chairman of the Chongqing Chamber of Commerce, he got his visa from the French Consulate, and on December 9, 1919, he boarded the ocean liner *Phoenix* for France to study on a work-study program. After he arrived, he worked at the Schneider & Cie Iron and Steel Factory in Creusot, and in 1922 he entered the Charleroi University of Labor in Belgium.

Since Nie Rongzhen came to France before the others, he was naturally regarded as a senior. During his stay in France, Father developed a revolutionary friendship with Nie Rongzhen. After liberation, between 1952 and 1957, our two families were next-door neighbors. The children of our family often went through a small door of the courtyard wall to play in Uncle Nie's house. From time to time, Uncle Nie would invite our whole family over for jellied bean curd, a local Sichuan favorite. Before Uncle Nie passed away in 1992, Father visited him occasionally, even though he seldom visited anyone else. Father would kindly greet Nie Rongzhen by calling him "old chap." The friendship that developed from that first meeting underwent seventy-two years of storm and stress, so profound, deep, and moving.

After a short stay in Paris, Father and his fellow students departed separately, as arranged by the SFEA, for Montargis, Fontainebleau, Saint Etienne, Flers, and some other places for study and French-language training in local middle schools.[4] Deng Xixian and Deng Shaosheng were sent to Bayeux Boy Middle School in Normandy, while their fellow villager Hu Lun was sent to Cambrai College.

The Chinese students had now obtained some rough ideas about the look of France and the elegance of Paris even before starting their studies. According to the description of Feng Xuezong:

> Paris is about fifteen kilometers in diameter and spreads out as far as some fifty kilometers. Most of the buildings along the streets are five- or six-story, without any dilapidated ones. Buses and trams speed by on the roads, while underground are layers upon layers of tunnels. One can travel from high above to down under in the twinkling of an eye. Isn't it convenient! The articles on display in the royal palaces are well preserved, and the museums have a whole variety of exhibits. The railway station can get you anywhere, both a tourists' attraction and a convenient facility for travelers. It is indeed the "flourishing capital of the world."

In the eyes of the newly arrived Chinese students, the landscape of the European continent and the bustling metropolis of the West, as well as the alien environment and customs, were a great curiosity and attraction that won their admiration. In contrast to their own poor, backward, feudal, and uncivilized motherland, this was like a different world. What they saw and heard in the first few days increased their confidence in, and their expections for, their new studies that were about to begin.

On October 21, 1920, Father and his uncle, Deng Shaosheng, together with twenty other Chinese students, began their study at Bayeux Middle School. Bayeux Middle School was in the Normandy region in northwestern France, over 200 kilometers away from Paris. On October 22, the *Bayeux Daily* carried a news report entitled "Chinese Students Arrive in Bayeux." It read, "Over twenty Chinese students led by two of their countrymen who speak French fluently arrived in Bayeux last night. These young people have been sent to France by their government and study the subjects that they are interested in at Bayeux Middle School, so that they can learn the French language and get familiar with the conditions and customs of France. They are resident students."[5]

The school opened a separate class for the Chinese students, mainly for the purpose of improving their French. They followed a normal regime of student life. Father once told us that the school treated them as children, and they had to go to bed very early every day. Father added that it was a private school, and they stayed there just for a few months. They did not learn much, and the diet was very poor.

Today in the national archives of France one can find an itemized account of the expenditures of the Chinese students in Bayeux.[6] The account shows that, for the month of March 1921, Ten Si Hien (Deng Xixian) spent a total of 244.65 francs for food and lodging. Of this sum, 200 francs was for living expenses, 7 for laundry, 7 for bedding rent, 12 for school fees, and 18.65 for miscellaneous expenses. More than 200 francs a month was by no means a small figure for the self-supporting students. When Father left home, his family was already hard up for money. To support his study in France, the family had sold some rice fields. He was therefore only too aware of the need to be frugal while in France.

According to that account, the miscellaneous expenses of the other Chinese students ranged from 15 to 50 francs, averaging about 25 francs per month. Father, at 18 francs, was relatively thrifty. Despite his thrift, Father soon used up all the money he had brought with him. So he had to leave the Bayeux Middle School. It was said in a report of the school in March 1921, "Of the 22 Chinese students, 19 left the school on the evening of March 13. They claimed to be heading for Creusot. I suspect that they were going to work."[7]

Father did not know that he would never enter another school in France after leaving this one. In less than five months, from the end of October 1920 to March 1921, Father wound up his studies in France. Unable to continue, Father had to look for a job. He recalled:

> Upon arrival in France, I learned from those students studying on a work-study program who had come to France earlier that two years after World War I, labor was no longer as badly needed as in the wartime (during which time the work-study program was started), and it was hard to find jobs. Since wages were low, it was impossible to support study through work. Our later experiences proved that one could hardly live on the wages, let alone go to school for study. Thus, all those dreams of "saving the country by industrial development," "learning some skills," etc., came to nothing.

The Chinese students came to France with no skills or technical know-how. When many of them wanted to work, they could take only ordinary odd jobs or miscellaneous work. In French, such work was called *malheureux*. The Chinese students jokingly called it *maolaowu*, meaning, "the fifth child of the Ma family." Odd-job men had no fixed jobs and had to go from one workshop to another for jobs. If they worked slowly, they would be dressed down by the foremen.[8]

By December 1920, the number of Chinese students studying in France had risen to over 1,500.[9] At that time, France was in economic recession; factories were reducing production or even closing. It was indeed difficult to find jobs. The SFEA somehow managed to get a hold of a large number of odd-job vacancies at the Schneider & Cie Iron and Steel Complex in Creusot and recommended more than 100 Chinese students to the factory. Nearly half of them were from Sichuan.[10]

On April 2, 1921, Father and Deng Shaosheng, together with several other students from Sichuan, were recommended for work in the Schneider & Cie factory. Father worked for four years as a laborer, a foreign worker.

Creusot was a city of heavy industry in southern France and headquarters for the Schneider & Cie Iron and Steel Complex—the largest French ordnance factory. It was then the second largest ordnance factory in Europe, only the German Krupp factory was larger.

The Schneider & Cie factory employed over 30,000 workers. During World War I, as large numbers of workers were drafted into the army and sent to the front, the factory employed many foreign workers. In 1917 several thousand Chinese laborers worked there as contract workers.[11] Twenty-one Chinese students worked there before August 1920, and the number increased to more than one hundred by the summer of 1921.[12] The factory handled rails, machinery, cannons, smeltery, construction, foundry, and electric engineering. Except for cannons, construction, and smeltery, all these branches employed Chinese students.[13]

Even today one can still find in the archives of the Schneider & Cie factory files on Father and his workmates. The employment registration card in the factory's personnel section clearly reads: "Deng Xixian, 16. Work Number: 07396. Date of Employment Registration: April 2, 1921. Sent by the Colombes Sino-Franco Workers' Committee, from Bayeux Middle School."[14]

Deng Xixian and Deng Shaosheng were assigned to the steel rolling workshop, where molten steel from the furnace was turned into steel ingots, then rolled into plates. It required no special technical training but was very labor-intensive and exposed workers to constant danger. The rolled steels (bars or plates) each weighed up to 100 kilograms. The workers had to pull them quickly along with long pincers, in the steaming hot workshop, whose temperature was above 100 degrees Fahrenheit, owing to the molten steel. Anyone who fell on the hot rolling steels was unable to escape being burned. Sometimes the rolling mill would malfunction and the steel bars would shoot out of it in all directions, causing accidents. Workers had to work in such conditions for more than fifty hours a week and sometimes had to work additional night shifts.[15] Father told us in our childhood that he had done odd jobs and pulled steel bars in France. One could well imagine how unbearable it was for a short, sixteen-year-old apprentice, hardly an adult yet, to do such hard work.

In the factory the Chinese students earned low wages, only twelve to fourteen francs a day as fixed wages. According to French law, those under the age of eighteen could only take up apprenticeship. The wage for an apprentice was even lower, only ten francs a day. Father was only sixteen then.[16]

The students lived in a hostel ten kilometers away from the factory. More than twenty shared a big room. There was a canteen in the hostel where they had breakfast and supper. As for lunch, they took only bread with them to work and drank some running water. There was no meat or vegetable for lunch. Even though the food in the hostel canteen was cheaper than outside, it still cost between forty and seventy centimes per meal. The students also had to buy their own work clothes, which cost twenty to thirty francs per suit.[17] The apprentices like Father were hard up, because they earned only ten francs a day. The Chinese students went to work in the hope of supporting their study. But they were exhausted by the strenuous work and could hardly make ends meet on the limited wage. Father once said that he had worked hard, pulling the red hot steel bars for a month in Creusot. What he earned in that month was not even enough to feed himself, and in the end he had to add more than 100 francs of his own to cover the shortfall.

On April 23, 1923, Father gave up his job at the Schneider & Cie factory and left Creusot.[18] A month later, Deng Shaosheng left, too. Father's first exposure to the dark side of capitalism came during his work in the French factory for almost a month. He personally experienced the plight of the working class. The oppression and exploitation by capitalists, the insults and abuse by foremen, and the miserable life greatly shocked his naive heart. Being young, however, he still longed for a wonderful life. He recalled later in Moscow that "in the first two years, I got an initial feel for the evils of the capitalist society. But with romantic ideas about life, I could not have a profound understanding."

He left Creusot, but he still hoped to find a job and save some money bit by bit, so as to fulfill his long-cherished desire to continue his study.

CHAPTER 15

FOR SURVIVAL AND STUDY

The Chinese students left their mark on some seventy factories and enterprises in all parts of France, mostly in Paris, Creusot, Saint Etienne, Saint Chamond, Le Havre, and Lyons. Some unforeseen events had occurred in the movement to study in France.

In November 1918, after World War I ended, the French economy was in recession, so factories were running under capacity and prices kept rising. The large number of demobilized soldiers made the problem of unemployment all the more acute. Beginning in the winter of 1919, the French economy took a turn for the worse. The French currency steadily depreciated. Formerly, one Chinese silver dollar could be converted into 8 francs, but now it was 14 francs, and eventually it would be 25. Bread was 25 centimes per kilogram in October and cost 50 centimes per kilogram in December, 105 centimes in March, then 130 centimes in September. The price of rice was 110 centimes per kilogram in the third quarter of 1919; it rose to 480 centimes per kilogram a year later. The price of staple foods kept rising, as did the price of other necessary items, non-staple foods, and transportation.

People found that they were living an increasingly tough life, but capitalists turned a blind eye to their difficulties. As a result, one strike after another occurred. On May Day 1920 the general trade union confederation of Paris launched the May Day general strike, which lasted for half a month. After the strike, some factories resumed production, but the economic situation as a whole remained unstable, with high unemployment.

More than 700 Chinese students came to France to study on a work-study program in 1919. The majority of them started to work after their arrival. At a time when even the French were haunted by the problem of unemployment, it was even harder for Chinese students to find jobs. Those who had jobs were under constant threat of being fired. By August 1920 some 1,000 Chinese students had come to France. Of them, only a few were able to continue studying at their own expense. Less than 300 had jobs, while more than 500 were out of work. The unemployed students gathered at the Overseas Chinese Association in Paris and lived on the five francs per day they got from the Sino-Franco Education Association. To find a means of livelihood, they frequently asked the association about job opportunities. As the students began to blame the association for inefficiency and neglect of duty, quarrels occurred from time to time. On one occasion, a student from Hunan gave Liu Dabei, secretary of the association, a hard thrashing.

When the association did happen to find one or two job vacancies, the situation was like "too many monks sharing too little porridge." The students scrambled for them and sometimes even exchanged blows over them. So in the end they had to decide by drawing lots.

Huang Lizhou recalled that there was a sense of pessimism and frustration among the students when they thought about their future in France. Some had to look for a different way out, returning home or heading for South Asia together with friends.

In November 1919 the Representation Agency wrote to the Global Student Union, asking it to act more prudently in recommending students to study in France on a work-study program.

> We have learned recently that among the students sent overseas, many are conscientious in their study. But there are also some who have come from poor families. Once in a foreign country, they are unable to support themselves and thus leave school to work in the factories. They do not know that France does not have a preferential policy toward foreign workers and the wage is very low. They often find themselves in utter destitution, lead vagrant lives, and are unable to return home even if they want to. Such students are not a small number. It pains one's heart even to talk about their dilemma. So it is better to be prudent in the first place than to have regrets later. Hope your union will pay special attention to it when it recommends students to France.[1]

Nonetheless, by the end of 1920, large numbers of Chinese students were still traveling far across the ocean to their ideal land. By then, quite a few Chinese students in France had awoken from the dream of study to the reality of work, and from then on to the distress of unemployment. The rich and beautiful Western country began to show its dark and miserable side. The much-discussed story of liberty, equality, and universal fraternity was replaced by stark cruel reality.

A student called Pei Zhen wrote:

> Now we actually work eight hours a day, plus two additional hours. So in total it is over ten hours a day, in addition to the energy loss and mental exhaustion. For so much input, we only get fourteen francs in return. Are all our time, energy, and mental labor only worth fourteen francs? The work is to produce goods for capitalists and has nothing to do whatsoever with the ordinary people. Some of us even help make guns for killing people. This is indeed worthless labor. Why do we do it? Why on earth do we take pains to do it! I don't know where our spiritual life lies, so dull and boring. Such numbness in life. Oh, life, what is your real value? We advocate the work-study doctrine and apply it, don't we? The work-study doctrine advocates jobs for all so as to maintain social equality and eliminate polarization; it also advocates education for all so as to create intellectual equality in society and dispel the dark shadow of the intelligentsia. But to produce goods for capitalists is something that could only partly satisfy the desire for material life. Is this in keeping with the work-study doctrine? Is this what we should do? In a nutshell, such factories of organized mass production are built on the basis of capitalist production, on top of workers. It is a method of plunder in every sense of the word. Such demonic plunder is even worse than that of the bandits engaged in killing, burning, and looting![2]

In December 1920 the students who were studying in France issued an appeal:

> Please note that we Chinese students who are studying in France on a work-study program spent all our money in getting out of China. While overseas, we are working hard for a living. The Chinese students in France have "experienced all kinds of hardship and thoroughly understand the hypocrisy of the society." As we cannot earn our living through work, are there any other alternatives for the working students? So please don't be mistaken. Don't take our suffering for something enviable. Please pay heed to our cries, complaints, groans, and wrath. Please show your sympathy and indignation and help us poor students studying here on a work-study program.[3]

Huang Lizhou recalled that from the end of 1920 to early 1921, the students could not find any jobs and had no money for education. Nor could they return home across the vast ocean. They were restless and at a loss as to what to do.[4]

Worse still, some students died of illness or in workplaces. Wang Biji and Li Zifen, two students form Hunan, died of gas poisoning. An Zichu, a student from Sichuan, died of arsenic poisoning in a chemical plant. Zhu Faxiang, a student from Jiangxi, disappeared in a Lyons factory. Four Chinese students in Creusot died of poisonous mushrooms. Pu Zhaohun, another student from Sichuan, who lost all hope for life, committed suicide by cutting his own throat and stomach.

While the Chinese students in France were faced with lack of access to education, unemployment, hunger, and even death, the Sino-Franco Education Association—the so-called family of students—was secretly engaged, under all sorts of pretexts, in malpractices for personal gain. Its members collectively embezzled public funds and lined their own pockets. Apathetic to the hardships of the students, they went so far as to accuse them of "lacking both the ability to work and the will to study" and threatened to send them back to China. What the Sino-Franco Education Association and the Chinese Legation in France did aggravated their acutely contradictory relationship with the students.

The students had always regarded the association as their parents and saviors. Yet the association issued a circular on January 12, 1921, declaring the de facto breaking off of its organizational ties with the students. On January 16, the association issued a second circular announcing that it would relieve itself of any financial liabilities and that beginning from the date of the issuance of the circular, it would continue to provide funds for only two months, until March 15, and no more thereafter. The circular stunned the students and triggered an uproar among them.

On January 23, the unemployed students from Sichuan, Hunan, Hubei, and Jiangxi who were living at the Overseas Chinese Association set up a preparatory committee for the Federation of Chinese Students Studying in

France on a Work-Study Program. The federation was formally established on February 14 and presented a petition to the Chinese Legation in France.

On February 27, a conference of Chinese students was held at the Overseas Chinese Association, and they decided to present a collective petition to the Chinese Legation the following day, designating the move an "Anti-hunger Campaign." Among the ten spokesmen elected were Cai Hesen, Zhao Shiyan, Li Weihan, Wang Zekai, Xiang Jingyu, Wang Ruofei, and Li Fuchun.

At around 8:00 A.M. on February 8, approximately 500 students led by Cai Hesen and others gathered in the vicinity of the Chinese Legation from all directions. Because of obstruction by the French police, only Cai Hesen, Zhao Shiyan, and eight other spokesmen entered the legation to make their representations. When Minister Chen Lu came to the square to meet the students, they angrily shouted, "Beat him, beat him!" Chen and his men retreated to the legation in a flurry. At that moment, the French police forced their way into the square and started to beat up the students. At about 9:00 P.M. that evening, armed French police made their way into the legation and arrested the ten spokesmen by force.

Thanks to the persistent struggles waged by the Chinese students, the French-Chinese Guardianship Society for Young Chinese Students in France was set up on June 1, and it was decided that a certain amount of money would be taken out of the Boxer Indemnity of 1901, paid by China to France, to be issued to the students for living expenses. Each student could receive five francs per day for no more than five months. Arrangements could be made to repatriate those students to China who wished to return.

Although the February 28th Anti-hunger Campaign was not a complete success, it forced the Chinese and French authorities to make some concessions. It was also the first struggle organized spontaneously by the Chinese students in France. Through the struggle, they became more united and began to defend their interests and their right to work, study, and have bread.

The Anti-hunger Campaign enhanced the morale of the Chinese students in France, and on June 2 of the same year they launched a protest against the secret loan deal between China and France.

In June 1921, to augment its military strength and intensify the civil war, the Northern Warlord government sent special envoy Zhu Qiqian and Deputy Finance Minister Wu Dingchang to Paris to discuss a secret deal with the French government to obtain a loan of 300 million francs for purchasing arms. When the news leaked, the Chinese workers, overseas Chinese, and Chinese students were all infuriated. Zhou Enlai, Zhao Shiyan, Cai Hesen, and others, in association with the organizations of Chinese residing in France, immediately organized a "reject-the-loan" committee.

On June 30, more than 300 Chinese residing in France showed their strong patriotism and held a "reject-the-loan" rally in Paris. At the rally one speaker after another vehemently denounced the Chinese traitors and the French government.

A second "reject-the-loan" rally, held on August 13 in Paris, was intended to bring Minister Chen Lu to account. Chen was too scared to show up. Wang Zengsi, secretary-general of the Chinese Legation, attended the rally on behalf of Chen and was beaten up by the angry participants. The mammoth "reject-the-loan" campaign scored a major victory, forcing the Chinese and French governments to abandon the loan scheme.

The just struggles of the Chinese students in France cut the Chinese and French governments to the quick. The French Foreign Ministry informed the Chinese Embassy of its decision to repatriate in two groups the chief plotters of the "reject-the-loan" campaign—the Chinese students. On September 15, the French government decided to discontinue the stipends for the Chinese students in an attempt to place them in a hopeless situation.

In September 1921, under the leadership of Zhao Shiyan, Zhou Enlai, and Cai Hesen, the Chinese students, who were in an extremely difficult position, launched the historic "back to the Lyons Sino-Franco University" campaign. The Sino-Franco University was established by the Sino-Franco Education Association with Wu Zhihui as its president. Its establishment was designed to make personal gains for a few bureaucrats and politicians in the name of organizing students to study on work-study programs. Once in operation, it refused to take in the Chinese students in France but instead tried to enroll students from within China. The pent-up anger of the Chinese students in France erupted like a volcano.

On September 20, an advance party composed of more than 100 students, including Zhao Shiyan, Cai Hesen, and Chen Yi was dispatched by the Conference of Representatives of the Chinese Students Studying on a Work-Study Program from Paris to Lyons. Zhou Enlai, Wang Ruofei, Li Weihan, and others remained in Paris to coordinate the action.

On September 22, the advance party reached Lyons, where it was surrounded by the French armed police. The police threw the students into prison vans by force and then took them to a barracks where they were locked up.

At the news of the imprisonment of the advance party, Zhou Enlai and others went about organizing a rescue operation. Public opinion in France showed deep sympathy with and support for the struggle of the Chinese students. The detained students carried on their struggle and even went on a hunger strike.

On October 13, the Chinese Legation, in collusion with the French government, escorted all the students of the advance party back to China. Among the 104 students sent back were Cai Hesen, Li Lisan, and Chen Yi. With the help of his fellow students, Zhao Shiyan cleverly escaped from the barracks and remained in France to continue the struggle.

Although the "back to the Lyons Sino-Franco University" campaign did not attain its goal, it further exposed the ferocious persecution of the Chinese students in France perpetrated by Chinese and French authorities and helped to raise the political consciousness of the students. Many of them went on to follow a thoroughly anti-imperialist and antifeudal revolutionary road. Most of the students who were escorted back to China threw themselves actively into the mighty torrent of the revolution. Cai Hesen, Chen Yi, Li Lisan, and others became the backbone of the Chinese revolutionary struggle and made immortal contributions.

Li Weihan wrote in his reminiscences of those struggles:

> This struggle broke out amid the internal contradictions of the work-study movement. As far as its specific aims were concerned, it was a failure. But its impact and achievements were of great historic significance. It showed to the French bourgeoisie the growing determination of Chinese youths after the May 4th Movement of 1919 to defy the warlord rule, yield to no foreign bullying, and strive to master their own destiny. It put an end to the surging work-study movement. All those who had advocated this movement with their good intentions came to see there was only a dead-end at the end of this road. . . . Most important of all was that out of this struggle emerged a new unity and a new awakening on an unprecedented scale among the work-study circles. Many people abandoned their unrealistic illusions of all kinds and accepted Marxism, taking the road of the October Revolution. Still more took an active part later on in the anti-imperialist and antiwarlord struggles.[5]

IN THE HUTCHINSON RUBBER PLANT

On April 23, 1921, Father left Schneider & Cie in Creusot for Paris. He discontinued his studies and also lost his job. He spent the money he had brought from China and was unable to find a job. He had to draw on relief funds from the Sino-Franco Education Association while waiting for a job opportunity.

According to the National Archives of France, Father received, under registration number 236, five francs worth of relief funds each day during the five months from May to October 1921.

The Chinese students who drew relief funds in Paris numbered some 500 at that time, and most of them lived in the building of the Overseas Chinese Association, located at Columbus in the western suburbs of Paris.

The Overseas Chinese Association was an ordinary three-story building that housed several overseas Chinese organizations, including the Sino-Franco Education Association, the Association of Chinese Students in France, the Association of Chinese Students Studying in France on a Work-Study Program, and the Society for Promoting Peace. The second floor of the building was a meeting room, and the third and first floors and the basement became the dormitories for Chinese students. As more and more students poured into Paris, the building became really crammed. A French senator's wife donated some tents, which were pitched in the vegetable garden behind the building. Later the building and the tents were all packed up. Each student had only five francs for each day's living expenses, so they could afford only two meals a day. Their food consisted mainly of bread and running water, sometimes with crude chocolate. They rarely ate vegetables. A few students with gas stoves could boil water, but the majority drank only running water. These young students had only one and a half kilograms of bread to eat each day because of the shortage of cooking oil and meat. They did not eat crude chocolate at times for lack of pocket money. The students, who crossed the oceans from thousands of miles away to study in France, had fallen into a hell of harsh realities far removed from their vision of paradise.

By August of that year, Father had reached the age of seventeen. His seventeenth birthday occurred at a difficult time with no apparent future or hope. Father did not take part in the two large-scale struggles waged by the students—the "reject-the-loan" and "back to the Sino-Franco University in Lyons" campaigns—that year, probably because he was young. On May 20 he only signed his name to the joint letter to Cai Yuanpei written by 243 Chinese students, including Wang Ruofei, Chen Yi, Liu Bojian, and Li Weinong, in which they demanded that the Sino-Franco University in Lyons and the Chinese-Belgian University be changed into engineering institutes so as to enable the students on work-study program to continue their studies.[1] At the forefront of struggle at that time stood a group of progressive youths who were older than

Father and had acquired fairly high political consciousness. Among them were Zhao Shiyan, Zhou Enlai, Cai Hesen, Chen Yi, Li Weihan, Wang Roufei, Xiang Jingyu, Li Fuchun, and Liu Bojian. Although Father had taken part in the May 4th Movement in China and had gone through the hardships of studying on a work-study program in France, he had acquired by then only preliminary political consciousness and progressive ideas and was still not under the influence of Marxist ideas. Thus he had failed to join the ranks of these struggles against the forces of darkness.

In September 1921 the French government decided to stop issuing subsistence expenses to Chinese students studying on work-study programs. In October, like other students, Father had no source of income at all and found himself in hopeless straits.

Probably "Heaven never cuts off a man's means." The Chambrelent, a factory on the edge of a canal in the Tenth District of Paris that made specialty fans and paper flowers, was recruiting workers. Father, his uncle Deng Shaosheng, and 103 other students were employed by this small factory on October 22, 1921, and Father was given the number 238.[2] The students were very lucky to get these jobs. A student named Luo Han recalled later that "being unexpectedly rescued from a desperate situation, we found the job of flower-making as Columbus discovered the New World. When the students studying on a work-study program were in extreme difficulties and suddenly discovered this new world, this was a good job to them regardless of the wage so long as they accepted it. Thus, over 100 students swarmed into this factory."[3]

The Chinese students there were making goods that were used for raising funds in the United States. They made flowers with gauze and silk, which were then tied up with an iron wire and a small label with the inscription "Made by the widows and orphans of those who have fallen in battle" stuck on. Generally, this job was done by women workers. The wage was very low. The workers earned only 2 francs for 100 flowers. Some skilled workers could make 600 to 700 flowers a day, earning a dozen francs.[4]

It seemed that these 105 students could eke out a living temporarily in this way. However, the good days did not last long. These jobs were not permanent but temporary. Soon they finished making all the flowers ordered. Two weeks later, on November 4, Father and his fellow students were fired by the factory.[5] They were out of work once again.

Father once said that he had done various jobs in France, and they were all odd ones. After he lost his job this time, he must have tried to find a new job everywhere and might have done some temporary odd jobs now and then. This unstable situation lasted for more than three months before he found a new job in the Hutchinson Rubber Plant near Montargis in February 1922.

The French economy started to improve in 1922. Some factories gradually restarted operations. The job opportunities increased, and advertisements recruiting workers often appeared in the newspapers. Some Chinese students found jobs, thereby managing to survive. In the meantime, thanks to the strong appeal of the Chinese students in France and the efforts of warmhearted people in China, some of the students' home provinces and counties raised funds to send to them in France. For instance, each student from Pengxian County, Sichuan Province, received 600 silver dollars in loans each year. Therefore, from 1922, all students in France who received loans or subsidies from China signed up for an entrance exam to a polytechnic or university so as to fulfil their long-cherished wish of studying on a work-study program in France. According to incomplete statistics, more than 200 students from Sichuan Province alone graduated from universities and schools of all descriptions in France.[6]

Montargis, which is situated in the south of Paris, is a small town in Louval Province. It was a place of residence for the French royal family in the Middle Ages. By the nineteenth century, it had a population of over 13,000 and was a town with well-developed industry, commerce, communications, culture, and education. Next to Montargis is the small town of Chalette. It had a population of only 3,000 and was the site of the old Hutchinson Plant, which manufactured a variety of rubber products. At the end of 1921, the plant started recruiting workers. Some Chinese students who were studying in France came to work there.

The owner of the plant was a British-American. In the middle of the nineteenth century, he had established some factories in France and other European countries, and the Hutchinson Plant in Chalette was said to be the sole rubber plant in Europe at that time. After World War I, the plant recruited large numbers of foreign workers, including Indians, Vietnamese, and Belorussians. As the personnel department of the plant had contacts with the French-Chinese Friendship Association, some Chinese students came to work there through the recommendation of the association. Their number reached 210 at most. In 1922 the Hutchinson Plant employed over 1,000 workers. Most of them were women and child laborers. The Chinese students numbered over 30. Some of them made tires, others raincoats, and the majority produced rubber overshoes.[7]

On February 13, 1922, Deng Xixian registered on the Chalette town council's list of foreigners. He wrote down the names of his parents and the date of his birth, giving his previous address as 39 rue de la Pointe, La Carenne-Colombe, and his identification card number as 1250394.[8]

On February 14, with the working number of 5370,[9] Father started a relatively stable period of work at the Hutchinson Plant. His time there was another important turning point in his life. He was assigned to the shoe-making workshop, where he made waterproof overshoes. The workday was ten hours long, with a half-day off on Saturdays. That is to say, they worked fifty-four hours per week. The system of payment by the hour was applied to new workers, and their wage per hour was one franc during apprenticeship, which was gradually increased afterwards.[10] When they became skilled workers, they were paid at a piece rate. The job was light manual labor but had to be done swiftly. It was suitable for clever and deft people. Zheng Chaolin also worked in this plant. He told me that he could only make ten pairs of shoes a day whereas Father could make some two dozen. A worker like Father could make fifteen or sixteen francs a day.

During a tour for the study of a project in France in 1988, I went to Montargis and, accompanied by the staff of the Hutchinson Plant, paid a visit there. With over 9,000 staff and workers, the plant remains a manufacturer of rubber products. On the right of the spacious courtyard, the workshop where Father worked has been kept intact. The building is no longer used as a workshop. Its second floor became a warehouse and the first floor is now packed with odds and ends. The workshop is big and bright with a high ceiling. One can well imagine that the working conditions there were far better than those in Schneider & Cie in Creusot. The staff who accompanied me told me that the plant had caught fire 100 years before. Later the well-known French architect Gustave Eiffel designed the building. Like the world-famous Eiffel Tower in Paris, the building was a steel structure. It is said that it was the first workshop in the world at that time with a metal structure. When I came back from the tour, I asked Father whether he knew that the workshop where he had worked had been designed by Eiffel. Father knew nothing about it.

As Zheng Chaolin, who had worked with Father, recalled, the plant vacated a wooden shed in the woods at a distance of five minutes' walk to the plant to accommodate over forty Chinese student-workers. When I was there, the staff of the plant told me that the shed had been dismantled long ago. But I could imagine how simple and crude it had been. The students prepared their meals together. They elected two students to be cooks whose wages were paid by others as timework in the plant. Mess accounts were made public. The money spent on meals per capita per day was about three francs. They had coffee and bread for breakfast and meat for both lunch and supper. No payment of rent was required. Therefore, those students, like Zheng Chaolin, who earned time wage could save more than 100 francs every month, and those like Father who earned a piece-rate wage could save over 200 francs every month.

Father spent his eighteenth birthday working in the Hutchinson Plant. He led a relatively stable life, and his work was not so strenuous as it had been in Cruesot, so perhaps he was relaxed at that time. Zheng Chaolin lived in the same shed with Father: "After supper there were two to three hours left before going to bed. We had a jolly time in the shed. Few and even none of us read books. We chatted, joked and abused each other. Fortunately, there was no scuffle. A boy from Sichuan Province, short and stout and aged only 18, always bounced here and there at this time every day. He cracked jokes here and then made fun there."[11]

It can be seen from this that Father had a lively and sanguine disposition in his youth. He retained an optimistic outlook his whole life, even in times of difficulty and hardship.

On October 17, 1922, Father and Deng Shaoshen resigned from the Hutchinson Plant.[12] On November 3, they left Chalette to go to the College de Chatillon-sur-Seine.[13]

My grandmother said that Father wrote home to ask for some money. My grandfather then sold some grain and fields and sent a small amount of money to his son in France. My grandfather was experiencing great financial difficulties then. However, he chose to sell his fields to raise the money needed. This proved that he gave his full support to his son's studies in France. Father probably did not receive the money until the end of 1922. Furthermore, he had earned some money in the nine months he worked in the Hutchinson Plant. It was quite natural then that he would be thinking about his purpose in coming to France from thousands of miles away. To achieve his wish of pursuing his studies, he had already endured untold sufferings. So he wanted to continue his studies when his financial resources permitted. Father went to Chatillon, but he failed to enter the College de Chatillon because he did not have enough money. Two months later, on February 1, 1923, he was back again in Chalette.[14]

His dream of continuing his studies was ultimately shattered by his failure to enter the College de Chatillon. Apart from his study later in Sun Yat-sen University in the Soviet Union, Father never again studied in a regular school. He joked with us, saying he had acquired the educational level of middle school only. Father accumulated his knowledge through self-study in later years. His wisdom originated from his own experiences in the revolutionary struggle.

Father is keen on learning for life. He has learned from books, from his work, from social life, and particularly from the revolutionary practice. What he has learned and gained from social and revolutionary practice is more significant and useful than what he learned in school.

Father is especially fond of reading books. He likes to read anything—famous Chinese and foreign classics, biographies of historical figures, special collections of current events, and even the complete volumes of the *Twenty-Four Histories*. His favorite historical book is *The Mirror of Governance*. Father has another hobby—consulting dictionaries. Since my childhood, I have always been asked by Father to look up a sentence, a phrase, or a word in *The Collection of Words*, *The Origin of Expressions*, or *The Kangxi Dictionary*. As a result, I have naturally acquired the habit of consulting dictionaries myself.

On February 2, 1923, Father returned to the Hutchinson Plant. On March 7, he left again after working in the shoe-making workshop for more than one month. The reason for his departure cited on his work card was "refusal to work."[15] In fact, Father left the Hutchinson Plant not because he did not want to work but because an important event occurred in 1922 that affected the rest of his life.

In June 1922 the most outstanding of the Chinese students who were studying in France organized the Chinese Youth Communist Party in Europe (renamed the Chinese Communist Youth League in Europe in the next year).

Between the summer and the autumn of that year, Father joined this organization and became a member of the Communist Youth League and a believer in Marxism and communism. He left the Hutchinson Plant in March 1923 and has been a career revolutionary ever since.

If Father had been only a progressive youth with patriotic ideas when he worked in the Hutchinson Plant in early 1922, he was a young revolutionary of some political consciousness and an advocate of communism by the time he left the plant. Large numbers of ardent Chinese revolutionary youths took the revolutionary road, established revolutionary organizations, and conducted revolutionary activities in France or other European countries earlier than or at the same time that Father did. I shall describe their revolutionary course in detail in the next chapter.

CHAPTER 17

ESTABLISHING COMMUNIST ORGANIZATIONS IN EUROPE

The Russian October Revolution in 1917 and the Chinese May 4th Movement of 1919 shook China tremendously. Many progressive Chinese youths were enlightened by new ideas and began exploring new ways. Among them were a number of outstanding people such as Li Dazhao and Chen Duxiu, who took the lead in accepting Marxism.

Members of the Xin Min Society (New People's Society) organized in 1918 by Mao Zedong and Cai Hesen—among them, Cai Hesen himself, Li Weihan, and Li Fuchun—went to France to study on work-study programs. These youths were quite progressive and revolutionary at home. Once in France, they continued their studies and explorations.

In February 1920, Li Weihan, Li Fuchun, Zhang Kundi, and others organized the Association for the Promotion of the Work-Study Program, which was devoted to the study of truth and the purpose of promoting the work-study program. In 1920, Cai Hesen arrived in France and lived in Montargis Public School. Soon thereafter, the members of the Xin Min Society in France were conducting their activities mainly in Montargis.

In August 1920 the Association for the Promotion of the Work-Study Program was renamed the World Society for Work-Study. In October the society called a three-day meeting in Montargis. Cai Hesen attended the meeting. After heated debates, most of the members agreed that the purpose of the society was to promulgate the need for a Russian-type social revolution, based on the tenets of Marxism.

The society was the earliest socialist organization among the students studying in France. With Montargis as the center of their activities, members of the society, together with those of the Xin Min Society, studied assiduously and disseminated Marxism and led the two large-scale mass struggles waged by the students studying in France in 1921.

In June 1920 a young man from Sichuan Province named Zhao Shiyan arrived in France. Zhao was born in Sichuan in 1901. In his childhood, he was influenced by new ideas. When he was studying at the middle school attached to the Beijing Advanced Normal School, he took an active part in the May 4th Movement of 1919 and the socialist youth work-study movement. After he arrived in France, he worked in several factories and went through the hardships of unemployment. This experience gave him many thoughts and feelings about the capitalist world.

In early 1921, Zhao Shiyan, Li Lisan, and others sponsored the Labor Society in Creusot where Chinese students and workers could gather. The Labor Society sincerely believed in the work-study doctrine and vowed to carry the work-study movement through to the end. Its members attached great importance to the workers' movement, went amid the Chinese workers, and thus became very popular with them.

The World Society for Work-Study based in Montargis and the Labor Society based in Creusot were the two progressive organizations among the Chinese students in France. Although at the beginning the two organizations held differing views and even opposed each other in their statements, they shared the principle of devotion to promoting the movement to study in France on work-study programs and to seeking truth.

In early 1921 Chinese students in France were faced with a crisis of survival caused by the discontinuation of their studies and their high rates of unemployment (as discussed in detail in chapter 15). They staged a spontaneous petition demonstration. At the critical moment of the struggle, the Xin Min Society and the World Society for Work-Study in Montargis reviewed the situation, under the leadership of Cai Hesen. They unanimously agreed that they should support the just struggle of the students in the suburbs of Paris. In late February, Cai Hesen, Wang Ruofei, Li Weihan, Zhang Kundi, Xian Jingyu, Cai Chang, and others went to Paris and led the struggle there. Although Zhao Shiyan and Li Lisan of the Labor Society in Creusot did not participate, they issued a statement in support of the struggle.

After the February 28th Anti-Hunger Campaign, the needs of the struggle were more apparent, and both the Montargis group, represented by Cai Hesen and others, and the Labor Society, represented by Zhao Shiyan and Li Lisan, were eager to clear up their misunderstandings and achieve unity among Chinese students.

In the summer of 1921 Zhao Shiyan made a special trip to Montargis to meet with Cai Hesen. They talked for three days and finally reached an agreement that their disputes were over, that they would come together to discuss problems and make revolution in the future, and that all advocated Marxism.[1]

It was at this time, in mid-February, that Zhou Enlai came to France from Britain. The name of Zhou Enlai is world-famous, known to almost everybody. He was born in the ancient city of Huai'an, Jiangsu Province, in 1898. He left home to study in northeastern China at the age of twelve and at the age of fifteen came to Tianjin, where he studied at Nankai School. Even at that time, his patriotism was growing, and in 1917 he went to Japan to pursue further studies. There he was greatly shocked to learn that the whole Chinese nation was at peril, and he was influenced by the progressive ideas of the journal *New Youth*. After he returned home in 1919, he threw himself into the May 4th Movement of 1919. In the struggle, he and twenty others, including Deng Yingchao, Guo Longzhen, and Liu Qingyang, organized the Awakening Society. The purpose of the organization was to seek awakening and self-determination in the spirit of reframing the mind and carrying out reforms. Later Zhou Enlai led a number of anti-imperialist patriotic struggles by the students in Tianjin. For this, he was arrested and jailed by the reactionary authorities.

After he was released, in November 1920, Zhou Enlai went to Europe via Shanghai to study on a work-study program. He stayed in France for two weeks before he went to Britain to study. But in Britain he did not get a chance to enter a school. What he saw was the serious social upheaval and unrest in Europe after World War I. At that time, the momentous workers' strike was sweeping the whole of England. He carefully studied the workers' movement in England, and in February 1921 he moved to France again. While learning French, he conducted social investigations. He earned a living by writing for newspapers and translating articles. In Europe there were all kinds of ideas after World War I. After repeated study and consideration, he chose to believe in communism.

In December 1920 Zhang Shenfu and Liu Qingyang also arrived in France. Zhang once worked with Li Dazhao to lay the foundations of the Communist Party of China. Chen Duxiu and Li Dazhao entrusted Zhang with the task of establishing an overseas branch. So in the spring of 1921, Zhou Enlai, with the introduction of Zhang Shenfu and Liu Qingyang, joined one of the eight sponsoring groups of the CPC—the Paris Communist Group. The group had five Party members: Zhao Shenfu, Liu Qingyang, Zhou Enlai, Zhao Shiyan, and Chen Gongpei. It was the predecessor of the European Branch of the Communist Party of China. At that time, it was still secret.

In February 1921, when Zhou Enlai arrived in France, the large-scale struggle waged by the Chinese students there had just come to a climax. In June, Zhou Enlai, together with Yuan Zizhen and others of the World Society for Work-Study, launched a struggle against the purchase of arms by the Chinese government (as described in detail in chapter 15). From then on, he had the leading position in the students' movement and revolutionary struggle.

In September, Zhao Shiyan and Li Lisan from Creusot and Cai Hesen and others from Montargis decided to wage a joint struggle for the right of the Chinese students to enter the Sino-Franco University in Lyons (as described in detail in chapter 15). On September 17, they set up first in Paris a "federation of Chinese students studying in France on a work-study program." Three days later, more than 100 Chinese students, including Zhao Shiyan, Cai Hesen, Li Lisan, Chen Yi, Chen Gongpei, Zhang Kundi, and Luo Xuezan, organized a "vanguard party" to march into Lyons. Zhou Enlai, Wang Ruofei, Li Weihan, and others coordinated the action in Paris. Although the struggle was suppressed by the French army and police and the core members, such as Cai Hesen, Chen Yi, Zhang Kundi, and Luo Xuezan, were expatriated to China, uniting and organizing themselves had become the common interest of the advanced elements of the Chinese students after this struggle, and they saw that it was time to establish a unified communist organization.

In July 1921 a big event—little known but very important to the destiny of China—took place in China: the Chinese Communists, headed by Li Dazhao and Chen Duxiu, founded the CPC.

In the early days of its founding, the CPC had only some fifty members. But this spark of communism began to spread across the Chinese land and soon became a prairie fire. The organization swelled its ranks not only at home but overseas. The establishment in Europe in 1922 of the Communist organization of students studying there was a historic moment.

At the end of 1921, Zhao Shiyan, who was living in hiding in northern France to escape the French army and police, maintained close contacts with Zhou Enlai, Li Weihan, Liu Bojian, Wang Ruofei, and Fu Zhong in France, Germany, and Belgium, and discussed with them the establishment of the Chinese Youth Communist Party. In early 1922, Zhao Shiyan and Zhou Enlai invited some people to have a full discussion at a Paris meeting, which reached an agreement. From then on, preparations were made separately by Zhao Shiyan and Li Weihan in France, Zhou Enlai in Germany, and Nie Rongzhen and Liu Bojian in Belgium. Zhao Shiyan also came into contact with Li Lisan, who was already back in China, to ask him for the constitution and other relevant documents of the Socialist Youth League at home.

In June 1922 a Communist organization among the youths studying in Europe was established. The organization held its first congress in Forest Bologne on the outskirts of Paris. Participating in the three-day congress were eighteen representatives from France, Germany, and Belgium. Among them were Zhao Shiyan, Zhou Enlai, Li Weihan, Wang Ruofei, Xiao Pusheng, Liu Bojian, Yuan Qingyun, Ren Zhuoxuan, Chen Yannian, Yin Kuan, Li Weinong, and Zheng Chaolin. The meeting was held at an open space in the forest. They rented eighteen chairs from an old French woman who opened the nearby outdoor coffee and tea house. Zhao Shiyan presided over the meeting, and he and Zhou Enlai made reports on the preparations for the establishment of the organization and its constitution. After discussion, it was decided to name the organization Chinese Youth Communist Party in Europe. The first congress elected Zhao Shiyan, Zhou Enlai, and Li Weihan members of the Central Executive Committee. Zhao was the secretary, Zhou Enlai was responsible for the propaganda work, and Li Weihan took charge of the organizational work. The office of the committee was located in a small hotel (17 rue Godfroy) near the place d'Italie in the Thirteenth District of Paris. Zhao Shiyan lived there, and Li Weihan and Chen Yannian often went there to work. Recruitment of new members started in Germany, then in Belgium.

At the same time, the French group of the CPC further developed into the European Branch of the CPC in the winter of 1922. The members of the Youth Communist Party became members of the CPC. At that time, the Party and the Communist Youth League were merged with a unified leading body, but the League was under the leadership of the Party. The League was open, while the Party was secret.

The establishment of the European Branch of the CPC and the Chinese Youth Communist Party showed that the Party was successfully conducting activities among the Chinese students in Europe and that the core of the leadership was in place to organize the political activities and struggles of the Chinese students, Chinese workers, and other overseas Chinese.

The Youth Communist Party had expanded rapidly since its founding. There were only some thirty members at the beginning. But barely six months later, the number had swelled to seventy-two. By 1924 the membership had grown to more than two hundred. Large numbers of prominent elements and progressive youths among the students studying in France and other European countries rallied around the Party and the League.

The Youth Communist Party conducted its activities under the absolute leadership of the European branch of the CPC. Zhao Shiyan was secretary of both the Central Executive Committee of the Youth Communist Party and the French branch of the CPC. He and Zhou Enlai, both leaders of the Youth Communist Party, were men of moral integrity and were held in great esteem and ardently loved by the members. Praising them in a poetic style, Cai Chang said that "Shiyan and Enlai are full of wisdom."

The Youth Communist Party reported its establishment to the Party Central Committee in China. On November 20, 1922, Li Weihan was sent back to China to contact and report to the Central Committee of the Youth League and to ask it to place the Youth Communist Party under its leadership.

February 17–19, 1923, the Youth Communist Party held a provisional congress in the auditorium of a police sub-bureau in a small town on the western outskirts of Paris. In addition to those who attended the inaugural meeting, forty-two delegates attended, including Chen Qiaonian, Nie Rongzhen, Deng Xixian, Fu Zhong, and others.[2] It was decided at the meeting to change the name of the Chinese Youth Communist Party into the Chinese Communist Youth League (CCYL) in Europe, also known as the European Branch of the Communist Youth League of China. Its leading body was renamed the Executive Committee of the Communist Youth League in Europe. Because the CPC Central Committee had decided to send Zhao Shiyan, Wang Ruofei, Chen Yannian, Chen Qiaonian, and eight others to the Soviet Union to study in the Communist University of the Toilers of the East, the meeting elected five members of the second executive committee, with Zhou Enlai serving as the secretary. Liu Bojian, Yuan Zizhen, and one other were made alternate members.

By then, the European Branch of the Chinese Communist Youth League had entered its second stage. It recruited more members, set up the organs, officially established its subordination to the CPC Central Committee and the CCYL, and came up with its own official name. It also launched its official publication, *Youth*. Its organization and other work began to be more complete and mature.

CHAPTER 18

THE STARTING POINT OF THE REVOLUTIONARY COURSE

Father recalled:

> I spent five years and two months in France, working in factories for about four years (doing office work for the Party and Youth League organizations in more than one year). From my working experience, with the help of progressive fellow students and under the influence of the French workers' movement, I started to make changes in my ideology. I began to read some Marxist works and attend some meetings at which some Chinese or French propagated communism. Then I also wanted to join revolutionary organizations. Finally, in the summer of 1922, I was admitted into the Chinese Socialist Youth League on the recommendation of Xiao Pusheng and Wang Zekai.

When Father began working in the Hutchinson Plant in Chalette near Montargis in February 1922, he came into contact with a number of students with progressive ideas. Although Father did not participate in their activities, what he saw and heard had a great impact on him. He began to read the *New Youth*, collections of discussions on socialism, and other books and newspapers.

At that time, a variety of ideological trends, especially anarchism, were in vogue among French youths. The two sons of Chen Duxiu—Chen Yannian and Chen Qiaonian—believed in anarchism for a time. Father, though very young, was never influenced by such ideas. He recalled, "Every time I heard people debating, I always championed socialism." From the very beginning, he accepted the ideologies of Marxism and communism and chose the road of proletarian revolution. He has remained steadfast despite the vicissitudes of more than seventy years.

He summed up his experience while studying in the Soviet Union:

> The sufferings of life and the humiliations brought upon by the foremen, the running dogs of capitalists, have had a great influence on me directly or indirectly. At the beginning, I only had some dim ideas about the evils of the capitalist society. Because of my romantic life, I was unable to have a deeper awareness. Later on, I acquired some knowledge about socialism, especially communism, and accepted the propaganda by some awakened elements, in addition to the sufferings I personally endured.

So he joined the European Branch of the Chinese Socialist Youth League. "In a word, I have never been influenced by ideas other than communism." Father made this summary in 1926 when he joined the Communist revolutionary rank in France; his was a true self-analysis of why a twenty-two-year-old young man would choose the communist ideal and take the revolutionary road at the age of eighteen.

Father told me that he joined the Youth League in Montargis and took the oath in Paris together with Mom Cai—Cai Chang. He said that everybody taking the oath was very excited, and the ceremony has remained fresh in their memories, even scores of years later.

Cai Chang was born in 1900 in Hunan. In her early years, she joined the Xin Min Society sponsored by Mao Zedong and, together with Xiang Jingyu, organized the Women's Work-Study Corps. In 1920 Cai Chang and her elder brother Cai Hesen went to France to study on a work-study program, taking their mother, Ge Jianhao, along. Cai Chang worked in Lyons, Paris, and other places. She took an active part in the discussions held by the members of the Xin Min Society in France and joined progressive organizations such as the World Society for Work-Study in Montargis. She also took part in the struggles waged by Chinese students in France. She joined the European Branch of the Chinese Socialist Youth League in 1922, and the following year she became a full member of the European Branch of the CPC.

Father did not know Cai Hesen well, but he was quite familiar with Cai Chang. As Cai was four years older than him, he called her "elder sister." Cai Chang fell in love with Li Fuchun, and they married in France. Father maintained close contacts with them. He called Li Fuchun "elder brother" and Li and Cai affectionately called him "little brother." Later Father lived with them for some time when they worked together in the Paris branch of the Youth League. Father told us that he often went to Cai Chang's home and ate noodles prepared by Mom Cai. They have remained friends for many years.

After 1957 our family moved to Zhongnanhai. There were four courtyards by the Huairen Hall, and together they were called the Qingyun Hall. Li Fuchun and Cai Chang lived in the number 1 courtyard, while our family lived in number 5. This made our two families even closer. My parents often took us to visit Uncle Li and Mom Cai. Uncle Li had a strong Hunan accent and pronounced the name of my brother Fei Fei "Hui Hui." We children respected Father's "elder brother" and "elder sister" very much. My mother and Mom Cai were also on very good terms. Mother had a high respect for her and often asked her for advice. Father and Uncle Li, two vice-premiers of the State Council, often traveled together for their work. The two families and other personnel often sat on a train for days on end going to northeastern, northwestern, southwestern, and eastern China.

When the Cultural Revolution came, Father became a target of attack and criticism and was put under house arrest. At that time, no one in the world dared to come close to him. One day the bodyguard of Uncle Li, Little Kong, stuffed two packages of cigarettes into the hands of an old orderly of our family, saying that they were sent by Comrade Fuchun. No sooner had he finished his words than he hurried away. The two packages of cigarettes were evidence of Uncle Li's political viewpoint and the value he put on the years of friendship he and Mom Cai had shared with their comrade-in-arms. Uncle Li died during the Cultural Revolution. Later Mom Cai was hospitalized for serious illness. My parents often went to see her. At her ninetieth birthday in 1990, Mother took the whole family to the hospital to congratulate her on behalf of Father. When Mom Cai died, Father presented a wreath and Mother attended the ceremony on Father's behalf to pay last respects to her.

The long friendship was a revolutionary friendship, like brotherly affection but deeper. Our generation saw this with our own eyes and were very much moved and educated.

At the beginning of 1923, Zhou Enlai became the leader of the European Branch of the Communist Party of China and the Chinese Socialist Youth League in Europe. Li Fuchun and Cai Chang were also active revolutionaries, while Father was only a young revolutionary who worked for the communist cause but remained immature. However, he was young, enthusiastic, active, and eager to make progress. He took firm revolutionary strides and, under the guidance of Zhou Enlai and other leaders, began his lifelong revolutionary career.

On June 11, 1923, Father left Chalette. The address he gave as his destination was 39 rue de la Pointe, La Garenne-Colombe. In accordance with the needs of the work of the secretariat of the Youth League Executive Committee, he became a professional worker of the European Branch of the Youth League in Paris while doing odd jobs.

After being admitted into the League, Father made much progress both in his thinking and in his mental outlook. When he first went to the Hutchinson Plant in early 1922, he was a lovely and mischievous boy. But when he joined the Youth League, he suddenly became much more mature. Wu Qi, one of the Chinese students studying in France at that time, recalled: "Among the students I knew, Comrade Deng Xiaoping was the youngest. In the second half of 1922, I happened to meet him in a restaurant on the outskirts of Paris. He was barely twenty at the time. Although he was young, he was experienced, capable, and talented. He was sturdy, full of grit. He was straightforward and spoke loudly and forcefully. All these are still fresh in my memory though half a century has passed."[1] This assessment shows that when a man has acquired an ideal and has something to pursue, clear beliefs, and an objective to fight for, it gives him a new life.

In 1922 the revolutionary situation in China developed rapidly. Dr. Sun Yat-sen chose a new road to revolution after suffering setbacks. Beginning in the summer of 1922, he began reorganizing the Kuomintang (KMT) and invited Chen Duxiu, Li Dazhao, and other Communists to guide the reorganization. Then Dr. Sun Yat-sen firmly established the Three Great Policies: alliance with Russia, cooperation with the Communist Party, and assistance to the peasants and workers.

In June 1922—that is, when the Youth Communist Party in Europe was founded—the Central Committee of the CPC published the "Propositions of the Communist Party of China on the Current Situation," which proposed the establishment of a democratic united front with the Kuomintang headed by Dr. Sun Yat-sen and other revolutionary democrats. In June 1923 the CPC called its Third National Congress in Guangzhou, at which it was decided that all its members would join the Kuomintang in their own names so as to set up the united front of all democratic classes. At that time, Dr. Sun Yat-sen sent Wang Jingqi to France to establish a branch of the KMT there.

Wang had been a student in France and was repatriated for his participation in the students' struggle to return to the Sino-Franco University in Lyons. After arriving in France, Wang immediately contacted Zhou Enlai and others. On June 16 of the same year, Zhou Enlai and others reached an agreement with Wang Jingqi that all members of the Chinese Communist Youth League in Europe would join the Kuomintang in their own names. In November the European Branch of the KMT was set up in Lyons, with Wang Jingqi serving as director of the Executive Department, Zhou Enlai head of the General Affairs Section, Li Fuchun head of the Propaganda Section, and Nie Rongzhen chief of the Paris Liaison Division. Along with other Youth League members, Father also joined the KMT in his own name in 1923. From then on,

> an excellent dynamic situation appeared in the united front work of the Kuomintang and the Communist Party in Europe. Facing this situation, the right-wingers of the Kuomintang were very frightened and launched fierce attacks on the Communist Party. They harbored extreme hatred toward Comrade Zhou Enlai, because he led Party and League members to wage a tit-for-tat struggle against them. At a meeting, a member of the right-wingers of the Kuomintang took out a pistol and pointed it at Zhou Enlai. Fortunately, one of our comrades was quick and wrested the pistol away, thus thwarting the assassination plot.[2]

In July 1923, while disseminating Marxism-Leninism and expanding the organizations, the European branches of the CPC and the Youth League led the Chinese students and workers in France in a struggle against the attempts of imperialist powers to "jointly administer" Chinese railways. Zhou Enlai, Xu Teli, Yuan Zizhen, and Xu Deheng proposed that twenty-two Chinese organizations in France meet in Paris and set up the Provisional Joint Committee of All Chinese Organizations in France, with Zhou Enlai as the secretary. On July 15, a mass rally of the Chinese in France, attended by more than 600 people, was held in Paris. Zhou Enlai delivered a speech entitled "Be United, Overthrow the Warlords at Home and Fight against International Imperialism." After the rally, it was decided to set up a federation of Chinese organizations in France. The struggle waged by the European branches of the CPC and the Youth League further united the Chinese workers and other Chinese in France and became more integrated with the political struggle against feudalism and colonialism in China.

After joining the Communist Youth League in 1922, Father matured quickly. He recalled that at that time he served as propaganda assistant in the

Bayeux branch for two terms and was instructed by the branch to co-edit with Fu Lie a workers' journal that was published every ten days.

In the summer of 1923, he became a member of the European Branch Committee of the Communist Youth League. Liao Huanxing recalled, "In June 1923 the European branch called the second congress and elected Comrade Deng Xiaoping into the Branch Committee."[3] Jiang Zemin also recalled that "in the summer of 1923, after the school was on vacation, I and Qiao Picheng went to Paris to look for temporary jobs. There happened to be the reelection at the second congress of the European Branch of the Communist Youth League. Both of us attended the congress as representatives. The congress elected the secretariat, with Zhou Enlai serving as the secretary, Li Fuchun as the director of propaganda department, and Yin Kuan as the director of organization department. Comrades Fu Zhong and Deng Xiaoping were also leading members. The congress decided to change *Youth* into *Chi Guang* (Red Light). But, in fact, the journal did not change its format until February 1924."[4]

It is clear that Father became an active member of the branch of the Communist Youth League, participating in the work of its leading body, beginning in the summer of 1923. Neither Liao Huanxing nor Jiang Zemin specified what post Father held. According to my analysis, Father only did some work for the leading comrades of the Branch Committee and could not be counted as one of its leading members. He once told us that when one participated in the leadership of the Youth League, one automatically became a full Party member. But he was only a Youth League member in 1923 and did not become a full Party member until 1924.

On August 1, 1922, not long after the Youth Communist Party was established in June of that year, it started publishing the journal, *Youth*. The editorial department and the organ of the Party were located in the same room at 17 rue Godfroy. The task of the journal was to "disseminate doctrines of communism." It carried translations of the works of Marx and Lenin as well as documents and news about the Communist International and the Youth Communist International. Zhao Shiyan, Zhou Enlai, and Zhang Shenfu all contributed to the journal, expounding on the nature and role of the Communist party, interpreting the fundamental tenets of Marxism-Leninism, and debating with the anarchists. Thirteen issues were published in all. Cai Chang recalled:

> The journal *Youth* was edited in turn. Comrades Deng Xiaoping and Li Dazhang cut mimeograph stencils, while Li Fuchun was responsible for distribution. Later, the journal was renamed *Red Light*. It was published irregularly, sometimes once every three days, sometimes every two days, and sometimes every month. The office was in an upstairs room of a coffeehouse at 5 rue S by the place d'Italie in Paris. I went there once in 1948. Comrades Deng Xiaoping and Li Fuchun worked in factories in the daytime and did Party work in the evenings, while Comrade Zhou Enlai worked as a full-time Party worker.[5]

Father has not forgotten the small coffeehouse near the place d'Italie. On his way to New York to attend the UN General Assembly in 1974, he told his entourage that he and his comrades-in-arms once lived near the place d'Italie and often drank coffee in a small coffeehouse. He asked the comrades from the Chinese Embassy in France to take him to the place d'Italie to have a look. After the visit, he said, with feeling, "Everything has changed!" As he could no longer drink the coffee of that small coffeehouse, Father asked the Embassy personnel to buy some coffee for him from other coffeehouses on the street every morning. He just likes the real coffee of the small coffeehouses of France and often compares them with the small teahouses in his hometown in Sichuan.

From 1923 Father began to work under the direct leadership of Zhou Enlai and became a professional revolutionary.

On February 1, 1924, the *Youth* was officially renamed the *Red Light*. If the *Youth* gave much weight to theory, the *Red Light* was even more militant. Changing theoretical *Youth* into practical *Red Light* shows that the work of the European Party and League branches at the time tended to be associated directly with the practical struggle to meet the demands of the revolutionary struggle.

The first issue of the *Red Light* carried a declaration:

> We want not only to review the current affairs in China but also to show the origin of such affairs and the way out. We would sincerely and honestly indicate the only road to national salvation, which cannot be achieved by other zigzag roads. In a word, the only objective we have set is: the national antiwarlord government alliance and the anti-imperialist international alliance.

The Chinese revolution had entered a new stage, the First Great Revolution, and the CPC European branch and the Youth League were making the militant call.

The journal published articles by Zhou Enlai, Li Fuchun, Xiao Pusheng, Ren Zhuoxuan, Deng Xixian, and Cai Chang. Zhou Enlai alone published nearly forty articles in the journal. Father said that he also published some articles, under the name of Xi Xian and other pen names.

Some of the many articles by Xi Xian exposed the evil acts of the Youth Party, an organization formed in December 1923 in France by a handful of fascist politicians who opposed the Communist Party and the Soviet Union; others refuted the brazen rumors spread by the Youth Party that Soviet Russia sent troops to the Sino-Russian border to pressure China; still others attacked the criminal attempts of the imperialist powers to interfere in China's internal affairs by organizing, at the supposed request of China, expert committees to examine China's political and economic situation and sort out China's debts. These articles used a pungent and forceful language and were very militant. However, they were simply exposé, not works of theoretical or political comment. They were compatible with Deng Xixian's experience, however: he was only twenty years old and had joined the revolutionary ranks not long before. As he commented himself: "I wrote a lot of articles for the *Red Light* under a number of names. There were no ideas to talk about. They merely advocated national revolution and the need to fight against the right wing of the Kuomintang and against Zeng Qi, Li Huang, and their like."

The Deng Xiaoping we see today is a great man of great talent, resourcefulness, and wisdom. We should know that every great man has to go through the process from immaturity to maturity and is tempered bit by bit and step by step.

The *Red Light* was a semimonthly journal published for sixteen months. Each issue had some ten pages. Its format was flexible and quick enough that more copies could be printed, and distributed more widely, than the former *Youth*. Its liveliness, diversity, and short, hard-hitting articles were very appealing to the Chinese students, workers, and other Chinese in France, and *Red Light* was well received by them. Chinese nationals in Europe praised it as "a vanguard of our struggle" and "a star of the Chinese in France."[6] By 1925 a total of thirty-three issues had been published. As it was the official publication of the European Branch of the CPC and the Youth League, it became the material for theoretical study and discussion among their members.

Zhou Enlai was responsible for editing and distribution and was the chief contributor. Li Fuchun was responsible for distribution, and Deng Xixian and Li Dazhang cut mimeograph stencils. They worked in a simply furnished room under harsh conditions. They had to go out to earn their living in the daytime and burned the midnight oil working on the journal. When they held a meeting, they crowded into the small room where Zhou Enlai lived and sat on the bed and table. They ate bread and drank water. Sometimes they had no vegetables to eat. It was under such harsh conditions that these cadres of the Party and the Youth League worked hard for their ideals. But they were optimistic and eager to make progress.

Turning the pages of the journal, one notices at once the beautiful and clear Chinese characters, which show that Father worked earnestly at his job of cutting stencils and printing. Because of his clear writing and simple and refined binding, he was praised as the "doctor of mimeography."[7]

The cover of the journal was a sketch of a teenager with nothing on. Holding a bugle in one hand and a flag in the other, with the boundless mountains and rivers under his feet, he was ready to take off against a background of red light. I do not know who designed it, but I think it is an excellent image that displays very well the features and temperament of the Party and Youth League members in France at the time. These youths in the land of France were in their prime and ready to reshape the world. Many of them later became the backbone of the revolutionary struggle that changed the face of China and the destiny of the Chinese people.

CHAPTER 19

TEMPERING IN THE PARTY

The Fifth Congress of the Chinese Communist Youth League in Europe was held July 13–15, 1924. The members of the new Executive Committee elected at the congress were:

Secretary: Zhou Weizhen

Members: Yu Zengsheng and Deng Xixian (the secretariat was composed of these three)

Director of Training Department: Li Junjie

Director of Propaganda Department: Xu Shuping[1]

According to the Party's regulations, if one held a leading position on the Executive Committee (branch) of the Communist Youth League in Europe, one became a Party member of the European Branch of the CPC. Therefore, when he was elected to the secretariat at the Fifth Congress, Father entered the second stage of his revolutionary career. He took up the leading post of the Chinese Communist Youth League in Europe and joined the Communist Party of China. He was a month away from being twenty years old.

The leaders of the CPC branch and the CCYL in France were often changed. Father told me that it was because the Party and the League organizations wished to train more cadres. The system of lifelong tenure was not followed; the longest tenure was one year. Another reason was that the battlefronts of the CPC were in China and so the CPC Central Committee often selected members of the CPC and the Communist Youth League in Europe to study in the Soviet Union before returning to work in China, or to return to China directly to take part in the struggles.

Fu Zhong recalled that the successive leaders of the Party and League branches in Europe were Comrades Zhao Shiyan, Zhou Enlai, Liu Bojian, Li Fuchun, Fu Zhong, and Hu Dazhi.[2] Fu Zhong did not mention Zhou Weizhen and Ren Zhuoxuan in his memoirs.

In 1924 the revolutionary movement in China, with Guangdong as its base, developed rapidly, and large numbers of cadres were urgently needed. Thus, after the Fifth Congress was concluded, the CPC Central Committee selected group after group of Party and League members to send either to study in the Soviet Union or directly back to China. In late July, according to the directive of the CPC Central Committee, Zhou Enlai and some other comrades came back to China.

Zhou Enlai went to Europe at the age of twenty-two and left France at the age of twenty-six. He had grown from a young student who sought truth into a professional revolutionary with a firm faith in Marxism and communism, some experience in revolutionary struggle, and strong organizational and leadership abilities. He and his comrades pioneered the communist cause carried on by the Chinese students in France. Under his leadership, guidance, and example, many revolutionary youths strengthened their belief and learned the art of struggle. When the comrades of the CPC and Youth League branches in Europe learned that Zhou Enlai was going back to China, they felt both excited and reluctant to part with him. Gathering around Zhou Enlai, they took a picture with him for remembrance. Besides Nie Rongzhen and Li Fuchun, Deng Xixian, the newly elected member of the secretariat of the Executive Committee, is in this picture, standing in the last row. He is in Western-style clothes with a cap on his head. Although his round face still seems a little childish, it shows some confidence and fortitude.

I once asked Father who he was closest to among the Chinese students in France. He thought it over and then answered that it was Premier Zhou: "I had all along regarded him as my elder brother, and we had worked together for the longest period of time."

It was true that during the two years when Father worked with Zhou in France, in the late 1920s and early 1930s when they did underground work in Shanghai, in the Central Soviet Area in Jiangxi, during the Long March and the revolutionary wars, when they worked in the highest leading bodies of the Party and the government after the founding of the People's Republic, and until the time when Premier Zhou, having given his all to the Party, the country, and the people, breathed his last, Father all along remained an able assistant and a loyal comrade-in-arms of Zhou Enlai. When the premier's health deteriorated, Father withstood high pressure, managed the state and army affairs, and shared some responsibility for the premier. When Premier Zhou became critically ill, Father kept watch day and night by his bedside. When Premier Zhou passed away, Father made the memorial speech on behalf of the Party and the people despite his deep sorrow. Tears are falling down my cheeks as I write these lines.

Because of Father's close relations with Zhou Enlai, we got to know Uncle Zhou in our childhood, and we all loved and respected him very much. Since his wife, Mom Deng, had the same family name as my father, for some time our parents told us to call her aunt. From this we can see the intimate relations between my parents and Uncle Zhou and Aunt Deng. Uncle Zhou was very kind to us, and he also made jokes with us, saying that I was the "foreign minister" of our family and my second elder sister Deng Nan was the "premier" and had a position as high as his!

We spent our childhood happily, having the care of many revolutionaries of the old generation like Uncle Zhou. During the Cultural Revolution, we saw with our own eyes the adverse circumstances that our seniors had experienced. When Uncle Zhou became seriously ill, we were as worried about him as our parents were. When he died, our parents took us to take part in the solemn memorial meeting held in the Great Hall of the People. After the meeting was over, we walked to the portrait of Uncle Zhou and made bows to him and we could not help crying. Whenever I think of that scene and the graceful bearing and likeness of Uncle Zhou, I cannot help feeling sad. I still keep the black armband I wore at the memorial meeting for Uncle Zhou.

Uncle Zhou and his comrades-in-arms will always be remembered by history and the people.

Zhou Enlai came back to China in late July 1924 to take part in the revolutionary struggles, but many Party and League members had already left France, among them, Zhao Shiyan, Li Weihan, Chen Yannian, Chen Qiaonian, Wang Ruofei, Yuan Qingyun, Xiong Xiong, Liu Bojian, and Zhang Shenfu. Father said that he did not know Zhang Shenfu but attended his farewell party in 1923.

Although these core members left France, the work and struggle of the CPC and League branches did not stop. The new branches continued to expand their organizations, organize Party and League members for study, propagate Marxism and Leninism, and brief the Chinese students and workers in France on the international and Chinese revolutionary situation. They also remained steadfast in their struggles against the KMT right-wingers and the Youth Party.

In the autumn of 1924 most of the organizations that supported etatism were dissolved and very few remained. The leading figures of the Youth Party, such as Zeng Qi and Li Huang, left Europe and gave up their positions there, whereas the Communist organizations in France became all the more active and energetic.

The Chinese Communist Youth League in Europe often organized mass meetings and lectures in Paris. Its members also attended the meetings of other organizations and used them as forums for propagating their political ideas.

On October 10, 1924, the various Chinese organizations in France held a rally to celebrate National Day. Ren Zhuoxuan led over 100 League members taking part in the celebration. Also attending the rally were members of the KMT, the etatist Youth Party, the Social Democratic Party, as well as Chinese from all walks of life and Chinese workers. The participants totaled several hundred. There was a dispute over which flag should be hoisted, the KMT's flag with a white sun against a blue sky, or the five-colored flag of the Northern Warlord government supported by etatists. The League supported the KMT and waged a resolute struggle against the Youth Party.[3]

Apart from the activities and protests carried out on public occasions, Li Fuchun and other leading League comrades also gave lectures on political economics in Paris. Some Chinese students and workers were admitted to the study groups, and the truth and new knowledge they were exposed to had a strong appeal to them.

In December 1924 the CCYL in Europe held the Sixth Congress, at which it was decided to set up a supervisory division under the League branch. The division was composed of seven to eight members, including Li Junjie and Deng Xixian, with Li Junjie as the director. Ren Zhuoxuan, Li Fuchun, and Deng Xixian, who had either joined the trade union or were familiar with the trade union movement, were responsible for organizing a trade union movement committee, with Yu Zengsheng serving as the director.

After the Sixth Congress ended, the CCYL in Europe made a decision to expand the Executive Committee: the Propaganda Department under the branch had six deputy directors. Fei Ziheng, Deng Xixian, and Xiong Jiguang were in charge of workers' movement, and Xiao Pusheng was in charge of publication of the *Red Light*.[4]

In the spring of 1925, after serving one term as the CCYL Branch Committee member, Father was sent to Lyons to work as a special representative of the CPC branch in Europe. Being the leader of the Party and League organizations in Lyons, he served as the deputy director of the Propaganda Department, as secretary in charge of training of the Lyons branch of the Youth League, and concurrently as the secretary of the CPC Lyons group.

In the same year, the Party and League organizations in France waged several large-scale mass struggles. On May 30, 1925, the May 30th Massacre occurred in China and shocked the whole world. In 1925 the Japanese imperialists and the Northern Warlord government wantonly suppressed the vigorously developing workers' movement in Shanghai. On May 15, Gu Zhenghong, the workers' leader and a member of the Communist Party, was shot dead. His death aroused the indignation of the public. The CPC Central Committee decided to continue the anti-imperialist struggle. On May 30, the students in Shanghai supported the workers, calling for the recovery of the foreign concessions. The British imperialists arrested over 100 people, killed a dozen, and wounded many more.

In response to the call of the CPC Central Committee, over 200,000 workers, 50,000 students, and a great many businessmen in Shanghai went on strike. Nearly 500 cities and towns across the country, such as Beijing, Nanjing, Hankou, and Guangzhou, immediately responded. The mammoth struggle lasted for three months. The May 30th Movement was a prelude to the upsurge in the Great Revolution.

After the May 30th Movement broke out, the European branches of the CPC and the CCYL immediately took action. They made a joint announcement with the General Branch of the KMT in France that an anti-imperialist rally of the Chinese residing in France would be held at 96 boulevard d'Auguste Blanqui in the central area of Paris on June 7. Among the over 1,000 people attending the rally—apart from the Chinese workers and students—were representatives of the French Communist Party and the Vietnamese Communist party group in France. The rally was chaired by Ren Zhuoxuan, then secretary of the CPC European branch. Representatives from various circles took the floor one after another to denounce the atrocities perpetrated by the imperialists who had invaded China and slaughtered the Chinese people. The rally adopted a resolution that firmly supported the May 30th Movement against imperialism, protested the dispatch of troops to Shanghai by the French government, demanded an immediate withdrawal of French troops and warships from China, called on the Chinese to unite to jointly oppose imperialism, and instructed the Office of the Red Light and other organizations to set up the Action Committee of the Chinese in France to Aid the Anti-Imperialist Movement in Shanghai. A demonstration was organized after the rally.

On June 14, the Executive Committee of the CPC European branch and the Executive Committee of the CCYL in Europe published a letter "to the Chinese who joined the demonstration." From the beautiful handwriting on the leaflet, we can easily tell that it was Deng Xixian who cut the stencil. The leaflet read:

> A serious and glorious action was taken in Paris—the center of the reactionary forces in Europe. We, the oppressed Chinese people, held for the first time a demonstration directed at the imperialist government! The Chinese taking part in the demonstration! Your spirit should be respected by us. No matter what doctrine you believe in, we pay tribute to you so long as you oppose imperialism in words and deeds, and wage from today uncompromising struggles against imperialism. We believe that all the struggle for overthrowing imperialism is more sacred than anything. The key to beginning to liberate the oppressed nations and all of mankind is to overthrow imperialism, to oppose the French imperialists that slaughtered the Shanghai people, and to oppose all imperialists who massacred the Shanghai people![5]

On June 21, uniting with the leftists of the KMT, the CPC branch and the CCYL in France organized a demonstration, and the demonstrators marched toward the legation of the Republic of China. Over 200 participants came from different directions. They blockaded the gate of the legation, cut off the telephone line, captured Minister Chen Lu, and forced him to sign a prepared document. This document stated that a telegram would be sent to the whole Chinese people, in the name of Minister Chen Lu, to support the anti-imperialism movement; that a diplomatic note in the name of Minister Chen Lu would be presented to the French government requesting it to withdraw its troops from China; that the minister would apologize to the Chinese in France and guarantee effective protection of the interests and freedom of the Chinese; and that Chen Lu had to donate 5,000 francs to be mailed to Shanghai to aid the workers on strike. This document was distributed to the relevant institutions and newspaper offices.

The actions taken by the Chinese in France to support the anti-imperialist May 30th Movement in China shocked all of Europe and greatly scared the French government. On the following day, June 22, the French government ordered the police to track down and arrest the Chinese Communists in France. In a few days, over twenty CPC and CCYL members who lived in the Paris area, such as Ren Zhuoxuan and Li Dazhang, were arrested and put into prison.[6] The French authorities then expelled forty-seven Chinese students.[7]

The arrest of Ren Zhuoxuan and other leaders caused great damage to the CPC branch and the CCYL branch in Europe. At that time, Father was in Lyons. He wrote:

> The leading comrades in Paris are expelled out of France on the ground of the anti-imperialist movement. Comrade Xiao Pusheng, the Party secretary, wrote me an urgent letter and appointed me special committee member of the area of Lyons and Cruesot responsible for giving guidance to all the work in the area. We were then quite ill informed of the situation in Paris and only knew that the central leading body of the organizations was no longer there and it was quite difficult to carry on any work. At the same time, urged by other comrades, I decided to resign from my job and go to Paris to work for the organizations. When I got to Paris, Comrade Xiao Pusheng was not expelled yet. So we held discussions on the establishment of a temporary executive committee. Before long, the executive committee was changed into the extraordinary executive committee. I was eventually appointed a member of both committees.

When the Party organization was sabotaged, Deng Xixian and his comrades-in-arms Fu Zhong, Li Zhuoran, and others returned to Paris from other places and took over the leadership positions of the Party and League organizations on their own.

On June 30, 1925, the Provisional Executive Committee of the CCYL in the European District was founded. The secretary was Fu Zhong (Xiao Pusheng acted in his place), and members were Deng Xixian and Mao Yushun. The Provisional Executive Committee decided that these three comrades formed the secretariat but that the living expenses of only one would be provided.[8]

According to a secret report received by the French police,

> On July 1, 1925, a meeting was held at 14 rue Taversiere in Billancourt with thirty-three participants. The chairman of the meeting spoke first. He said that since most of the members of the Chinese Action Committee in France were arrested, it was necessary to reorganize the committee. Besides, a protest statement will be printed both in French and Chinese soon and distributed in Paris. At the meeting, the radical people of anti-European capitalism expressed their firm opposition to the act of the French side to expel their Chinese compatriots, and especially their strong indignation at the decision to expel ten more Chinese on the coming Saturday. The meeting ended when the hotel owner came in and told them that the police had come.[9]

On July 2, the Provisional Executive Committee convened a meeting to protest imperialism. A secret French police report recorded that

> the Chinese Action Committee in France held a meeting yesterday afternoon at 23 rue Boyer to protest against international imperialism, with over seventy partici-

> pants. The chairman of the committee said that they had set up the Action Office, but its staff had not yet been reported to the congress and the members were to be elected at group meetings. Eight people spoke at the meeting. Among them, Deng Xixian proposed to unite with the government of the Soviet Union to oppose imperialism.[10]

On August 17, the first session of the Executive Committee of the Seventh Congress of the CCYL in Europe was held. The secretary was Fu Zhong; members were Deng Xixian and Shi Qubing; and these three formed the secretariat.[11] Fu Zhong, Deng Xixian, and Deng Shaosheng were also contributors to the Party and League publications. Father not only wrote during this period, but the Party organization appointed him concurrently the secretary of the supervisory committee of the KMT General Branch in France, responsible for all the work of the KMT.

Suppressed by the French government and the army and police, the CPC and CCYL organizations in Europe, instead of retreating, rapidly restored their organizations and carried on ever more positively their staunch and unyielding struggles. A secret report of the French police said:

> On September 6, a meeting was held at 23 rue Boyer in Bellevilloise with over forty participants. Since the incident in the Chinese Legation occurred, some of the Chinese Communists are now living in the Paris area and have taken emergent measures to prevent discovery. The purpose of this meeting was to commemorate Mr. Liao Zhongkai. Further investigation is to be made so as to find out who the meeting's organizer and participants were.[12]

On September 12, 1925, only two months after the large-scale arrest by the French army and police, the CPC European branch convened an enlarged meeting. The participants decided to organize another large-scale anti-imperialist rally of Chinese in France in the name of the General Branch of the Chinese KMT in France. At noon on September 15, over 1,000 Chinese held a mammoth, militant anti-imperialist rally in a meeting hall near the Seine in the central part of Paris. Shi Yisheng, a member of the CPC and vice-chairman of the General Branch of the KMT in France, spoke first. He said that since the May 30th Movement, the Chinese in France had taken an active part in the struggles, but that was not enough. By upholding the banner of proletarian internationalism, they should make persistent efforts to march forward heroically, direct the struggle at the British, Japanese, French, and American imperialists, and drive them out of Chinese territory so as to accomplish the great task of liberating the Chinese nation. After his speech, the French Communist Party representative, named Doriot, a French member of parliament, Malche, the Vietnamese Communist Party representative, and the representative of black Africans took the floor one after another. Finally, the CPC representatives Fu Zhong and Xiao Pusheng spoke. They pointed out that the May 30th Movement was part of the socialist revolution of the world proletariat, and the proletariat and laboring people of the whole world should be united to wage struggles against imperialism. Unless they won a complete victory, they would not stop their struggles. In the meeting hall, participants seethed with excitement. They shouted, "Down with imperialism! Down with warlords! Long live the victory of the Chinese national liberation movement!"

The convocation of this anti-imperialist rally brought the masses of Chinese in France closer to the CPC branch in Europe, and the revolutionary momentum was heightened. The rally also once again shocked the French government, which decided to immediately arrest its chairman and organizer. French police arrested Shi Yisheng and expelled him because the KMT right-wingers informed against him.[13] Although Father did not speak at the meeting, he took part in leading and organizing it as a leading member of the CPC branch.

The more militant the activities of the CPC branch in Europe and the CCYL became, the more nervous the French government became. The French police began to pay increasingly close attention to the movements of the Chinese and the Chinese Communists and conducted secret surveillance. The police were informed of even some small-scale meetings. However, the CPC members in France were not scared. They continued to hold meetings and make speeches as members of the CPC or of the KMT.

On October 25, 1925, the French intelligence agents reported: "From 20:00 to 21:30 yesterday [October 24], a meeting of Chinese Communists was held in a café on rue Charlot in Issy-les-Moulineaux City with twenty-five participants. The meeting was presided over by Deng Xixian. Wu Qi read out the lesson on education in communism and pointed out the necessity to reestablish the Chinese Communist group and start some publications."[14]

On November 16, 1925, the French intelligence agents reported that a mass meeting of the KMT was held in Paris chaired by Deng Xixian. The meeting commemorated Wang Jingqi, a leading member of the European branch of the KMT, and exposed the persecution of the progressives by the international imperialists and the French imperialists. The report said:

> The KMT convened a meeting from 15:00 to 17:00 on November 15 at 23 rue Boyer in Bellevilloise. The meeting was presided over by Deng Xixian and forty-seven people attended. The meeting commemorated Wang Jingqi, who had been expelled by France and died on the ship bound for China. Eleven representatives, including Chen Xi, spoke at the meeting, protesting against the arrest of Chinese by the French police. Finally, Deng Xixian said in his closing speech: "We hope that all of us present will always remember Comrade Wang Jingqi and continue to wage struggles against imperialism."[15]

On November 20, 1925, the French interior minister also mentioned this meeting in his letter to the French foreign minister. The letter read: "A meeting was held on November 15 at 23 rue Boyer in Bellevilloise to commemorate

Wang Jingqi, who had been expelled by France and died on the ship bound for China. Besides, the KMT side has appointed Chen Xi the representative of the KMT."[16]

By then, Deng Xixian was just twenty-one years old. He had gone from an ordinary League member to a full member of the CPC and had been elected to a leading position in the Party and League organizations in Europe. He had become a Communist of even firmer faith, was more mature in handling matters, and now had some experience in waging struggles and exercising leadership. He had served one and a half terms as the leader of the Party branch. His activities had drawn the attention of the French police, and they began secretly keeping watch on him and following him.

On July 30, 1925, Father was registered in the residents' book of the Boulogne police station in Billancourt; his registration number was 1250394.[17]

On November 6, 1925, Father went to work in the Renault Automobile Factory[18] and was assigned to work in the benchwork workshop. When I visited France in 1988, I went to the Renault Automobile Factory to look for Father's data. The factory personnel received me warmly and led me to visit the benchwork workshop where Father had once worked. They gave me some pictures of the factory and its workers in the 1920s. What was most valuable was that they found Father's work card.

On the work card, Father was registered as: Deng Xixian (Teng Hei Hien), Chinese, born on July 12 (according to the lunar calendar), 1904, in Sichuan; residing at 27 rue Traversiere in Bellancourt; a skilled bench worker, assigned to work in number 76 workshop. The wage for each grinding piece was one franc and five centimes.

On the bottom left corner of the card was a small photo of Father with the printed number of 82409A. He looked very young. Those who did not know him might have mistaken him for a teenager. But they did not know that Father, then twenty-one years old, had been a leading member of the CPC branch in Europe under the surveillance of the French police!

Father did not work long in the Renault Automobile Factory, but he was clever and mastered some benchwork technique. When he worked under surveillance in a factory in Jiangxi Province during the Cultural Revolution in the 1970s, this technique became very useful.

After Father went to work in the Renault Automobile Factory, the French police still kept close watch on him. I once asked him, "Why did the French police pay so much attention to you?" He replied, "Because I was quite active. The French police were quite clear about our activities!"

On January 7, 1926, the French police received a detailed report.[19] It said:

> According to the information received on January 5, the Action Committee of the Chinese Group in France held a meeting in the afternoon of January 3 at 23 rue Boyer in Bellevilloise. At this meeting, several speakers raised the slogan of "oppos-

ing imperialism" and demanded that the Chinese in France unite to support General Feng Yuxiang's pro-Communist and anti-Beijing government policy.

At the meeting the Action Committee also decided to ask the Chinese minister in Paris to declare his stand on the south-north conflict in China and to oppose any international intervention.

Although investigation has been made, since the Action Committee is well organized, the location and composition of the committee cannot be found. However, the several Chinese who spoke at the meeting on January 3 have been identified.

One of them is named Deng Xixian, who was born on July 12, 1904, in the family of Deng Wenming and Dan Shi in Sichuan Province of China. He has been living at 3 rue Casteja in Boulogne-Billancourt since August 20, 1925. He complies with the regulations of laws and decrees concerning foreigners. After arriving in France in 1920, he first went to work in Marseilles and then in Bayeux, Paris, and Lyons. When he came back to Paris again in 1925, he became a worker in the Renault Automobile Factory in Billancourt and stayed there until January 3. He attended meetings as a representative of the activists of the Communist Party and seemed to have made speeches at various meetings organized by the Chinese Communists. He especially advocated close ties with the government of the Soviet Union.

In addition, Deng Xixian also has a lot of booklets and newspapers of the Communist Party and has received many letters from China and the Soviet Union.

Two Chinese compatriots live with Deng Xixian. It seems that they are also in favor of Deng Xixian's political views. When they go out, they always accompany Deng Xixian. Fu Zhong (Fou-Tchang) was born in June 1903 [actually, 1900] in China. Ping-Suen-Yang, twenty years old, was born in Shanghai. They observe the French laws on foreigners. They claim to be students and are not engaged in any work.

Since the Chinese in Paris are quite exclusive, it is very difficult to find out their activities. To make everything clear, it seems necessary, with the permission of the commissioner of the General Police Bureau, to make visits and investigations at their residences in Billancourt. We might get some information from the house owners, and by so doing, we could, through checking their ID cards, find out who among them are the wanted Communists.

Three hotels should be kept under close watch: 3 rue Casteja; 14 rue Traversiere, and 8 rue Jules Ferry.

After receiving this report, the French police immediately decided to search the residences of Deng Xixian and others. On January 8, 1926, the French police made a surprise search of the three hotels in Billancourt and reported:

On the order of the commissioner of the police bureau, searches were made of three hotels from 5:45 to 7:00 this morning in Boulogne-Billancourt. The addresses of the three hotels are: 14 rue Traversiere; 3 rue Casteja; 8 rue Jules Ferry.

The purpose of searching the three hotels was to find out those Chinese engaged in propaganda of communism.

> All the rooms of these hotels were searched, and over 100 Chinese documents examined.
>
> In room 5 of the hotel at 3 rue Casteja, we found a lot of French and Chinese booklets disseminating communism (*Workers in China, Dr. Sun Yat-sen's Will, A.B.C. of Communism*, etc.), Chinese newspapers, and especially the Chinese Communist newspaper *Progress*, published in Moscow, as well as essential accessories of two mimeographs together with printing metal plates, rolls, and several packs of printing paper.
>
> The three men named Deng Xixian, Fu Zhong, and Ping Suen Yang lived in that room until January 7. They left suddenly yesterday, and the two men named Mon Fi Fian and Tchen Kouy who lived at 8 rue Jules Ferry also left in a hurry at the same time. These Chinese seem to be active Communists.
>
> It appears that these men had found themselves to be suspected and so they hurriedly disappeared. Their compatriots have taken precautions and have abandoned all the documents that might cause them trouble.[20]

Yes, the tenants who lived in room 5 of the hotel at 3 rue Casteja included Deng Xixian and Fu Zhong, leading members of the CPC branch in Europe. The French police were going to arrest them. They had obtained that information in advance and had resourcefully fled to distant places.

They had flown far and wide. Where and in what direction did they flee? The sacred place of the revolution at that time—the Soviet Union.

CHAPTER 20

ADIEU, FRANCE

As early as May 1925, the CPC European branch had decided to send some members, including Deng Xixian, to Moscow for further study.[1]

A letter, dated November 18, 1925, from Yuan Qingyun in Moscow to Fu Zhong and others mentioned again that "as soon as you receive our letter in the near future, you should set off at once."[2]

Another letter, dated December 9 from Moscow and addressed to Fu Zhong and others, said that

> we presume you have received our letter dated November 18. We ask Deng Xixian, Liu Mingyan, Fu Zhong, Zong Xijun, and Xu Shuping to come here as quickly as possible. If Zong Xijun cannot come, Li Junjie [Li Zhuoran] can take his place. The reason you have to come here has been explained in the previous letter. For the C.P. and the interests of the revolution, you have to come here at once.[3]

From this we can see that Father and others had received instructions from the CPC Moscow branch and were prepared to hurry to Moscow. Simultaneously, they were carrying out revolutionary activities, working in the factory, and preparing to leave France. So when the French police came searching, they had left without delay, and the police closed in on nothing.

On January 7, the Executive Committee of the European branch of the Chinese Communist Youth League issued a circular: "Twenty comrades who are leaving for Russia have decided to depart from Paris this evening (January 7). . . . Probably they will return to China soon. Comrades! When one group of our fighters after another advance to the front, we should keep in mind the slogan 'Go back home as early as possible.'"

On January 23, 1926, Liu Mingyan, a leading member of the Communist Youth League, wrote, "Twenty comrades left here for Russia on January 7." On his list were Fu Zhong, Deng Xixian, Li Junzhe [Li Junjie?], Deng Shaosheng, Hu Lun, and others.

On January 7, Father and his comrades-in-arms boarded the train that went north to the Soviet Union, the native place of the October Revolution. When boarding the train, they received the order of expulsion issued by the French police. This order actually prohibited them from setting foot on the soil of France ever again. The truculent French police authorities did not anticipate that the person they expelled in the 1920s would pay a visit to France in the capacity of a state guest fifty years later and would be accorded the warmest of red-carpet welcomes by the French government and the French people.

On their way, Father and his comrades-in-arms stopped in Germany and stayed in the houses of German workers. They were given a warm reception by the German working class and Communists. Father said their warm reception was genuinely proletarian and comradely.

From October 19, 1920, to January 7, 1926, Father lived in France for five years, two months, and nineteen days. When he came to France, he was a naive sixteen-year-old young man. After gaining extraordinary experience in his study, working, joining the Party and League organizations, and participating in revolutionary struggles, he left France at the age of twenty-two having grown into a career revolutionary who had a firm faith in communism and experience in revolutionary struggle.

In addition to disseminating Marxism and the ideal of communism and waging proletarian revolutionary struggles, a very important part of the revolutionary activities carried out by Father and his comrades in France was resolute and uncompromising opposition to imperialism, especially the piratical acts perpetrated by the imperialists against their motherland. China had suffered untold bullying by the Western powers, and its people had suffered oppression and plunder by international imperialism and the Chinese feudal forces. Since the very day they joined the Communist organization, Father and his comrades had resolved to link their own fates with the fate of their motherland and the people and to give their youth and even their lives to the cause of saving the nation. They lived under the powerful thumb of the Western powers, but they held high the banner of opposition to imperialism. They were not afraid of being arrested, expelled, or imprisoned. They did indeed have the heroic spirit and lofty ideals of young revolutionaries.

Some people have said that the Chinese are opposed to hegemonism, power politics, and external intervention because they have a strong national dignity. This national dignity stems from the brilliant 5,000-year history of their motherland and their splendid culture and, at the same time, from the bitter history of the national humiliation of subjection to bullying, insult, and plunder. Beginning from the very moment the Chinese were first subjected to aggression, many Chinese people realized that only by striving constantly to become stronger could there be a future for the Chinese nation, one maintained on an equal footing with all the other nations of the world.

However, in the last 100 years or so, only the Chinese Communists have led the Chinese people in standing upright and advancing from humiliation to national dignity, from poverty to prosperity, and have enabled the Chinese people to stop their suffering at the hands of foreign powers and to become masters of their own country. This objective was reached by the group of young Communists in France.

The European branches of the CPC and the CCYL trained, tempered, and nurtured a large group of fine revolutionaries who had faith in communism and practical revolutionary experience. They were a group of young intellectuals who cherished their country and their people. They had strong revolutionary enthusiasm. They lived simple and honest lives, and they had noble character and pure morals.

Father once told us:

> We lived a hard life then. After we became professional revolutionaries, our living expenses were provided by the organization. We could only eat some bread or cook some noodles. We did not institute the lifelong tenure system then, giving no consideration to position. For instance, while in France, Zhao Shiyan held a higher position than Zhou Enlai, and Zhou Enlai's position was higher than Chen Yannian's. But when they were back in China, Chen Yannian's position was the highest among them. Chen Yannian was very capable. He opposed his father (Chen Duxiu) and held better views than others did. His death was a deep regret. After coming back to China, Zhao Shiyan worked under their leadership, but he did not mind it. None of us minded the position and had no such ideas. We just carried out revolutionary activities. This was the characteristic of the Communists in the early period.

Father also said, with feeling, "It was hard to join the Communist Party. What an important matter it was to join the Communist Party at that time! We really gave our all to the Party, with nothing left over!"

Hearing these words of Father's, I was deeply moved and had many thoughts and feelings. As compared with the people of his generation, some youths today seem to lack something. Their enthusiasm seems to be not that strong, their faith not that firm, their character and morals not that pure, and even the blood flowing in their blood vessels not that red and hot.

As a Chinese saying goes, "The times produce the heroes." In the torrent of the Chinese revolution, each wave washed over the one ahead and each wave was higher than the last. In the early days of the Communist Party of China, men of talents gradually emerged. In France alone, a large number of pioneers who devoted themselves to the cause of revolution and to the liberation of the Chinese people arose from among the students who were studying on work-study programs and the Party and League members residing there.

Zhou Enlai: would be premier of the People's Republic of China

Deng Xiaoping: would be vice-premier of the State Council of the People's Republic of China and chairman of the Military Commission of the CPC Central Committee

Chen Yi: would be marshal and vice-premier of the State Council of the People's Republic of China

Nie Rongzhen: would be marshal and vice-premier of the People's Republic of China

Li Fuchun: would be vice-premier of the People's Republic of China

Li Weihan: would be a senior CPC leader and director of the United Front Work Department of the Party's Central Committee

Li Lisan: would be a senior CPC leader

Cai Chang: would be chairwoman of the All-China Women's Federation

There were also those martyrs who died a heroic death before the founding of New China:

> Wang Ruofei: a senior CPC leader who was killed in an air disaster at the age of fifty in 1946
>
> Zhao Shiyan: a senior CPC leader who was killed by the Kuomintang at the age of twenty-six in 1927
>
> Chen Yannian: a senior CPC leader who was killed by the Kuomintang in 1927, when he was less than thirty
>
> Chen Qiaonian: a senior CPC leader who was killed by the Kuomintang at the age of twenty-six in 1928
>
> Cai Hesen: a senior CPC leader who was killed by the Kuomintang at the age of thirty-six in Guangzhou in 1931
>
> Xiang Jingyu: a leader of the Chinese women's movement who died a heroic death in Wuhan in 1928
>
> Liu Bojian: a high-ranking general of the Red Army who was killed by the Kuomintang at the age of forty in 1935

There were many more names that have long been forgotten.

These people were born and bred in China but accepted progressive ideas from the West and eventually found truth. They sought truth boldly and heroically, and they shed their blood for their belief, determined as they were to save China and the Chinese people from peril.

They could have continued to study in France, worked in factories, engaged in trade, or made peaceful marriages and careers for themselves in the West. But they all came back to their poor, backward, and devastated motherland, to the land where they were born and bred, to their brothers and sisters, to their people who were undergoing untold suffering. They shed their blood and sweat on the Chinese land.

How many of them are still alive after several decades? Most of the Chinese students in France then either have been killed or have died of illness after working their hearts out for the revolution. They have gone to their eternal sleep on the land of their motherland and been integrated into the mountains, rivers, and plains of the Chinese nation. They were a generation who moved people to song and tears and were crowned with eternal glory.

We, the later generations, will always hold them in great respect from the bottom of our hearts.

Father had only four photos taken in France. One was a full-length picture, taken in March 1921 not long after his arrival in France, in which he is wearing a cap and Western-style clothes. He sent this photo to his fellow student Liu Puqing in June 1925. Fortunately, the venerable Liu kept it and gave it back to Father as a present nearly forty years later. This is the earliest photo of those we have and was taken when he was sixteen years old.

The second photo includes his uncle Deng Shaosheng and was taken a little later than the first one. The third is a group photo of all the delegates to the Third Congress of the European District of the Chinese Communist Youth League, taken in July 1924, the same photo on display at the memorial meeting for Zhou Enlai. In it, Father is standing on the right of the last row. The fourth is the one-inch photo on his work card for the Renault Automobile Factory.

Only these four photos are left. Some articles and reports have misidentified him in other photos.

While in France, Father contracted typhoid fever, probably between 1923 and 1925. He told me that he had contracted typhoid fever twice in his life, once in France and the other time during the Long March. He was on the verge of death on both occasions. Thanks to the fairly high level of medical treatment in France at that time, he was hospitalized for treatment and recuperated in a sanatorium for one month after being discharged, all free of charge. I think it was inevitable that Father would get ill considering how harsh his living conditions were at that time. Fortunately, he narrowly escaped death. Otherwise, I would not have the special honor of writing this book today.

Since Father lived in France for five years, he has always been described in some contemporary works as having some "outlandish" ways. For example, he was fond of Western classical music. Some works have capitalized on his present hobby of playing bridge and illustrated his disposition by showing him with playing cards in his hand, even when discussing important military and state affairs.

Father lived in the West for more than six years (including one year in the Soviet Union). Truly, he got used to some foreign habits. For instance, he likes to eat potatoes, to drink French wine, to eat cheese and bread, and to drink coffee. Some old comrades who studied with Father in France had such predilections as well. During his visit to France in 1975 and his stay in France on his way to New York for the United Nations General Assembly in 1974, Father bought some bread to send as a present to his old comrades who had studied in France, such as Premier Zhou, Nie Rongzhen, and Cai Chang. Whenever he received a present of good French wine or cheese from others, he never forgot to share them with these old friends.

Father acquired a lifelong hobby while in France—watching football matches. He did not have very much money in France, but once he spent five francs to buy the cheapest ticket to watch an international football match. In recollection now, he says that his food cost him five francs a day, and so it was hard for him to see the match. Furthermore, his seat was in the top row, and he could not even see the ball very clearly. He still remembers that Uruguay was the champion of that world tournament. Since liberation, he has been an enthusiastic spectator of football matches. He will go to any match, even one played by children's teams on the Xiannongtan Sports Ground. He also took us to watch the games. Although we did not know the rules, we had to go with him. When we were young, we spent most of the time in the lounge drinking soda water. But when we grew up, we all became enthusiastic lovers of football.

I was extremely moved on one occasion. In 1973, when the Cultural Revolution was not yet over, Father had just been freed from house arrest and had still not resumed his work. A foreign football team came for a match, and Father took us to watch the game. He had intended to sit quietly in the last row. Unexpectedly, he was discovered, upon his entry, by the audience sitting on the next bleachers over, and then more than 10,000 spectators in the stadium all stood up and applauded enthusiastically. Father had to move to the front row and also clap his hands repeatedly to greet the audience. The scene was truly exciting and has remained in my memory for years.

He forged an indissoluble bond with football while in France, but his desire to see it could not be met then. Today, conditions are much better. He can watch football games without leaving home. Whenever a world tournament is televised live, certainly he does not miss it. If he has no time, he has the match recorded for his later enjoyment. During the 1990 World Cup Tournament, he had just retired and had time to watch the game. He watched a total of fifty of the fifty-two matches broadcast on television—watching football to his heart's content.

As to Western classical music, Father is not interested in it, nor has he much knowledge of it. He likes Peking opera and knows a lot about it. He is an advocate of China's unique cultural features unadulterated by foreign flavors. What he likes most is the old male role of the Yan Jupeng school and the young female role of the Cheng Yanqiu school. Our home was close to the Huairen Hall. Whenever there was a performance of Peking opera, our whole family was always in the audience. My parents were the ones who really liked it, but we liked to join in the fun. It is indeed a wonderful artistic enjoyment to watch Peking opera. Listening to and watching it quite often, audiences can learn many historical stories and improve their cultural level. The words and phrases in the operas are really too beautiful!

Father learned to play bridge in Chongqing only after he led his troops to march to southwestern China. It was impossible for him to have such a normal hobby in the years of war before liberation.

CHAPTER 21

IN THE HOMETOWN OF THE OCTOBER REVOLUTION

The land is boundless yet bare, covered with silvery snow. Here is Russia.

The city is ancient but with a new look, surrounded by green forests. Here is Moscow.

The square is so vast and so solemn, with red flags fluttering. Here is Red Square.

This is the Kremlin, the site of Lenin's office and the people's regime of the first Soviet socialist country.

The first proletarian regime in the world was established in 1917 and had just spent eight years consolidating its political power and seeing to the recovery of its economy. Vladimir Ilyich Lenin, the leader of the October Revolution and teacher of the international proletarian revolution, had just passed away. While the newborn Soviet regime was going all out to heal the wounds of war and expand its economy, the Communist International under the leadership of Lenin was starting to fulfill its international obligations: by training cadres, it was helping the Oriental countries and regions where national and democratic revolution was cresting.

In 1921 the Communist University of the Toilers of the East was founded in Moscow, the capital of the Soviet Union, to train cadres for both the ethnic groups of the eastern part of the Soviet Union and the Oriental countries. Those trained in the university included Indians, Vietnamese, Japanese, Turks, Arabs, Persians, and Algerians. The Chinese students there numbered thirty-five in 1921 (most of them were Party or League members) and forty-two in 1922. The following year, the CPC branch in Europe sent twelve comrades, including Zhao Shiyan, Chen Yannian, Chen Qiaonian, and Wang Ruofei.

The revolutionary situation in China had developed rapidly since 1923. In June of that year, the CPC decided to form a united front with the Kuomintang. In the same year, Dr. Sun Yat-sen laid down the Three Great Policies—alliance with Russia, cooperation with the Communist Party, and assistance to the peasants and workers—and decided to reorganize the Kuomintang with the help of the Communist Party. In 1924 he founded the Whampoa Military Academy, which the CPC helped to lead, and the revolutionary army. As the First Revolutionary Civil War in China developed rapidly, both the CPC and the Kuomintang felt the need for more revolutionary cadres and asked the Soviet Union to train more people.

In response to that request, the Soviet Union founded Sun Yat-sen University for the Toilers of China solely for Chinese students. Its purpose was to use Marxism to train cadres for the Chinese communist mass movements and Bolshevik cadres for the Chinese revolution.

At the end of 1925, with the help of Vasili Constantin Borodin, the Soviet political adviser to the National Government in Guangzhou, the Kuomintang and the CPC jointly selected 310 students to be sent to Sun Yat-sen University for training. The first group of 118 students arrived in Moscow in November 1925; at least 103 of them were members of the CPC or CCYL, accounting for over 87 percent. Ten KMT members who had been studying in Germany entered the university in January 1926. Not long after, the CPC and CCYL branches in Europe sent twenty members who had been studying in France and other countries to Moscow for further study. Deng Xixian, Fu Zhong, and Li Zhuoran were among them. They went to the Communist University of the Toilers of the East first and soon entered the newly founded Sun Yat-sen University.

I visited the site of Sun Yat-sen University in 1990 when I led the group from the Chinese International Association for Friendly Liaison on a visit to the Soviet Union. It was a three-story building that was said to be the residence of an aristocrat of the old Russia. It was the Institute of Philosophy of the Soviet Academy of Sciences at the time of my visit. The rooms were furnished in the modern way, but in some large rooms one could still see the beautiful relief carvings on the ceilings and the intricate chandeliers left from the old times. All the rooms were light and spacious with high ceilings. There was a hall that had been turned into an auditorium. One could tell that the building had been magnificent and glamorous. A hall that used to be a ballroom in the times of aristocracy was said to have been used for the wedding of Pushkin, the great Russian poet. Most of the rooms had been used as offices or meeting rooms, and Lenin's busts were placed in the auditorium and all the meeting rooms.

When Father and other young CPC and CCYL members from France and other European countries came there in early 1926, they must have felt that they were in another world. When they were in France, they were foreign laborers and poor students living at the bottom of society. They were members of the secret Communist organizations tracked down by the French police. Yet when they were in the Soviet Union, they became guests of honor overnight who were accorded the warmest of welcomes. They became dignified students of the advanced Communist university. It was among the Soviet comrades and in the big family headed by Soviet laborers that they led a bright life free of oppression and darkness for the first time. There they could discuss communist ideals freely and carry out Party or League activities freely. They must have felt very happy and carefree.

The Soviet Union at that time had not completely recovered from the wounds of the civil war and the imperialist military intervention. In spite of that, the young Soviet state did its best to give support to the Chinese students in their lives and studies. The Promotion Association for Sun Yat-sen University was founded in the Soviet Union to raise funds for the university. The annual budget of the university then was around ten million rubles. To provide the foreign students with some necessary foreign currency (for example, to cover their travel expenses for their return home), the Soviet government had to use its extremely limited foreign currency. The Soviet government tried every means to provide a living to the foreign students, who were given more provisions than the Russian teachers and students. One Chinese student recalled, "We never ran out of eggs, poultry, fish, and meat, which were hard to get in 1926. Although the country was having economic difficulties, we had enough good food for three meals every day. We thought that we had better breakfast than the rich people."[1] Moreover, the university even provided the students with suits, overcoats, leather shoes, raincoats, winter clothes, and all other daily necessities. There was also a clinic for the students. The university organized students to go to ballet, opera, and other artistic performances, to the sanatoriums and summer camps for vacations, and to visit places of historical interest and scenic beauty in Moscow and Leningrad. Father said that he visited Leningrad in one of the trips organized by the university in 1926. Such a life formed a sharp contrast to the life Father had had in France.

Naturally, the main task of the Chinese students in the Soviet Union was to study. They had to learn Russian first. During the first semester, they had to spend four hours a day and six days a week on Russian. The compulsory courses of Sun Yat-sen University were: economics, history, contemporary world outlook, the theory and practice of the Russian revolution, the questions of nationality and colony, the question of social development in China, and linguistics. They had to take the following specific courses: history of the Chinese revolutionary movement and general history of China; history of the development of social formation; philosophy (dialectical materialism and historical materialism); political economy (mainly *Das Kapital*); economic geography; and Leninism. Another important course was military training.

The professors followed the teaching method of giving lectures first (in Russian but translated into Chinese), then answering students' questions; discussions and free debates by students followed, and professors concluded by providing summaries.

The basic unit of teaching was the group. There were over three hundred students in early 1926, divided into eleven groups consisting of thirty to forty students each. By early 1927 the number of students had exceeded five hundred.

The students of Sun Yat-sen University had reached different educational levels. Some had received secondary school education or higher education, while others had little prior education. There were also wide differences between the students in their understanding of the doctrines of Marxism. The university therefore divided the students into groups according to their educational levels. Preparatory courses were offered to help those at a low level to receive elementary education. There was a crash course on translation for those who had mastered the Russian language. One group in the university attracted special attention. It was the seventh group, known as the "group of theorists."

In this group gathered the important students from the CPC and the Kuomintang, among them, the CPC's Deng Xixian, Fu Zhong, and Li Zhuoran, and from the Kuomintang, Gu Zhenggang, Gu Zhengding, Deng Wenyi, Wang Jingwei's nephew and secretary, and Yu Youren's son-in-law Qu Wu. Father said that since the group had all top students from the two parties, it was quite well known.

Father, Fu Zhong, and Li Zhuoran were all leading members of the Executive Committee of the Communist Youth League in France and members of the CPC who had accepted the ideology of Marxism and had some experience in leading revolutionary struggle. They were already mature both in ideology and in action, and their experiences had been remarkable. They were in the same group with KMT members who were quite different in belief, viewpoint, and class stand, so they often had debates on various questions and the CPC students often even struggled against the Kuomintang students to some extent. Such a struggle found its expression especially in the trial of strength between them and the right-wingers of the Kuomintang. It was closely linked to the political struggle in China at that time.

Fu Zhong was the secretary of the CPC Branch of Sun Yat-sen University and Father was the leader of the Party group within the seventh group.

To get to know Deng Xiaoping at that time, I quote the following from the "Party Member Criticism Plan" of the CPC branch of Sun Yat-sen University dated June 16, 1926. The Party's appraisal of Father tells us something about him.

Name: Deng Xixian

Russian Name: Dozorov

Student Certificate Number: 233

Party Job: Leader of the Party group of the seventh group

Are all his behaviors compatible with his status as a CPC member? Yes. He has no non-Party tendency.

Does he observe discipline? Yes.

What about his knowledge of and interest in the Party's practical problems and other general political issues? Is he active or passive in raising questions for discussion in group meetings? Can he encourage other comrades to discuss all questions? He pays great attention to the Party's discipline. He has shown great interest in the general political issues, of which he has a quite clear understanding. He can actively participate in all discussions in the group meetings and can encourage other comrades to discuss all questions.

Does he attend the Party's conferences and group meetings? He was never absent.

Does he accomplish the work assigned by the Party? He can do so earnestly.

What is his relationship with other comrades? Close.

Is he interested in his lessons? Very much.

Can he set an example for others? He studies hard, and that has influence on others.

On his progress in the Party: He has made good progress in his understanding of the Party and has no non-Party tendency. As a Party member, he can set an example for the League members.

Has he eliminated the characteristics of a Party member within the Kuomintang? No.

Is he suitable for putting the Party's views into practice within the Kuomintang? Yes.

What job is most suitable for him? Propaganda and organizational work.

The above appraisal by the Party group can give us a general picture of the twenty-two-year-old Party member Deng Xixian.

When Father was in France, he read some works by Karl Marx and the works of the Communist Party of Russia, such as those by Karl Kautsky. He said that the Youth League group in France organized the study and discussion once every week. However, I do not think they could have studied systematically and in great depth in that way. The most important component of his study in the Soviet Union was learning the basic viewpoints of Marxism and related knowledge in a comprehensive and systematic way. If he had never had the chance to study in any institution of higher learning before, his study at Sun Yat-sen University was a good opportunity for him to receive higher education, especially education in an advanced school of the Communist Party. Since he and his comrades studied and lived together with the Kuomintang members, who had come directly from China, they learned a lot about all the factions of the Kuomintang and even had a test of strength with some KMT right-wingers. That laid a more solid foundation both in theory and practice for Deng Xixian's revolutionary activities and struggle after his return to China.

In his autobiography written and handed over to the Party in Moscow, Father wrote, "When I worked in the Party organization in Western Europe, I often felt that I was not competent enough. This was the cause for the mistakes I made from time to time. So I made up my mind long ago to come to Russia for study." "I have felt even more strongly that my study of communism was much too superficial." "Therefore, I will spend all my time in Russia on my study until the day I leave this country so that I could have a fairly good understanding of communism."

In that valuable autobiography, this young Communist who was just over twenty years old wrote further, "When I came to Moscow, I had already made up my mind to give my whole life to our Party and our class more resolutely. Henceforth, I am ready to absolutely receive the Party's training, obey the Party's command, and always fight for the interests of the proletariat."

That was the oath made by Deng Xixian, the young fighter for communism. In the decades of revolutionary struggle since then, he has kept his oath.

Among Father's schoolmates in Moscow, two people are worth mentioning. One is Chiang Ching-kuo, the son of Chiang Kai-shek. He was not in the same class with Father and was younger. He was not well known in the university.

The other is a young woman Communist named Zhang Xiyuan who was sent to Moscow from China for training.

Zhang Xiyuan, born in 1907, was just nineteen years old. Her hometown was Liangxiang Town, Fangshan County, Hebei Province. Her father, Zhang Jinghai, once served as the chief of the Liangxing railway station and participated in the Great Strike of February 7, 1923. Zhang Xiyuan used to be a student of the No. 2 Girl Normal School of Zhili Province. In 1924 she, as a key member, participated in the student strike for the educational reform of that school. She joined the Communist Youth League there. She met Li Dazhao, Zhao Shiyan, and other Party leaders after she came to Beijing in 1925. She joined the CPC in the same year and participated in the activities of the Communist-led Association for the Convocation of a National Assembly. She was sent by the Party to Moscow to study in Sun Yat-sen University probably in the second half of 1925. It was at the university that she and Father met and became very close.

Zhang Xiyuan appeared with over twenty fellow girl students in a group photo taken in a sanatorium on the outskirts of Moscow in 1926. In that photo, she looks dignified, pretty, and bright in her short hair. Standing together, she and other girls looked quite intimate. Yet who can tell from that photo that this girl was a young Communist who had been much tempered in the revolutionary struggle? She sent that photo to her family in China. It was as late as 1978 that the Shanghai Longhua Revolutionary Cemetery obtained that precious photo from her kinsfolk. That is the only photo of Zhang Xiyuan, whose short life ended when she was merely twenty-four. The photo has now been laid straight in her tombstone in the Shanghai Longhua Revolutionary Cemetery.

Father and Zhang Xiyuan were nothing but schoolmates and comrades-in-arms in Sun Yat-sen University in Moscow. They were not exactly in love yet. Nevertheless, it was in that university that they started their personal relationship.

The course of study at Sun Yat-sen University was two years long. Father did not complete his studies. He was instructed to return to China to participate in the Great Revolution at the end of 1926. He studied at Sun Yat-sen University for less than one year.

He embarked on the road home. After he had been away for six years, he was to return to his motherland, to the battlefields that were filled with the smoke of gunpowder.

Oh, China, your sons have come back.

CHAPTER 22

VICISSITUDES OF THE FIRST REVOLUTIONARY CIVIL WAR

In 1926 the First Revolutionary Civil War rapidly developed.

On March 12, 1925, Dr. Sun Yat-sen, the great forerunner of the bourgeois democratic revolution, unfortunately passed away. His death was undoubtedly a tremendous loss to the vigorously developing democratic revolutionary movement. But the people's revolutionary movement was like an arrow on the bow and an erupting volcano. In July 1925 the Guangzhou National Government was established, with Wang Jingwei serving as the chairman, Liao Zhongkai as the minister of finance, and a Russian named Vasili Constantin Borodin as the adviser, so as to continue to exercise leadership over the Chinese revolutionary movement. At the same time, the army under the National Government was reorganized into the National Revolutionary Army. After the reorganization, the army had Party representatives and political departments. Zhou Enlai, Li Fuchun, Lin Boqu, and other Communists all served as Party representatives in various corps. The Three Great Policies formulated by Dr. Sun Yat-sen were still carried out. After the victories in its eastern and southern expeditions, the National Revolutionary Army went on to consolidate the Guangdong Revolutionary Base Area.

In January 1926, after the Kuomintang had held its Second National Congress, the conditions for the Northern Expedition were ripe. In July of the same year, the Guangzhou National Government, influenced, urged, and helped by the Communist Party, launched the Northern Expedition.

After setting off from Guangdong, the National Revolutionary Army defeated the troops under the warlords Wu Peifu and Sun Chuanfang in less than ten months. It fought all the way from Guangzhou to Wuhan, Nanjing, and Shanghai, and the revolution swept across half of China. Although the reactionary Northern Warlord government, which was supported by the imperialists, attempted to put up resistance and launch a counteroffensive, its scheme fell through and the reactionary rule of the Northern Warlords was on the verge of collapse. The Northern Expedition was an unprecedented revolutionary war against imperialism and feudalism. It was an earthshaking people's war. It undermined the reactionary rule by the imperialists and the feudal warlords and launched the large-scale people's revolution.

In northern China, in Xi'an, an ancient capital, Gen. Feng Yuxiang joined the national revolution, thus changing forever the entrenched state of the various warlords in the north. Feng Yuxiang was born in Anhui. He joined the army in his early years and served as a brigade commander and then as a military governor in the Northern Warlord government. Impressed by the upsurge of the revolution, he began to incline toward the cause. Later he launched the Beijing Coup. He reorganized his army into the National Army and established the provisional mixed cabinet in Beijing, with his clique at the core. He also drove the last emperor of China and the imperial family of the Qing Dynasty out of the Forbidden City. Then his army fought the army of the Fentian warlord clique and occupied Tianjin, thus extending the area of control of the National Army in the north from Henan to the whole of Zhili.

While in Beijing, Feng Yuxiang received support from the Communist Li Dazhao. Under the influence of the Communist Party, Feng Yuxiang had patriotism and the revolutionary will. In May 1925, as arranged by the Communist Party, Feng Yuxiang went to the Soviet Union for a study tour. During his visit, General Feng was warmly welcomed by the Soviet government and people from all walks of life. The CPC Central Committee sent Liu Bojian, secretary of the Moscow branch of the CPC, to accompany him throughout the tour. With the help of the Communist International and the CPC, Feng Yuxiang gained a clear understanding of the Chinese revolution and began to support Dr. Sun Yat-sen's Three Great Policies of alliance with Russia, cooperation with the Communist Party, and assistance to the peasants and workers. Feng Yuxiang's experience in the Soviet Union had a great influence on his ideology.

When Feng Yuxiang was in the Soviet Union, the Chinese domestic situation underwent many changes. With the support of Wu Peifu and Sun Chuanfang of the Zhili warlord clique and Zhang Zuolin of the Fengtian warlord clique, the Northern Warlord government headed by Duan Qirui suppressed the masses of people in an attempt to stamp out the National Revolutionary Army and the revolutionary forces in the north. In view of the changes in China, Feng Yuxiang, accompanied by Liu Bojian, Yu Youren, and Usmanov, an adviser sent by the Communist International, started the journey back home in August 1926. When Feng Yuxiang arrived in Shaanxi in mid-September, he was greeted with the news that the National Revolutionary Army had captured Hankou. On September 17, Feng Yuxiang held an oath-taking rally of his army in Wuyuan at which he declared that his troops would join the Kuomintang and vowed to eliminate the traitorous warlords and overthrow imperialism. After that, with the help of Liu Bojian and other Communists, Feng Yuxiang began to reorganize his army. To increase his strength, he received from the Soviet Union military aid in the form of large quantities of weapons and matériel. At the same time, the CPC agreed to send more personnel to help his work in his army. After the oath-taking rally in Wuyuan, the CPC sent more than 200 members to work in Feng's army. Among them were Liu Bojian, Xuan Xiafu, Chen Yannian, and Deng Xixian. They were all fine, capable Communists who were transferred from Moscow, the Whampoa Military Academy, and the Northern Bureau of the CPC.

The Party member Deng Xixian, who was then studying at Sun Yat-sen University, was among the first group of some twenty personnel sent to work in Feng Yuxiang's army. They set off from Moscow toward the end of 1926. They first traveled by train and then by truck from Udinsk until they reached Kulun (now Ulan Bator) in Mongolia. After staying for a short period in Kulun, they set off again. Owing to the truck's limited carrying capacity, they sent three people—Deng Xixian and two League members—ahead as an advance party with the trucks carrying the ammunition sent by the Soviet Union to Feng Yuxiang's army. There were three vehicles in all with Soviet drivers.

It was only some 800 kilometers from Kulun to Baotou. It would have taken about an hour by air today. But at that time, it was very hard to find roads in the vast, sparsely populated wilderness and deserts. In the boundless wilderness, three Soviet trucks bumped along. When the passengers and drivers were hungry, they ate some solid food, and when they felt cold, they collected some dry cow dung for heating. The grassland of Mongolia at that time was absolutely different from the present-day grasslands of plentiful water, lush grass, and herds of cattle and sheep. Instead, it was a vast wasteland. Furthermore, it was midwinter; the ground was frozen and covered by snow and a piercingly cold wind prevailed. The journey was extremely difficult. There was no road in the wilderness. Sometimes they had to push the trucks along and could cover only a couple of kilometers a day.

They at last came out of the wilderness, only to find that a desert was lying ahead. The desert conditions were even harsher than in the wilderness. There was no grass, no water, no trees, and no human beings. When the wind blew, they could hardly see the way ahead. The sand was burning hot under the sun. When night fell, they could see the stars sparkle in the sky, but below were no human beings but only a sea of sand. Although there was no road in the wilderness, the trucks could still move. But in the desert they could not. So they had to ride camels. After eight days and nights, they at last walked out of this seemingly boundless sea of death.

More than a month had passed before they arrived in Yinchuan of Ningxia in northwestern China. Father told us that for a month or more, none of them had washed their faces. After a rest for a short time in Yinchuan, they began to ride horses and traveled day and night. At last they arrived in Xi'an in February 1927 by taking the road between Shaanxi and Gansu.

Father once said that the more than twenty comrades were dressed in rags when they arrived in Xi'an after the journey from Moscow via the Mongolian grassland and the desert in northwestern China. Once Feng Yuxiang received these personnel sent by the Communist Party of China, almost everyone he saw was dressed in this way.

As soon as Father arrived in Feng Yuxiang's army, he met Liu Bojian, with whom he was very familiar because they had studied together in France. Of course, they were very happy about meeting again. Soon Wang Chongyun and Zhu Shiheng, who came there together with Father as the advance party, were appointed chiefs of corps political division, and Father was appointed director of the Political Department of the newly founded Sun Yat-sen Military School in Xi'an.

Father recalled:

> The School was run by Yu Youren, commander-in-chief of the National Revolutionary Army stationed in Shaanxi. Yu belonged to the left wing of the Kuomintang at that time. All the principal posts at the school were held by those sent by our Party. The president of the school was Party member Shi Kexuan (he laid down his life later) and the vice president was Comrade Li Lin who had returned from the Soviet Union. (We knew each other when in France, and he laid down his life later in the Central Soviet Area.) I concurrently served as the Party secretary of the school. The school was set up after a very short period of preparations. Many of the trainees were Party or Youth League members. Apart from military training, the school mainly conducted political education and improved and developed Party and League organizations. The main subject of political education was the revolution and Marxism-Leninism. In Xi'an, it was a red school, which became the base for the Weihua Uprising in Shaanxi in 1928.

While in Xi'an, Father mainly did school work, and for a short time he also gave lectures at the Party-run Sun Yat-sen College in Xi'an. He also attended some Party and League meetings and rallies of the revolutionary masses in Xi'an. The atmosphere of Xi'an was permeated with mass revolution, so parades and rallies must have been held frequently.

During this period, Father received his living expenses from Feng Yuxiang's army. For a military life, there was, of course, not much money. So those comrades sent by the Party to the army often sought extra treats. Father told me that almost every week they would go to the Drum Tower area of Xi'an, where they would ask Shi Kexuan, president of the military school, to give them a treat—pancake boiled with beef. He says that pancake boiled with beef was very good food at that time!

Father and the other comrades worked in Feng Yuxiang's army for three to four months. By June tremendous changes had taken place in the political situation of the First Revolutionary Civil War. On April 12, 1927, Chiang Kai-shek betrayed the revolution and launched the counterrevolutionary coup.

After founding the Kuomintang, Dr. Sun Yat-sen put forward wisely the Three Great Policies, one of which was cooperation with the Community Party. The first cooperation between the Kuomintang and the Communist Party resulted in the rapid development of the Great Revolution and the brilliant achievements of the Northern Expedition. But during the period of the First Revolutionary Civil War, there was always an acute struggle between the left and right wings of the Kuomintang, and within the revolutionary united front, there was an acute struggle between revolution and counterrevolution.

Not long after Dr. Sun Yat-sen passed away in August 1925, Liao Zhongkai, a staunch left-winger of the Kuomintang, a close comrade-in-arms of Dr. Sun Yat-sen, and the minister of finance of the Guangzhou National Government, was murdered in cold blood by the reactionaries. This was a sordid scheme by the counterrevolutionary forces to eliminate the left wing of the Kuomintang.

After the May 30th Movement of 1925, the struggle for leadership became increasingly acute between the advocates of revolution and counterrevolution and between the proletariat and the bourgeoisie within the revolutionary united front. While further mobilizing the masses of the people and expanding its organizations, the Communist Party of China launched a struggle against the right wing of the Kuomintang. At that time, the right wing of the Kuomintang had organized many counterrevolutionary organizations to oppose the Three Great Policies, cooperation between the Kuomintang and the Communist Party, and the left wing of the Kuomintang. Chiang Kai-shek of the new right-wing clique overtly accepted the will of Dr. Sun Yat-sen but covertly persisted in his anti-Communist stand. He had been nurturing and expanding his own forces and carried out clandestine activities against the left wing of the Kuomintang and the Communist Party of China.

To seize the revolutionary military power, Chiang Kai-shek first had to remove the Communist Party, the biggest obstacle to his scheme. In March 1926 Chiang Kai-shek engineered the *Zhongshan* Warship Incident. He ordered Li Zhilong, then acting director of the Naval Bureau and a Communist Party member, to dispatch the *Zhongshan* warship to Whampoa Harbor, awaiting assignment. Then, after making a false charge that the Communist Party had plotted to wage a rebellion, he gave an order to arrest and detain Li Zhilong and other Communists in the National Revolutionary Army. Zhou Enlai questioned Chiang Kai-shek face to face, denouncing in strong terms his sabotage of the cooperation between the Kuomintang and the Communist Party. Feigning repentance, Chiang Kai-shek took further steps to attack and split the left wing of the Kuomintang.

As the Northern Expedition unfolded, the CPC organized and launched many large-scale workers' strikes and workers' armed uprisings against imperialism and feudalism. It also organized the mighty peasant movements. These struggles effectively coordinated the Northern Expedition and promoted its victory. In the face of the growing revolutionary forces of the people, the imperialists dispatched troops to step up their intervention in China while drawing the right-wingers of the Kuomintang to their side. At the beginning of 1927 Chiang Kai-shek had secret consultations and collaborated with the pro-Japanese Fengtian warlord Zhang Zuolin and, at the same time, openly asked for British and American "aid" and the support of Japan. After the National Revolutionary Army captured Nanchang, Chiang Kai-shek made Nanchang his base to expand his sphere of influence and increase his strength and opposed the decision to move the capital of the National Government to Wuhan. Meanwhile, he mustered the counterrevolutionary forces in Jiangxi to sabotage and suppress the peasant revolutionary movement, slaughtering members of the peasant associations.

While Chiang Kai-shek was trying to attain his goal of dictatorship by taking advantage of the revolution, the right-wing capitulationist line within the

Communist Party of China, represented by Chen Duxiu, also jeopardized and obstructed the development of the revolution. Chen Duxiu was one of the principal founders of the Communist Party of China. He was held up as an exemplar in the new cultural movement. But after he became general secretary of the Communist Party, he pursued an erroneous right-deviationist line, taking a passive attitude toward the Northern Expedition and making compromises with and concessions to the right wing of the Kuomintang. In December 1926, at a special meeting of the Central Committee in Wuhan, Chen Duxiu even made serious right-wing mistakes in order to take the predominant position on the committee. Because of these concessions, and with the instigation and support of imperialist and feudal forces, Chiang Kai-shek was reassured and emboldened and started his large-scale counterrevolutionary massacre in March 1927.

Toward the end of March, Chiang Kai-shek first sent He Yingqin to disarm the three regiments of the National Revolutionary Army in Nanjing and then went on to attack the general trade union headquarters in Hangzhou, massacring the workers' leaders and the revolutionary workers. On April 12, Chiang Kai-shek completely disarmed 2,700 picketing workers in Shanghai, banned the activities of all trade unions, and arrested and shot leaders of the workers' movement and the revolutionary masses. On April 13, workers in Shanghai went on a general strike and held a mass rally attended by more than 100,000 people. Chiang Kai-shek sent troops to suppress them, killing over 100 workers and wounding many others. In just three days, more than 300 workers in Shanghai were killed, more than 500 others were arrested, and more than 5,000 workers were missing.

On April 15, Chiang Kai-shek ordered Li Jishen in Guangdong to engineer another massacre in Guangzhou and to arrest and shoot a large number of Communists and worker activists. After that, large-scale slaughters of Communists and the revolutionary masses were carried out in Nanjing, Wuxi, Ningbo, Hangzhou, Fuzhou, and other cities. Fine CPC leaders, such as Chen Yannian, Zhao Shiyan, and others, died heroic deaths under the butcher knives of Chiang Kai-shek. At the same time, the warlord Zhang Zuolin killed Li Dazhao, another CPC outstanding leader in Beijing, and other Communists and left-wingers of the Kuomintang in the Beijing-Tianjin area.

The counterrevolutionary crimes perpetrated by Chiang Kai-shek aroused great indignation among the revolutionary masses. Mao Zedong and other Communists and left-wingers of the Kuomintang—Deng Yanda, Soong Ching Ling, He Xiangning, and others—signed a joint telegram to denounce Chiang Kai-shek. The Central Committee of the Kuomintang and the National Revolutionary Government in Wuhan also issued orders to denounce Chiang Kai-shek, expelled him from the Kuomintang, dismissed him from all his posts, and ordered his arrest.

One trouble followed another. The year 1927 was the most tragic in the history of the Chinese revolutionary movement. In May the counterrevolutionary military officers in Ji'an, Jiangxi, killed many of the revolutionary masses. Xia Douyin, commander of the 14th Division in Wuhan, collaborated with the Sichuan warlord Yang Sen to betray the revolution and attack Wuhan. The reactionary officer Xu Kexiang in Changsha betrayed the revolution and killed more than 100 people. In June, Xu Kexiang slaughtered more than 10,000 people in Hunan. The reactionary officer He Jian in Wuhan declared a split with the Communist Party and arrested Communists. On June 10, Wang Jingwei, Sun Ke, Tan Yankai, and Feng Yuxiang called an anti-Communist meeting in Zhengzhou. On July 15, Wang Jingwei officially declared his break with the Communist Party and frenziedly slaughtered Communists, progressive youths, and revolutionary workers and peasants in the Wuhan area.

CPC organizations were seriously sabotaged, large numbers of outstanding Party leaders were killed, and the ever-surging anti-imperialist and antifeudal revolutionary movement was bloodily suppressed as a result of the betrayal of the revolution and the large-scale massacre of Communists and revolutionary masses by Chiang Kai-shek and Wang Jingwei. Chiang Kai-shek gradually established a new warlord rule that prevailed against communism, democracy, and the people. The blood of the Communists and the revolutionary workers and peasants reddened the rivers and mountains and land of China. The dynamic Great Revolution ended in failure.

In this changed environment, Feng Yuxiang, who had participated in the revolution, transferred his allegiance to Chiang Kai-shek. On June 10, he attended the anti-Communist meeting convened by Wang Jingwei in Zhengzhou. On June 19, he ordered all the chiefs of the political divisions of his army to come to Kaifeng under the pretense of assembling for training. There he had the CPC member Shi Kexuan, president of the Sun Yat-sen Military School, arrested and killed. Besides eliminating the Communists in his army, Feng Yuxiang also imprisoned CPC members in a concentration camp in Kaifeng to carry out brainwashing.

Father was then the director of the Political Department of the San Yat-sen Military School. The startling domestic situation that arose after the April 12th Incident had long been clear to him, and through direct contact with Liu Bojian, he was well informed. After he got the news about the assembling of all CPC members, he consulted with Liu Bojian, Shi Kexuan, and Li Lin, and they all agreed that Deng should go to Wuhan to look for the Central Committee instead of going to Kaifeng to "receive training." Therefore, toward the end of June, Father left Xi'an and soon reached Wuhan via Zhengzhou.

Although Feng Yuxiang went over to the anti-Communist movement waged by Chiang Kai-shek and Wang Jingwei, he turned out to have been influenced by progressive ideas. He showed mercy to most of the CPC members without killing them, and in the end he "courteously sent away" the more than 200 Communist Party members, including Liu Bojian.

Feng Yuxiang abandoned the revolution, severed his ties with the Communist Party, and embarked on an uneven road in life. He suffered one setback after another, an experience that awakened him, and in the end he threw himself into the mighty torrent of the revolution of resistance against Japan. He joined hands once again with the Communist Party in the struggle against dictatorship. In September 1948, at the invitation of the Communist Party, he returned from abroad to take part in the preparations for a new political consultative conference. Unfortunately, he died on his way back when his ship caught fire.

The life of General Feng Yuxiang was not peaceful but it was extraordinary. He might have participated in building New China and made greater contributions to the people and the development of his motherland, but he died too early. His wife, Madam Li Dequan, carrying out his behest, devoted all her energy to the construction of New China. While she was serving as the minister of public health of the Central People's Government, she treated people honestly and set an example of kindness and industriousness that left an unforgettable impression on others. Father once worked in Feng Yuxiang's army, so he was quite cordial and respectful to Madam Li Dequan. She died in Beijing in 1972 at the age of seventy-six.

CHAPTER 23

OUT OF BLOODBATH

Between June and July 1927 Father arrived in Wuhan and reported to the Central Military Commission.

Father then transferred his Party member registration to the Central Committee and was assigned to work as a secretary of the Central Committee. The secretary-general of the Central Committee was Deng Zhongxia. Zhou Enlai, whom father had known in France, also came to Wuhan at this time to serve as a member of the Political Bureau and director of the Military Department of the Central Committee. Father was mainly responsible for the safekeeping of the documents of the Central Committee and for liaison and confidential work. He was also responsible for taking notes at important meetings of the Central Committee and for drafting some of the less important documents.

Father changed his name to Deng Xiaoping, as required by the secret work.

After Wang Jingwei betrayed the revolution, Wuhan, where the National Revolutionary Government was located, was in White terror, the terror created by large-scale massacre and arrest under the counterrevolutionary regime. The Communist Party was forced to go underground. The whole nation was in a state of White terror. Many local Party organizations were seriously sabotaged, and a large number of CPC members were slaughtered. The Central Committee lost contacts with most of the Party organizations throughout the country. There was little work in the central organs because they maintained minimal contact with local Party organizations.

Father once told us that Chen Duxiu had planned to appoint eight key political secretaries under the secretary-general—he did appoint Liu Bojian, Deng Xiaoping, and several others—but because of the change in the situation, there were still vacancies, and even those secretaries already named did not assume office. So under the secretary-general of the Central Committee, only Father took up his post.

Father attended various meetings of the Central Committee in the capacity of secretary. One meeting, to discuss the problems in Henan, was presided over by Chen Duxiu himself. Father had the impression of Chen Duxiu that he alone had the say. He presided over the meeting in a quite simple way. Not long after the meeting opened, Chen only said, "Land to the tillers," and adjourned the meeting arbitrarily. After the meeting Father was asked to draft a document addressed to the Henan Provincial Party Committee. Because Father worked in the Central Committee for just a short time and knew little about Henan and its problems, and because there was not much discussion at the meeting, the document he drafted contained about 300 characters only. After reading it, Deng Zhongxia said that it was too simple and that it would be all right this time but the next one should be longer. Chen Duxiu was a typical big intellectual, so the red-tape style was quite prevalent at the time. As

Father was a new hand on the Central Committee, this left a deep impression on him.

In mid-July the Central Committee called a meeting of its Political Bureau. At the meeting, it was decided that Chen Duxiu was to be sent to the Communist International to discuss the problems of the Chinese revolution, and that in China a five-member Standing Committee of the Political Bureau was to be set up to exercise the functions and powers of the Political Bureau. The five members were Zhang Guotao, Zhou Enlai, Li Weihan, Li Lisan, and Zhang Tailei. Thereafter, Chen Duxiu no longer attended to Party affairs.[1] The shuffling of the Party's central leadership was a turning point in the struggle to eliminate the right-capitulationist line.

In late July the Central Committee decided to stage the Nanchang Uprising; Zhou Enlai, Li Lisan, Zhang Tailei, and Deng Zhongxia all went to Nanchang. Li Weihan took up the post of secretary-general concurrently. Although Father had no chance to meet Li Weihan while studying in France, the name had long resounded in his ears, as Li had been one of the senior students there. Father continued to work under the leadership of Li Weihan and lived with him on the upper floor of a French merchant's building in Hankou.

After the April 12th Incident of 1927, the right wing of the Kuomintang, represented by Chiang Kai-shek, established the new warlord rule. They relied on the support of imperialists, feudal forces, and financial magnates to suppress and exploit the broad masses of workers and peasants ever more cruelly and to suppress the revolutionary forces. The Chinese revolution was then at a low ebb, and the organizations of the Communist Party went underground with their activities. In the bloody White terror, some people in the revolutionary ranks were killed or arrested, some wondered where to go, some were panic-stricken, some left the Party and the revolutionary ranks, and some even went over to the counterrevolutionary camp.

Persecuted by new warlords, the Chinese revolutionary forces suffered a great deal. By 1932, about one million people had been slaughtered by the counterrevolutionaries. Between January and August 1928 alone, more than 100,000 people lost their lives. The Party organizations suffered serious damage. By the end of 1927 Party membership had been reduced from more than 50,000 to some 10,000.

But in those bloody and dark days, "the Communist Party of China and the Chinese people were neither cowed nor conquered nor exterminated. They picked themselves up, wiped off the blood, buried their fallen comrades, and went into battle again."[2]

Although Chen Duxiu left the Central Committee, the new Standing Committee of the Political Bureau of the Central Committee did not stop working. On August 1, 1927, Zhou Enlai, secretary of the Party Central Military Commission, and He Long, Ye Ting, Zhu De, and Liu Bocheng led a historically important event, the Nanchang Uprising. This was the first shot fired at the

Kuomintang reactionaries by the revolutionary armed forces led by the Communist Party of China.

The Nanchang Uprising was staged when the White terror prevailed and the Communists were surrounded by the powerful enemy. Although it ended in failure, it created a precedent for the seizure of political power by the proletariat armed forces. The independent people's armed forces, led entirely by the Communist Party of China, emerged after this uprising.

Not long after the Nanchang Uprising, on August 7, the Central Committee held an emergency meeting in Hankou. Qu Qiubai and Li Weihan presided, and participants resolutely corrected the right-capitulationist line followed by Chen Duxiu and established the general principles of agrarian revolution and armed resistance to the Kuomintang reactionaries. Meeting participants called on all Party members and the people to continue their revolutionary struggle and decided to send experienced cadres to organize peasant uprisings in some major provinces and regions. The meeting elected the Provisional Political Bureau of the Central Committee. On August 9, the Political Bureau meeting elected Qu Qiubai, Li Weihan, and Su Zhaozheng members of the Standing Committee of the Political Bureau, with Qu Qiubai serving as the principal leader.

Although the August 7th Meeting was characterized by an incorrect tendency toward encouraging adventurism and commandism, it did, at a critical moment, correct the mistakes of right-capitulationism, stabilize the situation, strengthen revolutionary conviction, and point the Party toward the correct orientation to the revolution. So it was a meeting of undeniable historical significance. As a secretary of the Central Committee, Father attended the meeting as an observer.

After the August 7th Meeting, the Central Committee moved from Wuhan to Shanghai between late September and early October to avoid the grave situation in Wuhan and meet the needs of the developing revolutionary movement. Father also went to Shanghai.

Bordering on the East China Sea, Shanghai is the largest city of China. Owing to its geographical location and easy communications, it had become the most important economic center of China at that time. National industry had begun to develop in Shanghai, as well as the beginnings of a banking industry, and commercial and trading systems were relatively well developed. It was from Shanghai that large numbers of Chinese students left their motherland for France to study on work-study programs in the early 1920s.

Since Shanghai had highly developed industry and commerce, the city was home to a great number of workers, many of whom had tempered themselves in the May 4th Movement of 1919, the May 30th Movement of 1925, and the successive revolutionary struggles against imperialism and feudalism.

In 1926 and 1927, when the Northern Expedition achieved great momentum, the workers in Shanghai, under the leadership of the Communist Party of

China and under the direct command of Zhou Enlai, Luo Yinong, and Zhao Shiyan, launched three large-scale armed uprisings. With their flesh and blood, the workers in Shanghai opened the gate of Shanghai to the Northern Expeditionary Army. The workers and revolutionary masses in Shanghai had a high level of revolutionary awareness, rich experience in struggle, and a glorious revolutionary tradition. They were a major revolutionary force on which the Party relied.

As a birthplace of capitalism in China, Shanghai had become a base for the colonial aggression waged by the imperialists against China over the past 100 years. All the imperialist forces flocked together in Shanghai, making the city their sphere of influence. There were many foreign concessions and foreign police bureaus. Foreign troops, officials, merchants, and missionaries ruled and were Shanghai's first-class citizens at the time. On a warning board at the gate of a park was written: "No admittance for dogs and Chinese."

Shanghai was a city of great resources where the bureaucrat-capitalists, assistants to foreigners, various forces of feudal secret societies, and elements of all political factions gathered. In the high buildings and large mansions lived the financial magnates and roaming down below in the lanes were hooligans and local ruffians. These "gentlemen" and demons mingled and collaborated, forming a vast net that was hard to break. On the surface, they lived lives of luxury and dissipation in the sprawling, foreign-dominated metropolis, but in reality their Shanghai was a place of filth and mire with an extremely foul and rotten smell.

Shanghai was once the revolutionary base of the Northern Expeditionary Army, but now it had become the sphere of influence of the reactionary new warlords. Large numbers of troops and police were there, and special agents rode roughshod. They searched for and arrested the revolutionaries everywhere and bought out traitors. Collaborating with foreign police, they suppressed the revolutionary people and the Communists. People were arrested and killed almost every week. Shanghai was also a bloody battlefield where the White terror prevailed. Chen Yannian, Zhao Shiyan, and other noted Communists were killed there in cold blood.

Why did the Central Committee move from Wuhan to Shanghai? Precisely because of the special nature of the changing social environment in Shanghai, the committee was able to gain a firm foothold, set up its organs, and conduct work amid the various enemy factional activities and under the very noses of the reactionary forces.

Before and after the reshuffling of the Central Committee and its move to Shanghai, the Party continued to organize armed uprisings while resuming Party building. In September 1927 Mao Zedong was entrusted by the Central Committee to go to Hunan as its special representative to organize the world-renowned Autumn Harvest Uprising and to found the 5,000-strong 1st Division of the 1st Corps of the Chinese Workers' and Peasants' Revolutionary Army.

This army arrived in the Jinggan Mountains in the Jiangxi-Hunan border area in October and established there the first revolutionary base of the Chinese Workers' and Peasants' Red Army.

In the wake of the Autumn Harvest Uprising, the Communist Party of China launched the Guangzhou Uprising, which shocked the world.

The Nanchang Uprising, the Autumn Harvest Uprising, and the Guangzhou Uprising dealt heavy blows to the Kuomintang reactionary forces and showed that the Communists were not daunted. They had the heroic spirit and feared neither death nor hardships. They stood up and fought again and refused to give up before they achieved their aim!

These three major uprisings marked the great beginning of the CPC's establishment of the revolutionary armed forces under its own leadership, founding of the Red Army, and eventual seizure of political power by the revolutionary armed forces.

Somebody once told me that the Communist Party of China had only fifty members and not a single rifle or gun at the beginning, but in the short span of twenty-eight years, it had a powerful army and seized political power. What great strength! The Communist Party of China is such a special party, in that it was nurtured by the Chinese nation, which had gone through all manner of hardships, and its laurel of victory was woven with the blood and lives of countless pioneers.

Yes, the Chinese Communists are special people. They came from the laboring people, who suffered most, and they seized political power through heroic struggle and by arms. They emerged from bloodbath and marched fearlessly along the road toward even more arduous struggles.

CHAPTER 24

THE TWENTY-FOUR-YEAR-OLD SECRETARY-GENERAL OF THE CENTRAL COMMITTEE

After the Central Committee moved to Shanghai, it had to work under the very pressing political conditions of the grave nationwide White terror.

Through various secret channels, the Central Committee rapidly relayed the guidelines of the August 7th Meeting to the whole Party and sent a number of people to Hunan, Hubei, and Guangdong to provide guidance to those working there. The Central Committee organized workers' and peasants' revolutionary uprisings in Hunan, Hubei, and Jiangxi, launched a powerful counterattack against the frenzied slaughter and bloody suppression by the Kuomintang, and steered the revolution toward rural areas. In these military struggles, the Soviet banner was unfolded and the Chinese Workers' and Peasants' Red Army led by the Party was organized. In the areas under the rule of the new warlords of the Kuomintang, the Central Committee actively launched movements of workers, students, and women, set up secret trade unions and student associations, and organized workers' struggles in some urban areas.

After moving to Shanghai, one of the major tasks of the Central Committee was the difficult organizational work of restoring, rectifying, and rebuilding Party organizations and improving the chaotic situation caused by the grave White terror. The Central Committee soon established a secret organizational system and secret working organs, set up a national secret liaison network, and published a secret Party journal.

In January 1928 the Central Committee decided that Zhou Enlai, a member of the Standing Committee of the Political Bureau, would take up the post of director of the Organization Bureau concurrently. From then on, he assumed important responsibilities for handling the day-to-day work of the Central Committee.

In December 1927, not long after Father moved to Shanghai, he was appointed secretary-general of the Central Committee to assist Zhou Enlai and other committee leaders in handling routine affairs. Apart from attending various meetings of the Central Committee as either an observer or a participant, Father was also in charge of such affairs as documents, telegrams, secret liaison, Central Committee funds, and arranging meetings.

Because Shanghai was in the firm grip of the Kuomintang, the comrades of the Central Committee were working in an extremely dangerous environment. They had to change their residences and names frequently, especially Zhou Enlai, who was then very important and famous. Sometimes he had to change his residence once a month or even every two weeks. In light of the requirements of their secret work, the leading comrades did not even know each other's addresses. But Father, as the secretary-general, had to know the addresses and locations of all responsible comrades and secret organs of the Central Committee. He was the only person who knew all these top secrets.

The Central Committee organs were mostly in the foreign concessions, and the leading bodies were mostly in the Central District in the International Settlement. At Fourth Avenue (Fuzhou Road) in the downtown section of Shanghai, there was a Tianchan Theater, and a house with the number 447; at the back of the theater was the secret organ of the Central Committee. The Shengli Hospital was on the ground floor, and upstairs Xiong Jinding and his wife, Zhu Duanshou, rented three rooms to use as the offices of the Political Bureau of the Central Committee. Xiong Jinding disguised himself as a businessman from Hunan dealing in handwoven cloth and yarn. By the gate was hung a plate that said "Fuxing Firm." He was called "Boss Xiong." Between November 1928 and April 1931, almost all the meetings of the Political Bureau were held there.

In 1990 I went to see the veteran revolutionary Zhu Duanshou. She was eighty-two then but still full of spirit and able to walk in steady and firm steps. Mom Zhu told me:

> I came to Shanghai in the summer of 1928, and at that time I came to know your father. He was only twenty-four at the time. Our organ was in the International Settlement. It had not been discovered, and I and Boss Xiong did not leave there until 1931, when Gu Shunzhang turned a traitor. Your father was the secretary-general of the Party Central Committee. He often came to this office and stayed for half a day or at least one or two hours before leaving again. The meetings of the Political Bureau and its Standing Committee were all held there. Your father was responsible for the agenda. The first meeting would decide the date of the next. The Standing Committee had a few members, and the meeting of the Political Bureau was held in one room. When an enlarged meeting was held, two rooms were occupied sometimes. Your father often spoke at the meetings. There was one speech I still remember clearly: he opposed the idea of Li Lisan, who advocated winning victory in one or several provinces first. Your father said that the Kuomintang had several million troops, but our Party was organized without armed forces, so how could we defeat them with crudely made guns? At that time, the secretary was Xiang Zhongfa. He was a good-for-nothing. Your father and Comrade Zhou Enlai had been to France and the Soviet Union and knew a lot of things.
>
> I was a special secret messenger of the Central Committee organ, under the direct leadership of your father. The reports from various places and the Soviet areas were all written in invisible ink on paper or cloth, and I developed them in alum water. Then I copied them. They were top secrets and never went out of the doors of the Political Bureau. Boss Xiong was the special accountant of the secretariat, also directly under your father. I also did odd jobs for the organ, boiling water and preparing meals for the comrades who came there to work or to attend meetings. The comrades of the Central Committee all liked the dishes I prepared. Comrade Zhou Enlai liked the meatball, and your father liked pepper. Your father has a good character and is amiable and easy of approach. He is four years older than I and calls me "little sister." Your father loves talking and joking, but in a gentle way.

At that time, the comrades doing underground work had to dress up as wealthy men. They wore long gowns and ceremonial hats. Your father also dressed like that.

Mom Zhu joined the Party in Hunan. When she came to Shanghai to work on the organ of the Central Committee, she was only twenty. Introduced by Zhou Enlai, both she and Boss Xiong remained in the office, posing as a phony couple. Later, in August 1928, they really married. They invited two tables of guests, including Xiang Zhongfa and other leading comrades as well as my father. Boss Xiong was more than twenty years older than Zhu. After leaving the Central Committee organ in Shanghai, the couple moved to the Soviet area in Hunan and western Hubei and were once put in prison. After liberation, when Boss Xiong was hospitalized in the Beijing Hospital, Father went to see him. Boss Xiong died, but Zhu Duanshou is still alive. The great contributions of this revolutionary couple earned quite a name for them when they did underground work in Shanghai.

In July 1991 I visited the revolutionary veteran Huang Jieran in the Office of the All-China Association of Industry and Commerce in Beijing. His original name was Huang Wenrong. He joined the Party in 1926 and served as a division chief of the secretariat in Shanghai. When Father was sent by the Central Committee to Guangxi in 1929, he succeeded Father as secretary-general of the Central Committee. He related many of the events at that time to me in detail:

When I was in Wuhan, I served as the secretary of Chen Duxiu. When the Central Committee moved to Shanghai, I first worked in the office of the Party journal *Bolshevik* and later in the office of the Central Committee at Yong'anli, Northern Sichuan Road. In 1928 I was transferred to take up the post of division chief of the secretariat. I first came to know your father in 1928. At that time, there was a two-story building with two sitting rooms at 700, Baideli, Tongfu Road, Shanghai, and that was an organ of the Central Committee. In fact, the organ handled the routine affairs of the Central Committee, and so we all called it the General Office of the Central Committee. Comrades Enlai and Xiaoping came there every day. Comrades of all departments and units under the Central Committee came for instructions. As secretary-general, Comrade Xiaoping handled administrative and technical matters. Such problems as asking for personnel, funds, and instructions and reporting on work were solved by Comrade Enlai on the spot, and problems that could not be solved or were important ones were submitted to the meeting of the Political Bureau for discussion and decision. Comrade Enlai was actually like a general manager and handled all the routine affairs. At that time, I worked in the office of the Party journal and also went to him for instructions. I met Comrades Enlai and Xiaoping there for the first time. They were very busy. There were so many people asking for instructions that sometimes they had to queue up outside.

In 1929 Comrade Xiaoping was to be transferred to Guangxi, and I was ready to take over his job. So I also attended some of the Political Bureau meetings held in the office of Boss Xiong. During that two to three months, we had more contacts. The Political Bureau meetings were presided over by General Secretary Xiang

Zhongfa. The meetings mainly discussed the topics decided upon in advance, including workers' movement, the international situation, the domestic situation, economic problems, the situation in the whole country as well as in local areas, tactics, countermeasures, guiding principles, and methods of work and struggle. All these were major problems. In every discussion, the leading comrade responsible for the work under discussion made a speech first and then others aired their ideas, opinions, or differing views. Each person's time on the floor was limited, and the speeches were short. Xiang Zhongfa could speak a lot sometimes, but what he said was not to the point. The person who spoke most was Zhou Enlai. He knew more, did more work, and often prepared outlines for his speeches. The subjects he spoke about most were related to the work of the Soviet areas and military work. Comrade Xiaoping was the secretary-general responsible for keeping the minutes. (Sometimes others took minutes.) He also spoke because as the secretary-general he had the right to speak and to raise questions, which was useful for him in handling the affairs decided upon by the Political Bureau meetings. He shouldered the big responsibility of linking the higher organs with the lower ones. As the secretary-general knew a lot and handled many affairs, his work had a direct bearing on the safety of the Central Committee. Although Comrade Xiaoping did not talk much, once he did, and it was rather weighty. Although he was a bit tacit, what he talked about was simple but meaningful and easy to understand. On the contrary, some people talked much but were unable to make themselves understood. After the meetings, the secretary-general was responsible for drafting and processing documents. The secretary-general was also in charge of the work of the secretariat of the Central Committee. We can say that the secretary-general was responsible for everything and assumed big responsibilities.

After the Political Bureau meetings, sometimes we dined with Boss Xiong. The beef boiled in chicken soup prepared by Zhu Duanshou was our favorite. While eating, we all talked and laughed. Comrade Xiaoping also loved talking and laughing and sometimes was very jocular. He made a deep impression on me. He was very calm, prudent and amiable.

I asked Old Huang to give a brief account of the personnel and work of the secretariat.

The secretariat had five sections: clerical section, internal liaison section, external liaison section, accounting section, and translation section.

The clerical section was responsible for cutting stencils, mimeographing, receiving and sending documents, distributing documents, and writing in invisible ink. These aspects of work were very secret and done separately. The documents and minutes of meetings of the Central Committee were triplicated; one copy was kept by the Central Committee, another was sent to the Communist International in the Soviet Union, and the third one was sent by the Special Branch to the countryside for preservation. It was said that those copies sent to the countryside were well kept and all were retrieved after liberation. Some organs were discovered and documents were seized and kept in the archives of foreign police stations. After liberation, we recovered them. It was hard to keep the documents safe. In the clerical section, there was also a place for leading comrades of the Central Committee to read documents. As soon as we received a document, the secretary-general was always the first to read it.

The internal liaison section was chiefly responsible for sending documents, some notices, and information to various sections and units. The tasks were very heavy. So long as the wives of the leading cadres were calm, bold, meticulous, and good at watching people's every mood and handling sudden emergencies, we often entrusted them with the work. The wives of all the leading cadres had done this job.

The external liaison section was responsible for the liaison between the Central Committee in Shanghai and Shunzhi, Manchuria, Hunan, Hubei, Guangdong, and Guangxi provinces. It was divided into southern, northern, and Yangtze River routes. Each route had branch routes. They formed a national liaison network to link up with the Central Committee in Shanghai. All the documents, cash, cadres, and other personnel who were coming and going had to be received by this network. We had liaison offices across the country under the cover of shops and restaurants where the underground messengers could live. Around the Soviet areas, there were many shops that functioned as liaison offices, as it was convenient to make contacts in the name of purchasing goods. Comrades who acted as underground messengers were carefully selected, as they had to be very firm and cool-headed and could do whatever was required.

The internal and external liaison seemed an ordinary job, but it was hard and required techniques and responsibility. Each person had to rack his brains in doing anything. Documents might be hidden in books, cotton quilts, or thermos bottles. Some microphotos might be put into fountain pens. Other things might be put into pastry or cloth. The funds from the Soviet areas included paper money, bullion, and personal gold and silver ornaments. How to send them? They were put into bamboo shoulder poles or stuffed into fish stomachs to avoid discovery by the enemy. In a word, you had to think of many ways.

The accounting section was staffed by Boss Xiong only. The place where he lived was the most secret and the safest. Only those with the prior approval of the Political Bureau could go there. That organ was under the direct administration of the secretary-general. Boss Xiong was responsible for managing Party funds. The Political Bureau made the decisions, and Boss Xiong issued the funds. The Political Bureau designated some people to examine the expenditures. The bullions and personal ornaments and even the paper money sent in from the Soviet areas had to be converted into cash, but that could not be done in ordinary banks for fear of exposure. So we made use of the connections with Zhang Naiqi, who was then the vice-chairman of the board of directors of Zhejiang Industrial Bank in Shanghai. He had good relations with Comrade Chen Yun, but he did not know we were Communists. Comrade Chen Yun and I often invited him to amuse ourselves and seized the opportunity to ask him for help. He was very pleased to do so because his bank could collect some service charges.

Although the translation section was a component of the secretariat, it was, in fact, under the direct leadership of the Political Bureau.

In February 1990 I visited another revolutionary of the older generation in Shanghai, Zhang Ji'en. He started working for the organs of the Central Committee in 1928. At the beginning, he worked in an organ at 135 Yong'anli and then was transferred to another organ at Qinghefang of the Fifth Avenue. Downstairs was a grocery that sold cigarettes, soap, and matches.

Old Zhang told me:

> This shop was opened by Deng Xiaoping. At that time, we opened many shops as our cover. Upstairs was the place where Li Weihan, then a member of the Political Bureau, lived. After Li was transferred to Jiangsu as secretary of its Provincial Party Committee, he could not live in the organ of the Central Committee anymore and had to move to the house in the western district of Shanghai where the Jiangsu Provincial Party Committee was located. So I and my wife were moved to this place. Several Political Bureau meetings were held in my place. Xiang Zhongfa, Zhou Enlai, and Qu Qiubai all attended the meetings, at which the problems in Zhejiang and Yunnan were discussed. We also received many other people who passed by. Zhou Enlai paid particular attention to confidential work. He insisted that women comrades wear their hair in a bun and wear embroidered shoes. People living in the office had to be a couple and were prohibited from talking about revolution. My office was under the administration of the secretariat. I once worked in the clerical section.

Old Zhang was later transferred to a confidential department. He said: "When the Political Bureau held a meeting, Deng Xiaoping kept minutes. After he left, I took his place. Many of the leading comrades of the Central Committee came from Hunan. This made it very difficult for me to keep the minutes as I could not understand the Hunan dialect."

Comrades who did underground work in Shanghai had to change their residences and meeting places frequently, and to escape the enemy's searches, the houses had to have exits in several surrounding lanes. Father lived with Li Weihan for some time and spent half a year with Zhou Enlai and his wife. At that time, those who maintained the closest working and personal relations with Father were, first, Zhou Enlai and his wife, and then Li Weihan. Because he was then working on the Central Committee, Father also had many other connections with its leading comrades, such as Zhao Shiyan, Chen Yannian, Li Suoxun, Deng Zhongxia, Luo Yinong, Qu Qiubai, Guan Xiangying, Su Zhaozheng, Li Lisan, Gu Shunzhang, and Xiang Zhongfa. He was quite familiar with all of them.

In his memoirs, Li Weihan said that during the Sixth National Congress held in 1928,

> I and Ren Bishi were instructed to stay in the office of the Central Committee. The secretary-general at the time was Deng Xiaoping. From April to September 1928, when the leading members of the new Central Committee returned, the meeting place was still the two upstairs rooms of the house at the back of the Tianchan Theater on Fourth Avenue in Shanghai. This house was rented in the winter of 1927 or in early 1928 to be used secretly as the meeting place of the Standing Committee of the Political Bureau of the Central Committee. The house was kept by Xiong Jin-

> ding and his wife Zhu Duanshou. Comrades coming to the meetings could go straight to the meeting room from a flight of stairs at Yunnan Road to the west of the Tianchan Theater. In the meeting room, there was a small desk under the window facing the west. It was by this desk that Xiaoping kept minutes of the meetings. This organ was never discovered by the enemy from its establishment until after the Fourth Plenary Session of the Sixth Central Committee held in January 1931. Later on, the Central Committee decided to discontinue the use of this house, probably only after Gu Shunzhang was arrested and turned traitor.
>
> Deng Xiaoping and I went to this old place to have a look on our way to Hangzhou in 1952 when Mao Zedong summoned us while he was presiding over the meeting to draft the constitution of the people's republic. At that time, I, Bishi, and Xiaoping met at 9:00 A.M. every day to handle routine matters, not at this place but on the upper floor of a shop just a block away. Among those present were also Xiong Jinding, chief of the internal liaison section, and other leading cadres, such as Li Fuchun, who was the leader of the comrades of Jiangsu Provincial Party Committee, stayed behind and attended the meetings occasionally.[1]

The Sixth National Congress was held in Moscow. Most of the leading comrades of the Central Committee went there. Those who stayed behind in Shanghai continued their work energetically. They waged the anti-Japanese movement and the struggle against selling out Shandong and Manchuria to Japan by the Kuomintang government in collusion with Britain and the United States. They stepped up their efforts to organize workers' movements in urban areas, develop the work in rural areas, and win over and destroy the enemy forces. They also wasted not a single minute in rectifying and expanding the Party organizations and strengthening the Party's underground work.

While in Shanghai, Father worked all along in the organ of the Central Committee. As a cover, he disguised himself as an owner of a grocery and then an antique shop. As the secretary-general of the Central Committee, he knew very well all the locations of the committee's organs and its meeting places and times. He was quite familiar with all the roads, streets, and lanes, especially the zigzag lanes that extended in all directions and where the secret Party organs were located. As most urban districts in Shanghai were foreign concessions, many streets and roads had foreign names, such as Petain Road and Foch Road. What is interesting is that Father can still remember the old names but is unable to match them with the new. In 1991, when Father went to Shanghai, he drove around the city in the company of some comrades. Father was in high spirits and related something about their underground work in Shanghai before liberation. He said, Why was there a road called Foch Road? Because this road was in the French concession. The road was named after Ferdinand Foch, a famous French general. Father mentioned the old names of some streets and asked the comrades from Shanghai what they were called now. The young local officials who accompanied him on the visit seemed to know very little about these things of the past and could only look at each other in blank dismay.

CHAPTER 25

MOM ZHANG XIYUAN

In the summer of 1989 I went to visit Mom Liu Ying, a Red Army woman and the wife of Zhang Wentian, in the summer resort of Beidaihe not far from Beijing.

Mom Liu Ying told me, "I knew your father very early." In 1928, when the Hunan Provincial Party Committee was sabotaged in the White terror and its Party secretary laid down his life, the Party organization there sent her to Shanghai to contact the Central Committee. She arrived in Shanghai by going through numerous hardships. There she found Zhou Enlai and contacted the Central Committee.

Liu Ying was admitted into the Party during the May 30th Movement of 1925 on the recommendation of Li Weihan, who was then secretary of the Hunan Provincial Party Committee. After she arrived in Shanghai, Li Weihan asked her to stay in his home and disguise herself as his cousin.

Mom Liu Ying said, "At that time, Zhou Enlai often went to Li Weihan's home, and each time he took Comrade Xiaoping along. They held meetings at Li's home. At that time, the general secretary of the Central Committee was Xiang Zhongfa. Qu Qiubai, Li Lisan, Zhou Enlai, and Li Weihan also worked on the Central Committee. Li Weihan was responsible for the work in Hunan."

Mom Liu Ying told me with a smile, "That was how I came to know Comrade Xiaoping. In 1928 he was twenty-four and I was twenty-three. We all liked joking, and we became familiar. I did not even ask about his job and simply called him Xiaoping. He was very lively and liked to talk and laugh. I remember that he was responsible for writing things, a very easygoing person." But in our eyes, Father is introverted and talks little, is kind but very serious. But he does talk loudly, sometimes with hearty laughter, when he meets his old comrades and friends. It is very hard for us to imagine the lively, cheerful, and talkative young Deng Xiaoping.

I assumed that there were two reasons for his cheerfulness. One was that after he arrived in Shanghai, the Party's work was restored, and the other was that he married in the spring of 1928.

In chapter 21, I mentioned Zhang Xiyuan, one of Father's schoolmates in Moscow. She was a nineteen-year-old Communist coming directly from China, and he was a twenty-one-year-old Chinese Communist coming to study in Moscow from France. They were not in the same class, but they quickly became acquaintances, leaving on each other a very good impression. But at that time, their tasks were to study attentively and fight against the right wing of the Kuomintang. They remained no more than schoolmates, with no love developing between them.

In 1927, not long after Father moved from Xi'an to Wuhan to take up the post of secretary-general of the Central Committee, he met, to his astonishment, Zhang Xiyuan, who had just returned from Moscow. She returned via

Mongolia around August or September 1927. After her return, she led a railway workers' strike in Baoding. All the preparations for the strike were made at her home, as even her eight-year-old brother remembered. After the strike, Zhang Xiyuan went to Wuhan, where she worked in the secretariat of the Central Committee. When Father met his old schoolmate, he must have been very happy. Not long afterwards, the Central Committee moved from Wuhan to Shanghai. Father went to Shanghai, and so did Zhang Xiyuan, who worked in the secretariat under Father's leadership.

Not long after New Year's Day 1928, Father and Zhang Xiyuan married. Father was less than twenty-four, while Zhang Xiyuan was about twenty-two. To celebrate the wedding, the leading comrades of the Central Committee gave a special dinner party in a Sichuan restaurant called Ju Feng Yuan, in Guangxi Zhonglu, Shanghai. Present were Zhou Enlai, Deng Yingchao, Li Weihan, Wang Ruofei, and most of the other comrades working in the Central Committee, totaling more than thirty.

Zheng Chaolin, who attended the dinner party, told me, "At that time, the situation in Shanghai was not very tense [before October 1928], so a dinner party could be given. The dinner parties were given in celebration of the weddings of several couples working on the Central Committee." Zheng Chaolin was over ninety-one when he was telling me this, but he could still remember very clearly.

> Zhang Xiyuan was very pretty, not very tall. She was a student of Baoding No. 2 Normal School. Together with Li Peizhi [wife of Wang Ruofei], she took part in a student strike. She also did secret work in Wuhan. Zhang Xiyuan had many friends, and some of them also courted her, but in the end she married your father. Later on, when I lived in the home of Wang Shaoxing, your father often took Zhang Xiyuan to visit Wang because he had known Wang in the Northwestern Army. So I often met them.

The old revolutionary Zhu Yueqian also told me,

> While in Shanghai, my husband Huo Buqing worked in the Central Military Commission, and I also worked in the commission's office. At that time, I and my husband, your father and Zhang Xiyuan, Zhou Enlai, and Deng Yingchao were in the same Party group. We held group meetings once every week and often changed the meeting place. We mainly met to study. Your father was a very good and capable cadre. Zhang Xiyuan was a Beijinger, speaking in pure Beijing dialect. I still remember her look. She was about your height [5'2"], pretty, with a fair and clear face, delicately made and gentle and soft. She spoke gently and softly and had a good affection for your father.

When I saw Zhu Yueqian in 1990, she was eighty-one. Born in 1909, she was only nineteen, three years younger than Zhang Xiyuan, in 1928 when she was in the same Party group with Father. Uncle Zhou was just over thirty, while Mom Deng was of the same age as Father, also twenty-four. Their average age was only a little over twenty—what a young Party group! But it was a mature, firm, and lively Party group.

Mom Zhu Duanshou also told me,

> Of course, I knew Zhang Xiyuan! She came to our office, and we got along very well. She had a good temperament and was pretty, lively, just like me, very frank and straightforward. She was tender, very lovely, and very kind to others. We were of the same age at the time, and so we had a lot of things in common to talk about. At the time when we did underground work, we disguised ourselves as the monied people, wearing close-fitting gowns and high-heeled shoes, with hair cut short. Comrade Enlai and your father also wore long gowns and doctor's hats.

Father and Zhang Xiyuan once lived for more than half a year with Zhou Enlai and Deng Yingchao in a house in the International Settlement. Uncle Zhou and Mom Deng lived on the upper floor, and Father and Zhang Xiyuan on the ground floor. Mom Deng once said that they often heard Father and Zhang Xiyuan talking and laughing downstairs.

I asked Father about that, and he said, "We were all young people at the time, and naturally we talked and laughed!"

Father once said thoughtfully, "Zhang Xiyuan was a rare beauty." He and Zhang Xiyuan were schoolmates and comrades-in-arms. Moreover, they were a young couple with a deep affection for each other.

How pleasing it must have been to see the pure and beautiful true feelings of this couple in Shanghai during the White terror and in the concessions heavily guarded by foreign police.

I have often imagined what Zhang Xiyuan looked like. How beautiful was she? What was her character? Around her there seemed to be a mysterious and obscure light that aroused reverie and reminiscence.

When she died, she was only twenty-four, and the newborn child died, too. It was very sad. But in our minds, she is young forever.

She was not my own mother, but she was still my mother.

CHAPTER 26

FIGHTING IN THE DRAGON'S POOL AND THE TIGER'S DEN

Although in Shanghai good people coexisted with bad people, making it possible for the Party to do underground work, the city was under the strict control of the imperialists and the reactionary forces. The enemy tried every means to destroy the Party's underground organizations and arrest or kill the Party leaders. They used foreign police for suppression, special agents for surveillance, and traitors for information against the CPC. In the three years after the Great Revolution failed, many important Party leaders were betrayed, arrested, or killed, one after another. They included Chen Yannian, secretary of the Jiangsu Provincial Party Committee; Zhao Shiyan, acting secretary of the Jiangsu Provincial Party Committee; Luo Yinong, a member of the Political Bureau of the CPC Central Committee; Peng Pai, alternate member of the Political Bureau of the Central Committee; and Yang Yin, director of the Military Department of the Central Committee. In November 1928 the Central Committee decided to set up the Special Task Committee, which was composed of Xiang Zhongfa, Zhou Enlai, and Gu Shunzhang. It was in charge of leading the Special Branch of the Central Committee and protecting the Party's work and its leading comrades.

As early as the spring of 1928, the Party's Special Branch established the first antiespionage connection, Yang Dengying, under the direct leadership of Zhou Enlai, with Chen Geng and others taking charge of the work.

Yang Dengying, also known as Bao Junfu, had studied in Japan. He had complex social connections and came into contact with all parties and their factions, with various people in the foreign concessions, and with the underworld. Some people called him "veteran of four dynasties," simply because he had contacts with the KMT, the Japanese, the Chinese collaborators, and the CPC. In 1928 Chiang Kai-shek started to set up his secret services in Shanghai, and Yang Dengying was chosen to set up a detective agency. Yang was sympathetic to the revolution and gave this information to the Party. Then Chen Geng, head of the Special Branch of the Party, was personally responsible for recruiting Yang Dengying to be the first antiespionage connection within the enemy's detective agency.

On the Party's instructions, Yang Dengying soon established contacts with Chen Lifu and Zhang Daofan, prominent figures in the KMT intelligence system, and became their trusted follower. Furthermore, he tried to win the trust of Xu Enzeng, chief of the KMT's special agents in Shanghai. He also established close ties with some foreign police stations in Shanghai, especially the British police station. The enormous amount of information provided by Yang Dengying played an important role in preventing the Party's organizations from being sabotaged by the enemy, in rescuing the arrested personnel, and in eliminating the traitors. Making use of Yang Dengying's connections, and on his recommendation, the Special Branch dispatched the CPC members Li Kenong, Qian Zhuangfei, and Hu Di to infiltrate the KMT's high-level secret services in 1929. Qian Zhuangfei even served as the confidential secretary of Xu Enzeng, director of the Party Affairs Investigation Section of the Organization Department of the KMT Central Committee.

Under the direct leadership of Zhou Enlai, the Special Branch of the Central Committee did its work smoothly and effectively. The Central Committee's first secret radio station in Shanghai was established in 1928. In the same year, the Party's second secret radio station was established by the Southern Bureau, based in Hong Kong, and the two radio stations started to communicate with each other in 1930.

Because of the enemy's sabotage after 1931, the Central Committee in Shanghai found itself in a more difficult and dangerous working environment. At that time, two terrible incidents caused enormous damage to the committee. One was the betrayal of the revolution by Gu Shunzhang, an alternate member of the Political Bureau of the Central Committee. Being a worker from Shanghai, Gu Shunzhang was once a member of the Green and Red Bands and had a lot of connections with all sorts of people. According to some veteran comrades, he was quite capable and had broad connections; he worked effectively to organize the workers' movement in Shanghai and led the Special Branch, especially in eliminating traitors. However, he was a man of bad character. He had some lumpenproletarian habits and led a corrupt life. He smoked opium, for which he was criticized by Zhou Enlai many times. On April 25, 1931, after he had escorted Zhang Guotao to the Hubei-Henan-Anhui Base Area, Gu Shunzhang openly did magic tricks without permission in a recreation center in Wuhan. He was recognized by a traitor and arrested. He betrayed the Party the very day of his arrest.

Gu Shunzhang had been in charge of the Party's security work for a long time. He knew not only many of the Party's secrets but also the addresses of many organs of the Central Committee and leading cadres. So his betrayal could have caused a fatal disaster to the Party.

After catching such an important CPC leader, the KMT secret service in Wuhan cabled its headquarters in Nanjing the same day. The cable was intercepted by Qian Zhuangfei, a CPC member and Xu Enzeng's confidential secretary. Knowing the urgency of the matter, Qian himself went to Shanghai at once to report to Zhou Enlai.

In view of this extremely dangerous and grave situation, Zhou Enlai promptly decided to move elsewhere all the organs and personnel that Gu Shunzhang knew and abolish all the secret working methods that he knew. With little regard for his own safety, Zhou Enlai even went personally to tell some responsible comrades to move elsewhere. When the enemy started the large-scale search and arrest according to information given by Gu Shunzhang, the organs and personnel of the CPC Central Committee had been moved safely, thus avoiding a big disaster.

But the betrayal of the revolution by Gu Shunzhang did cause immeasurable damage to the underground work system and the work on which the Central Committee had spent so many years. Some organs were withdrawn, and some leading comrades and staff had to leave Shanghai. Gu Shunzhang also betrayed Yun Daiying, a leader of the Central Committee who was arrested by the enemy but did not expose his true identity. Yun Daiying was killed in the end. Gu Shunzhang also betrayed Yang Dengying, but Yang was able to stay out of danger because of his special relationship with Zhang Daofan. Being guilty of the most heinous crimes, Gu Shunzhang could not escape punishment in the end. He was executed by the Central Statistics Bureau, a secret service of the KMT, in 1935 for trying to set up his own faction in the KMT's secret service. He is a man who should be condemned through the ages, and even his death did not expiate all his crimes.

The second incident was the arrest of Xiang Zhongfa, the general secretary of the Central Committee, and his betrayal of the revolution. Xiang Zhongfa was originally a worker. His political consciousness was low and his leadership skills undeveloped. Simply because he had the support of the Communist International, he became the general secretary of the Central Committee. During his tenure, it was Zhou Enlai and other comrades who handled the practical day-to-day work of the committee.

Xiang Zhongfa had a poor record of conduct and was not disciplined. A prostitute became his concubine. Mom Chen Congying, the wife of Ren Bishi, said:

> When Comrade Enlai discovered this, he asked my mother to live with the concubine of Xiang Zhongfa so as to keep an eye on them. Enlai intended to transfer Xiang Zhongfa to the Soviet area. So he moved the concubine to a hotel first and then Xiang Zhongfa to the place where Zhou himself lived, telling him not to go out. But Xiang Zhongfa, breaching the discipline, sneaked out to see his concubine when Comrade Enlai was not around. Worse still, he would not leave and stayed there overnight. My mother knocked on their door to warn them, and they just turned a deaf ear to the warning. The next day, he took a cab and was recognized by the driver, who reported to the police bureau. Xiang Zhongfa was arrested.

That happened on June 22, 1931, and Xiang Zhongfa betrayed the Party on June 24.

Huang Dinghui, also known as Huang Mulan, an underground worker at that time, told me:

> I was in a café with a lawyer. Also with us was a friend who served as an interpreter in a foreign police station. He told us that a ringleader of the Communist Party, for whose arrest the KMT had offered a reward of 100,000 silver dollars, was captured. He was a native of Hubei Province. He had some teeth inlaid with gold and had

> only nine fingers. He was over sixty years old and had a brandy nose. He was a coward and could not stand the electric chair. Hearing this, I told myself that this must be Xiang Zhongfa. I immediately came back and reported to Kang Sheng through Pan Hannian. At 11:00 P.M. that night, Zhou Enlai, Deng Yingchao, and Cai Chang moved quickly to a French hotel. At 1:00 A.M. that night, the personnel of the Special Branch, who disguised themselves as peddlers selling dumpling soup near Enlai's residence, saw the police coming with Xiang Zhongfa. Xiang Zhongfa had the key of Enlai's house. They saw Xiang in handcuffs open the door only to find it empty. That was a near thing!

Being the general secretary of the Central Committee, Xiang Zhongfa would have caused serious damage to the Party because of his betrayal of the revolution. Luckily, thank God, this traitor of unpardonable evils got his due punishment!

As soon as Xiang Zhongfa was arrested, the KMT's secret service in Shanghai immediately cabled its headquarters in Nanjing, reporting that the ringleader of the "bandits" was captured. Chiang Kai-shek was in Mount Lushan at that time. Reading the cable, he ordered a summary execution. After Xiang Zhongfa's betrayal of the CPC, the secret service in Shanghai sent another cable to Nanjing at once but soon received the summary execution instruction. So it executed Xiang Zhongfa, according to the instruction. When Chiang Kai-shek learned about Xiang's betrayal of the revolution, it was too late to change his mind, as Xiang had already been shot.

Such vicious incidents happened one after another, making it more and more difficult for the Central Committee to carry out its activities in Shanghai. Under these circumstances, some leading comrades of the Central Committee were transferred to the Soviet areas. Zhou Enlai left Shanghai in December 1931 for the Central Soviet Area in Jiangxi Province.

Father recalled:

> We did secret work in Shanghai under very difficult conditions. Doing revolutionary work there was like holding your head in your hand. I had never taken any photo and never went to any cinema. I never got arrested while doing the underground work. Although I was in the army so long, I was never wounded. That was rare. Nevertheless, there were quite a few risky occasions, of which two were quite bad.
>
> One was caused by Ho Jiaxing, who had betrayed Luo Yinong. I went to have a secret contact with Luo Yinong. As soon as we finished talking, I left through the back door just before the policemen came in through the front door. Luo Yinong was arrested. When I came out of the house, I saw a member of the Special Branch, who was disguised as a shoeshine boy, near the front door pointing his finger at the house quietly. I immediately knew what had happened. It was a matter of less than one minute. Luo Yinong was shot later.
>
> The other happened when I was living with Premier Zhou, elder sister Deng, and Zhang Xiyuan in the same house. The work of our Special Branch was done remarkably well. They learned that the policemen had found out Zhou's residence and were coming to search the place. They were informed of this, and all comrades

> at home moved out immediately. I did not know this, as I was away from home at the time. I knocked on the door when the policemen were searching the house. Luckily, we had a member of the Special Branch infiltrating the enemy secret service who was inside and answered that he was coming to open the door. I found it was not the right voice and immediately left, thus avoiding a disaster. I didn't even dare to walk in that lane in the following six months.
>
> These were the two most dangerous occasions I came across. It was so dangerous! The difference of even a few seconds could have had grave consequences!

The most detestable people in the Party's secret work in Shanghai were those traitors. They sold out not only their own souls but also the lives of other Communists.

Ho Jiaxing, whom Father mentioned, was one of them. After coming back from the Soviet Union, he and his wife were dissatisfied with the poor life in China and betrayed the revolution. On April 15, 1928, Ho betrayed Luo Yinong, a member of the Political Bureau of the Central Committee and the director of the Organization Bureau.

In August 1929 Bai Xin, originally a secretary of the CPC Central Military Commission, betrayed the revolution and sold out five comrades who were attending the meeting of the Military Commission, resulting in the tragic death of four of them. In the same month, a hidden traitor, named Dai Bingshi, informed against some CPC members, and seven comrades of a secret organ of the Party were arrested on the spot by the Shanghai-Wuson Garrison Headquarters of the KMT.

In addition, Huang Dihong, who was among the first group of students of the Whampoa Military Academy, betrayed Zhou Enlai by passing information about his meeting with Zhou. Other leading members of the Central Committee who were betrayed by the traitors included Li Weihan and Li Lisan. In the end, these traitors, guilty of the worst crimes, were unable to escape the net of justice, and all of them got the severe punishment they deserved.

It was extremely difficult to do underground work in the White area. But it was only in the difficult and dangerous situations that the scums could be exposed and the heroism of the Communists displayed.

CHAPTER 27

IN THE POLITICAL ARENA OF GUANGXI

Between July and August 1929 the Central Committee sent Deng Xiaoping to work in Guangxi Province. As the representative of the Central Committee, he led the Party's work in the province and prepared and organized armed uprisings. He was sent there by the Central Committee at the request of Yu Zuobai, the governor of Guangxi Provincial Government.

Why did Yu Zuobai invite the Communists to Guangxi when there was a White terror throughout the country? We have to explain from the beginning.

After the betrayal of the revolution by Chiang Kai-shek, Wang Jingwei, and other right-wingers of the KMT in 1927, the Great Revolution ended in failure and the target of the revolution, the Northern Warlord regime, was not overthrown. In addition, the KMT itself was facing a split into different factions. These events made it impossible for Chiang Kai-shek to fulfill his dream of autocratic rule over the country. The political situation in China at that time was characterized by a tripartite balance of forces. The warlord regime ruled in Beijing, the Wang Jingwei regime in Hankou, and the Chiang Kai-shek regime in Nanjing. China was still in a state of chaos caused by the separatist warlord regime, which inevitably led to new wars.

In October 1927 the war between the warlords in Nanjing and Wuhan broke out first. Having formed an alliance with the warlords in Guangdong and Guangxi provinces as well as the Northern Warlord Feng Yuxiang, Chiang Kai-shek launched a punitive expedition against Tang Shengzhi in Wuhan. Taking advantage of the situation, the Guangxi warlord clique was able to expand its forces, aggravating in turn the conflicts between it and Chiang Kai-shek.

In 1929 Chiang Kai-shek killed two birds with one stone by making use of the conflicts between Wang Jingwei and the Guangxi clique. He was able to strike a blow at the Guangxi clique while driving out Wang Jingwei, thus taking quick military and political control of the KMT. However, having resolved the conflicts between the Nanjing and Wuhan cliques and between himself and the Guangxi clique, Chiang had not brought the whole of China under control. At the time, Feng Yuxiang controlled the areas of Henan, Shaanxi, Gansu, and Ningxia; Yan Xishan controlled the areas of Shanxi, Hebei, Suiyuan, Chahar and Beiping, and Tianjin; the Guangxi clique controlled the areas of Guangxi, Hunan, and Hubei; and Chiang Kai-shek controlled only Shanghai, Nanjing, Hangzhou, and the areas of Jiangsu and Zhejiang. The cliques of Chiang, Feng, Yan, and Guangxi each occupied some areas, and they were on their guard against one another.

To achieve his goal of dictatorial rule, Chiang Kai-shek tightened his control over all the cliques. As a result, the conflicts between Chiang Kai-shek and the forces of all the other cliques were intensified. To display their strength to Chiang Kai-shek, Feng Yuxiang accelerated his preparations for war in northwestern China; Yan Xishan's troops conducted operational exercises in Shanxi Province; Li Zongren carried out the policy of arms production in Wuhan; and Bai Chongxi staged large-scale joint military drills in Hebei Province. It looked as if war would break out at any minute, and the whole country was panic-stricken.

As old warlords disappeared and new ones appeared, the warlord forces still played the leading role in China's political arena. The wars between the old warlords were followed by the wars between the new warlords, and such internecine fighting between warlords went on year after year, bringing untold misery to the Chinese people. How people wished that such chaos, such disaster, such bitterness and hatred would come to an end!

Yet the wars between the new warlords befell the Chinese in the end. In March 1929 the war between Chiang Kai-shek and the Guangxi clique broke out. In November of the same year, the war between Chiang Kai-shek and Feng Yuxiang broke out. In the same month, the second war between Chiang Kai-shek and the Guangxi clique broke out. In March 1930 Yan Xishan joined forces with Wang Jingwei, Feng Yuxiang, Zhang Xueliang, and Li Zongren and Zhang Fakui of the Guangxi clique. They sent an open telegram denouncing Chiang Kai-shek, thus launching a major war in the Central Plains.

During seven months of fighting, approximately one million troops fought on a 500-kilometer front in the heartland of China, with 300,000 casualties. It was an unprecedentedly large-scale and costly war. Many warlords joined the war—the largest internecine war among them in the modern history of China.

Through these wars, Chiang Kai-shek repeatedly made use of the conflicts between and within various cliques and tried to maneuver among them by employing massive troops for suppression, offering their members huge sums of money to turn traitor, and promising high positions. Finally he succeeded in defeating the several hundred thousand troops of the powerful cliques of Yan Xishan, Feng Yuxiang, and the Guangxi warlord one after another, thereby establishing his autocratic rule and achieving at least the formal unification of the country.

In these internecine wars among the warlords, the war between Chiang Kai-shek and the Guangxi warlords broke out first. Like a thorn in the throat, the forces of the Guangxi clique represented a major obstacle to the establishment of Chiang Kai-shek's autocratic rule, so he was determined to remove it. In March 1929 Chiang Kai-shek ordered a punitive expedition against the Guangxi warlords, who were also in full battle array to fight back. When the swords were drawn and bows were bent, Li Mingrui, a general in the Guangxi army, declared suddenly that he would accept the leadership of Chiang Kai-shek's "Central Government." Li's action shocked the Guangxi clique.

On April 4, Chiang ordered his troops to continue to pursue and suppress the Guangxi troops. At the same time, he continued to buy out more and more of the Guangxi troops with money and promises of high positions. Although the Guangxi warlords Li Zongren, Bai Chongxi, and Huang Shaohong tried hard to fight back, they were hopelessly outnumbered. They abandoned their troops and left for Hong Kong. Then Chiang Kai-shek appointed Yu Zuobai governor of the Guangxi Provincial Government.

The war between Chiang Kai-shek and the Guangxi clique thus ended, with Chiang the victor. The important figures who caused the defeat of the Guangxi warlord were Yu Zuobai and Li Mingrui. Who were these two people?

Yu Zuobai was from Guangxi Province. When he was young, he joined the Guangdong army against Yuan Shikai and for the protection of the republic. Although he was Li Zongren's subordinate, he was always on bad terms with Li.

Li Mingrui was also from Guangxi Province. He was Yu Zuobai's cousin. Being young and courageous, valiant and skillful in battle, he always charged at the head of his troops. Together with his cousin Yu Zuobai, he decided to support Dr. Sun Yat-sen and to throw himself into the national revolution.

In 1926, Guangdong and Guangxi provinces were unified, and cooperation between the KMT and the CPC surged. At the climax of the Great Revolution, the revolution developed in Guangxi Province. Trade unions, peasant associations, student associations, and women's associations were established one after another. Yu Zuobai and Li Mingrui were quite excited about the revolutionary situation. They supported wholeheartedly Dr. Sun Yat-sen's Three Great Policies. They were known as the KMT leftists in the Guangxi army.

After the oath-taking rally for the Northern Expedition in July 1926, the whole country witnessed an upsurge in the national revolution. The Northern Expeditionary Army marched north directly and defeated all enemy forces.

Having rendered meritorious service many times in unifying Guangdong and Guangxi provinces, Li Mingrui repeatedly performed outstanding services again during the Northern Expedition. He was well known in the Northern Expeditionary Army as a general who was skillful in battle.

Instead of giving him awards for his outstanding performances, Li Zongren and Huang Shaohong went so far as to hate him, out of jealousy. The conflicts between Li Zongren, Huang Shaohong, and Li Mingrui thus became extremely serious.

Such a conflict among the Guangxi warlords was no secret to Chiang Kai-shek. Actually, he was extremely worried about the rapid growth of Guangxi forces. To get rid of Li Zongren, Huang Shaohong, and Bai Chongxi, Chiang Kai-shek had no choice but to use Yu Zuobai and Li Mingrui.

Chiang Kai-shek sent his men secretly to see Li Mingrui and Yu Zuobai, who was in Hong Kong then, to convey Chiang's desire to draw them over to his side. Rumors spread everywhere quickly, even in the Guangxi army, where people felt puzzled.

Yu Zuobai's military power to command troops was removed and he was forced out of Guangxi by Li Zongren. There was nothing he could do but live in exile in Hong Kong. Chiang Kai-shek tempted him with promises: first, he would be appointed adviser to the commander-in-chief of the navy, army, and air force, with the rank of general, and concurrently governor of the Guangxi Provincial Government; second, he would receive two million silver dollars for soldiers' pay and military supplies; and third, Yu Zuobai and Li Mingrui would share the authority to govern Guangxi and to reorganize the Guangxi army and discharge surplus personnel.

Chiang Kai-shek was known for being sinister, suspicious, and crafty. He could not tolerate anyone who posed a threat to him. Yu Zuobai and Li Mingrui had long been aware that Chiang Kai-shek tolerated them not because he favored them but because he was using them to get rid of the forces of the Guangxi clique, such as Li Zongren. Even so, it was a good opportunity for them to get rid of Li Zongren and Huang Shaohong at Chiang's hands.

After the war between Chiang Kai-shek and the Guangxi clique broke out, Chiang's troops mounted large-scale attacks on the troops of the Guangxi clique on all fronts. The Guangxi warlord Li Zongren was truculent and determined to fight to the end. Knowing that the time was ripe, Li Mingrui suddenly declared independence and announced his decision not to join the war. This move dealt a major blow to the Guangxi troops. Before long, the Guangxi troops suffered crushing defeats. The once powerful Li Zongren, Huang Shaohong, and Bai Chongxi fled to Hong Kong. Overwhelmed with joy, Chiang Kai-shek sent an open telegram to the whole nation: "Wuhan is under control without firing a single shot." Right after that, Li Mingrui commanded his army to march south and stabilized the situation in Guangxi Province with one fell swoop.

Having achieved their goal, Yu Zuobai became the governor of the Guangxi Provincial Government, and Li Mingrui the deputy commander-in-chief of the 8th Route Army of the Punitive Army (Chen Jitang was the commander-in-chief), the special military commissioner to Guangxi for reorganizing the army and discharging surplus personnel, and the Guangxi pacification commander.

When Yu Zuobai and Li Mingrui controlled the military and administrative power in Guangxi, their foundation was extremely weak, and it was difficult to raise funds for military spending. There were economic difficulties in the province. Chiang Kai-shek did not keep his promise of offering two million silver dollars. Although Huang Shaohong's remaining troops accepted Yu and Li's command, they were still strong, and not willing to be controlled by them.

Chiang Kai-shek defeated the Guangxi clique by making use of Yu and Li. However, no one knew when Chiang would make use of someone else to defeat Yu and Li. Yu Zuobai and Li Mingrui absolutely could not trust Chiang Kai-shek and had to deal with their sworn enemy, the old Guangxi clique. Consequently, they approached the Communist Party through Yu Zuoyu, Yu Zuobai's younger brother, hoping that the Party would send people to help them control the situation there.

It was not the first time Yu Zuobai and Li Mingrui had made contact with the Communist Party. As early as the first cooperation between the Kuomintang and the Communist Party, a group of CPC members joined the Guangxi army and held some important positions in it. Under the influence of the progressive ideological trend of the Great Revolution and progressive persons, Yu Zuobai supported the movement of the workers, peasants, and students.

Yu Zuobai had always had a good impression of the Communist Party and harbored no ill feeling against it. Li Mingrui was open and upright and inclined toward the revolution. After the April 12th Incident of 1927, he was very indignant over the persecution of the Communists, in the name of "purging the party," by Chiang Kai-shek and the Guangxi warlords, and he sent off quite reluctantly the CPC members in his army.

Coincidentally, a CPC member worked secretly for Yu Zuobai and Li Mingrui, influencing them imperceptibly with his words and deeds. He was Yu Zuobai's younger brother, Yu Zuoyu. Yu Zuoyu studied in the Yantang Military School of the Guangzhou Army for Upholding the Provisional Constitution and took part in the Northern Expedition during the Great Revolution as a regiment commander under Li Mingrui. He was a brave officer. After the Great Revolution failed, Yu Zuoyu joined the CPC in Hong Kong in October 1927 and participated in the Guangzhou Uprising in December of that year. He was sent back to Guangxi later to carry out revolutionary activities, lead the peasant movement, and organize peasant self-defense corps. In the spring of 1929 he was instructed to do secret work in the Guangxi army led by Yu Zuobai and Li Mingrui. When his brother Yu Zuobai and his cousin Li Mingrui expressed a willingness to consult with the Communist Party, Yu Zuoyu took it as his duty to contact the Party at once.

The Central Committee thus decided to send cadres to work in Guangxi. Deng Xiaoping was selected as the representative of the Central Committee.

CHAPTER 28

GOING TO GUANGXI

Between July and August 1929, when southern China was in the grip of the intense heat of midsummer, Father carried out the instructions of the Central Committee and the Central Military Commission and bid farewell to his wife, boarded a southbound ship, and headed for Guangxi via Hong Kong.

Deng Xiaoping then was no longer the young man Deng Xixian who had just returned to China after studying in the Soviet Union. The revolutionary activities in China of the preceding two years or more, especially the hardships that he endured after the failure of the Great Revolution and the revolutionary work under the White terror, had added to his experience of revolutionary struggle. Since he was transferred to the Party's central organs before the August 7th Meeting of 1927—and especially since he had become the secretary-general of the Central Committee—he had had opportunities to attend as an observer various top-level meetings of the Central Committee, read the reports on the Party's work from all over the country, and take part in the technical work for some major policy decisions of the Party. All this exposure enriched his work experience, enhanced his understanding of political and policy matters, and increased his knowledge of and experience in Chinese revolutionary activities. When he went to Guangxi, therefore, Deng Xiaoping was very experienced in revolutionary struggle and had a high level of political awareness.

Father got off the ship in Hong Kong and got in touch with the Southern Bureau of the Communist Party immediately. Located in Hong Kong, the Southern Bureau of the Central Committee was in charge of the Party's work in both Guangxi and Guangdong provinces. Like some districts of Shanghai, Hong Kong was a leased territory and thus provided better cover for the Party's activities.

The secretary of the Southern Bureau was He Chang, and the secretary of the Military Commission of the Guangdong Provincial Party Committee was Nie Rongzhen. The two couples—He Chang and his wife Huang Mulan (or Dinghui), and Nie Rongzhen and his wife Zhang Ruihua—lived close to Fenghuangtai by the Hong Kong race course. As soon as he arrived in Hong Kong, Father got in touch with them. Huang Dinghui recalled:

> At the time my husband and I lived together with Rongzhen and Ruihua. After he arrived in Hong Kong, Comrade Xiaoping lived in a hotel. He came to our residence once, discussing the Party's work in Guangxi with both Comrades He Chang and Nie Rongzhen. He had dinner at our residence, which was prepared by me and Ruihua. Later on He Chang went to Guangxi and attended the meeting held by the Guangxi Provincial Party Committee, at which both he and your father made speeches. A few days later He Chang came back from Guangxi.

To maintain contact with the local Party committee, the Central Committee sent Gong Yinbing of the Special Branch to accompany Deng Xiaoping to Guangxi. Gong brought with him the cipher code and took charge of the confidential work there.

At the same time, the Party sent dozens of political and military cadres who, through various channels and connections, tried to work in the provincial government led by Yu Zuobai and the troops under Li Mingrui. They did so either with the help of various connections or by using different names. None of them revealed their true identity as Communists. In this way, Communists came to Guangxi quietly, one after another, on the instructions of the Party.

In September of that year Father and Gong Yinbin reached Nanning. Once in Nanning, Father first got in touch with Lei Jingtian, secretary of the Guangxi Special Party Committee. On September 10, Deng Xiaoping, representative of the Central Committee, presided over the First Congress of the Communist Party in Guangxi. More than thirty delegates from Nanning, Wuzhou, and the Zuojiang and Youjiang areas attended the congress, at which Father spoke about the current situation and tasks. Participants adopted a series of important resolutions on unfolding the agrarian revolution, establishing workers' and peasants' armed forces, and preparing for armed uprisings.

Father recalled, "After we arrived in Nanning, I met with Yu Zuobai several times and did some united front work according to the guiding principles formulated by the Central Committee. Meanwhile, I carefully assigned those cadres sent by the Central Committee to suitable posts in Yu's government." After he arrived in Guangxi, Father operated in the open as secretary of the Guangxi Provincial Government, under the assumed name of Deng Bin, while as the representative of the Central Committee he led the Party work in Guangxi.

Father was soon cooperating closely with Yu Zuobai in Guangxi. Under the influence of the Communist Party, Yu Zuobai and Li Mingrui first released a group of "political prisoners." These political prisoners—members of the Communist Party, members of trade union and peasant associations, and progressive youths during the Great Revolution—had been arrested by the Guangxi warlord clique after the April 12th Massacre of 1927. This group, especially the cadres of the Communist Party and the Youth League, later became mainstays in the Red Army in Guangxi.

Drawing on past experience, Li Mingrui was very eager to expand his army and military strength after he seized military power in Guangxi. The troops he brought back from Wuhan numbered only some 30,000, whereas a comparable number of troops of the former warlords in Guangxi remained. The latter had undergone a nominal reorganization, but they were still not ready to accept Li Mingrui's command. Because Li Mingrui wanted to establish an army under his own control, the Guangxi Garrison Detachment, composed of two newly reorganized groups, the 4th and the 5th, was established.

In view of the situation, Father and others proposed to Li Mingrui through Yu Zuoyu that the General Training Corps be set up to train junior officers. The Party sent more than 100 Communist cadre trainees to this corps and trained and educated nearly 1,000 progressive youths from Li Mingrui's army. Some trainees were admitted into the Party.

After consultations with the Communist Party, a group of Communists were assigned to the newly established Garrison Detachment. Zhang Yunyi, a Party member, served as the commander of the 4th Group, and Li Qian, another Party member, as the deputy commander. Yu Zuoyu and Shi Juran, both Party members, were made commander and deputy commander of the 5th Group, respectively. As instructed by Deng Xiaoping, Zhang Yunyi and others appointed more than 100 Party members as company or platoon commanders. They also strengthened the revolutionary ideological education, punished the bitterly hated officers of the old army, and recruited large numbers of workers, peasants, and progressive students, thus bringing about a change in the composition and ideology of the army. Such volunteers expanded the 4th Group from 1,000 to 2,000 men in a short time. In the meantime, Yu Zuoyu reinforced his 5th Group with some activists in the peasant movement, and his group also swelled rapidly, to 2,000 men. The General Training Corps was also commanded by a Party member. The newly established army was later to become the basis of the 7th and 8th Corps of the Red Army.

Under the influence of the Communist Party, Yu Zuobai lifted the ban on the progressive mass movements in Guangxi. As a result, the Provincial Peasant Association could resume its work and convened a representative conference. The progressive organizations, such as trade unions, women's associations, and student associations, resumed their operations one after another, thus immediately bringing about a vigorous revolutionary upsurge across the entire province of Guangxi—a situation similar to that before the Great Revolution.

Also under the influence of the Communist Party, Yu Zuobai eliminated the old forces of Huang Shaohong in the rural areas and appointed large numbers of the peasant movement leaders who had emerged during the Great Revolution as county magistrates. Among them was Wei Baqun from Donglan County. On the recommendations made by the Communist Party, Yu Zuobai also appointed a group of Communists and progressive persons as county magistrates in the Zuojiang and Youjiang areas. As a result, more than twenty counties in these areas were controlled by Communists and progressive personages.

With the support of the Communist Party, Yu Zuobai and Li Mingrui backed the peasants' armed forces in Donglan County, gave that military body the formal name of the Youjiang Group for Protecting Businessmen, and gave them a few hundred guns to show their support.

As early as the period of the Great Revolution, the Donglan peasant movement in the Youjiang area was well known both inside and outside the province, thanks to a prominent leader of the Guangxi peasant movement, Wei Baqun.

Born into a wealthy family, Wei Baqun was patriotic from an early age and joined the Communist Party of China in 1925. In 1926 the peasant movement led by Wei Baqun in Donglan and Fengshan in the Youjiang area of Guangxi Province spread like a raging fire, and the area became one where the peasant movement was the most highly developed in the whole country. Wei Baqun not only established peasant associations but also organized and expanded peasant armed forces for self-defense. After the April 12th Massacre of 1927, the revolutionary struggles in Guangxi went underground. Only the peasant armed forces of Donglan led by Wei Baqun persisted under the White terror in open armed struggle.

Wei Baqun was the hero whom the vast number of peasants in Guangxi were proud of. The solid foundation laid by him in the Youjiang area provided very favorable conditions for the Red Army to establish the Youjiang Revolutionary Base Area.

Besides their work in the army, Father and the Guangxi Special Party Committee made great efforts to restore and expand Guangxi's disorganized local Party organizations. As a result, the local Party organizations at all levels established their contacts. Meanwhile, study classes were organized for Party members, and Party journals were published.

Nanning was different now! A new revolutionary upsurge had developed rapidly in Nanning and other parts of Guangxi. At a time when the revolution was at a low ebb and the White terror prevailed across the country, Guangxi alone was in a new revolutionary upsurge. The new developments encouraged the revolutionaries and the progressives. Li Zongren, the head of the Guangxi clique, cried out in alarm: Yu Zuobai and Li Mingrui, after returning to the south, put wings onto the tiger, giving rise to Communist peril. Guangxi Province has now become the southwestern base of the Communist Party.

After Father and the other comrades sent by the Central Committee arrived in Guangxi and took charge of the Party's work there, their work quickly provided results. However, people in Guangxi then did not know Deng Xiaoping. On the instruction of the Central Committee, and according to his experience in secret work over the years, Father did not appear in public in Guangxi. He worked only in the Party and maintained contacts with only a very few people. Besides his Party work, Father only met with Yu Zuobai a few times. Father maintained close contact with the Central Committee so as to report his work or ask for instructions.

The developments in Guangxi had long drawn the attention of many different kinds of people. Although the Communist Party did not operate openly, the lively revolutionary atmosphere attracted the close attention of the counterrevolutionaries. Some alarmed outsiders were already shouting, "Yu Zuobai and Li Mingrui have been creating troubles, resulting in the red upsurge across the Zuojiang and Youjiang areas and in the rebirth of the former Communist bandits in Donglan." Chiang Kai-shek, who had supported Yu Zuobai and Li Mingrui in overthrowing the Guangxi clique, naturally paid close attention to the developments in Guangxi.

Chiang Kai-shek was known for his subtleness and adeptness in scheming. When he cooperated with Yu Zuobai and Li Mingrui to overthrow the Guangxi faction, he sent one of his men, Zheng Jiemin, to become the director of the Political Department of Li Mingrui's 15th Division, so as to keep watch on him. Naturally, all the events that took place in Guangxi were reported to Chiang Kai-shek by Zheng Jiemin. Chiang Kai-shek, of course, was not happy with what was happening in Guangxi, but at the moment he had no brilliant scheme to contain Yu Zuobai and Li Mingrui.

The developing situation, however, soon gave Chiang Kai-shek an opportunity to eliminate Yu Zuobai and Li Mingrui. In August 1929 Wang Jingwei, who had always been at odds with Chiang Kai-shek, attempted to unite with Feng Yuxiang, Yan Xishan, Tang Shengzhi, and some other military groups discontented with Chiang Kai-shek's conduct in a joint effort to oppose and topple him. Wang Jingwei sent his man to Nanning to persuade Yu Zuobai and Li Mingrui to oppose Chiang Kai-shek jointly. At the time, Yu Zuobai and Li Mingrui felt that, although it was their goal to overthrow Chiang Kai-shek, their military and political power was not yet solid enough, and they were very short of funds for military expenditures and soldiers' pay and provisions. So they hesitated to oppose Chiang Kai-shek at that time.

Knowing that Yu Zuobai and Li Mingrui remained undecided, Wang Jingwei repeatedly sent telegrams to urge them to launch an anti–Chiang Kai-shek campaign quickly. At the same time, Wang sent his man to give them a large amount of Hong Kong dollars as a bait.

When the news reached Nanjing, Chiang Kai-shek was shocked. Chiang stepped up his efforts to persuade Yu Zuobai and Li Mingrui and personally sent a telegram to Li Mingrui in September denouncing Wang Jingwei.

Urged repeatedly by Wang Jingwei and under the threat and coercion by Chiang Kai-shek, Yu Zuobai and Li Mingrui resolved to oppose Chiang. On October 1, 1929, Yu Zuobai and Li Mingrui organized in Nanning an oath-taking rally against Chiang Kai-shek and sent an open telegram announcing that they were serving as the commander-in-chief and deputy commander-in-chief, respectively, of the Southern Route Expeditionary Army against Chiang Kai-shek. Soon after that, Yu and Li conducted operational deployments of their troops. Li Mingrui himself went to the front to direct the attack on Guangdong.

After Yu Zuobai and Li Mingrui sent the anti–Chiang Kai-shek open telegram, the *Central Daily* in Nanjing cried out in alarm that Yu and Li were colluding with the Communist Party to oppose Chiang Kai-shek. Chiang analyzed the situation in Guangxi and decided to disintegrate the forces of Yu Zuobai and Li Mingrui by buying them out. Chiang Kai-shek was indeed an unscrupulous and ambitious man. His adeptness at analyzing a situation, making use of contradictions, and using any means to deal with his opponents enabled him to maneuver among various political groups and realize his ultimate goal of dictatorship.

Among Li Mingrui's troops, only the 15th and 57th were divisions he could trust. The remainder formerly belonged to the Guangxi clique and were difficult to control. Chiang Kai-shek made a move first to buy out the men of the former Guangxi clique, and even a brigade commander trusted by Li Mingrui, by offering huge amounts of money and high posts to them.

Li Mingrui had just arrived at the front and was preparing to direct the battle when he learned unexpectedly that his three main divisions and one brigade had turned to the side of Chiang Kai-shek. Li Mingrui lost almost all his military strength in an instant.

The anti–Chiang Kai-shek expedition failed in less than ten days. Finding that nothing could be done to reverse the situation, Li Mingrui had to go back hurriedly to Nanning with only a few of his followers.

When Yu Zuobai and Li Mingrui decided to join in the anti–Chiang Kai-shek campaign by sending the open telegram, the Communist Party made an objective analysis of the situation and came to the conclusion that the campaign was doomed to failure, because Li Mingrui had only three divisions, all handicapped by internal disunity; he had been in Guangxi only a short time; and his political and economic foundation was very weak.

Since Yu Zuobai and Li Mingrui sent the open telegram without heeding the advice of the Communist Party, Father decided to leave behind the 4th and 5th groups, as well as the General Training Corps that had been controlled by the Party, to protect the rear, so as to preserve the strength of the revolutionary forces and guard against eventualities. Thanks to the persuasion and insistence of Father and others, Yu Zuobai and Li Mingrui finally agreed to this plan.

As soon as Yu Zuobai and Li Mingrui set off with their troops, Father and Zhang Yunyi started to prepare for contingency. They dispatched to the Zuojiang and Youjiang areas two battalions, one each from the 4th and 5th groups, to get everything ready. In Nanning, because Zhang Yunyi was also commander of the Nanning Garrison, the troops controlled by the Party took over the provincial ordnance depot and some other organs in his name and seized 5,000 to 6,000 rifles as well as mountain guns, mortars, machine guns, transceivers, and piles of ammunitions. At the same time, steamboats were ready by the river banks. So everything was ready. Had it not been for the farsightedness of Father and other Communists, Yu Zuobai and Li Mingrui would have completely lost their troops in the anti–Chiang Kai-shek campaign and, instead of staging a comeback later, would have had nowhere to shelter themselves.

Soon after their hasty flight back to Nanning, Yu Zuobai and Li Mingrui planned to flee to the Zuojiang area. Father, in consultation with other comrades, promptly decided to stage a mutiny. They planned to pull the troops out of Nanning for the Zuojiang and Youjiang areas with a view to establishing a new base with Bose and Longzhou as its centers. Their decision was reported through a secret radio station to the Central Committee in Shanghai, which approved it.

The situation was pressing and brooked no delay. One mid-October day, at nightfall, gunshots were heard in Nanning proper. The mutinous troops went into action suddenly, opening the ordnance depot and seizing all the arms and ammunitions. After that, the 4th and 5th groups and the General Training Corps quickly withdrew from Nanning.

The 4th Group and a small part of the General Training Corps, led by Zhang Yunyi, withdrew up the Youjiang River toward the Bose area in the northwest, while the 5th Group, led by Li Mingrui and Yu Zuoyu, headed for the Longzhou area in the southwest along the Zuojiang River.

Leading the comrades of the local Party organizations engaged in secret work, Father directed the steamboats loaded with arms and the security forces up the Youjiang River toward the Bose area. Xu Fengxiang, who took part in the mutiny, recalled:

> One misty morning in mid-October at the pier of the Nanning customs, our comrades were busy loading all the big and small boats with weapons, ammunitions, and other matériel in an orderly way and then boarded the boats themselves. I boarded a steam boat. The last to get on the boat was a comrade who was short but strong, in his twenties and full of energy. As he boarded the boat, he greeted with a smile those comrades already aboard. I was about to ask the other comrades about him, for I didn't know him, when someone cracked a joke: "There comes the secretary. Let's make room for him!" So that was the top leader of our mutiny—Comrade Deng Xiaoping. His public capacity was secretary of the Guangxi Provincial Government.[1]

Yuan Renyuan, a former political instructor of the General Training Corps, also recalled:

> Comrade Xiaoping directed the steamboats loaded with arms and the security forces to advance toward the Youjiang area by water. Coincidentally, She Hui and I were on the same boat with Xiaoping. We had only known that our leader was Xiaoping, but we had never met him. That was the first time I saw him. He had the assumed name of Deng Bin then. That first encounter left an unforgettable impression on us. Xiaoping treated unexpected events cool-headedly and calmly and was a resourceful and resolute man. He was approachable, easygoing, talkative, and sometimes humorous.[2]

The fleet led by Father sailed upstream, pressing ahead and sending ripples across the river. The soft October breeze of the south touched the faces of the revolutionary fighters. In their hearts, they were as clear as bright daylight about the significance of their action. Their thoughts were as thrilling as the surging river. The journey was smooth, and they soon reached the Youjiang area.

Zhang Yunyi had long heard about Deng Xiaoping, the representative of the Central Committee, but had never met him because of the secret nature of their work. After the troops arrived in Tiandong, Zhang Yunyi saw Deng Xiaoping for the first time. He wrote in his memoirs:

> The boats with ordnance arrived not long after we got there. A moment later, I caught sight of Ye Jizhuang, who was accompanying a comrade whom I didn't know toward the headquarters of the group. That comrade was of medium build, in his twenties, glowing with health, and radiating vigor and had an easy manner. We hurried to meet them. Introducing him to us, Ye Jizhuang said: "This is Deng Xiaoping."—"So you are Deng Xiaoping!" I cried out. In the past three or four months I had been getting many valuable instructions on our work, which enabled us to solve many difficult problems we faced. But I had never had the chance to meet him. Deng Xiaoping was very excited, too, and grasped my hands. Our hearts were filled with such comradely feelings that we forgot to talk for quite some time. When we sat down, we were joined by Lei Jingtian and several other comrades of the Special Party Committee. We introduced one another, talking and laughing excitedly. At this moment, Deng Xiaoping suggested that we go to Bose the following day, take with us most of the ordnance, and conceal those heavy weapons and ammunitions not needed for the time being in the mountains of Donglan and Tiandong. All of us agreed to this plan. So we began to implement it immediately. We continued marching for two more days before arriving in Bose. From then on, Deng Xiaoping and I lived together.[3]

Thus, under the careful arrangements and organization of the representative of the Central Committee Deng Xiaoping and other Communists, the local Party organizations in Guangxi successfully preserved revolutionary strength by moving the troops to the Bose and Longzhou areas. As a result, the basic conditions were created for the Red Army and the revolutionary base areas to be established under the red banner in the near future.

THE BOSE AND LONGZHOU UPRISINGS

On a map, Guangxi resembles a flat and broad mulberry leaf. Its water comes from the Hongshui River, and its soil is red. On the red soil, all the mountains and fields are green, the exuberant green of mulberry leaves.

The capital, Nanning, is situated in southwestern Guangxi. Liuzhou, a city of strategic importance in the north, leads to Hunan and Guizhou provinces. Wuzhou, the passage to Guangdong Province, lies in the east, and the Zuojiang and Youjiang areas are to the west.

To the west of Nanning, the Yongjiang River is divided into two rivers, one in the south and the other in the north. The river in the northwest leading to Bose is named the Youjiang River, and the other in the southwest leading to Longzhou is named the Zuojiang River. The delta between the two rivers is called the Zuojiang and Youjiang areas.

Bose is over 210 kilometers from Nanning. It is not the hinterland of Guangxi. There are no big towns nearby. Yunnan Province is in its west. Longzhou is about 150 kilometers from Nanning. It adjoins Zhennanguan (now Pingxiang City), a fort in southwestern China. The distance between Longzhou and Vietnam on the opposite side is only about 10 kilometers.

The Youjiang area is where the three provinces of Guangxi, Yunnan, and Guizhou border. It is a multinational area where the Zhuang, Han, Yao, and other nationalities live in compact communities. During the Great Revolution, Wei Baqun established peasants' revolutionary armed forces in the two counties of Donglan and Fengshan in this area. Here the revolutionary struggle led by the Communist Party of China never stopped, not even during the White terror after the failure of the Great Revolution. Wei Baqun and his comrades-in-arms established a solid mass base in the Youjiang area, thus creating favorable conditions for the arrival of the revolutionary armed forces led by Deng Xiaoping and Zhang Yunyi.

Father recalled, "The Youjiang area in Guangxi had a fairly solid mass base. In that area there was the outstanding and prestigious leader of the peasants, Comrade Wei Baqun. In the areas of Donglan and Fengshan, Comrade Wei Baqun worked for a long time, and the areas were a very good revolutionary base area. It was greatly conducive to the establishment and activities of the 7th Corps of the Red Army."

As soon as Father and his comrades arrived at Bose, they immediately began political and organizational work and the work of planning armed uprising. Yuan Renyuan, Wei Guoqing, and others recalled:

> Comrade Deng Xiaoping called a meeting of the Party committee. It decided to further mobilize the masses and propagate the propositions of the Sixth National Congress, to reorganize and expand the army units, to establish the system of political work, to organize soldiers' committees, to practice equality between officers

and men, to arm the peasants, and to wage struggles against local tyrants and the evil gentry through local organizations. The revolutionary activities in the Youjiang area developed day by day. In early November the Central Committee approved the plan of armed uprising in the Zuojiang and Youjiang areas, gave the designations of the 7th and 8th corps of the Red Army, and appointed leading cadres. In accordance with the directive of the Central Committee, Comrade Deng Xiaoping immediately drew up an overall plan and prepared for the armed uprising.[1]

Yuan Renyuan recalled:

After we arrived at Bose, we started our work, in accordance with the directive and plan of Xiaoping. First, we propagated the propositions of our Party among the army units and the masses to mobilize the masses. Second, we controlled the political power and dismissed some reactionary county magistrates, who were replaced by our own men in the areas where our power could reach. Third, we reorganized the army and expanded the armed forces. Our army units were orginally the old-style units of Li Mingrui with a complex class status. To transform this old-style army into a revolutionary army, we purged, first of all, some reactionary officers. We neither arrested nor killed them; instead, we gave them traveling expenses and sent them away in a courteous way. Then in the army units we carried out a democratic revolution, set up soldiers' committees, practiced equality between officers and men both politically and in daily life, prohibited hitting or cursing, and established the system of political work. Fourth, we established Party organizations and recruited new Party members. Fifth, we wiped out the armed forces of landlords and bandits and consolidated the base area. Sixth, we trained cadres. Xiaoping attached great importance to this work. He gave lectures himself. I remembered that he gave us lectures on such topics as the Resolution of the Sixth National Congress, the Ten-Point Program, and Soviet political power. In his lectures, he explained profound knowledge in simple terms. His lectures were easy to understand and integrated theory with practice. They were well received.[2]

Xu Fengxiang recalled that in Bose

I worked with Comrade Deng Xiaoping. At that time, we constantly received letters and telegrams addressed to "Secretary Deng" by our comrades who had been sent to different places in the Youjiang area, which told us about the cruel oppression by the old regime, all kinds of outrages perpetrated by local tyrants and evil gentry, and the warm welcome extended to our army by the people. Whenever I sent these letters and telegrams to Comrade Deng Xiaoping, he paid great attention to the news transmitted back from various places. To prepare for the uprising, Comrade Xiaoping worked day and night. In the daytime, he talked with comrades, attended meetings, and gave instructions about the work, while at night he discussed important questions and planned matters concerning the uprising with Zhang Yunyi and other cadres. Soon our army wiped out the reactionary 3d Garrison Group, mobilized the masses of workers and peasants, and continued to reorganize and transform our army.[3]

Right at this time, Gong Yinbin, who had returned to Shanghai to ask the Central Committee for instructions on the work, came back to Bose and communicated to Father and others the instructions of the Central Committee. The Central Committee approved the proposal put forth by Father and others and asked them to establish a base in the area of the Zuojiang and Youjiang rivers and to found Red Army units. The designation given was the 7th Corps of the Red Army. Zhang Yunyi was appointed commander, and Deng Xiaoping political commissar. The units in the Zuojiang area were organized into three columns constituting the 8th Corps of the Red Army.

Father and his comrades must have been very inspired by the decision of the Central Committee. They immediately sent Gong Yinbin back to Shanghai to report to the Central Committee on the withdrawal of the troops of the Youjiang area and stated:

> We will carry out the directive of the Central Committee resolutely and finish all preparations in about forty days. Then we will announce the uprising at once.
>
> Deng Xiaoping called a meeting of the Party committee immediately, communicated the directive of the Central Committee, and decided to step up preparations, announce the uprising, and establish the 7th Corps of the Red Army and the Youjiang Soviet on December 11—the day of the second anniversary of the Guangzhou Uprising.[4]

On December 11, 1929, the red flag of armed uprising was hoisted high on top of the city wall of Bose, and the establishment of the 7th Corps of the Chinese Workers' and Peasants' Red Army was officially declared.

The First Conference of Representatives of Workers, Peasants, and Army-men of the Youjiang Area was held at Pingma on the next day. Through election, the Youjiang Soviet Government was established, with Lei Jingtian as chairman and Wei Baqun, Chen Hongtao, and others as members.[5]

After the 7th Corps of the Red Army was founded, a front committee was set up, with Deng Bin, representative of the Central Committee, as secretary and Zhang Yunyi, Chen Haoren, Lei Jingtian, and others as members. The committee exercised a unified leadership over the work of both the army and the localities.

On the eve of the Bose Uprising, that is, in late November, Father suddenly received a telegram from the Central Committee in Shanghai asking him to go to Shanghai to report to it on the work. After he planned the work with Zhang Yunyi and others, Father left Bose several days before the Bose Uprising, that is, in early December. Accompanied by a guide and disguised as a businessman, Father planned to go to Longzhou first to give instructions and check on work there in preparation for the Longzhou Uprising and the founding of the 8th Corps of the Red Army, and then to leave Longzhou and to sail to Hong Kong via Haiphong, Vietnam, and then on to Shanghai. Father started off from Bose with Yuan Renyuan and She Hui and stayed in Tiandong that night. They had a chance to meet with Li Mingrui on the way the next day.

Li Mingrui and Yu Zuoyu had arrived in Longzhou to raise soldiers' pay and provisions and reorganize the troops at the same time. Yu Zuobai had already left Guangxi for Hong Kong by then. Toward the end of November, Li Mingrui intended to launch a counterattack on Nanning by taking advantage of the political chaos in Guangxi and the vulnerability of Nanning. He had ordered Yu Zuoyu to lead the troops in the Zuojiang area to Chongshan to await orders, and he crossed the Youjiang River personally to discuss the joint attack on Nanning with the troops stationed in the Youjiang area. The day after Deng Xiaoping's stop overnight in Tiandong, Li Mingrui and the Central Committee representative met.

Deng Xiaoping and Li Mingrui had known of each other for a long time, but this was their first meeting. After they became acquainted, Father and Li Mingrui forged a profound comradeship and started their long side-by-side fight in revolutionary struggle.

Yuan Renyuan recalled that when they met, Xiaoping talked with Li Mingrui for a short while, and they decided to return to Bose together. Father had found out from talking with Li Mingrui that Li Mingrui and Yu Zuoyu were still hesitating about raising the red flag, though they had no alternative.

He Jiarong recalled,

> When Li Mingrui had a chance to meet with Comrade Deng Xiaoping on the way, they went back to Bose. Comrade Deng Xiaoping did political and ideological work on Li Mingrui. He told Li the revolutionary truths and pointed out the harms of internecine warfare among warlords. Furthermore, he explained to Li that the Party planned to establish the Zuojiang and Youjiang revolutionary base areas, prepare for the uprisings in Bose and Longzhou, and found the 7th and 8th corps of the Red Army. . . . [Deng] then asked him to be the commander-in-chief of the two corps and hoped Li would follow the Party to take a revolutionary road. Comrade Li Mingrui accepted with pleasure the advice of Comrade Deng Xiaoping and took the revolutionary road resolutely.[6]

After making up his mind, Li Mingrui went back to Longzhou at once. Deng Xiaoping started again on his journey to Shanghai.

When Li Mingrui was on his way back to Longzhou, he found that the deputy group commander, Meng Zhiren, had rebelled in Longzhou when the troops led by Yu Zuoyu marched to Chongshan. After Li Mingrui joined with Yu Zuoyu, they decided immediately to recover Longzhou. After their troops blockaded and encircled Longzhou, Meng Zhiren was defeated and fled. Longzhou was recovered by the 5th Group on December 3.

Huang Yiping, an old soldier of the 7th Corps of the Red Army, recalled that when Li Mingrui arrived at Longzhou after the failure in the struggle against Chiang Kai-shek,

> under the pressing situation, he had to support the revolution. Deng Xiaoping and Zhang Yunyi talked with him many times, recommended revolutionary books like the *Communist Manifesto* to him, pointed out the future of the revolution and his

way out, encouraged him to raise the red flag and join the revolutionary ranks, and trusted him politically by making him the commander-in-chief of the 7th and 8th corps of the Red Army. Under the influence of the Party's policy, Li Mingrui witnessed the increasing growth of the Red Army forces in the Youjiang area and the warm support given by the people of all nationalities to the Party and the Red Army and realized consequently that his only way out was to follow the Party and to join the revolutionary ranks. During this period Chiang Kai-shek sent his trusted special agents many times to Longzhou and other places to lure Li Mingrui and his kinsfolk with certificates of appointment as the governor of Guangxi Province and the commander of the 15th Corps and a huge sum of money. But Li Mingrui was not lured and gave a flat refusal.[7]

He Jiarong recalled,

> After the recovery of Longzhou [according to Wu Xi, it was the next day after the recovery of Longzhou], Comrade Deng Xiaoping immediately led Yan Min, He Shichang, Yuan Zhenwu, and others to come to Longzhou. There he studied with Li Mingrui, Yu Zuoyu, and other comrades the work plan and specific arrangements for the Longzhou Uprising and then left for Shanghai via Vietnam to report work to the Central Committee. The revolutionary enthusiasm ran high in the Zuojiang area. Efforts were made to restore order rapidly, establish local governments, found the workers' and peasants' red guards, organize the masses, and conduct propaganda and education. In accordance with the important instructions given by Comrade Deng Xiaoping on army transformation and the lessons drawn from the past in particular, efforts were stepped up to reorganize the 5th Group and transform it into reliable revolutionary forces.[8]

On February 1, 1930, the revolutionary uprising by the people of the Zuojiang area of Guangxi broke out in Longzhou. A red flag with the hammer and sickle was hoisted on the city wall of ancient Longzhou.

Longzhou was seething with excitement. At every corner of the city, red flags could be seen, the sound of firecrackers could be heard, and songs were sung loudly and clearly. The "Internationale" resounded through the air, and the procession of parading workers, peasants, students, and Red Army units was full of power and grandeur.

The 8th Corps of the Chinese Workers' and Peasants' Red Army was officially founded. The corps commander was Yu Zuoyu, the corps political commissar was Deng Bin (Xiaoping), and the corps encompassed two columns. The Zuojiang Revolutionary Military Committee, the Committee for the Suppression of Counterrevolutionaries, and committees for workers, peasants, and women were all set up.

After the Longzhou Uprising, the 8th Corps of the Red Army dispatched units to wage the struggle against local tyrants and suppress bandits, exterminated the despots who had been guilty of the most heinous crimes and incurred the greatest popular indignation, and established democratic regimes of workers and peasants in eight counties.

The people of Longzhou, while establishing the red political power, waged a heroic struggle against French imperialists under the leadership of the Communist Party. Longzhou had been a French sphere of influence since the late Qing Dynasty. After the Longzhou Uprising, the French imperialist forces were in a great panic. They went so far as to slander the red political power in a note they sent and dispatched aircraft into Longzhou airspace to pose armed threats. The 8th Corps of the Red Army denounced sternly the French slanders and threats, stating clearly its determination to "eliminate all privileges of imperialism in China."

The unjustifiable acts of the French imperialists enraged the people of Longzhou, who had suffered much under imperialist aggression. Tens of thousands of people held demonstrations, encircled the French Consulate in Longzhou, the customhouse, and the Catholic Church controlled by the French, confiscated the weapons, transceivers, and other military matériel the French had planned to use in unleashing a riot, captured 150,000 silver dollars, and expelled the French consul and missionaries. Li Lisan, then the leader of the Central Committee, gave this appraisal of the anti-imperialist patriotic struggle waged by the people of Longzhou: "Within a few days, the regime achieved what the government of the Kuomintang warlords could not and did not dare to achieve in decades. It carried out the political program of the Communist Party of China of fighting imperialism and ushered in a new era of the Chinese revolution."[9] Through the Bose and Longzhou uprisings, Chinese Communists heroically held high the fluttering red banner in southern China, thereby shocking the reactionary forces and heightening the morale of the revolutionaries.

After the Bose and Longzhou uprisings, the twenty counties in the Zuojiang and Youjiang areas with a population of over one million became one of the red revolutionary base areas and attracted nationwide attention.

In accordance with the appointments of the Central Committee and the Central Military Commission, Li Mingrui served as the commander-in-chief of the 7th and 8th corps of the Red Army and Deng Bin (Xiaoping) as the general political commissar of the two corps and concurrently as secretary of the Front Committee.

Li Mingrui was thirty-three years old at that time, and Deng Xiaoping was twenty-five.

STATE AFFAIRS, FAMILY AFFAIRS, AND PERSONAL GRIEF

When Father, as instructed by the Central Committee, returned to Shanghai in January 1930, he first reported to the Central Committee and the Central Military Commission on the work in Guangxi.

An article entitled "A Discussion on the Work Arrangements of the Red Army in Guangxi" was published in the second issue (March 15, 1930) of *Military Bulletin*, which is now kept in the Central Archives. All those in the article have been given assumed names. I think the reporter must be Deng Xiaoping and the participants in the discussion must be the leaders of the Central Committee and the Military Commission. *Military Bulletin* added an editorial note to the article:

> Originally we did not intend to publish all the records of the discussion. It is precisely because this change in Guangxi is the best-organized and best-intentioned mutiny in the country that we published the records, so that mutiny could be launched across the whole country. It is very important to disseminate the lessons and experience gained from this mutiny to all local Party committees. Therefore, we publish all records.

The reporter discussed at great length the work of the previous period in Guangxi and proposed the future work there. His report stated the need to continue the agrarian revolution, to set up Soviet revolutionary regimes through direct election by the masses, and to strengthen the combat effectiveness of the army. He proposed that the military strategy should be to establish ties between the Zuojiang and Youjiang areas and advance toward the border area between Hunan and Guangdong provinces so as to try to join forces with Zhu De and Mao Zedong.

During the discussion, the participants made many comments and suggestions on the work in Guangxi, some of which apparently showed a strong tendency toward "left" deviation. For instance, some held the view that the Red Army in Guangxi should advance to Liuzhou and Guilin (meaning to launch attacks on big cities). Several speakers showed an utter lack of trust in and rejection of Li Mingrui. After all the speeches were made, the reporter's additional remarks considered the views of the participants, reiterated some focal points of the work, and explained their views and misunderstandings.

As to Li Mingrui, he was an officer of the old army, but he was also a famous general in the Northern Expedition, a brave opponent of Chiang Kaishek, and one of the leaders of the Bose and Longzhou uprisings. Moreover, under the influence of the Communist Party, he had made a firm and resolute commitment to the revolutionary ranks. Father thought he knew Li Mingrui better than anyone and had the greatest trust in him. Since the opinion of some leaders of the Central Committee and the Military Commission differed from his own, Father explained very sincerely, "Of course, it is right to place the emphasis of our work on the mobilization of the local people, but we cannot neglect the establishment of the ties of work at the higher level!" Father told me later, "The purpose of the Central Committee to send me to Guangxi was exactly to do the united front work!"

I am convinced that this unnamed reporter was Father because daring to uphold the truth in the face of fallacy and to air views in the presence of superiors who hold wrong views are acts that exactly conform with his conduct and character.

Toward the end of the discussion, a leading comrade of the Military Commission (Hao) made a concluding statement. Although his statement failed to deviate from the then "leftist" policy of the Central Committee, it made an objective and realistic analysis of the situation in Guangxi by and large and gave instructions on the future work of the Red Army in Guangxi. This statement was quite different in style and content from those of the previous speakers. At the time, Zhou Enlai had taken charge of the work of the Military Commission, so I would guess that it was he who made this concluding statement.[1] The question of Li Mingrui was not mentioned in the statement. Father said that he proposed to the Central Committee that the recruitment of Li Mingrui as a member of the Communist Party of China be approved, and that this proposal was approved by the committee.

On March 2, soon after Father reported to the Central Committee and the Military Commission, the Central Committee issued a directive to the Front Committee of the 7th Corps of the Red Army, in care of the Guangdong Provincial Party Committee. It read:

> Comrade Xiaoping has come here and made a detailed report on the previous work and changes in the army in Guangxi. In addition to many specific questions that were discussed at length with Comrade Xiaoping, who would tell you personally, more instructions are as follows.
>
> As for the current situation, the crisis of imperialist attack on the Soviet Union is more serious, and chaos has resulted from the internecine warfare among warlords across the country, which has accelerated the crisis of the ruling class. The 7th Corps of the Red Army was founded under the objective conditions of the whole country and under the influence of the struggle waged by the masses of Guangxi. Although it emerged in remote Guangxi, its great role and importance cannot be belittled.

The directive summed up the strong and weak points of the previous work of the 7th Corps of the Red Army and pointed out clearly the main line of its future work: to carry out agrarian revolution thoroughly, to extend guerrilla warfare, to destroy feudal forces completely, and to set up Soviet regimes on the basis of the trust of the masses. The future of the 7th Corps "is to advance in the direction of the border area between Hunan and Guangdong provinces and the central part of Guangdong Province and to establish contacts with the Red Army units led by Zhu and Mao as well as the rebellion in Beijiang so as to try to win victory in Guangdong or several provinces first."

The directive approved the list of seven members of the Front Committee of the 7th Corps, with Deng Xiaoping serving as secretary. Zhang Yunyi was appointed commander of the 7th Corps, and Deng Xiaoping political commissar.[2]

This Central Committee directive pointed out the future tasks and orientation of the 7th Corps. It contained many very important points, but its contents were still imbued with "left"-adventurism—especially the propositions that the 7th Corps should defend the Soviet Union, attack big cities, and win victory in one or several provinces first—as well as some "leftist" policy measures. It was precisely these "leftist" instructions and their spirit that set a most arduous and dangerous path for the 7th and 8th corps of the Red Army to traverse.

After he finished reporting to the Central Committee in Shanghai, Father was busy with his family affairs because he had suffered a misfortune in his private life.

As soon as Father finished reporting on the work, he hurried to see his wife. At that time, Zhang Xiyuan was in Baolong Hospital in Shanghai and was going to give birth to a child. The birth should have been a happy event, but, unexpectedly, it was a difficult labor. Not only did Zhang Xiyuan have a hard time giving birth to the child but she contracted puerperal fever. Although she was in the hospital, medical conditions were very poor. Father, suffering the greatest anxiety, stayed in the hospital to be with his wife day and night. Unfortunately, Zhang Xiyuan passed away several days later.

After the baby was born, it was entrusted to the care of the family of Xu Bing and Zhang Xiaomei. The baby, a girl, died a few days later, too, probably because of the difficult labor.

I heard from Mom Deng Yingchao that Father was deeply grieved by the death of Zhang Xiyuan. However, even the greatest misfortune was personal misfortune at most, and even the greatest grief had to be buried deep in the heart. The military situation on the front was pressing and required urgent action.

After the unfortunate death of Zhang Xiyuan, Father had to stay in Shanghai for several more days. In late January he went back to Guangxi hurriedly, before he had had time to bury his wife. The Central Committee had already approved their plans, and the troops and comrades in Guangxi were waiting for him to give instructions and provide leadership.

When Father passed through Hong Kong once again, he made contact with the underground messenger of the Party there, Li Qiang, who was establishing a secret radio station in Hong Kong. Father asked him about how he could make radio contact with Shanghai when he was back in Guangxi. Li Qiang told Father about the call sign and other things. Li Qiang recalled that at that time they "also talked about entrusting the burial of his wife to my care. That was the first time I was acquainted with Comrade Xiaoping."

Li Qiang was a staff member of the Special Branch and was then in charge of burying some comrades of the Party. In the spring of 1930, Li Qiang returned to Shanghai and took charge of burying Zhang Xiyuan in accordance with the direction of the Military Commission. After Father was sent to work in Guangxi, he was under the leadership of the Military Commission. The secretary of the Military Commission was Zhou Enlai. Uncle Li Qiang told me:

> We buried Zhang Xiyuan in the cemetery at Jiangwan, Shanghai. The name used on her tombstone was Mrs. Zhang, nee Zhou, but the name used for cemetery registration was the true name, Zhang Xiyuan. Pseudonyms were mostly used for the burial of such comrades at that time. For example, the pseudonym used for Luo Yinong was Bi Jue, and that for Su Zhaozheng was Yao Weichang. Among those who took part in the funeral procession were Comrade Deng Yingchao and her mother, as well as a girl. After the burial, we held a memorial ceremony in accordance with the custom of the times. Later I got to know that the girl who went with us was Zhang Xiaomei, the younger sister of Zhang Xiyuan.

My second uncle, Deng Ken, recalled that he went to Shanghai for study in 1931 and in May found his elder brother there. Father led his younger brother to visit the grave of Zhang Xiyuan in the Jiangwan Cemetery. Deng Ken remembered that the name on the tombstone was not Father's real name. That was the requirement of their underground work.

After Shanghai was liberated in 1949, Father immediately made efforts to find the grave of Zhang Xiyuan. The graves of many martyrs could not be found as a result of the chaos caused by war and the construction of an airfield by the Japanese in the area around the cemetery. But thanks to the help of Li Qiang and his good memory, the grave of Zhang Xiyuan was found at last. When my parents went to see the grave, they found it was flooded. So Father had the bones of Zhang Xiyuan taken out of the grave and put into a small coffin. This coffin and another with the bones of Su Zhaozheng were placed downstairs in the house where Father lived in Shanghai. That was the house used by the Li Zhi Society (a club of senior KMT officials). Father soon left Shanghai to lead his troops to march southward and westward and into southwestern China until the last remnants of Chiang Kai-shek were driven out of the mainland of China.

The coffins of Zhang Xiyuan and Su Zhaozheng were stored at the site of the Li Zhi Society for a long time, and no consideration could be given to them at all when the Cultural Revolution broke out. But curiously enough, the bones of these revolutionary martyrs and others were buried at last in the Shanghai Cemetery of Revolutionary Martyrs in 1969.

When the Cultural Revolution was spreading, Father was overthrown as the "No. 2 Capitalist Roader." I presume that those in charge of the construction of the Shanghai Cemetery of Revolutionary Martyrs must have known nothing about Zhang Xiyuan and that they buried her coffin with that of Su Zhaozheng, as they were seen to be placed together. If they had known that this Zhang Xiyuan was the wife of Deng Xiaoping, they might not have buried her there. Furthermore, in their deep class hatred, they might have disposed of the bones of Zhang Xiyuan in such a way as to indicate their thorough criticism of Deng Xiaoping. Perhaps some supernatural forces protected the bones of Zhang Xiyuan during that crazy and chaotic period.

Today, the Shanghai Cemetery of Revolutionary Martyrs has been renamed Longhua Revolutionary Cemetery. On the plain tombstone of Zhang Xiyuan are engraved the characters "The Tomb of Martyr Zhang Xiyuan," and it is also inlaid with her photograph, taken when she was in Moscow. She is now lying serenely and quietly amid fresh, green pines and cypresses with Su Zhaozheng, Yang Xianjiang, Gu Zhenghong, and other martyrs. When we paid a visit to her grave, we presented quite a few fresh flowers in the hope that their beauty and splendor would accompany her in restful peace there.

Father had not yet reached the age of twenty-six in January 1930. Before going to Shanghai, he had thought he would have a reunion with his wife and greet his first child, but both his wife and child were dead when he left Shanghai. One can imagine how deeply Father was grieved by this unexpected and tremendous misfortune.

However, he could not be sunk in his personal grief, nor could he stay in Shanghai long enough to bury his wife and child. He had to hurry back at once to Guangxi. The urgent and volatile revolutionary situation in Guangxi was waiting for him! Many revolutionary comrades in Guangxi were waiting for the return of the representative of the Central Committee and their political commissar!

Communists also had emotion and tears. Countless Communists lost their kinfolk and comrades-in-arms, but they buried their emotions deep in their hearts and forced their tears to dry up, because in their minds the revolutionary interests were above all else.

These revolutionaries could give their lives and blood for the cause of liberation of the people and their motherland. What other sufferings could they not endure, and what difficulties could they not overcome? What forces could overwhelm and defeat them?

CHAPTER 31

THE RISE AND FALL OF THE 8TH CORPS OF THE RED ARMY

On February 7, 1930, Father returned to Longzhou, Guangxi Province, via Hong Kong and Vietnam. Before he left the territories of Vietnam, he saw from a distance the red flags flying high over the Zhennan Pass. He knew instantly that the Longzhou Uprising must have been launched and that the 8th Corps of the Red Army must have been established.

As soon as he arrived in Longzhou, he found that the troops of the 8th Corps had been sent to various counties to suppress bandits and local tyrants. Only Wan Danping, commander of the 2d Column, had stayed behind in the headquarters of the 8th Corps. Wan Danping reported to Deng Xiaoping in detail the work of the 8th Corps and the situation in Longzhou. Deng Xiaoping found that there were some problems in the work in the 8th Corps and in Zuojiang.

The Zuojiang Revolutionary Committee had been set up, but it did not actually operate and its political power was still unstable. The newly formed 8th Corps was small in number and equipped with only 1,000 rifles. The troops were entirely controlled by the officers of the old army, and not many CPC members led troops. The main force of the corps had left, and those who had stayed behind for garrison duty were the unreliable troops of the old army that had been incorporated into the corps. Meanwhile, since the Longzhou Uprising, the contradictions between the red political power and the reactionary forces had intensified, and the original mass base in the Zuojiang area had not been solidified.

The situation in the Zuojiang area was increasingly difficult. Father learned that right at that moment, the 7th Corps, led by Li Mingrui, commander-in-chief of the 7th and 8th corps, and Zhang Yunyi, commander of the 7th Corps, was marching toward Nanning. On the way, it fought a fierce battle with nearly four regiments of the Guangxi army in Long'an. At the same time, the 8th Corps, led by its commander Yu Zuoyu, was preparing, according to plan, to attack Nanning in coordination with the 7th Corps.

Father held that, judging from both subjective and objective conditions, the attack on Nanning was doomed to failure and the attacking troops were likely to be destroyed. According to the instruction of the Central Committee to not attack Nanning, Father immediately sent an urgent telegram to Li Mingrui and Zhang Yunyi, asking them to stop the attack on Nanning and, at the same time, asking Yu Zuoyu to lead his troops back to Longzhou. Receiving the instruction from Political Commissar Deng, Yu Zuoyu hurried back to Longzhou.

Soon after that, Father called a meeting of cadres from the Guangxi Military Commission and local Party committees. At the meeting, Father delivered several detailed speeches in line with his report to the Central Committee in Shanghai, pointing out that all aspects of the work in the Zuojiang area were failing to concentrate on the central task. Through the discussions at several meetings, participants decided on principles and policies regarding the agrarian revolution, expansion of the Red Army, and other questions. They also decided that the general objective for the 8th Corps was to join forces with the 7th Corps and then concentrate their strength to advance toward the border area between Hunan and Guangdong so as to join forces with the 4th Corps, commanded by Zhu De and Mao Zedong.[1]

Because the meeting clarified the Party's objectives, principles, and policies, significant success was soon achieved in the work in the Zuojiang area. The integration of the agrarian revolution with the movement against French imperialists (they were as close as Vietnam), in particular, received warm support from the people.

Political Commissar Deng told Yu Zuoyu to heighten vigilance and to move toward the 7th Corps in the Youjiang area when he was unable to gain a firm foothold. But because the 8th Corps relied entirely on the tax revenue from the Longzhou area for its military expenses, Yu Zuoyu hesitated to move close to the Youjiang area.

Soon Father learned that the 7th Corps had suffered a defeat in the battle at Long'an in the Youjiang area and withdrawn its main force from the area to an unknown place. The Guangxi clique, which thus returned to local military and administrative power, sent four regiments to invade Longzhou. The 8th Corps had realized that it was impossible to hold on to Longzhou, so it decided to capture the important town of Jingxi, between the Zuojiang and Youjiang rivers, so as to maintain its contact with the 7th Corps in the Youjiang area.

Having assigned tasks in Longzhou, Father hastily set off for the Youjiang area. On March 7, he arrived at Leiping, where the 1st Column of the 8th Corps was stationed. Zhou Zhi, an old soldier of the 8th Corps, recalled:

> Under the direct command of General Political Commissar Deng Bin (Xiaoping), our 1st Column of the 8th Corps of the Red Army, totaling over 1,000 men, left Longzhou for Jingxi. I was then an orderly in the 2d Battalion. During our march, Political Commissar Deng crossed over mountain after mountain and shared weal and woe with us. His straightforwardness, amiableness, and revolutionary optimism left a deep impression on us. During the journey, he inquired about the ideology and living conditions of the troops with great interest. He encouraged us to be good at fighting and doing propaganda work among the people and to persevere in the struggle no matter how difficult the situation might be. Political Commissar Deng's teaching greatly educated and inspired us.[2]

On March 11, the 1st Column surrounded the enemy in Jingxi. He Jiarong recalled:

> Political Commissar Deng came to the front and directed the battle together with me at our position outside the south gate [close to the present-day Jingxi Bridge]. Four days later, we were still unable to take the besieged town. As Political Commissar Deng could not stay in Jingxi for too long, I sent the 8th Company, led by its commander Tan Jin, to escort him across the Youjiang River. Before his departure, he instructed the commanders of the 1st Column to spare no effort to take Jingxi so as to remove the obstacle to the contact between the Zuojiang and Youjiang areas and at the same time, keep an eye on the developments in Longzhou. Political Commissar Deng, escorted by the soldiers of the 8th Company, arrived safely at the place where Comrade Wei Baqun was.[3]

While in Jingxi, Father learned that along the banks of the Youjiang River, Guohua was still controlled by the 7th Corps of the Red Army. So he left the 1st Column of the 8th Corps to go to the Youjiang area to convey the instructions of the Central Committee. Before he left, he sent a telegram instructing the troops in Longzhou to move rapidly toward the Youjiang area in accordance with the established principle of establishing liaison with the 7th Corps.

After Father left the 8th Corps, its 1st Column mounted repeated attacks but was unable to capture Jingxi. When it withdrew toward Longzhou, it fought a fierce battle with the enemy at Tieqiao, and more than 400 officers and men heroically laid down their lives.

At this moment, the enemy attacked Longzhou with massive numbers of troops. Outnumbered by the enemy, the 8th Corps of the Red Army abandoned Longzhou after putting up a brave resistance, and led by Corps Commander Yu Zuoyu, it retreated to Pingxiang. The enemy pursued it. The commander of the 2d Column and others laid down their lives, and Corps Commander Yu had only 700 men left. The morale of the 2d Column of the 8th Corps was shaken. This column suffered heavy losses after fighting several fierce battles with the enemy. Some political workers had to leave the column, and many Red Army and Peasants' Army soldiers died in action. Regiment Commander Liu Xiding betrayed the revolution, and as a result, the 2d Column of the 8th Corps was destroyed.

The occupation of Longzhou by the enemy marked the failure of the 8th Corps of the Red Army and the revolutionary regime in Longzhou. Yu Zuoyu, commander of the 8th Corps and a CPC member, went to Hong Kong to contact the Party organization. But unfortunately, he was betrayed by a traitor, arrested, and then escorted back to Guangzhou. On August 18, 1930, Corps Commander Yu Zuoyu was executed at Honghuagang, Guangzhou, on the order of Chen Jitang. Yu Zuoyu was only thirty.

After the 8th Corps was defeated, its 1st Column, which had attacked Jingxi and was under the leadership of Commander He Jiarong and Chief of Staff Yuan Zhenwu, a CPC member, made several attempts to establish contact with the 7th Corps of the Red Army. But it failed to do so because of strong converging attacks by the enemy. The 1st Column retreated to the border area of Guizhou but continued to fight.

This troop of the Red Army suffered innumerable hardships and fought in the border areas between Yunnan and Guangxi, and between Guizhou and Guangxi, for several months. After half a year, in September, the remaining 300 or so soldiers, led by Chief of Staff Yuan Zhenwu, eventually joined forces with the 7th Corps, commanded by Li Mingrui and Zhang Yunyi, in the Hechi area in Guangxi. When Yuan Zhenwu grasped Commander-in-Chief Li Mingrui's hands, the eyes of the two fighters brimmed over with warm, excited tears, as if they were kinsfolk meeting again after a long separation.

The remaining troops of the 1st Column of the 8th Corps of the Red Army were incorporated into the 7th Corps and participated in the reorganization at Pingma.

CHAPTER 32

THE RISE OF THE 7TH CORPS OF THE RED ARMY AND THE YOUJIANG RED REVOLUTIONARY BASE AREA

In March 1930 Father and a company of the 1st Column of the 8th Corps rushed to the Youjiang area from Jingxi. By then, the entire area along the Youjiang River had been occupied by the enemy, and the 7th Corps had retreated to Donglan. Father managed to break through the enemy's tight encirclement and arrived in Donglan in a roundabout way in late March.

Great changes had taken place in the Youjiang area since the Bose Uprising in December 1929, and the situation there was excellent. Despite the favorable revolutionary situation, however, the Front Committee, instead of focusing its work on in-depth mobilization of the masses to carry out agrarian reform, decided to attack Nanning and suffered a defeat on the way there, with heavy casualties. As they retreated to Pingma, the troops again suffered heavy casualties at the hands of the pursuing enemy. The main force of the 7th Corps broke away from the enemy's pursuit before entering the areas of Donglan and Fengshan for rest and reorganization in mid-February. About half a month later, the Front Committee decided to conduct guerrilla warfare outside their areas. The 1st and 2d columns, led by Corps Commander Zhang Yunyi, were to move to Hechi in the north, while the 3d Column, led by Wei Baqun, stayed in Donglan of the Youjiang area.

In early April the 1st and 2d columns arrived at Si'en (now Huanjiang). They suffered a minor setback when the Guangxi army launched a surprise attack on them; afterwards, they climbed over the Miaoshan Mountains and arrived in the Rongjiang area in Guizhou Province. Toward the end of April 1930, the 7th Corps took Rongjiang (then called Guzhou) and captured large quantities of weapons, ammunition, and other matériel, thus enhancing its morale. After a short rest and reorganization in Rongjiang, the 1st and 2d columns of the 7th Corps returned to the Youjiang area of Guangxi Province. One day in April 1930, Father arrived in the Wuzhuan district of Donglan County. There he found Huang Meilun, who was working in the County Women's Association.

Huang Meilun recalled that "it was drizzling. At dusk, a smart and vigorous young man, who was holding a walking stick, wearing a bamboo hat, having his trouser rolled up over his straw sandals, and followed by a Red Army soldier, came to my door." That was Deng Bin (Xiaoping), the political commissar.

Huang Meilun immediately led Political Commissar Deng to meet Wei Baqun. When the two men met, they greeted each other warmly. Wei Baqun "arranged for Political Commissar Deng to change his wet clothes. After supper, he sat beside the brazier, which was characteristic of the Zhuang nationality and began chatting for a long time." Early the next morning, Wei Baqun led Political Commissar Deng to the Kuixinglou Building.

The Kuixinglou Building, formerly a place for worshiping the god of literature, Kuixing, had now become the office building for the peasant association and the Soviet government. "After Political Commissar Deng came to Wuzhuan, Baqun put up a bamboo bed and an old square table on the second floor for him as his office and study. From then on, the light in the Kuixinglou Building was often on late into the night."[1]

Father had two tasks in Wuzhuan: trying to get in touch with the main force of the 7th Corps, which was marching north, and, together with Wei Baqun, investigating the agrarian revolution and experimenting with it in selected places.

Deng Xiaoping and Wei Baqun often called meetings attended by military and administrative cadres and key members of the Party leadership to discuss and formulate principles and policies concerning the agrarian revolution. He briefed the leading Party cadres in the 3d Column on the agrarian revolution conducted by the 4th Corps and asked them to hold serious discussions. He often went with Wei Baqun to the countryside to propagate the policies guiding the agrarian revolution and arrange for the work in that area. A lack of competent cadres made it difficult to do the work in the Youjiang area. So Father not only provided guidance on the specific work but also organized training classes for poor peasants and farm laborers. Jiang Maosheng, an old soldier of the 7th Corps, recalled: "In April 1930, Comrade Deng Xiaoping, secretary of the Front Committee, organized a training class for party members in Wuzhuan, Donglan County. He compiled teaching materials and gave lectures himself, explaining, in clear and simple language, the fundamental tenets of Marxism-Leninism and the Party's principles and policies."[2]

Father once said that some of the methods he applied to the agrarian revolution in the Youjiang area arose from the experience he had gained working on the Central Committee in Shanghai, specifically, from the reports of the 4th Corps, led by Mao Zedong and Zhu De, and from the oral reports made by the comrades from the 4th Corps to the Central Committee. Yuan Renyuan, Wei Guoqing, and others said in their recollections of this period that

> Comrade Deng Xiaoping, together with Comrade Wei Baqun and others who commanded the 3d Column, made great efforts to carry out the agrarian revolution in the base area of Donglan County. Comrade Deng Xiaoping briefed the cadres on the situation and experience in the agrarian revolution in the Jinggang Mountains and elaborated on the importance of the agrarian revolution to expanding the Red Army and consolidating and developing the worker-peasant democratic regime. Comrade Wei Baqun was a well-known leader of the peasants in Guangxi. Their work proceeded very smoothly.

Father stayed in the Kuixinglou Building in Wuzhuan for about two months. Toward the end of May, he decided to go to the Hechi area to look for the 7th Corps, which was believed to have moved there.

Ya Meiyuan, who was sent by Wei Baqun to escort Political Commissar Deng, recalled that, as they came to Deng's place, they saw the twenty-five-year-old Deng "dressed in a gray uniform, wearing a Red Army cap and a pair of sandals. His face was aglow, and his eyes very kind and intelligent."[3] In the early morning sunlight, Father and his party bid farewell to Wei Baqun and set out on horseback to look for the 7th Corps.

On the way, they climbed mountains and crossed rivers and had only solid food to eat and spring water to drink. They spurred their flying horses to full speed for the search. On the fourth day, they learned that an army unit that held a huge red banner bearing the hammer and sickle had arrived in Hechi. On the sixth day, Father hurried to Hechi and met with Li Mingrui and Zhang Yunyi at last.

Once in Hechi, Father called a meeting attended by the leaders of the 7th Corps to convey to them the directives of the Central Committee. They also discussed whether to lead the troops back to the Youjiang area.

In Hechi, Father also called a meeting attended by party members, at which they decided to move the troops back to the Youjiang area, to carry out the agrarian revolution, to transform the Red Army, and to take it as the general objective to expand the base area rapidly. They also decided that the 7th Corps should exploit the victories it had won on the way back from Guizhou, march toward Bose, and recover it without let-up. After the meeting, the morale of the 7th Corps was very high, and all the officers and men were ready to set off.

When he returned to Guangxi, Father did one more important thing: he admitted Li Mingrui into the Communist Party, with the approval of the Central Committee and on Li's own request. As soon as Father arrived in Longzhou, the decision was announced. Thus, Li Mingrui had turned from an old-style commander with patriotic and democratic ideas into a revolutionary fighter who had firm faith in communism.

Li Mingrui's transformation was extraordinary as well as rather common. In the tortuous course of Chinese history, many people had gone through struggles, setbacks, hesitation, and perplexity before awakening and choosing the road to truth—the only road that could save the Chinese people from the hell of their sufferings. Only the revolution led by the Communist Party of China could truly break away from the influence of the warlords and foreign forces, could be carried out for the real benefit of China and the Chinese people, and could be called a genuine people's revolution. Many people like Li Mingrui took the revolutionary road delineated by the Communist Party of China. Zhu De, Peng Dehuai, He Long, Ye Jianying, and Liu Bocheng—the future marshals of the People's Republic of China, all traversed this course, as did countless others from the day the Chinese Workers' and Peasants' Red Army was established to the time of the three-year War of Liberation. They joined the revolutionary ranks neither for high positions nor for handsome pay. So what did they strive for? They knew only that they had found the truth. They were proud to devote their lives to the future destiny of China and the 400 million Chinese people. In so doing, they became heroes of the revolution and of the people.

Although he had known Li Mingrui for hardly one year, Father established a true friendship and trust with him. Many people distrusted Li Mingrui because of his background as an old-style military officer. The "leftist" Central Committee, led by Li Lisan, had even gone so far as to ask Party members repeatedly and clearly to have no illusions about Li Mingrui. It issued a clear directive on "being firmly opposed to admitting him into the Party" and even gave an order to "drive him away from the base area."[4]

Because of poor communication between Guangxi and the Central Committee, Father was not aware of its attitude. Even if he had known of it, however, Father would have withstood the pressure and firmly, warmly, sincerely, and courageously welcomed Li Mingrui into the revolutionary ranks, because he had the fullest understanding of the situation in Guangxi and of Li Mingrui. He knew that Li Mingrui needed the revolution, and the revolution needed Li Mingrui.

Father has always been like that. As a Party member, he has a very strong sense of organization and discipline, and at the same time, he never gives in to mistakes. He dares to uphold the truth, to seek it from the facts, and to assume responsibility. In early June 1930, the 7th Corps, led by Commander-in-Chief Li Mingrui, Political Commissar Deng Xiaoping, and Corps Commander Zhang Yunyi, advanced toward Bose. On June 8, its 1st and 2d columns launched an attack on Bose and succeeded in recovering it.

After Bose was recovered, the Front Committee decided to exploit the victory. The 7th Corps carried on fighting and took the county towns along the Youjiang River, including Fengyi, Enlong, Silin, and Guode, thus recovering the entire lost Soviet areas along the Youjiang River. While the 7th Corps won this series of military victories, it continued to conduct the agrarian revolution, expand the mass base, and consolidate the red revolutionary base area in the Youjiang area. "A thriving scene prevailed in the entire Youjiang area."[5]

In early April the major war of the Central Plains between Feng Yuxiang, Yan Xishan, Li Zongren, and Zhang Xueliang on the one hand and Chiang Kai-shek on the other started. The Guangxi clique was on the anti–Chiang Kai-shek side. Li Zongren, Bai Chongxi, and Huang Shaohong led the main forces of the Guangxi army up into Hunan Province so as to join the decisive battle on the Central Plains. It was when the warlords were involved in internecine war and the main forces of the Guangxi army had moved north that the 7th Corps was able to achieve these victories and consolidated the Youjiang Revolutionary Base Area.

However, the raging red flames in the Youjiang area had long perturbed

Chiang Kai-shek. In early July, when the situation on the battlefields of the Central Plains had just begun to be favorable to Chiang Kai-shek, his government in Nanjing immediately ordered the Yunnan army commanded by Long Yun to attack Nanning along the Zuojiang and Youjiang rivers via Longzhou and Bose. This scheme was designed to kill two birds with one stone and to reap the spoils of the contest fought by others.

The attack on Guangxi mounted by the Yunnan army was designed to assault the rear of the Guangxi army, which was moving north, so as to contain its expedition against Chiang Kai-shek and at the same time to attack the red political power in the Youjiang area. Sending the Yunnan army into Guangxi to attack both the Guangxi army and the Communist Party was a well-devised, cunning scheme that, by playing off one power against another, would not diminish Chiang Kai-shek's strength. In view of this situation, the 7th Corps withdrew from Bose and retreated to Silin.

The Yunnan army was well known for its bravery and combat effectiveness. The three divisions led by Commander-in-Chief Lu Han marched into Guangxi, with Division Commander Zhang Chong leading his division to spearhead the attack. The powerful, 20,000-strong Yunnan army advanced menacingly. As soon as they arrived in the Youjiang area, they took Bose and became all the more arrogant. But to their surprise, on the way to Nanning, Gen. Zhang Chong's troops were ambushed by the 7th Corps at Guohua near Pingma. A fierce battle broke out, and between 500 and 600 Yunnan troops were killed or wounded. This battle dealt a heavy blow to the truculent Yunnan army, which did not dare to attack the red areas anymore.

In 1986 we accompanied Father on his trip to Guilin, Guangxi Province. On an excursion on the Lijiang River, Father mentioned this ambush of the Yunnan army when he was recalling his combat experience in Guangxi:

> The Yunnan troops were good at fighting calmly. But each soldier had two guns—a rifle plus a pipe for opium smoking. Opium smokers couldn't walk for a long distance, so the Yunnan troops were only good at fighting defensive battles. I fought a battle against Zhang Chong near Pingma east of Bose. Zhang Chong was a combat commander. The Yunnan troops actually intended to attack Guangxi with only three divisions! Later Zhang Chong joined the revolutionary ranks. He belonged to the Yi nationality. He died after liberation.

After fighting the Yunnan troops, the 7th Corps also suffered over 200 casualties, so it moved to Pingma for reorganization and training.

From the very beginning of the founding of the 7th Corps, importance was attached to transforming the troops, recruiting Party members, and establishing Party organizations. Meanwhile, soldiers' committees were set up and the warlord style eradicated in the army units.

While conducting political, ideological, organizational, and military education and training, the 7th Corps cooperated with the local Party committees to carry on the agrarian revolution. Because of that revolution, the poor peasants were delighted and inspired and showed enthusiasm for sending one group of young peasants after another to join the army. As poor young peasants became new recruits in the army, the composition of the 7th Corps changed greatly, bringing about a new general attitude.

The 7th Corps stayed in the Youjiang area for about three and a half months. While undergoing reorganization and training, it also fought the armed forces of local tyrants and bandits. Father said that hardly a day passed without military operations during that period.

To safeguard the fruits of the agrarian revolution and protect the autumn harvest, the 7th Corps decided to wind up its reorganization and training and move to the Hechi area in early October. Mo Wenhua, an old soldier of the 7th Corps, recalled:

> When the reorganization and training ended, the agrarian revolution in the base area had been basically accomplished successfully. Politically, the poor peasants had become masters of their own fate and had stood up economically, thus greatly arousing their enthusiasm for revolution and production. Many emancipated peasants asked to join the Red Army or the Red Guards. Several thousand peasants signed up for joining the Red Army, thus expanding it from three to four columns, with a total of 8,000 men. The Red Army was then well trained and powerful and was fully prepared to fight new battles and open up new prospects.[6]

In Father's words, it was the "heyday of the 7th Corps of the Red Army."

In September 1930, after six months of fighting from one place to another, the remaining troops of the 8th Corps, led by Yuan Zhenwu, eventually arrived in Hechi and successfully joined forces with the 7th Corps. From then on, the 7th and the 8th corps were combined into a valiant armed force in the revolution in Guangxi and started a new journey.

After Father reported on the work in Guangxi to the Central Committee in Shanghai in February 1930 and left to return to Guangxi, for various reasons the Central Committee lost contact with Guangxi. On June 16, the committee sent a letter of instruction to the Southern Office of its Military Commission and to the Front Committee of the 7th Corps:

> With respect to the 7th Corps, the Central Committee has not received any report since Comrade Xiaoping returned to the 7th Corps. The Central Committee learned from some comrades who had come to Shanghai to report work about the defeat suffered by the troops in Longzhou since they withdrew from Longzhou and Bose, but heard nothing about the whereabouts of the main forces after they withdrew from Bose. Some foreign newspapers in Shanghai recently reported that they were now in the vicinity of Liuzhou, but no detailed information was available either![7]

In this letter, the Central Committee reiterated the viewpoint of the Li Lisan line: that "it is most likely that the world revolution will begin in China," and demanding that "victory in the revolution be won first in one or several major provinces," that in southern China efforts be made to achieve victory in Guangdong, and that "resolute actions be taken to attack Liuzhou and Guilin held by the enemy, with a view to advancing to the Xijiang and Beijiang areas in Guangdong." The letter also criticized the Front Committee of the 7th Corps for its approach to Li Mingrui and Yu Zuoyu, noting that it was "a very erroneous and dangerous case" that was "contravening the correct directive of the Central Committee, resulting in the opportunistic defeat!"

We can see from the letter that the Central Committee in Shanghai, dominated by Li Lisan, was not only unhappy about some practices of the Front Committee of the 7th Corps but very worried about having lost contact with it. To make sure that its "leftist" policies were implemented in Guangxi, the Central Committee sent a special representative to Guangxi, Deng Gang, to guide the work there.

On September 30, Deng Gang, representative of the Southern Bureau of the Central Committee, came to the 7th Corps of the Red Army. On October 2, the Front Committee of the 7th Corps held a meeting at Pingma at which Deng Gang briefed the participants on the guidelines of the Central Committee. The Front Committee of the 7th Corps decided to reorganize its four columns into three divisions, of which the 19th and 20th divisions were to move north to Hechi, where a military parade of the whole corps would be held for the representatives of China's Soviets so as to boost morale and where a conference for all Party members would be convened. The 21st Division, commanded by Wei Baqun, was to remain in the Youjiang Base Area to persist in the struggle there.

On October 4, the main force of the 7th Corps, 7,000 strong, marched north in a mighty formation toward the Hechi area in the Guangxi-Guizhou border region. These Red Army troops were full of vigor and high in morale, but they did not know that their political commissar and the secretary of their Front Committee, Deng Xiaoping, was very worried.

Deng Gang had relayed the new guidelines of the Central Committee, that is, that a new revolutionary high tide had come and victory must be won first in one or several provinces so as to proceed to establish national revolutionary political power. The 7th Corps was ordered to attack Liuzhou and Guilin, eventually to take Guangzhou, and to capture Wuhan in cooperation with the 3rd Corps. The objective was to "join forces in Wuhan and to water the horses in the Yangtze River." In addition to making arrangements for military operations, Deng Gang also conveyed the Central Committee's criticism of the agrarian reform policies implemented by the Youjiang Special Party Committee in the Youjiang Base Area of Guangxi Province, stating that those policies followed a right-wing rich-peasant line.

The Central Committee's new strategic military arrangements and its criticism of agrarian reform policies in the area concerned Father greatly. His concern was based in his different view of the revolutionary situation and the revolutionary tasks at hand. Moreover, he was worried about the fate of the more than 7,000 soldiers of the 7th Corps.

Victory in one or several provinces first, then establishment of national revolutionary political power and the arrival of a revolutionary upsurge—this was Li Lisan's strategic concept. How did it take shape and develop? A detailed explanation needs its own chapter.

CHAPTER 33

THE ORIGIN OF LI LISAN'S "LEFT"-ADVENTURISM

During one of my interviews with veteran revolutionaries, an old revolutionary grandma sighed after telling me so much about the history of the Party: "At that time, the Party was still in its infancy, and it vacillated now to the left and now to the right!"

That is quite true. A human being has to go through the ignorance of infancy before he becomes an adult. Likewise, a party necessarily grows from small to big, from weak to strong, from vacillation between the left and the right in its infancy to a firm, strong, mature position.

From 1921, when the Communist Party of China was founded, to 1935, when the Zunyi Meeting was held, the Party went through several historical periods characterized by the leadership of, among others, Chen Duxiu, Qu Qiubai, Li Lisan, and Wang Ming. In the course of the fourteen years of struggle, the Communist Party and Chinese Communists, through heroic and indomitable fighting, achieved great and brilliant revolutionary results. However, because of repeated changes in the Party's lines and policies, the mistakes that some Party leaders consistently made in their decision-making and understanding, and the improper influence of the Communist International, the Chinese revolution experienced many twists, turns, and dangers and suffered deplorable losses.

Father once said that the Party never formed a genuine nucleus of leadership before the Zunyi Meeting.

Between 1921 and 1934 the ranks of the Communist Party steadily swelled. It founded red armed forces and established red revolutionary regimes. It also suffered the mistakes and setbacks caused by Chen Duxiu's right-capitulationism, Qu Qiubai's "leftist" putschism, Li Lisan's "left"-adventurism, and Wang Ming's "leftist" opportunism.

From July 1921, when the CPC was founded, its membership increased from some 50 to approximately 60,000 at the time of the Great Revolution. In 1927 the Great Revolution ended in failure because of the betrayal of the revolution by the right-wingers of the KMT and the erroneous right-capitulationist line followed by Chen Duxiu, the general secretary of the CPC. At that time, the Party membership dropped sharply to a little more than 10,000. Chen Duxiu's right-capitulationist line was discarded, and the CPC reshuffled its Central Committee. It fought doggedly under the harsh conditions of the White terror. It rapidly and effectively restored Party organizations to conduct its work. By the Sixth National Congress in June 1928, its membership had grown to over 40,000.

Under the leadership of the CPC, many armed uprisings were launched. In those uprisings, red flags were hoisted openly, the Chinese Workers' and Peasants' Red Army led by the CPC was founded, and the Soviet red regimes and red revolutionary base areas were established. While waging the revolutionary armed struggle, Chinese Communists launched the agrarian reform movement in the red areas.

The emergency meeting of the Central Committee held in Wuhan on August 7, 1927, corrected in time Chen Duxiu's right-capitulationism and established the guiding principle on agrarian revolution and red armed struggle. However, when the Central Committee opposed the right-deviationist mistakes, it opened the way for the "leftist" mistakes.

In early November 1927, the Central Committee held an enlarged meeting of the Provisional Political Bureau presided over by Qu Qiubai. The meeting held that the revolution at that time was still on the rise. The Party should therefore oppose retreat and continue to attack. Its general tactic was carrying out armed uprisings and integrating rural and urban uprisings.[1]

At the enlarged meeting of the Central Committee, the "leftist" putschism represented by Qu Qiubai dominated the leading bodies of the Central Committee and received the firm support of B. Lominadze, the representative of the Communist International. Under the leadership of wrong "leftist" putschism, the main task of the Central Committee was implementing the general tactic of staging a general uprising throughout the country. The Central Committee organized one uprising after another in the cities, including Yixing, Wuxi, Shanghai, Wuhan, and Changsha. Because of the lack of a mass base, each of those uprisings failed, causing severe damage and losses to Party organizations and revolutionary forces. Qu Qiubai's wrong guiding principles met with criticism and resistance from many comrades in the Party. The Communist International also criticized the CPC and its own representative who made the mistakes. As a result, Qu Qiubai's "leftist" putschism came to an end in less than six months, lasting only from November 1927 to April 1928.

To steer clear of the dangerous conditions under the White terror in China, the Communist Party of China held its Sixth National Congress in Moscow between June and July 1928. The Sixth National Congress made basically correct policies. It correctly summed up the experience and lessons of the revolutionary work, opposed both right and "left" tendencies, correctly analyzed the nature of Chinese society and the Chinese revolution, and correctly assessed the political situation in the Chinese revolution. It also decided on the various tasks of the Party. However, the congress lacked a correct understanding of the protracted nature of the Chinese revolution and the importance of the rural base areas. Since it did not correct the "leftist" mistakes once and for all, the possibility of making "leftist" mistakes in the future remained.

The Sixth National Congress elected a new Central Committee and Political Bureau. Under the influence of the Communist International, the congress placed excessive emphasis on the composition of workers in the Party leadership. As a result, Xiang Zhongfa, a man of working-class origin who was incompetent for the job, was elected chairman of the Standing Committee of the Central Committee (equivalent to the general secretary).

After the Sixth National Congress, Party organizations became less disorganized and expanded further. By September 1930 CPC membership had increased to over 122,300. The work in the White area underwent rectification and development. The Party's work in some key cities, including Wuhan and Guangzhou, was restored. The Party further extended its influence among the masses, and the membership of the Red Trade Union grew to over 100,000.

By mid-1930 the Workers' and Peasants' Red Army, under the leadership of the CPC, had expanded rapidly and established and consolidated revolutionary base areas in the countryside. The regular Red Army troops throughout the country had more than ten corps and totaled some 100,000 men, including the local armed forces. They had established a dozen revolutionary base areas in the countryside.

The warlords set up separate regimes and were involved in internecine warfare among themselves, so they had no time to attend to other matters. This gave the CPC an opportunity to expand the red revolutionary base areas and the Chinese Workers' and Peasants' Red Army throughout the country, especially in the border areas of the southern provinces, where the warlords' strength was weak. A single spark was going to start a prairie fire, which would soon light up the vast land of China.

As soon as the revolutionary regimes were set up, the agrarian revolution was conducted. China at that time was a poor and backward agricultural country; its national industry was rather weak. Therefore, to win victory in the revolution, it was imperative to first emancipate the tens of millions of poor peasants and to mobilize them to form the backbone of a massive revolutionary army. In effect, the Chinese revolution was first and foremost a peasant revolution. After the poor peasants got the land and gained their personal freedom, their revolutionary enthusiasm was considerably aroused. They supported the red revolutionary regime by vying with one another to join the Red Army. The excellent revolutionary situation in the base areas greatly strengthened the Red Army and consolidated those base areas.

When the revolutionary situation developed and the revolutionary forces expanded, some Central Committee leaders became conceited. The "leftist" ideas that had not been completely eliminated in the Party began to prevail again. Moreover, those ideas developed further into "left"-adventurism.

In June 1930 the Political Bureau of the Central Committee held a meeting that was presided over by Xiang Zhongfa. With Li Lisan playing the predominant role, the meeting adopted "The Resolution on the Current Political Tasks," drafted by Li Lisan. This resolution overestimated the situation and strength of the Chinese revolution, denied the unevenness of the development of the Chinese revolution, insisted on the "theory of making cities the center of the Chinese revolution," and put forward the "Party's general tactical line," that is, to start "a great workers' struggle" to push the revolution immediately to an upsurge, to stage armed uprisings so as to win "victories first in one or several provinces" around Wuhan, and further, to establish a national revolutionary regime. The "left"-adventurism represented by Li Lisan was thus born.

Taking the "left"-adventurist line, the Central Committee made a series of plans for armed uprisings in key cities, with Wuhan as the center, and ordered the Workers' and Peasants' Red Army to attack those key cities. The order of the Central Committee was: the 3d Corps was to attack Wuhan; the 1st Corps was to attack Nanchang and Jiujiang so as to seize the whole of Jiangxi Province; the 2d Corps was to coordinate the attack on Wuhan and Changsha; the 1st Corps was to cut off the Beiping-Hankou Railway and to approach Wuhan; the 10th Corps was to attack Jiujiang; and the 7th Corps was to attack Liuzhou and Guilin and finally seize Guangzhou. The ultimate objective was to realize the adventurist plan of "joining forces in Wuhan and watering horses in the Yangtze River." At the same time, plans for staging general strikes and armed uprisings in the major cities in the White area were drawn up.

Those comrades in the Party who held different views were suppressed and criticized. Some were even punished by expulsion from the Central Committee and dismissal from their posts. These practices led to the growth of factionalism in the Party.

To guarantee effective implementation of the general tactics of Li Lisan's adventurism, the Central Committee sent many special representatives to all Soviet areas and Red Army units to direct and supervise the work. The representative sent to the 7th Corps of the Red Army in Guangxi was Deng Gang.

CHAPTER 34

THE EXPERIENCE OF THE 7TH CORPS

In the autumn of 1930 more than 7,000 fighters of the 7th Corps of the Red Army, led by Commander-in-Chief Li Mingrui, Political Commissar Deng Xiaoping, and Corps Commander Zhang Yunyi, arrived at the Hechi area in the Guizhou-Guangxi border region in northern Guangxi.

On October 10, the Front Committee of the 7th Corps held a Party congress of the corps. At the congress, Deng Gang, representative of the Southern Bureau of the Central Committee, insisted on implementing the instructions of the Central Committee and demanded that the 7th Corps first attack Liuzhou. Among 7th Corps personnel, both Chief of Staff Gong Chu (Hecun) and the director of the Political Department, Chen Haoren, strongly supported this view. Father clearly expressed his disagreement.

Father once said that he was very excited to hear the view of the Central Committee, conveyed by its representative, that there was a national revolutionary upsurge. But he was also soberly aware that Li Zongren and Bai Chongxi of the Guangxi clique had restored their rule in Guangxi. Although the 7th Corps had established base areas in the Youjiang area and expanded its forces, it had only several thousand men. It was difficult for the 7th Corps to capture Bose. It had no guarantee of victory in capturing any larger cities, such as Guilin, Liuzhou, or even Guangzhou.

But the Central Committee ordered the 7th Corps to attack Liuzhou, so it had to carry out the order, with resolution. What was to be done? Father exchanged views with some comrades and found that a consensus had not been reached on the question. Some comrades did not believe such a plan was feasible; others did not express their views at all. In the end, Father suggested at the meeting that since it was difficult to capture Liuzhou in the southeast from Hechi with a wide river in between, it would be better to attack Guilin in the northeast first before capturing Liuzhou. The suggestion was approved by the majority of the delegates. Although Father and other comrades expressed different views, the Hechi meeting dominated by Deng Gang accepted the Li Lisan line in its entirety and defined the tasks of the 7th Corps as to take Liuzhou, Guilin, and Guangzhou, so as to accomplish the revolution in the south. The meeting also criticized the previous "mistakes" made by the Front Committee of the 7th Corps.

On November 5, all the officers and men of the 7th Corps held a military parade in Hechi. More than 7,000 Red Army men, in high spirits and full of vitality, were reviewed. On November 9, the 7th Corps got ready and advanced eastward.

When this proud and enthusiastic revolutionary force was marching to the front to implement the "leftist" resolutions of the Central Committee, it was not aware that the Third Plenary Session of the Party's Sixth Central Committee had been held in Shanghai at the end of September, that is, two months before the 7th Corps set off and not long after Deng Gang had arrived in Guangxi to implement the "leftist" line of the Central Committee. At that plenary session, the "left"-adventurist mistakes made by Li Lisan and others were criticized and corrected. The convocation of the meeting stopped many erroneous adventurist actions that were still under way and lessened in time the losses caused by "left"-adventurism. It was indeed a crucial meeting for the Chinese revolution.

Nevertheless, the 7th Corps, far away in southwestern China and cut off by one mountain range after another, did not know anything about this crucial Party meeting and the important change in the principles and policies of the Central Committee. It set off to carry out the instructions of the former "leftist" Central Committee—two months after the committee had corrected its "leftist" mistakes.

Dressed in army uniforms, wearing army caps with a red star, and holding high the red flag, they valiantly set out. They did not know that a series of unwinnable battles were waiting for them. They did not know that they were going to encounter untold hardships along the way. They were not being deliberately mocked by history. History is history, and no one can rewrite it.

On the second day after it set off, the 7th Corps captured a small town, Huaiyuan.

There were two different opinions in the Front Committee of the 7th Corps. Some people proposed to cross the river to attack Qingyuan in the south, while Political Commissar Deng Xiaoping and Commander-in-Chief Li Mingrui held that, as Qingyuan was a strategic town of the enemy, it would be difficult to take it, so they should proceed to Dukou in the east at once.

After the 7th Corps abandoned its plan to attack Qingyuan, it skirmished with the enemy in Siba before coming to a stalemate with the enemy for three days near Tianhe. By now, it had been several days since the 7th Corps left Hechi, and it had moved forward only fifty or sixty kilometers, won some battles, and suffered defeat in others. The 7th Corps turned north and reached Sanfang.

Sanfang was located in the Damiaoshan Mountains and was free from enemy harassment. Because it was raining, the 7th Corps had a rest for a few days. It was at this time that Deng Gang, Gong Chu, Chen Haoren, and others accused Deng Xiaoping of violating the order of the Central Committee and insisted on their plan to attack Liuzhou first. A meeting of cadres at and above the battalion level was held in Sanfang to discuss the matter. A heated argument followed, and in the end most of the participants expressed their obedience to the instructions of the representative of the Southern Bureau of the Central Committee. Deng Xiaoping felt deeply isolated and thus offered his resignation from the post of secretary of the Front Committee and proposed that the representative of the Central Committee or someone else take his post.

But Deng Gang, Gong Chu, and Chen Haoren objected to this proposal, and Deng Xiaoping had to comply with the decision of the meeting.

In the long and hard journey, arguments never ceased within the Front Committee of the 7th Corps, and the differences were never ironed out. In Father's words, they quarreled every day and all the way. Soon after the meeting in Sanfang, the 7th Corps moved southeast and prepared to attack Liuzhou.

While advancing on Chang'an by the Rongjiang River halfway to Liuzhou, the 7th Corps discovered that Chang'an was guarded by two enemy divisions. On December 15, the main force of the 7th Corps, led by Li Mingrui, launched an attack on the town of Chang'an. But they were unable to storm the town because of the enemy's strong fortifications and heavy fire. The battle lasted for five full days, and the 7th Corps suffered several hundred casualties. The 7th Corps decided to pull out of the battle.

After it withdrew from Chang'an, the 7th Corps was compelled by reality to give up its plan to attack Liuzhou. However, Deng Gang and his followers abandoned neither the Li Lisan line nor their plan to take Guilin. Owing to the enemy's defense preparations, the 7th Corps could move nowhere but north, so as to maneuver back to Guangxi from Hunan to attack Guilin. So the 7th Corps moved straight north, crossed over mountain after mountain, and captured the small town of Suining in southwestern Hunan on December 21.

The news of the 7th Corps' entry into Hunan and occupation of Suining was soon reported in the enemy's newspapers and in the Communist Party's official organ, *Red Flag*. But the 7th Corps read none of these news reports. Nor did it know anything about the military deployments of the enemy. After a short rest in Suining, the 7th Corps advanced northeast again. Since it had left Chang'an, the 7th Corps had trudged more than 200 kilometers to reach the strategic town of Wugang in the southwestern border region of Hunan.

The 7th Corps originally had planned only to raise some funds in the Wugang area, not to capture Wugang. It learned, however, that only a local armed corps, not regular troops, guarded the town. So it decided immediately to take Wugang. To the surprise of the 7th Corps, it was unable to capture the town after several days of repeated attacks. By the fourth day, it had suffered considerable casualties and continuing the attack was becoming untenable. When it was about to make readjustments, the Hunan enemy reinforcements arrived unexpectedly. So the attempt by the 7th Corps to capture Wugang failed, and it had to withdraw.

During the battle of Wugang, more than 200 men of the 7th Corps were killed or wounded. Not only was troop morale lowered but quite a few got lost in the retreat. The 7th Corps reassembled its troops and decided to move southeast immediately. Eventually, it returned safely to Guangxi and arrived at the town of Quanzhou in the border region between Hunan and Guangxi.

The leading cadres of the 7th Corps held a meeting in Quanzhou to discuss the future of the corps. Since it started off from Hechi in early October, in only two months the strength of the 7th Corps had dropped from more than 7,000 to somewhere between 3,000 and 4,000. Defeatism prevailed among the troops, and many deserted. It was already winter, and the troops were still short of food and clothing. Moreover, the Hunan enemy forces in the north were eying their prey covetously, while the Guangxi enemy troops in the south were fully prepared to wait for the 7th Corps. It was by no means easy to attack Liuzhou and Guilin and then to "water their horses in the Yangtze River" under such difficult conditions. Grim reality forced the 7th Corps to thoroughly abandon the "left"-adventurism inspired by Li Lisan, and the plan to take Liuzhou and Guilin was not mentioned again.

After the meeting, Deng Gang asked to return to Shanghai and report on the work to the Central Committee. He then left the 7th Corps. Soon after, his supporter Chen Haoren, director of the corps' Political Department, also left. Father said that Chen Haoren left quietly after a battle without telling anyone, and no one knew where he had gone. After the Li Lisan line lost its influence in the 7th Corps, Deng Xiaoping, Li Mingrui, and Zhang Yunyi regained command of the corps.

Deng Gang and Chen Haoren had an unshirkable responsibility to implement in the 7th Corps the adventurist directive of Li Lisan's Central Committee—as well as responsibility for the series of defeats the corps suffered. They did not formulate the erroneous line, they only resolutely carried it out. Therefore, although they could be held responsible for the setbacks of the 7th Corps, the main responsibility lay with the Central Committee.

After the meeting in Quanzhou, the 7th Corps felt an urgent need for rest and replenishment. So it stayed there for three days and raised some pocket-money funds for the officers and men. Then, having learned that the Guangxi enemy troops were advancing toward Quanzhou, the 7th Corps decided to move southeast and enter the area of southern Hunan.

After capturing Daozhou, the 7th Corps organized a mass rally to educate the masses of the people and to heighten its morale. The 7th Corps had stayed in Daozhou for barely two days when it learned that the Hunan enemy troops were approaching from three directions. So it had to advance southward to Jianghua in the Hunan border region. It was midwinter and exceptionally cold all the way, with heavy snowfalls and a chilly wind. The troops plodded along on their forty-five-kilometer journey. They lacked clothing and pay, and some wore only unlined clothes and sandals or even shorts. They were miserably cold and hungry. In one day quite a few Red Army soldiers died of cold.

When they arrived at Jianghua, they found that, with no Party organization or mass base, the local conditions were even worse than in Daozhou. So they had to move again toward Lianzhou, Guangdong. Once more they crossed over the Laomiaoshan Mountains in the border region of the three provinces of Hunan, Guangxi, and Guangdong and fought the armed forces of the landlords before eventually arriving at the Guiling Mountains in the Guangxi border region in mid-January 1931.

The 7th Corps stayed in the Guiling area for four days of rest and reorganization. By now the 7th Corps was down to less than 4,000 men. After consulting with Li Mingrui and Zhang Yunyi, Deng Xiaoping decided to reorganize the troops by changing the two former divisions into two regiments. To boost army morale, the chief commanders were to serve concurrently as regiment commanders. Thus, the former 19th Division was changed into the 55th Regiment with two battalions each 1,200 strong. Gong Chu was its commander, and Deng Xiaoping was concurrently its political commissar. The former 20th Division was changed into the 58th Regiment with Commander-in-Chief Li Mingrui as its commander. It had two battalions totaling over 1,300 men. The corps also had units of more than 800 men under its direct command. It had six heavy machine guns and three mortars.

With such a reduced establishment, the Corps was well organized. The corps commanders became regiment commanders, the former regiment commanders became battalion commanders, and battalion commanders became company commanders. So the corps had enough officers and restored to a certain extent its combat effectiveness. It also got some food and clothing, heightening the morale of the troops.

The Guiling area was located in the important juncture of the three provinces of Guangxi, Hunan, and Guangdong, and the enemy's situation was rather complex. The 7th Corps decided to move to the Lianzhou area of Guangdong Province.

Not long after its departure, the 7th Corps crossed the border between Guangxi and Guangdong. When it arrived at Dongpi, some thirty kilometers away from the city of Lianzhou, on January 19, it discussed whether to attack Lianzhou. After weighing the advantages and disadvantages, it decided not to attack Lianzhou. Instead it decided to move north into Hunan, as suggested by Gong Chu.

The 7th Corps learned on January 17, 1931, when it arrived in Xingzi, forty kilometers away from Lianzhou, that 1,000 Hunan enemy troops had organized a defense on the route to Hunan, so the corps turned back to Lianzhou.

When the 7th Corps arrived, the enemy in Lianzhou was extremely frightened and set fire to the city in an attempt to stop the advance of the 7th Corps. On hearing the news, the 7th Corps rushed into the city to put out the fire, thus preventing the city from being completely destroyed. The Red Army's courageous efforts to put out the fire greatly moved and educated the residents of Lianzhou. Residents and businessmen raised funds and grain for the 7th Corps of their own accord and also arranged for medical treatment for some 100 wounded fighters of the Red Army.

While in Lianzhou, the 7th Corps still wanted, according to its general plan, to go up north into southern Hunan so as to gain a firm foothold there, establish the Soviets, carry out agrarian revolution, and make new Red Army recruits. So it once again turned north. Toward the end of January, the 7th Corps passed through Xingzi, Guangdong Province, and arrived at Meihua Village, Ruyuan County, Guangdong Province, on January 30. It was still midwinter, the season when wintersweet was in blossom.

Right after the corps was settled at Meihua Village, a messenger sent by the Lechang County Party Committee in Hunan brought to the corps the circular of the Third Plenary Session of the Party's Sixth Central Committee and other documents. It was only after reading these documents that Father and the Front Committee of the 7th Corps learned that the Central Committee had repudiated "left"-adventurism and put an end to the Li Lisan line as early as September 1930, when the 7th Corps was still in Hechi![1] Between September 1930 and January 1931, the 7th Corps had plodded hundreds of kilometers and fought one battle after another. It lost not only its revolutionary base area but also two-thirds of its strength! The enormous responsibility, the great losses, and the grave mistakes could not be more shocking in retrospect.

What was lost was lost, and the 7th Corps had no time to retrieve the losses. It had been at Meihua Village for only a few days when it learned, on February 3, that one regiment of Guangdong enemy troops was coming from Xingzi. The Front Committee of the 7th Corps, considering it an excellent opportunity to wipe out the enemy since only one regiment would be engaged, prepared the troops for battle.

As soon as the battle began, the 7th Corps found that the enemy strength was by no means only one regiment but three, two of which had come from Lechang. Incorrect information led to incorrect decisions, and incorrect decisions led to defeat.

In the fierce and brutal five-hour battle, the 7th Corps wiped out more than 1,000 enemy troops. At dusk, it pulled out of the battle because it also had suffered heavy casualties: more than 200 men were wounded; the commander of the 20th Division, Li Qian, and the commander of the 55th Regiment, Zhang Jian, died heroically; and the corps chief-of-staff, Gong Chu, the commander of the 59th Regiment, Yuan Zhenwu, and a battalion commander of the 58th Regiment, Li Xian, were wounded. With over half of its cadres killed or wounded, the 7th Corps now had only 2,000 men left.

The battle at Meihua has remained deeply fixed in Father's memory. When he went to Guangdong in 1992, he mentioned with emotion his combat experience there. He said that, regrettably, in the battle at Meihua many important cadres had laid down their lives.

The battle at Meihua Village ended on February 5, 1931. The 7th Corps buried its dead comrades-in-arms and then retreated into the mountains. After suffering heavy losses in the battle, the 7th Corps was both exhausted and dispirited.

The Front Committee of the 7th Corps decided to abandon its plan to establish a base area in the border region between northern Guangdong and southern Hunan and to leave Lechang immediately for Jiangxi Province to join forces with the Central Regular Red Army in the Central Soviet Area.

Having made arrangements for the more than 200 wounded soldiers, the 7th Corps, led by Deng Xiaoping, Li Mingrui, and Zhang Yunyi, marched west and crossed the Lechang River (now Wushui River).

At noon, Deng Xiaoping and Li Mingrui, leading the 55th Regiment, crossed the Lechang River first. To their surprise, the enemy had transported troops by truck from Lechang and Shaozhou to prevent the 7th Corps from crossing the river. Only one company of the 58th Regiment, led by Zhang Yunyi, was able to cross the river before it was cut off by enemy artillery fire.

In the battle at the Lechang River, the 7th Corps was cut in two—the 55th Regiment advanced toward Jiangxi Province while the 58th Regiment continued to make detours in the Hunan-Guangdong border region before eventually crossing the Lechang River and heading also for Jiangxi. The two regiments lost contact and heard nothing about each other until mid-April, when they finally rejoined forces at Yongxin, Jiangxi Province.

The 55th Regiment broke through the enemy's encirclements. The 1,000 troops advanced northward into Jiangxi via northern Guangdong. On February 8, they arrived at Neiliang, where, to their pleasant surprise, they met the Congnan guerrilla forces under the leadership of the Hunan-Jiangxi Border Region Special Committee of the Central Committee.

Jiangxi was indeed the birthplace of the first revolutionary base area of the Communist Party. There were Party organizations there, a solid mass base, local armed guerrilla forces, and the Central Regular Red Army. It was so different from Guangxi. The troops of the 7th Corps met their own comrades—even just a small local guerrilla force—as soon as they entered Jiangxi. Being with their own comrades warmed their hearts.

Although it had suffered setbacks and lost its strength, the 7th Corps still had a widespread reputation as a valiant, battleworthy force. Therefore, as soon as it entered Jiangxi, the enemy newspapers carried a series of news reports: "Li Mingrui Leads the 7th Corps of the Red Army to Enter the Hunan-Jiangxi Border Region"; "Li Mingrui Leads the Red Army to Advance Toward Lechang and Renhua by Paths"; and, "Red Army's 7th Corps Enters Hunan to Attack Jiangxi in Two Directions." By then, the enemy was still keeping a close watch on the 7th Corps of the Red Army and did not dare to take it lightly!

With the help of the guerrilla forces, the 7th Corps made appropriate arrangements for the wounded and was briefed on the situation in Jiangxi. It learned that the place was still a guerrilla area with a relatively poor mass base, that the county town of Chongyi thirty kilometers away was occupied by the enemy, but that the enemy's strength was rather weak there. The 7th Corps decided to advance thirty kilometers northward and take Chongyi!

Chongyi, located in southwestern Jiangxi and in the juncture of the three provinces of Hunan, Guangdong, and Jiangxi, was about 100 kilometers away from the Jinggang Mountains and in the periphery of the Red Army's Hunan-Jiangxi Base Area. With strong combat effectiveness and overwhelming force, the 7th Corps quickly captured the county town. In light of local conditions at the time, the 7th Corps decided to conduct revolutionary work in Chongyi.

Although it was cut into two troops when it crossed the Lechang River, the 7th Corps did not engage in any battles in its eastward advance and thus suffered little loss. When the 55th Regiment arrived at Chongyi, it had more than 1,000 men (some 40 percent of them were Party members), nearly 800 guns, one mortar, and five machine guns. During its stay in Chongyi of more than twenty days, the 7th Corps rapidly established several district Soviet regimes, thus creating a red area in the county. At the same time, the question of distribution of land was raised, and the relevant work went on smoothly. With a short rest, the troops largely restored their combat effectiveness.

During this period, New Year's Day 1931 of China's lunar year fell on February 17. All the households in Chongyi rang out the old year and rang in the new and celebrated a peaceful new year. All the officers and men of the 7th Corps, who had gone through many hardships in the old year, shared with the local people the joys of a happy festival.

After the New Year holiday, the 7th Corps learned that the enemy troops were ill at ease with the Red Army in Chongyi. The secretary of the Front Committee of the 7th Corps, Deng Xiaoping, discussed the situation with Li Mingrui and Xu Zhuo, who were also on the committee, and decided to move the troops to the Xinfeng area as suggested by the Party Committee of the Administrative Region of Southern Jiangxi.

While in Chongyi, Father learned from the comrades of the Party Committee of the Administrative Region that the Party's Sixth Central Committee had held its Fourth Plenary Session and Wang Ming (also known as Chen Shaoyu) had taken over the leading position of the Central Committee.

The news somewhat shocked Father because he had never had a good opinion of Wang Ming. Father had had no contact with the Central Committee since he left it to travel to Guangxi in 1930. Now that the 7th Corps had arrived in Jiangxi, the enemy situation was not serious and the local Party Committee of the Administrative Region had a reliable communication route to the Central Committee in Shanghai, so he could go to Shanghai to report on the work to the Central Committee. Father, Li Mingrui, and Xu Zhuo held a Front Committee meeting, which unanimously agreed that Deng Xiaoping should go to Shanghai to report on the work and ask for instructions. Deng Xiaoping nominated Xu Zhuo to be the acting secretary of the Front Committee during his absence. He exhorted the 7th Corps again and again not to operate in isolation and said that only by securing the support of the people in areas with a solid mass base could the 7th Corps gain a firm foothold and accomplish its tasks, and that if necessary, it could move toward the revolutionary base area of the Jinggang Mountains.

Then Father left Chongyi after bidding farewell to Li Mingrui, the comrade-in-arms with whom he had founded the 7th Corps and established the Zuojiang and Youjiang revolutionary base areas. Father never thought that his parting with Li Mingrui would be their final parting.

Father and Xu Zhuo first went to a town fifteen to twenty kilometers away from Chongyi to see over one hundred wounded fighters and to discuss with the comrades of the Party Committee of the Administrative Region and make arrangements for establishing base areas. There they decided to have a messenger from the committee escort Father to Shanghai.

When Father said good-bye to Xu Zhuo, shots could be heard far off in the distance. Father told Xu Zhuo once again to move the 7th Corps toward the Jinggang Mountains if necessary. Then, disguised as a mountain products merchant and accompanied by the messenger from the Party Committee of the Administrative Region, Father walked for several days before reaching Nanxiong, Guangdong Province.

Nanxiong was then a major liaison station of the Communist Party managed by a couple named Li. Father stayed there overnight, then another messenger from Guangdong, sent by the couple, escorted him on foot to Shaoguan, where he boarded a train for Guangzhou. After spending half a day in a hotel in Guangzhou, Father, with a steamer ticket bought by the messenger for the sea journey from Guangzhou to Hong Kong and then to Shanghai, left Guangzhou the same night for Hong Kong. Once in Hong Kong, he boarded a ship and arrived in Shanghai safely.

CHAPTER 35

ETERNAL GLORY TO THE 7TH CORPS

After Father left the 7th Corps for Shanghai to report on the work to the Central Committee, Li Mingrui and Xu Zhuo were ready to set off for Xinfeng in the east.

At that moment, the enemy troops were quickly, and unexpectedly, approaching the city of Chongyi. Because of poor intelligence, the 7th Corps did not know about the enemy's three-pronged attack until they approached the city. There was a dense fog at the time. Deploying a deceptive battle array to mislead the enemy, Li Mingrui ordered the 7th Corps to withdraw from the city. It attacked now northward and now southward and withdrew from the city while fighting. Very soon, the 7th Corps disappeared in the dense fog, like divine troops, while a bloody battle was fought between the enemy troops. When they found that they were fighting themselves, it was already too late. The 7th Corps had advanced northward to Suichuan near the Jinggang Mountains long before.

In March 1931 Li Mingrui led the 7th Corps even further north and reached Yongxin. There the corps met a unit of the 3d Corps of the Red Army led by Teng Daiyuan. That was the first time the 7th Corps had met friendly Red Army units. The 7th Corps joined forces with the Red Army units in Jiangxi. In Yongxin, while doing its work among the masses, the 7th Corps continued to inquire about Corps Commander Zhang Yunyi and the 58th Regiment.

One bright spring day in April, at about 4:00 P.M., Li Mingrui and Xu Zhuo were leading an army unit to cross a river, hoisting a red flag. An army unit suddenly approached. It turned out to be the lost 58th Regiment. Grasping each other's hands, Li Mingrui and Zhang Yunyi were so excited that tears ran down their cheeks like two boys. The 55th and 58th regiments of the 7th Corps of the Red Army joined forces in Yongxin at last.

Two months before, in February 1931, most units of the 58th Regiment, the corps training unit, the special task company, and some noncombatant units of corps detachments, had failed to cross the Lechang River. Led by Corps Commander Zhang Yunyi, they at last crossed the river elsewhere and came to Lingxian County, Hunan Province, in the Hunan-Jiangxi Revolutionary Base Area. In April, after fighting many bitter battles, they had arrived at Yongxin, Jiangxi Province. During a battle, they found a copy of the Kuomintang's newspaper, the *Central Daily*, on which a line of large characters caught their attention: "Remnants of Communist Bandit Li Mingrui's Troops Fleeing toward Suichuan." They had learned the orientation of the main forces of the 7th Corps at last. Zhang Yunyi and his troops were very happy and immediately moved toward Suichuan.

On a small wooden bridge at the river mouth to the east of Suichuan, the 55th and 58th regiments and Commander-in-Chief Li Mingrui and Corps

Commander Zhang Yunyi met at last after suffering innumerable hardships and being separated from each other for more than two months.

On top of the city wall was flying high the red banner that read, "Headquarters of the 7th Corps of the Chinese Workers' and Peasants' Red Army." A rally of more than 2,500 officers and men of the 7th Corps was held there to celebrate their triumphant rendezvous. Everywhere there was laughter. Every face revealed unadulterated joy.

On the rostrum, Li Mingrui and Zhang Yunyi were sitting upright jubilantly. Suddenly there was the deafening sound of fireworks and the sound of a bugle call.

In his speech, Corps Commander Zhang Yunyi reviewed the arduous course of struggle waged by the 7th Corps under the leadership of the Party. He pointed out that after undergoing harsh struggle, the 7th Corps had become even stauncher and its triumphant rendezvous showed that it could stand any test. Commander-in-Chief Li Mingrui and leading members of the Yongxin County Soviet also spoke at the rally. The whole meeting was like a jubilant ocean.

Not long after, the 7th Corps received an order from General Political Commissar of the Central Red Army Mao Zedong and Commander-in-Chief Zhu De to establish the Hexi General Headquarters, with Li Mingrui serving as the commander-in-chief.

After days of rest in Yongxin, the men of the 7th Corps, full of energy and with high morale, went into action to smash the second counterrevolutionary campaign of "encirclement and suppression" launched by Chiang Kai-shek's Kuomintang troops against the Central Soviet Area. The 7th Corps operated in the areas of western Jiangxi and eastern Hunan, took part in many battles, cooperated with the Central Red Army to crush the second counterrevolutionary campaign, and expanded the Hunan-Jiangxi Revolutionary Base Area.

In mid-June 1931, on the order of the Central Red Army, the main forces of the 7th Corps crossed the Ganjiang River in the east and advanced to Xingguo County to join forces with the Central Red Army.[1] Old soldiers of the 7th Corps emotionally recalled:

> During the more than ten months from September 1930 when the 7th Corps of the Red Army left the Youjiang area to July 1931 when it joined forces with the Central Red Army, it fought its way through four provinces of Guangxi, Hunan, Guangdong, and Jiangxi, heroically crushed the enemy's encirclement, pursuit, blocking, and interception, overcame unimaginable hardships, and realized its dream of joining forces with the Central Red Army headed by Zhu and Mao at last. From then on, the 7th Corps became part of the Central Red Army and fought south and north under the direct command of Chairman Mao and Commander-in-Chief Zhu.[2]

Ten months of tortuous fighting and a march of 3,500 kilometers—that was the crooked but glorious path traversed by the 7th Corps, the dauntless unit of the Chinese Workers' and Peasants' Red Army. In November 1931 the

First Conference of the Representatives of Workers, Peasants, and Soldiers of the Chinese Soviet Republic was held in Ruijin, the "Red Capital" of the Central Soviet Area. The 7th Corps sent five representatives to the conference. To commend the revolutionary spirit and outstanding merits of the 7th Corps, Mao Zedong, chairman of the Provisional Central Government, personally presented a silk banner to the 7th Corps at the closing ceremony. On the banner were written four large Chinese characters: "Zhuang Zhan Qian Li" (meaning, fighting over the course of 500 kilometers). Father recalled that even in the 1970s, Mao Zedong told him several times that "the 7th Corps of the Red Army was really good at fighting."

The history of the 7th Corps is moving and tragic, a history of momentous revolutionary fighting. It contains both chapters of glory and victory and chapters of tragedy and defeat. The 7th and 8th corps of the Red Army started from scratch and developed quickly, and the Zuojiang and Youjiang revolutionary base areas expanded from small to large. For all this, the Chinese Workers' and Peasants' Red Army gained great fame and prestige throughout Guangxi. Like other brotherly troops of the Red Army, they will go down in the history of the revolution.

The revolutionary and military practices of the 7th and 8th corps gave Father good opportunities to temper himself. Experience and lessons, victories and defeats—all of it enriched his experience and was valuable for his revolutionary career in the following years, making him more mature.

In the ranks of the 7th and 8th corps, many revolutionaries fought heroically, advanced wave upon wave, and even laid down their lives for the revolution. Many others later became high-ranking officers of the revolutionary army. Li Mingrui was a militant general during the Northern Expedition. He gave up his superior social status. He rejected the high post and handsome salary offered by Chiang Kai-shek and resolutely chose the arduous revolutionary road. He joined the Communist Party of China. In the face of rigorous trials, he never wavered. He had both courage and cunning as well as the art of command and made indelible contributions to the 7th and 8th corps of the Red Army.

Unfortunately, Li Mingrui was killed in cold blood under the rule of the "left"-opportunist line followed by Wang Ming in October 1931. He was only thirty-five. Later, his good name was restored and he was made a revolutionary martyr by the Party Central Committee at the Seventh National Congress in 1945.

The old soldiers of the 7th Corps felt both joyful and grievous when they learned that their commander-in-chief had been exonerated at last. If Li Mingrui had not been killed, how many more meritorious deeds he would have performed in his revolutionary career! A vigorous and courageous general, he would also have received his due honor and position from the Party and the people.

In 1986 when Father went to Guangxi, he recalled the 7th and 8th corps. He especially mentioned Li Mingrui.

> I first met Li Mingrui on the way from Bose to Longzhou. I was the one who went to Shanghai to ask the Central Committee to approve the admission of Li Mingrui into the Party. We two went all the way to Jiangxi. Li Mingrui was the commander-in-chief of the 7th and 8th corps of the Red Army, and I was the chief political commissar. Lei Jingtian was the chairman of the Soviet. The 8th Corps was destroyed, but the 7th Corps was good at fighting. Yu Zuobai fled to Hong Kong, but Li Mingrui remained firm!

Father also inquired about Li Mingrui's kinsfolk and descendants.

He has always missed Li Mingrui over the decades. He said that in the 1970s he told Mao Zedong on several occasions that Li Mingrui had been killed wrongly. To this day, Father feels aggrieved whenever he mentions Li Mingrui.

Wei Baqun was born in a family of the Zhuang nationality in Guangxi. After the May 4th Movement of 1919, and under the influence of progressive ideas, he participated in the revolution led by Dr. Sun Yat-sen. In 1929 he joined the Communist Party of China. The following year, when the 7th Corps was ordered to march northward, he served as the commander of the 21st Division. He said good-bye to the main forces and undertook the arduous task of staying behind in the Youjiang Revolutionary Base Area. He led the people and the revolutionary armed forces in the Youjiang area to persist in struggling under very harsh conditions.

Many of the fighters of the 7th Corps laid down their lives in the Youjiang area. Many of the revolutionary people in the Youjiang Base Area were killed. Wei Baqun's younger brother and son were killed, and his mother was driven by the reactionaries up the mountain and starved to death. The enemy offered a reward of 14,000 silver dollars for the capture of Wei Baqun. He not only persisted in struggle but also led the Red Army to crush two campaigns of encirclement and suppression launched by the Guangxi warlord clique against the base area. However, on October 19, 1932, he was murdered by a traitor during the third countercampaign against encirclement and suppression. He was then thirty-eight. The people in the Youjiang area risked their lives to bury him in the land of Guangxi.

Wei Baqun was a famous peasant leader in Guangxi and a hero in the eyes of the people of Guangxi. He once wrote these lines: "In the Red Army and as men of the Red Army, all are fighting for the salvation of the people; you die first, I die later, and all of us lay down our lives for the revolution!" What a fearless revolutionary spirit to fight on in that cruel and harsh environment! That is the hero of the people's revolution and a Chinese Communist!

Father never forgets the mountains and rivers in Guangxi, and he will never forget the revolutionary comrades-in-arms of the 7th and 8th corps. In Nanhu Park in Nanning, capital of the Guangxi Zhuang Autonomous Region, a monument was erected in memory of the martyrs Li Mingrui and Wei Baqun. On the monument, Father wrote: "In memory of Comrades Li Mingrui and Wei

Baqun and others. Eternal glory to the revolutionary martyrs who laid down their lives in the Bose Uprising!"

The 7th and 8th corps were a heroic people's army. In their course of struggle, they trained many outstanding fighters who became commanders and leading cadres in the later periods of the anti-Japanese war, the War of Liberation, and the reconstruction of New China. The former commander of the 7th Corps, Zhang Yunyi, once said, "There are no fewer than fifty high-ranking commanders from the 7th Corps of the Red Army. How many battles made one high-ranking commander? How many battles altogether made those fifty?"[3]

In 1955 the Chinese People's Liberation Army introduced the military rank system. Among those upon whom was conferred the rank of general, nineteen were from the former 7th and 8th corps of the Red Army, including one senior general, two colonel generals, four lieutenant generals, and twelve major generals.

Although they had a revolutionary history of only several years, the 7th and 8th corps were indeed staunch contingents of the Chinese revolutionary army and made tremendous contributions to the Chinese revolution. The Bose Uprising and the Longzhou Uprising will go down in history as two important revolutionary uprisings of the many led by the Communist Party of China during the period of the agrarian revolution.

CHAPTER 36

CHANGES IN THE EARLY 1930s

In about February 1931, Father returned to Shanghai from Jiangxi through the secret communication line of the Party.

According to the address given him by the liaison station, he soon came into contact with the underground messenger of the Central Committee and reported for duty to the committee. With the help of the underground messenger, he first lived in the old Huizhong Hotel for some days and later moved to a garret.

After arriving in Shanghai, Father requested through the underground messenger that the leading comrades of the Central Committee hear his report on the work of the 7th Corps. By April 29, while waiting to be received by the Central Committee, he had completed a written report, entitled "The Report on the Work of the 7th Corps." In this work report, Father gave a detailed account of the course of struggle of the 7th and 8th corps, the Party work, and the work of agrarian revolution done by the 7th Corps in the Youjiang area of Guangxi, throughout the 3,500-kilometer march and in Chongyi, Jiangxi. He earnestly analyzed the experience and lessons of the work of the 7th Corps during this period.

He held that the primary weakness of the work of the 7th Corps was an overemphasis on military questions instead of on the work among the masses. As it neglected the work among the masses, it was in a passive position throughout the march and unable to gain a foothold, so it had to march to southern Jiangxi.

Second, in his analysis, the 7th Corps should have left the Youjiang area quickly, because it had played a very small role there, and when it arrived at Meihua Village, it should not have tried to gain a foothold there and should have quickly moved to Jiangxi. If it had, it would not have fought the costly battle at Meihua Village.

Third, under the influence of the "leftist" line, the 7th Corps made mistakes and suffered setbacks on several occasions by assaulting heavily fortified positions, so the attack on Liuzhou, Guilin, and Guangzhou became empty talk. Moreover, in fighting operations the 7th Corps was weak in its intelligence work and tended to take the enemy lightly. It also did not do enough work to transform the officers and men from the old army. The 7th Corps was composed of these old army members and a number of newly recruited peasants. The weak foundation plus the implementation of the Li Lisan line made the 7th Corps suffer many setbacks. There were also many weaknesses in its Party and political work, and its work did not proceed well because it failed to carry out the agrarian revolution thoroughly.

The report included more than 16,700 Chinese characters and was a serious summary of the political and military work by a leading member of the

Party. It should have drawn great attention from the Central Committee. But six months passed. The Central Committee did not hear Father's report. Living in Shanghai, Father received funds for living expenses from the underground messenger only once a month. His only contacts with the Central Committee were the periodic visits from the underground messenger.

In Shanghai, Father soon met with some acquaintances, including Li Weihan, He Chang, Li Fuchun, and Nie Rongzhen. He also stayed overnight in the homes of Li Weihan and He Chang.

Gradually, Father learned that considerable change had taken place in the Central Committee and the Party work from the summer of 1929, when he left the Central Committee for Guangxi, up to his return to Shanghai. The implementation of the "left"-adventurism followed by Li Lisan had caused not only serious damage to the 7th Corps but also losses to the Party's cause and armed struggle in other areas. The mistakes of "left"-adventurism aroused great resentment and opposition from vast numbers of cadres in the Party. The Li Lisan line lasted for only three months before it ended in failure. In September 1930, according to the instructions of the Communist International, Qu Qiubai and Zhou Enlai presided over the Third Plenary Session of the Sixth Central Committee. It criticized Li Lisan and others who made "leftist" mistakes, stopped their adventurist plans, and reorganized the Central Committee. Li Lisan admitted his mistakes and left his leading post on the Central Committee. This marked the end of the Li Lisan line.

In its early development, the Communist Party of China was not very lucky. It suffered time and again from interference and sabotage by erroneous lines. At that time, the young CPC was not mature enough to control its own destiny. It needed the support of the Communist International, which often interfered in its work and gave wrong orders.

After Li Lisan's "left" putschism ended in September 1930, Pavel Mif was sent by the Communist International to directly handle the affairs of the Communist Party of China and the Chinese revolution. By taking advantage of Wang Ming, Mif issued arbitrary instructions and acted autocratically. This led to a more theoretical and more perfect "left"-adventurist line through Wang Ming.

Wang Ming, whose original name was Chen Shaoyu, was born in Jinzhai, Anhui Province. He joined the Communist Party of China in 1925. The same year, he went to study at Sun Yat-sen University in Moscow. In 1927 he returned to China with Mif, the vice president of the university. He then took up the posts of secretary of the Propaganda Department of the Central Committee and editor of the Party organ, *The Guide*. After the Great Revolution failed, he returned to the Soviet Union to work in Sun Yat-sen University. While there, he became a favorite of Mif. When he led the Party branch at the university, he carried out factional activities with the support of Mif and attacked and persecuted those Chinese comrades who opposed him. In 1929 he returned home and engaged in underground work. Originally, he was a supporter of Li Lisan's "leftist" views, but when he heard that the Communist International criticized the Third Plenary Session of the Sixth Central Committee (the plenary session that corrected Li Lisan's mistakes), he turned around to attack the plenary session and the new Central Committee. He openly unfurled the banner of "supporting the line of the Communist International" and demanded a thorough reshuffling of the Party central leadership.

Why did the Central Committee tolerate their wanton acts? Because it was being manipulated behind the scenes by Mif from the Communist International. At that time, the CPC still lacked a strong leading nucleus and its own leader and backbone. Therefore, it had to comply with and obey the views of the Communist International.

On January 7, 1931, the Fourth Plenary Session of the Sixth Central Committee of the Communist Party of China was held secretly in Shanghai. The meeting was still presided over by General Secretary Xiang Zhongfa, but in name only. Actually, it was Mif who ran the show. Heated disputes prevailed throughout the meeting. Mif and others criticized the Li Lisan line, but their real purpose was to reorganize the central leadership of the Communist Party of China and prop up Wang Ming to assume power. Manipulated by Mif, the meeting held that "the main danger in the Party now" was "right deviation," and it decided to reelect the Central Committee and its Political Bureau.

The list of the new Political Bureau was drafted by the Communist International in advance. The general secretary was still Xiang Zhongfa, who played only an insignificant role. Wang Ming became a member of the Political Bureau and in fact led the Central Committee.

Thus, Wang Ming, who "could only play petty tricks" and "lacked practical work experience,"[1] became the master of the Communist Party of China, with the support of Mif.

Wang Ming called himself "100 percent Bolshevik," but in reality he dogmatized Marxism and deified the experience of the Communist International and the Soviet Union. He continued to stress nationwide "revolutionary upsurge" and the Party's "offensive line" across the country. He practiced factionalism within the Party and often waged excessive struggles against people who held different views or even resorted to ruthless struggle and merciless blows.

With the support of the Communist International, and because of the overarching dogmatism and "leftist" tendencies within the Party, very soon Wang Ming's "left"-adventurist line took shape, a line more serious and more arrogant than the Li Lisan's "left"-adventurism.

In April 1931 Gu Shunzhang, a member of the Political Bureau of the Central Committee, was arrested and turned traitor. In June General Secretary Xiang Zhongfa was arrested and turned traitor, too. As the Central Committee in Shanghai was destroyed, a provisional Central Committee was formed, with Bo Gu as the general leader. Wang Ming went to Moscow to act as the CPC representative to the Communist International. But, in fact, he still controlled the Party through Bo Gu. Wang Ming's "left"-adventurist line lasted for four years, causing tremendous losses to the Communist Party of China and the Chinese revolution.

While Wang Ming's "left"-adventurist line was holding sway in the Party, another major event took place in China in the early 1930s, an event that almost threw the Chinese nation into an abyss of national subjugation and genocide. That was the aggression against China perpetrated by its eastern neighbor, Japan.

In the second half of 1929, the capitalist world was plunged into an unprecedentedly large-scale and grave economic crisis. It started in the United States and very soon spread to the whole of the capitalist world. From 1930, the total industrial output of all major capitalist countries dropped by 44 percent, and the number of unemployed came to about twenty million by the end of 1933. This economic crisis aggravated the domestic contradictions of all capitalist countries and the contradictions between imperialism and the colonies because the capitalists tried to shift the losses caused by the economic crisis onto the people in the colonies.

In September 1931, when Britain, the United States, and other countries were busy with their own internal affairs and China's Chiang Kai-shek was going all out to conduct the campaign of encirclement and suppression against the Red Army, the Japanese militarists stepped up their war of aggression against China. On September 18, the Japanese invaders launched an attack against northeastern China. But the Chiang Kai-shek government in Nanjing went so far as to issue a "no clash" order, prohibiting the Northeastern Army, led by Gen. Zhang Xueliang, from putting up resistance. As a result, the hundreds of thousands of troops of the Northeastern Army retreated to areas south of the Shanhaiguan Pass without firing a single shot. Zhang Xueliang had to bear the blame and was cursed by the Chinese people as "a general of nonresistance." In less than five days, the Japanese army occupied almost the whole of Liaoning and Jilin provinces. And in the short span of three months, all the three provinces in northeastern China, Liaoning, Jilin, and Heilongjiang, had fallen into the hands of the Japanese troops.

After occupying northeastern China without fighting a battle in January 1932, the Japanese troops became more brazen and even went so far as to attack Shanghai. On March 3, they captured the city. On May 5, Chiang Kai-shek signed the Truce Treaty of Wusong and Shanghai, permitting Japan to station troops in Shanghai, promising to ban the nationwide anti-Japanese movements, and ordering the anti-Japanese 19th Route Army, which resisted the Japanese troops in Shanghai, to leave for Fujian to "suppress the Communist Party."

To Chiang Kai-shek, Japanese imperialists were not his principal enemy, and the fall of the vast land into Japanese hands meant nothing. The enemy in his heart was the Communist Party and the Red Army it led!

On November 30, 1931, when the country was in calamity, Chiang Kai-shek proposed that "to resist foreign aggression, it is essential to achieve stability at home first," meaning, the Communist Party had to be eliminated before China could resist foreign invaders! What a strange logic!

The September 18th Incident astonished the whole nation and aroused widespread condemnation. Workers and students went on strike, demanding that the government resist the Japanese invaders. On September 28, students in Nanjing were so indignant that they destroyed the building of the Foreign Ministry of the Kuomintang government. Toward the end of the same year, students from Beiping, Tianjin, Shanghai, Hankou, and Guangzhou demonstrated in Nanjing and presented petitions to the government. But Chiang Kai-shek dispatched troops and police to suppress them, killing more than thirty students and seriously wounding more than one hundred. The perverse acts of Chiang Kai-shek's Kuomintang government—suppressing the students in cold blood instead of resisting the Japanese invaders—aroused even greater indignation among the people.

In September 1932 the Provisional Central Committee of the Communist Party of China called on the people of China to arm themselves and resist the aggression by the Japanese imperialists. Under the leadership and influence of the Communist Party of China, the mass movements to demand resistance to Japan were waged in Shanghai and other places one after another.

When the national contradictions became the principal contradiction, the CPC should have seized the opportunity and done everything possible to mobilize support and unite all available forces to wage the anti-Japanese struggle. However, under the control of Wang Ming, the Central Committee headed by Bo Gu wrongly sized up the situation and put forward the slogan of "Defend the Soviet Union with arms." At home, it advocated "Down with all." It held that the conditions had been ripe for seizing key cities, and it put forward the adventurist proposition of attacking cities and waging general strike. Because of these erroneous propositions and slogans, the Party lost the opportunity to maintain close contact with the anti-Japanese masses, its supporters and sympathizers.

On the other hand, Chiang Kai-shek took the opportunity to suppress large numbers of Party members. They were arrested and killed, and the Party was greatly weakened. In the White area, Party organizations were almost completely destroyed. By early 1933 even the Provisional Central Committee controlled by "leftists" was unable to stay in Shanghai and had no choice but to move to the Central Soviet Area in Jiangxi. Later, its radio station was destroyed and—unfortunately?—the Central Committee even lost contact with the Communist International. In the long history of China, this chapter, the early 1930s, was marked by national humiliation and Party mistakes.

In early 1931, when the country faced calamity and the Party made mistakes, Father returned to Shanghai, under these complex circumstances.

For six months after his return to Shanghai, the Central Committee and its

Military Commission did not hear his report or meet with him. But he did not know that before he completed his report on the work of the 7th Corps, two others from the 7th Corps—Chen Haoren, former director of the Political Department of the 7th Corps, and Yan Heng, an officer of the 7th Corps—had already finished their reports to the Central Committee, on March 9 and April 4, respectively.[2] Apart from giving a detailed account of the experiences of the 7th Corps, they analyzed the successes and failures of the corps from a "leftist" point of view, especially Yan Heng.

The Central Committee controlled by Wang Ming was more "leftist" than Li Lisan's committee. While paying no attention to Deng Xiaoping, the political commissar of the 7th Corps who had come to report on the corps' work, it sent a letter to the Front Committee of the 7th Corps on May 14, making arbitrary criticisms of the work of the 7th Corps, in an overbearing and extremely stern tone.[3]

From its "leftist" point of view, the letter attributed the defeat of the 7th Corps chiefly to its lack of class analysis and denounced the line followed by the 7th Corps as "an obviously putschist adventurist line of the Li Lisan-ism. . . . It fully shows the right-opportunist and rich-peasant line under the cover of that line." The letter was sent to the 7th Corps in Jiangxi. Political Commissar Deng Xiaoping was in Shanghai, consigned to limbo and with no knowledge of the Central Committee's severe criticisms of the work of the 7th Corps and discontent with him.

But Deng Xiaoping felt in his heart the obvious coolness shown toward him by Wang Ming's Central Committee. Father once said that he occasionally met with his old friends and they voiced their discontent, and that this period was the most difficult one in his political career. How could he, a Communist, stay idle, doing nothing all day long? He requested permission from the Central Committee, through the underground messenger, to go back to the 7th Corps, but the request was turned down on the grounds of lack of communication and liaison. Later, he again requested permission to work in the Soviet area, and in about June 1931 he received that permission from the Central Committee.

Before leaving for the Soviet area, between May and June, the Central Committee instructed Father to inspect the work of the Anhui Provincial Party Committee in Wuhu. So he went, together with an underground messenger who was a native of Anhui. After arriving in Wuhu by ship, Father was at a restaurant while the underground messenger went to contact the Anhui Provincial Party Committee. The messenger soon returned, telling Father that the secret sign of the Anhui Provincial Party Committee had disappeared, indicating that it had been discovered by the enemy. As the situation was very dangerous, they immediately bought tickets and returned to Shanghai by ship that very day. Father reported their experience to the Central Committee, having "accomplished his task," in his words.

In mid-July Father left Shanghai by ship for Jiangxi via Guangdong. He was accompanied by a woman comrade, named Jin Weiying, or Ah Jin, as she was called by all. Jin Weiying, whose original name was Jin Aiqing, was born in Daishan, Zhejiang Province, in 1904; she was the same age as Father. She joined the Communist Party of China in 1926 and had been engaged in the workers' movement. In 1930 she was the leader of the United Action Committee of the Trade Unions in Shanghai. Father and Ah Jin came to know each other in Shanghai in 1931. Both were sent to work in the Central Soviet Area in Jiangxi, and later they married.

That was Father's third departure from Shanghai. The first was in the summer of 1929, when he was twenty-five years old. With lofty ambitions, he was instructed by the Central Committee to go to Guangxi to organize armed uprising. The second departure was at the end of January 1930, when he came and left all in a hurry. He had come to Shanghai to report to the Central Committee, only to be confronted with the loss of his wife and daughter. His grief, however, had to wait while he hurried back to the battlefront, where the military situation was very pressing.

The third departure was in July 1931. When he went south again by ship, the situation was quite different. The tortuous 3,500-kilometer march of the 7th Corps still lingered in his mind, and he was doubtful about the future and destiny of the Party. But he longed to work in the Central Soviet Area.

Only two years had passed. It seemed to him that they were both very short and very long. Those two years had witnessed army life and battles, both victories and defeats.

Time was passing, and men were changing every year, and each year they became more mature. Deng Xiaoping at this time was about twenty-seven. He had gained rich revolutionary experience and become deeper and more mature.

His future would be filled with even more militant enthusiasm that would rage like a fire.

CHAPTER 37

RUIJIN AND THE CENTRAL SOVIET AREA

In mid-July of 1931 Father and Ah Jin arrived in Shantou from Shanghai by ship and found the liaison station. A comrade from Guangdong led them northward and they smoothly entered Yongding, Fujian Province, via Dapu at the Guangdong-Fujian border, the territory of the Central Soviet Area. They continued their journey northwestward through Shanghang and Dingzhou (Changding) and turned westward across the border of Fujian and Jiangxi and arrived in Ruijin, Jiangxi Province. By then, it was August 1931.

Ruijin was the center of the Central Revolutionary Base Area. In the month after the Nanchang Uprising failed in August 1927, Mao Zedong had led the Autumn Harvest Uprising in Hunan. After that uprising, Mao Zedong thought it was impossible to capture key cities at that time and that the armed forces should move to rural areas, where enemy rule was relatively weak, so as to preserve military strength, persist in struggle, and expand the revolutionary forces. He decided to march the troops into the mountains on the Hunan-Jiangxi border and establish a revolutionary base in the Jinggang Mountains.

Mao Zedong held that the CPC and its army should combine armed struggle with the agrarian revolution and the establishment of revolutionary base areas. The Jinggang Mountains are located in the border area between Jiangxi and southern Hunan, far away from the key cities controlled by the enemy, and being steep, they harbor many advantageous strategic positions, for both defense and attack. Under the influence of the First Revolutionary Civil War, the people there were politically sound, and local Party organizations and armed forces had already been established. It was the most suitable place for establishing a revolutionary base area.

The establishment of the revolutionary base area in the Jinggang Mountains kindled the sparks of an "armed independent regime of workers and peasants," a great turning point in the seizure of power by revolutionary armed forces. From the end of 1927 to 1928 the troops that had participated in the Nanchang Uprising led by Zhu De, the Guangzhou Uprising, and the Pingjiang Uprising led by Peng Dehuai, all came to the Jinggang Mountains, one after another, to join forces with the Autumn Harvest Uprising forces led by Mao Zedong. Together, they succeeded in establishing a vast revolutionary base area there and red revolutionary political power.

In April 1928, with the approval of the Central Committee, Mao Zedong, Zhu De, and Chen Yi founded the 4th Corps of the Chinese Workers' and Peasants' Red Army, with Zhu De serving as the corps commander, Mao Zedong as the Party representative, and Chen Yi as the director of the Political Department. The corps had three divisions and nine regiments, totaling more than 10,000 men.

The establishment and expansion of the Jinggang Mountain Revolutionary

Base Area terrified Chiang Kai-shek. Toward the end of 1928 the enemy set up the "Bandit Suppression Headquarters" for Hunan and Jiangxi and mustered twenty-five regiments, about 20,000 strong, to launch fierce, five-pronged attacks on the Jinggang Mountains.

To thwart enemy "suppression," in January 1929 Mao Zedong and Zhu De led the main forces of the 4th Corps, 3,600 strong, toward southern Jiangxi, while Peng Dehuai stayed in the Jinggang Mountains with the 5th Corps of the Red Army. The 4th Corps advanced to western Fujian first and then turned to southern Jiangxi, capturing a number of counties on the way. In June the 4th Corps conducted guerrilla warfare in the border area between southern Jiangxi and western Fujian and, at the same time, mobilized the masses to carry out an agrarian revolution and expand the Red Army and local forces. It established red political power in more than ten counties.

In early 1930 the Western Fujian Revolutionary Base Area expanded to cover an area a couple of hundred kilometers in circumference, with a population of 850,000. Land was distributed to 800,000 peasants in the agrarian revolution, and Soviet red power was established in all counties.

In June 1930, according to the directives of the Central Committee, the Front Committee of the 4th Corps reorganized the 4th and 6th corps into the 1st Corps, with Zhu De serving as the commander-in-chief and Mao Zedong as the political commissar.

In a year and a half, the Red Army had established new base areas in southern Jiangxi and western Fujian, forming a solid central revolutionary base area, or the Central Soviet Area.

In October 1930 Chiang Kai-shek ended his major battle in the Central Plains with Yan Xishan and Feng Yuxiang and turned around to encircle and suppress the Red Army with concentrated force. In November the 100,000 Kuomintang troops started their first campaign of encirclement and suppression in the Central Soviet Area.

Adopting a flexible strategy and tactics, Mao Zedong and Zhu De first lured the enemy troops deep into the area, then broke through their middle and defeated them one by one. By December 30 the Red Army had wiped out more than 9,000 enemy troops and captured the enemy commander-in-chief, Zhang Huizhan, winning the first countercampaign against the enemy's encirclement and suppression effort.

Even after this disastrous defeat, Chiang Kai-shek did not stop and arrogantly boasted that he would "wipe out the Communist troops within three months." In February 1931 he ordered He Yingqin to assemble 200,000 troops to launch the second campaign of encirclement and suppression against the Central Soviet Area. This time, Mao Zedong formulated the strategy of concentrating forces to attack the weak enemy troops and annihilate them one by one in mobile warfare. Over 30,000 Red Army men marched over 350 kilometers, won five major battles in fifteen days of fierce fighting, and captured more than

20,000 weapons, thus crushing Chiang Kai-shek's second campaign very swiftly.

Flustered and exasperated, Chiang Kai-shek immediately assembled 300,000 troops. With a number of military advisers from Germany, Japan, and Britain, he assumed personal command in Nanchang, the capital of Jiangxi Province, to launch the third campaign of encirclement and suppression against the Central Soviet Area, in July 1931. Facing a grave situation, Mao Zedong adopted the strategy of avoiding the enemy's main force and striking at its weak forces. The Red Army effectively annihilated the enemy by circling around, making a feint to the east but attacking in the west, and maneuvering skillfully. The enemy was so confused that it did not know where the main forces of the Red Army were. The exhausted enemy troops pursued the Red Army but were attacked repeatedly by the main forces of the Red Army and local armed forces. After three months, the Red Army had wiped out more than 30,000 enemy troops and captured more than 20,000 guns, forcing the enemy to retreat in the end. The enemy's third campaign, led personally by Chiang Kai-shek, also ended in failure.

After these three successful countercampaigns against the enemy, the revolutionary base areas in southern Jiangxi and western Fujian were combined to form the Central Soviet Area, which included twenty-one counties with a population of two and a half million. Chiang Kai-shek's smug calculation that he would "achieve stability at home before resisting foreign aggression," regardless of public feelings and the humiliation of the nation, had come to nothing before the invincible Red Army.

Father arrived in Ruijin, Jiangxi Province, in August 1931, during the third countercampaign. Ruijin was at the rear of the Central Soviet Area. After Father arrived, he found that the Party and government leadership of Ruijin County had been usurped by counterrevolutionaries. Many revolutionary cadres and people had been killed. On just one hill opposite the county town, more than 100 people had been killed. The local people and the cadres were in low spirits. All aspects of revolutionary work had stopped in the county. At that time, Xie Weijun from the Red Army was in Ruijin, and Yu Zehong, sent by the Central Committee in Shanghai to work in the Central Soviet Area, had also arrived. They had no contacts with the higher authorities then.

They met with Deng Xiaoping and elected him secretary of the Party Committee of Ruijin County. After he became secretary, Deng Xiaoping, together with Xie Weijun and Yu Zehong, first quickly punished counterrevolutionaries and rehabilitated the revolutionary cadres. Then he called the Soviet congress of the county, established red revolutionary political power, and mobilized the masses. The enthusiasm of the cadres and people in Ruijin was soon aroused, especially that of local peasant cadres. With the support of the local people and the cadres who had close connections with the masses, a great change took place in the work of the county.

After the enemy's third campaign was crushed, the center of political power in the Chinese Soviet moved to Ruijin, which became renowned throughout the whole country as the "Red Capital."

A mass rally of 50,000 was held in the Red Capital of Ruijin to celebrate the brilliant victory of the third countercampaign. Father told me that, because of the very poor material conditions, there were no loudspeakers and the rally had to be held separately at four or five sites. Father presided over the celebration rally and accompanied Mao Zedong to all the sites to deliver a speech. All of Ruijin County became a sea of red flags and streamers. It was a scene of jubilation as people shouted slogans and cheered. Since he had gained work experience in the Guangxi revolutionary base areas, Father worked skillfully in Ruijin. Everything went well.

He often mentioned his experience in Ruijin, though he lived there for only a short time. One day in 1992, when our whole family was sitting together at a dinner table, he saw the six-year-old son of my younger brother eating ravenously because he was too hungry. He chuckled, saying, "When we were formulating policies for land distribution during the agrarian revolution in Ruijin, some people said that children should not get land. I told them that, as a Sichuan saying goes, a three-year-old son may eat more than his father can afford! Children could eat very much, and so they, too, should get shares of land. Then they accepted my view. Look, Xiao Di is just like what I said." We all laughed. I told Father, "It should be that a grandson of six may eat more than his grandfather can afford!" Father must have many stories in his mind, but he is reluctant to tell about his past. If he could tell more stories like this one, I would have had more "rice straws" to clutch, and this book would have been more vivid.

For less than a year Father served as the secretary of the Ruijin County Party Committee. In May 1932 Jiangxi Provincial Party Secretary Li Fuchun transferred Father from Ruijin to the post of Party secretary of the Huichang County Party Committee in the south of Ruijin.

In the autumn of 1972, when Father had become a target of attack during the Cultural Revolution, he was banished to Jiangxi. My parents got permission to make "investigations and studies" in Jiangxi. They set off from Nanchang and went south to the Jinggang Mountains, where they paid a visit to the revolutionary ruins of the Jinggang Mountain Revolutionary Base Area. Then they turned southeast to Ruijin, Huichang, and Xunwu. When Father arrived in Ruijin, it had been forty-one years since he was Party secretary there. Although he was still the "No. 2 capitalist roader" in the country and had not been reinstated, the cadres and people of the old revolutionary base area in Jiangxi warmly received him. The comrades of Ruijin County told him, "You are our old secretary of the Ruijin County Party Committee." Father was deeply moved by these words. He was still remembered by the people of the old base area when he suffered injustice.

While the third countercampaign was being launched in the Central Soviet Area, Wang Ming was gradually implementing his "left"-adventurist line in the

Central Soviet Area and other revolutionary base areas. In June 1931 the Central Committee, dominated by the erroneous "leftist" line, gave instructions to the Red Army and all local Party organizations, stating that the pressing task of the Party was to win victories in one or more provinces and to link up all the Soviet areas in Hunan, Hubei, and Jiangxi, and asking the Red Army to attack the much stronger enemy troops, regardless of the conditions.

After the September 18th Incident, the Central Committee made the wrong decision that the Red Army should concentrate its forces to pursue the retreating enemy troops and capture one or two key cities or smaller ones. The Central Committee controlled by Wang Ming even pointed out bluffingly that the main danger in the Party was still right-opportunism. To practice "left"-adventurism, the leaders who made "leftist" mistakes developed factional organization. On those who suspected or resented "leftist" policies or did not actively support and implement them, they imposed the labels of "right-opportunism," "rich-peasant line," "line of conciliation," or "double-dealer." Comrades so labeled suffered "ruthless struggle" and "merciless blows" organizationally and personally, and even in the way the Party dealt with the enemy.

In April 1931 the Central Committee in Shanghai sent a delegation to the Central Soviet Area. As soon as they arrived, they ordered people about like overlords and made arbitrary criticisms. They charged at the very start that the Front Committee of the First Front Army and the Central Soviet Area had made many serious mistakes.

In November 1931 the First Congress of the Central Soviet Area of the Communist Party of China was held in Ruijin, Jiangxi Province. Under the control of the Central Committee delegation, the congress criticized those people in the Central Soviet Area who had made "mistakes" in many areas. The congress attacked Mao Zedong's line of agrarian revolution as a "rich-peasant line" and a "right-opportunist mistake" and criticized those in the Red Army who did not discard the "traditions of guerilla-ism." The congress erroneously urged the Party to wage a struggle between the two lines and "concentrate firepower on the current main danger in the Party—right deviation." On the question of the Red Army, the congress stressed the need "to build" the Red Army "on the basis of large-scale operations." The congress excluded in the end the correct leadership of Mao Zedong and others over the work of the Party and the Red Army in the Central Soviet Area.

At the same time, in November 1931, the First National Congress of the Chinese Workers' and Peasants' Soviet was held in Ruijin. That congress elected Mao Zedong chairman of the Provisional Central Government of the Chinese Soviet Republic.

The First National Congress should have been a meeting of tributes to heroes and celebrations of victories, but because it was controlled by the Central Committee delegation, high-handed "leftist" pressure cast a deep shadow over the meeting.

Father attended the congress. There he came to know many comrades from the Soviet areas, including Wang Zhen—"the Bearded Wang," as he was called—a famous commander of the Red Army from a railway worker's family. Father said that before the congress, he caught sight of Li Mingrui at a distance. The two old comrades-in-arms who had shared weal and woe should have been very excited to meet again. But Father knew that the Central Committee was not pleased with Li Mingrui, and in the prevailing "leftist" atmosphere, he did not even walk up to greet Li, so as not to implicate him. The two just had a look at each other and departed. That was the last time Father saw Li Mingrui before he was killed at the evil hands of the "leftists."

The series of erroneous "leftist" policies did the greatest harm to the revolutionary work in the Central Soviet Area, in all its aspects, and they were resisted from the very beginning by Mao Zedong and others. The officers and men of the Red Army in the Central Soviet Area and the cadres and masses in the base areas faced two tasks: to repulse the powerful military attacks by the enemy, and to fight the erroneous "leftist" line.

But the Central Soviet Area and the Central Red Army stood the major tests presented by this series of difficulties and dangers.

CHAPTER 38

THE FIRST SECRETARY OF THE PARTY COMMITTEE OF THE CENTRAL COUNTY OF HUICHANG

In May 1932 Father took up the post of secretary of the Huichang County Party Committee, on the order of the Jiangxi Provincial Party Committee.

Huichang borders on Ruijin, fifty kilometers south of Ruijin. In April 1930 Mao Zedong and Zhu De had led the 4th Corps to Huichang, where they expanded Party organizations and established the County Party Committee. After the enemy's third campaign of encirclement and suppression was smashed, many of the rural areas in the county became red areas, but the county town was still under White rule. In November 1931 the 3d Corps of the Red Army seized the county town, and the following month, the Provisional County Party Committee was set up. In May 1932 Father came to Huichang to lead the Party, government, and army work there.

Within a few days after he arrived in Huichang, Father had stopped the harassments by the enemy "Peace Guarding Corps," thus stabilizing the situation and ensuring the safety of the people. While discussing work with Luo Pinhan, director of the Organization Department of the County Party Committee, Father said that, as Huichang was an important passage to Jiangxi, only fifty kilometers away from the Red Capital of Ruijin, and was also a big county, with fourteen to fifteen districts, it was necessary to establish a military department to meet the needs of struggle. On the recommendation of Luo Pinhan, Zhong Yaqing, the former deputy commander of the Independent Regiment of the 11th Corps of the Red Army, was appointed director of the Military Department. As Zhong Yaqing was an illiterate, he was afraid that he would be unable to do the job and did not go to Huichang to take up his post until Deng Xiaoping urged him by telephone several times. Zhong Yaqing recalled:

> When I walked to the Soviet of Santang District of Huichang County, I just happened to meet with Comrade Deng Xiaoping. Upon seeing me, he asked very loudly, "You, this comrade, where do you come from, and where are you going? What is your name?" "I come from Chengjiang. My name is Zhong Yaqing. I am going to Huichang," I replied.
>
> "So you are Zhong Yaqing. Good. I am Deng Xiaoping. Come along to the Soviet of Santang District and have a chat."
>
> I saluted Comrade Deng Xiaoping and followed him to the Soviet of Santang District on a dike. The chairman of the Soviet received us with tea. Comrade Xiaoping told me in a criticizing tone, "You are just too unruly. Old Luo phoned you many times, and you insisted on not coming. I criticized Luo Guibo again in my call, and you've come now at last. Look!" He pointed to the documents that were hanging on the wall, saying, "We have issued documents on your appointment as director of the Military Department. How dare you not come!" Then he said, "Just stay here and follow me to Luotang District today."

> I followed Comrade Xiaoping to the Soviet of Luotang District. In the evening Comrade Xiaoping called a meeting attended by the leaders of the district Party committee and the district Soviet, assigning tasks such as the expansion of the Red Army. The following day, at the breakfast table, we ate pork given by the leaders of the district Party committee. I remember that Comrade Xiaoping told the secretary of the district Party committee, "Pork is good, but one thing is missing." The Party secretary asked, "What is missing?" Comrade Xiaoping said frankly, "Pepper." The secretary rose and immediately disappeared. In no time, he returned with a handful of new pepper. Comrade Xiaoping took one and put it into his mouth, saying, "It is not too hot, but just right." Everybody ate pleasantly.
>
> In Huichang, our comrades in the Military Department often went down to various districts to organize the Red Guards. Those between eighteen and twenty-five were the core members of the Red Guards, and the others were ordinary members. We had to make lists of them and report to Comrade Xiaoping. We discussed many tasks with Comrade Xiaoping, including the expansion of the Red Army. Comrade Xiaoping was always ready to provide enthusiastic guidance for my work. He was very easy of approach.[1]

The reminiscences of Zhong Yaqing give a vivid picture of the work and life of Deng Xiaoping, the twenty-eight-year-old secretary of the Huichang County Party Committee.

In 1932 Chiang Kai-shek was still following his policy of nonresistance to the aggression of the Japanese militarists and vigorously suppressing the anti-Japanese democratic movements, regarding the Chinese Communists and the red revolutionary base areas as a serious hidden danger. After the previous three campaigns against the Central Red Army failed, he reassembled over 100,000 troops and launched an attack on the Soviet area in Hunan and western Hubei in June. Because of the erroneous "leftist" command, the 3d Corps of the Red Army was ordered to attack Jingzhou and Shashi. As a result, the 3d Corps suffered heavy casualties and also lost the revolutionary base area.

After this success, Chiang Kai-shek started again to assemble more troops to advance toward the Central Soviet Area. He was determined to annihilate the Chinese Workers' and Peasants' Red Army.

In the south, the Guangdong troops under Chen Jitang occupied the areas of Shanghang, Fujian, and Meixian, Guangdong. The three adjacent counties of Huichang, Xunwu, and Anyuan to the south were the borders of the Central Soviet Area, bordering on Fujian and Guangdong. They encompass chains of mountains, and the terrain there is strategically situated and difficult of access. The three counties constituted an important southern passage to Jiangxi and an important border area of the Central Soviet Area.

To meet the needs of the war situation, strengthen the work in the border area of the Central Soviet Area, and more effectively smash the enemy attack from the south, the Central Committee and the Jiangxi Provincial Party Committee decided to integrate Huichang, Xunwu, and Anyuan and establish at

Junmenling, Huichang, the Huichang Central County Party Committee to lead the revolutionary struggle in the three counties.

Junmenling was situated at the juncture of Huichang, Xunwu, and Anyuan, fifty-five kilometers from the county town of Huichang. It was the key junction of Fujian and Guangdong, a militarily contested place throughout history. It was only 100 kilometers from the Red Capital of Ruijin and could be called the southern gate for defending the Central Soviet Area. In July 1932 a meeting attended by Party activists of the counties of Huichang, Xunwu, and Anyuan was held at Junmenling. More than 100 Party cadres at and above the district level were present. A representative of the Central Committee, Luo Mai (Li Weihan), attended the meeting. Deng Xiaoping, Luo Pinhan, and others spoke, and the Huichang Central County Party Committee was formally set up, with Deng Xiaoping serving as the secretary. The office of the committee was at Junmenling.

In 1932, in accordance with the resolutions adopted by the Central Committee and the Jiangxi Provincial Party Committee on expanding the Red Army, the three counties, under the leadership of the Huichang Central County Party Committee, discussed in detail the very important task of expanding the Red Army and made assignments related to this task to various districts. At that time, the Red Army mainly recruited peasants and workers between the ages of sixteen and twenty-five. The principle was to conduct political mobilization with patience and heighten the awareness of new recruits without resorting to coercion or commandism, deception or bribery. Cadres and Party members were encouraged to take the lead. Propaganda was conducted widely, and relatives were encouraged to persuade relatives, neighbors to persuade neighbors, parents to persuade sons, and wives to persuade husbands.

In September a drive of Red Army expansion was launched. Thanks to the Party's thoroughgoing work and the growing awareness of the masses, young people asked to join the Red Army one after another. In just three months, from July to September 1932, more than 1,000 people joined the Red Army in Huichang County alone.

While expanding the Red Army, the Huichang Central County Party Committee paid full attention to recruiting Party members to expand the Party organizations. Party members in the three counties totaled nearly 6,000. Under the leadership of the county Party committee, Party membership increased rapidly. Party organizations were very sound and active.

To consolidate Soviet political power, Secretary Deng Xiaoping of the county Party committee called and presided over many meetings attended by the chairmen of the Soviets and leading members of the departments in the three counties.

The main tasks of the local Soviets were to expand the Red Army through organization and mobilization, sell government bonds, distribute land completely, improve conferences of representatives in urban and rural areas, fight corruption and slackness in work, and start to hold formal elections in some places, according to the election regulations promulgated by the Central People's Government.

Under the leadership of the Huichang Central County Party Committee, the people showed great enthusiasm, particularly for production. In May 1932 the county Party committee set up spring ploughing production committees at the county, district, and township levels to organize the season's work. That year the peasants in the county reaped a good harvest. The county Party committee also restored and developed the handicraft industry and ran small ordnance factories in Huichang and other places. To increase revenues, the county Party committee also established a "tariff office" at Junmenling and abolished all the exorbitant taxes and levies while practicing the system of unified progressive taxation. It also set up an "external trade bureau" to deal on behalf of the county government in salt, cloth, medicine, tobacco, paper, cereals, and other important goods. It found various ways to break through the blockade imposed by the enemy so as to ensure a steady flow of materials.

The Huichang Central County Party Committee paid great attention to expanding local armed organizations. It attached importance to setting up the Communist Youth League, the trade union commission, and the committee of guidance for women. It also gave much consideration to education. Over 90 percent of the children in the three counties went to school.

Father told us that when he was in the Soviet area, he traveled all over the vast area of Ruijin and Huichang on horseback with a bodyguard who also acted as the horse tender. His horse died before climbing the snow-capped mountains during the Long March, and his bodyguard was also changed before the march. Father does not like ostentation and extravagance and opposes hairsplitting. His habit of keeping one horse and one bodyguard lasted until the beginning of the anti-Japanese war. Even after he took on much higher positions, he still lived a simple life. During the whole period of the anti-Japanese war and the War of Liberation, he had no private secretary. From postliberation days up to the start of the Cultural Revolution, he had only one secretary. To him, the number of people one employs does not matter; what is important is efficiency.

Father is a man of action and a man of daring and resolution. Wherever he worked, he could soon open up new prospects, whether in the Zuojiang and Youjiang areas, Guangxi, or in Ruijin and Huichang, Jiangxi, because he was resolute and firm and did his work in a well-organized manner. His boldness and resolution and his ability grew with his experience. These qualities, plus his extraordinary foresight and sagacity, enabled him to create a completely new situation for China decades later when the destiny of the entire nation was in his hands.

Between 1932 and 1933, when Father worked in this area of Huichang, he not only thoroughly changed the face of this border of the red area but also left a deep impression on his comrades-in-arms. Zhong Yaqing, who was the former commander of the Third Operational Subarea of the Jiangxi Military Area in Huichang, recalled:

> In September 1932 I fought a battle with the several hundred troops led by Zhong Shaokui in Dongliu, Fujian. I was wounded again in action. Then the troops were led by my chief of staff, and I was taken to Guikeng. Because I was bleeding so much, I was taken again to Luotang District in Huichang for medical treatment. The following day I was moved again to the medical collecting center of the Guangdong-Jiangxi Military Area at Junmenling.
>
> When Comrade Deng Xiaoping (who was then the political commissar of the Third Operational Subarea) learned from the battlefield report that I was seriously wounded, he himself immediately made a call to inquire about my condition. As I was unable to receive the call, the head of the center, Luo Tianguan, did it for me. He told me that Comrade Xiaoping urged me time and again to go by boat to Huichang for medical treatment the next day. I asked Luo to tell Comrade Xiaoping that I was so seriously wounded that the wound bled whenever I moved and I could not go to Huichang, at least not for a few days. During that period, Comrade Xiaoping made a call every day to ask about my condition. I felt very poorly, and so after four days in the collecting center, when I was a little more energetic, I was taken to Huichang by a small boat. The afternoon of the following day, Comrades Deng Xiaoping and Luo Pinhan came to the hospital to see me. Comrade Xiaoping cordially comforted me, "I read your report from the battlefield. The affairs on the front have been handled by somebody else, and you need not worry. Have a good rest!." No sooner had he finished saying that than he took out fifty silver dollars and gave them to me to buy food. I took the money and was unable to calm down for a long time. I was moved to tears because Comrade Xiaoping was very busy yet he took time to see me and, furthermore, gave me so much money under very difficult financial conditions.
>
> I stayed in the Huichang hospital until March 1933. Right after I left the hospital, I went to the office of the Huichang County Party Committee, where I met Comrade Luo Pinhan but not Comrade Deng Xiaoping. I was so disappointed. Fifty years have elapsed in an instant since then. What a great change in these fifty years! Comrade Luo Pinhan heroically laid down his life in battle long ago. Fortunately, Comrade Deng Xiaoping is still in good health and is leading us in the drive for four modernizations. Whenever I recall the past and how Comrade Deng Xiaoping devoted himself to the revolution, with his foresight and industriousness and his consideration for other comrades, I feel very excited.[2]

Indeed, Zhong Yaqing did not meet Deng Xiaoping, secretary of the Huichang Central County Party Committee, when he left the hospital in March 1933, because Deng had already been transferred from Huichang after being criticized by the "leftist" element.

CHAPTER 39

THE "DENG, MAO, XIE, AND GU INCIDENT"

Wang Ming's "left"-adventurism and factionalism prevailed in the Central Soviet Area, in the Central Red Army, and in all the revolutionary base areas.

In October 1932 the Central Bureau of the CPC in the Central Soviet Area held a meeting in Ningdu, Jiangxi Province. At the meeting, the "left"-adventurists strongly insisted on attacking big cities. They accused Mao Zedong of being "slack in work" and "disrespectful" of their leadership and of making the right-deviationist mistake of "waiting for enemy attack." Because of his resolute struggle against the "leftist" mistakes at the meeting, Mao Zedong was dismissed from his military post of general political commissar of the First Front Army of the Red Army and was transferred to a post where he was to do "special government work." As a matter of fact, he was deprived of his military authority.

By January 1933, owing to the wrong policies of the "left"-adventurists, the Party work in the White area had suffered a nearly total failure. The Provisional Political Bureau of the Central Committee had to move from Shanghai to Ruijin in the Central Revolutionary Base Area, thus enabling the Provisional Central Committee, which had been implementing Wang Ming's line, to exercise even more direct leadership over the work in the Central Soviet Area.

After the September 18th Incident, Chiang Kai-shek's government in Nanjing had adopted an absolute policy of nonresistance to the Japanese imperialists. As a result, the Japanese aggressors brazenly occupied China's three northeast provinces and in 1932 established the "Manchukuo," the puppet state under Japanese control. The Japanese army established ruthless colonial rule over northeastern China, frenziedly suppressed anti-Japanese organizations, and slaughtered innocent civilians, thus driving the Chinese there into an abyss of misery. In January 1933, intensifying their aggression, Japanese troops forcibly occupied Shanghaiguan, the communication center of northern China, killed Chinese soldiers and civilians wantonly, and began to target their aggression at China's Rehe Province. In early March Japanese troops seized Chengde, the capital of Rehe Province, arrived at various passes along the Great Wall, and prepared to launch a large-scale attack on China's Central Plains.

In the face of the national calamity caused by these powerful enemy aggressors, Chiang Kai-shek, in spite of the strong opposition of all kinds of Chinese people, went so far as to assemble a total force of 650,000, which included 81 divisions, 29 brigades, and 39 regiments, and launched the fourth campaign of encirclement and suppression against the Red Army. He assumed personal command in Wuhan and appointed himself commander-in-chief of the "Bandit Suppression Headquarters" and concurrently commander-in-chief of "bandit suppression" in the three provinces of Henan, Hubei, and Anhui.

Between February and March 1933, Chiang Kai-shek arrogantly employed half a million troops to attack the Central Soviet Area. By this time, Mao

Zedong had already been pushed out of the Red Army. While resisting the "leftist" obstructions of the Central Bureau of the CPC of the Central Soviet Area, Zhou Enlai and Zhu De were adhering to correct strategies and tactics. Using ingenious tactics, the Red Army smashed the enemy's fourth campaign. In these battles, three divisions of the enemy's 1st Column were totally wiped out, and the commanders of the 25th and 59th divisions, as well as over 10,000 guns and enemy troops, were captured.

The victory won by the Red Army in its fourth countercampaign was described by Mao Zedong as "unprecedentedly glorious and great," whereas Chiang Kai-shek, in his letter to Chen Cheng, lamented it as "the first secret anguish in my lifetime."

As a result of the battles of the fourth countercampaign, the Central Soviet Area expanded to cover areas in the four provinces of Hunan, Jiangxi, Fujian, and Guangdong and was linked up with the Soviet areas in Fujian, Jiangxi, and Zhejiang. The Central Red Army had swelled to 100,000 men. The Red Army men and members of the Communist Party numbered 300,000, respectively, throughout the country.

The victories won in the first, second, third, and fourth countercampaigns were possible because the Red Army adopted flexible, correct, and suitable strategies and tactics. Victory was also possible because, while red revolutionary political power was being established in the Central Soviet Area, the agrarian revolution was being carried out. The correct land policies of Mao Zedong and others resulted in the distribution of land to vast numbers of poor peasants. People of all strata showed unprecedentedly great enthusiasm for production. With the support and cooperation of the people, revolutionary base areas were increasingly consolidated, and the Red Army grew daily and won one victory after another.

However, there is no easy path to victory under heaven. Wherever there is truth, there is falsehood. The two are as inseparable as negative and positive numbers. Truth can show its radiance only in the struggle against falsehood. The history of mankind has always been tortuous and is full of ups and downs. Sometimes truth stays on, and sometimes falsehood prevails.

In early 1933, when the Provisional Political Bureau of the Central Committee moved to the Central Soviet Area, some "left"-adventurists, represented by Bo Gu, opposed the policies followed by Mao Zedong and others in the Soviet area. They not only deprived Mao Zedong of his leadership over the Red Army but also tried to squeeze out and attack other comrades who resisted "leftist" policies. At the same time, practicing large-scale factionalism, they dispatched representatives to all Soviet areas to carry out the so-called antirightist struggle and to "reform the Party's leadership at all levels."

In February 1933 Luo Ming, acting secretary of the Fujian Provincial Party Committee, did not agree to erroneous "leftist" policies and was accused by the "leftist" leaders of making right-opportunist mistakes and of taking pessimistic views on the revolution. This was the so-called Luo Ming's line. Luo was dismissed from his post and suffered many disciplinary actions. In March the struggle waged by the Provisional Party Central Committee was directed at Jiangxi Province.

On March 12 the Jiangxi Provincial Party Committee sent letters of instruction to Huichang, Xunwu, and Anyuan counties, accusing the Party and League organizations there of making mistakes of "opportunism similar to Luo Ming's line and the purely defensive military line." It had all started with the "Xunwu Incident."

In 1932, while the fourth countercampaign was under way in the Central Soviet Area, Wang Ming and other "left"-adventurists who placed undue emphasis on expanding the Central Red Army organized some of the local armed forces of the three counties into the regular Red Army, thus greatly weakening the local armed forces in the southern border district of the Soviet area. Then they ordered the Red Army units that were defending the southern front of the Soviet area to move north, thereby further weakening the military strength in that area. In November 1932 the enemy troops launched a large-scale attack. Owing to the great disparity in strength, the county town of Xunwu, which was situated in the border area of Jiangxi, Guangdong, and Fujian provinces, fell.

Leaders of Wang Ming's "leftist" policies made use of the Xunwu Incident. They accused the Huichang Central County Party Committee of "panicking and attempting to run away before the enemy attack" and implementing the "purely defensive military line." This became the prelude to the campaign against the "Jiangxi's Luo Ming's Line" in the counties of Huichang, Xunwu, and Anyuan.

The Xunwu Incident was only a pretext. The campaign was actually the result of the struggle between "leftist" and anti-"leftist" policies. It was a strategic move by Wang Ming's "leftist" leadership to carry out factional attacks on those comrades in the Party who held different views.

Although Mao Zedong's "rich-peasant line" was criticized and he was removed from military leadership during the First National Congress and at the Ningdu Meeting of the Political Bureau of the Central Committee in the Soviet area, many Party members and cadres in the Central Soviet Area and the Red Army did not agree with Wang Ming's "leftist" policies and firmly resisted them. The struggle was represented by Luo Ming in Fujian and Deng Xiaoping in the Central Soviet Area.

On a series of questions, the county Party committee, with Deng Xiaoping serving as secretary, earnestly implemented the correct policies proposed by Mao Zedong, which conformed to the actual conditions in the border areas. The committee resolutely resisted, in both theory and practice, the mistakes of Wang Ming's dogmatism in an attempt to reduce the losses those mistakes inflicted on the Party. Thus the committee had become a major obstacle to the general implementation of "leftist" policies in the Central Soviet Area by Wang Ming and other "left"-adventurists.

When Father was in the 7th Corps of the Red Army, he resented but had passively implemented "left"-adventurist policies. This time, in the Central Soviet Area, he unhesitatingly took the lead in offering clear-cut resistance to "leftist" mistakes.

The experience of the 7th Corps and the losses incurred by the Party and the revolution had given large numbers of Communists like Father a sober understanding of the mistakes of the "left"-adventurism. They were not scared by Wang Ming's labels of dogmatism and factional arrogance. They started to fight, and to fight conscientiously.

An intensive campaign was now launched to criticize Deng Xiaoping, Mao Zetan, Xie Weijun, and Gu Bai. In February 1933 these four were criticized by name in an article entitled "Oppose Luo Ming's Line" in *The Struggle*, the official organ of the Central Bureau of the Central Soviet Area. They were described as "the leaders of Jiangxi's Luo Ming's line." In another signed article entitled "What Is the Offensive Military Line?" the Huichang Central County Party Committee was criticized by name for making the mistake of holding a "purely defensive military line." The article accused the counties of Yongfeng, Ji'an, Taihe, Huichang, Xunwu, and Anyuan of "getting stuck in the mud of a purely defensive military line for a long time" and pointed out the need to "fight against all opportunist wavering, flightism, and purely defensive military lines, as well as the compromise between these lines."

The "leftist" leaders instructed the Jiangxi Provincial Party Committee to issue repeated directives to the regions and work units where these four people worked to mobilize grass-roots cadres and Party members to directly criticize and denounce them. On March 12, again in accordance with the intention of the Central Bureau, the Jiangxi Provincial Party Committee published instructive documents from the Party on the three relevant counties within the Party in the Jiangxi Soviet area, accusing the Huichang Central County Party Committee led by Deng Xiaoping of being "panic-stricken," "retreating and running away," and making "the mistake of taking a purely defensive military line." Therefore, the committee was "opportunist," in the same way as "Luo Ming's line."

In late March Deng Xiaoping, secretary of the county Party committee, was sent to Wantai, Gonglue, and Yongfeng to solve certain problems. At the end of March, at a meeting held at Junmenling and attended by Party activists from the counties of Huichang, Xunwu, and Anyuan, Deng Xiaoping came under converging attacks. After the meeting, Deng was removed from the post of secretary of the county Party committee and transferred to the post of director of the Propaganda Department of the Jiangxi Provincial Committee.

In April 1933 the "leftist" factionalists continued to wage "ruthless struggles" against the four, dealing them "merciless blows" and instructing them to make "statements" and "self-criticism." But Deng and the three others did not yield. They did not give in on matters of principle, and they stood up strongly to the "leftist" factionalists. By writing two statements that explained their views and actions, they seriously challenged the abuses, attacks, and slanders inflicted on them. Their uncompromising position further enraged the "leftist" leaders and led to an even fiercer and larger attack on them.

At the meeting of the Jiangxi Provincial Party Committee held on May 5 to sum up its work—presided over by the representatives from the Provisional Central Committee and the Central Bureau—the Jiangxi Provincial Committee adopted "The Resolution of the Jiangxi Provincial Committee on the Two Statements Made by the Four Comrades of Deng Xiaoping, Mao Zetan, Xie Weijun, and Gu Bai." Party disciplinary actions were taken against them. All or part of their responsibilities were removed, and their pistols were handed over in the presence of the participants. They were instructed to go to the grass-roots units to remold their thinking. They were asked to further "state" and "expose" their mistakes and to make new self-criticisms, and they were warned that "no cover-ups will be permitted any more."

Deng Xiaoping was dismissed from his post as director of the Propaganda Department of the Provincial Committee and was given a "final serious warning" within the Party. Mao Zetan was removed from all his posts in the Red Army. Xie Weijun was transferred to another post, and Gu Bai was removed from his post and given a "final serious warning."

In this struggle against "Jiangxi's Luo Ming's line" not only were Deng, Mao, Xie, and Gu criticized and attacked but all cadres at the provincial, county, and branch levels who held correct views as representatives of "Luo Ming's line" were denounced. This was done in the name of "waging the struggle against opportunism in an in-depth way down to the lower levels and in all practical work." Large numbers of cadres were replaced everywhere, and Party members were in panic, thereby causing heavy losses to the work in the revolutionary base areas.[1]

Although Deng, Mao, Xie, and Gu were criticized, denounced, and even dismissed from their posts and punished, they did not give in because they were staunch Communists and seasoned revolutionaries. Instead, they adhered all along to their correct position and the truth. For a very long period, they endured humiliation and bore a heavy load for an important purpose. They continued to fulfill the duties that a Chinese Communist should perform and advanced bravely along the rough and difficult road of revolution until their last breath.

Some people might ask, Why did they act like that? Why did they not give in to the wrongdoing? Why were they not swayed by personal feelings? Why did they stay in the revolutionary ranks in the face of such unfair treatment?

Because they were Chinese Communists who had firm faith. They believed that despite occasional twists and turns, their cause was just. They believed that although the Party could sometimes be misled, their Party was always great. They believed that although there were many dangers and difficulties on the road to save the country and the people, bright prospects always lay ahead. Indeed, they were the truly outstanding CPC members and revolutionaries.

After the Central Red Army began to move to western Hunan and went on the Long March in October 1934, Mao Zetan stayed in the Central Soviet Area to persist in guerrilla warfare and serve as a member of the Branch Bureau of the Central Soviet Area and commander of the Independent Division of the Red Army. In early 1935 he led some units of the Independent Division to Changting, Fujian Province, and served as a leading member of the Military District Headquarters in the border area between Fujian and Jiangxi provinces. In April 1935, when he led guerrillas in a march, he died a heroic death in the Honglin Mountains in Ruijin, Jiangxi Province, at the age of thirty.

After being attacked by "leftism," Xie Weijun was appointed an inspector and took charge of collecting grain and recruiting Red Army soldiers. In this difficult situation, he endured humiliation to carry out an important mission, worked hard, and was not upset by criticism. He took part in the Long March in 1934. After the Zunyi Meeting of 1935, he served as secretary of the local work department of the General Political Department of the Red Army. When he reached northern Shaanxi, he was appointed secretary of the Sanbian Special CPC Committee. While leading troops to advance to Bao'an County, he encountered a surprise attack mounted by bandits and was killed in an intense fight at the age of twenty-seven.

Gu Bai went on to take charge of grain collection. After the start of the Long March, he stayed behind and served as the commander of the guerrilla column of the Red Army in the Fujian-Guangdong-Jiangxi area. Between the spring and summer of 1935, while leading his units to Longchuan, Guangdong Province, he was surrounded by the reactionary civil corps because of information provided by a traitor. He died a heroic death in the fighting at the age of twenty-nine.

All these three comrades joined the revolutionary ranks in their youth, endured the attacks by "leftists" in their twenties, and gave their young lives for the revolution before reaching the age of thirty.

Youth—the twenties and thirties—is the brightest moment in one's life. Everyone cherishes youth and loves life. However, these three revolutionaries were resolutely thrown into the revolution and gave their all to it.

Setbacks and difficulties are inevitable. Some people are scared in the face of setbacks or shrink from difficulties. To a revolutionary like Deng Xiaoping, suffering setbacks and difficulties is a common experience. In his long and legendary revolutionary career, every time he overcame them, he took a step forward.

More than 2,000 years ago, the famous thinker Lao Zi of the Spring and Autumn Period said that good fortune lies within bad, and bad fortune lurks within good. In other words, the relationship between good and bad fortune is dialectical and interchangeable. An event is rarely absolutely good or bad, and the degree of good or bad fortune in it varies from person to person and from time to time.

The attack on Father by "leftists" in the Central Soviet Area was a heavy political burden to him at that time. However, this event of the 1930s became quite an important factor in his political career forty years later—and moreover, a good and positive factor.

In 1966 the Cultural Revolution broke out in China. One year later Father was overthrown as the "No. 2 capitalist roader." In 1971 Lin Biao, handpicked by Mao Zedong as his successor, was too eager to usurp power and failed in his attempt to assassinate Mao. Lin fled and was killed in a plane crash. When Father heard of the crimes of Lin in Jiangxi, where he was under house arrest in 1972, he was most excited and wrote a letter to Mao Zedong, expressing his views on the incident. On August 14, 1972, Mao Zedong wrote the following instruction on Father's letter:

> Comrade Deng Xiaoping has made serious mistakes, but he should be differentiated from Liu Shaoqi. First, he was under attack in the Central Soviet Area. He was one of the four guilty persons—Deng, Mao, Xie, and Gu—the head of the so-called Maoist faction. The materials about him can be found in the history of struggle between two lines in the two books after the Sixth National Congress. . . . Second, there are no questions of a political nature in his history. That means he did not surrender to the enemy. Third, he assisted Comrade Liu Bocheng in fighting battles and rendered meritorious military service. Besides, after we entered the cities, we cannot say that he did nothing good. For instance, when he was the head of a delegation to Moscow for negotiations, he did not bow to the Soviet revisionists. I have said these things many times. Now I repeat them once again.

This was Mao Zedong's written instructions, which were then regarded as the supreme guide.

Starting from these instructions, the downswing in Father's political career began to reverse. He finally returned to Beijing in March 1973 and was reinstated in his former position of vice-premier of the State Council of the People's Republic of China, assisting Premier Zhou Enlai in the day-to-day work of the State Council. In 1975 Father was reappointed vice-chairman of the Military Commission of the Central Committee and, concurrently, chief of the General Staff of the People's Liberation Army. After these appointments, Father began to stem the frenzied "leftist" tide and make general rectifications in every field throughout the country.

As for Mao Zedong's reasons for reinstating Father after his second downfall, apart from those stated in his instructions and his view of Father as a "rare talent," the "Deng, Mao, Xie, and Gu Incident" in the 1930s was an important factor to be reckoned with. Deng was attacked precisely for implementing Mao Zedong's policies and practices and for being "the head of the Maoist faction."

"It is imperial ideology to say that there are no other parties outside the party; it would be most odd if there were no factions within the party." This is a famous quotation from Mao Zedong.

The struggle in the 1930s put Deng in the Maoist faction. That Mao Zedong would remember that fact for forty years was not anticipated by Deng Xiaoping at the time.

CHAPTER 40

THE EDITOR-IN-CHIEF OF *RED STAR*

In May 1933, after Father was subjected to the factionalist criticism of Wang Ming's "left"-adventurism and dismissed as director of the Propaganda Department of the Jiangxi Provincial Party Committee, he was assigned to the post of inspector at Nancun, Le'an County. Less than ten days after he arrived at Le'an, he was ordered back to the Provincial Party Committee. It was said that since Le'an was a border area, the "left"-adventurists were afraid he might cause trouble.

Soon Father was transferred to the General Political Department of the Workers' and Peasants' Red Army to serve as its secretary-general.

At the time, the director of the General Political Department was Wang Jiaxiang, and the deputy director was He Chang. Because Wang Jiaxiang had been wounded in action, he was in poor health. He Chang was actually in charge of the work of the General Political Department.

When He Chang was the leader of the Southern Bureau of the Central Committee, he went with Father to Guangxi to prepare for the Bose Uprising. Later, they often met in Shanghai. Father lived in He Chang's residence for some time. So they were very familiar with each other. Many of Father's comrades resented the attacks on him by Wang Ming's "left"-adventurists. He Chang was very sympathetic to Father's bitter experience and freed him from the predicament by trying to transfer him to the General Political Department to serve as its secretary-general.

A woman comrade called Zhu Yueqian worked in the General Political Department at the time. Her husband, Huo Buqing, had once worked in the Military Commission of the Central Committee in Shanghai. While in Shanghai, Huo Buqing and his wife, Zhou Enlai and his wife, Father and Zhang Xiyuan belonged to the same Party group of their branch. Later, when Huo Buqing went to the Central Soviet Area in Jiangxi Province, he often met with Father. Both were natives of Sichuan Province and were secretaries of county Party committees, so they maintained relatively close relations. Zhu Yueqian told me that

> Huo Buqing was a native of Qijiang County, Sichuan Province. He and your father were from the same province. While in Jiangxi, they both carried Mauser pistols and wore straw sandals with leg wrappings. When they sometimes met in Ruijin, they went together to eat noodles. At the time, noodles with shredded meat were good food, as they could not have chicken or meat. Although your father was young, he was always cheerful. Many cadres were persecuted under Wang Ming's line then, and your father took objection to Wang Ming's ultra-"leftist" line. So Wang Ming attacked your father, punished him by giving him the most serious warning within the Party. Huo Buqing was also given the punishment of warning within the Party. But your father did not mind the punishment in the least. He continued to chat

> and laugh and remained optimistic without looking worried. Both of them were responsible to the Party and the people.
>
> In September 1933 Huo Buqing died of illness. Chen Tanqiu, secretary of the Fujian Provincial Party Committee, transferred me to the General Political Department. Your father was the secretary-general of the General Political Department, taking charge of everything. At the time, Huo Buqing had just died, and I was pregnant. I felt very unhappy. Your father often comforted me. In view of the lesson of Zhang Xiyuan's death, your father let me inform him early of the childbirth so as to make proper preparations. When I was about to give birth to the child, your father sent three stretcher-bearers to carry me on a stretcher to the hospital more than ten kilometers away and had his own bodyguard escort me all the way. After giving birth to the child, I had only two suits of clothes. I had neither clothes nor diapers for the child, so I had to wrap the child with my own clothes. I wrote a note to Comrade Xiaoping to the effect that I had no clothes or diapers for the child and asked him to help me draw my share from the Red Army's public field so as to buy some clothes. At the time, the Red Army had public fields, and each soldier could draw a share of harvests. Your father replied with a note, saying that we cadres should not draw our shares of the Red Army's public fields, which should be given to the soldiers. He drew for me ten silver dollars for child-bearing expenses and four for child-care expenses, sent his bodyguard to give me the money, and solved my problem. While adhering to principles, he cared for the comrades under him. But he did not take care of himself. Your father is a very good cadre.

Father is just such a man. He worked conscientiously and never gave in on matters of principle. He cared for other comrades very much and did not mention what he had done for them. He was very strict with himself and did not show his feelings too openly, whether they stemmed from joy or misfortune.

At the time, Father suffered setbacks not only in his political career but also in his private life. After he had been subjected to the "leftist" criticism, Ah Jin left him in 1933.

The moon waxes and wanes, and humans have their joys and sorrows, their partings and reunions. Ever since ancient times, no complete perfection can be expected. People do not know how many joys and sorrows, partings and reunions, or hardships in their careers and lives they have to experience to complete their asceticism and attain nirvana.

Father experienced vicissitudes both in his career and in his life. He lost two wives (for two different reasons). Politically, the "leftist" criticism was the first and in fact the smallest hardship he would be faced with. He was to suffer two greater political setbacks thirty years later.

He was overthrown three times and reinstated three times, and each of his reinstatements was more glorious and more shocking than the previous one. As many people have commented, his extraordinary rise and fall three times has made Deng Xiaoping's life legendary, and people admire him for it. Uli Franz, his German biographer, wrote that Deng Xiaoping

> has overcome his political rise and fall three times and numerous schemes and intrigues with his extraordinary ability and has come one step closer to the goal of his life each time. In our century, I have not seen in either the East or the West such a stateman as Deng Xiaoping who has traversed so tortuous a path in life and accomplished so much.[1]

Suffering hardships and setbacks is by no means relaxing, but not experiencing them can lead to an insipid life. Father was about thirty when he suffered setbacks in both his political career and his private life in 1933. Others may be just learning to stand on their own feet by the age of thirty. But Father at thirty had already experienced a great deal. As a result, he did not fear even great political storms, and he also did not take the unhappiness of his private life too much to heart.

Ah Jin was a capable woman cadre of the Red Army. After she arrived with Father in the Central Soviet Area, she served as secretary of the Yudu and Shengli county Party committees. She led the Party, the government, the army, and the people of the two counties in carrying out economic construction, expanding the Red Army, and supporting the front. After she separated from Father, she was transferred to the Organization Department of the Central Committee to serve as the chief of the Organization Section, and as the deputy director of the Armed Forces Mobilization Department of the Revolutionary Military Commission of the Central Committee the following year. In October 1933 Ah Jin was instructed to serve as the general leader of the Red Army Expansion Shock Force of Ruijin County. She accomplished her task brilliantly, and the Central Committee made a favorable comment on her. When the Central Red Army started the Long March in 1934, Ah Jin and over twenty women comrades of the First Front Army were organized into the 2d Column of the Central Committee, the "Red Armband Column." She worked with the local work department and mobilized the masses on the way. Later on, she was transferred to a cadre recuperation company directly under the Central Committee to serve as the secretary of its Party branch. After she arrived in northern Shaanxi after the Long March, she served as the chief of the Organization Section of the Organization Department of the Central Committee and held other posts. Many women comrades, including Ah Jin, contracted diseases because of the long years of hard wartime conditions. In 1938 the Party organization decided to send her to the Soviet Union for medical treatment. The war between the Soviet Union and Germany broke out in 1941. The flames of war soon spread to Moscow. Ah Jin was receiving medical treatment in a hospital in the suburbs of Moscow, and unfortunately, she died, at the age of thirty-seven, in the chaos caused by the war.

Father did not hold the post of secretary-general of the General Political Department for very long. Because he had little work to do, he requested, after two or three months, a transfer to another post so as to do more practical work.

The General Political Department assigned him to the post of assistant in its Propaganda Department. In addition to ordinary propaganda work, Father edited *Red Star*, the official organ of the Revolutionary Military Commission of the Workers' and Peasants' Red Army during the Second Revolutionary Civil War. It was launched on December 11, 1931, and published by the General Political Department. Father held this post until the eve of the Zunyi Meeting during the Long March.

Mom Liu Ying, wife of Zhang Wentian, told me:

> I came back in 1933 after studying in Moscow and met your father in the Central Soviet Area. At the time, he had made mistakes and had been dismissed from his posts. He was editing *Red Star* in the General Political Department. I was assigned to the post of director of the Propaganda Department of the Central Committee of the Youth Communist Party (Youth League). Our office was very close to that of the General Political Department. There were three houses—one each for the Central Bureau, the General Political Department, and the Youth Communist Party. They were all in the same village and very close to each other.
>
> There was a group of youths, including us. We had few recreational activities in the countryside, so we paid neighborly visits to each other after supper. We liked to go to He Chang's house to amuse ourselves and to chat with your father because he had much knowledge. He was very optimistic. He Chang told us how your father had been criticized and divorced [from Ah Jin]. Actually, He Chang took charge of the work of the General Political Department. He was very sympathetic to Comrade Xiaoping. He said Comrade Xiaoping was very capable and had suffered many injustices. Later, I became the leader of the Red Army Expansion Force and was overzealous about my tasks. Your father told me jokingly, "You don't speak unless you can say something sensational. You were reported by *Red Star*!"

The readers of this newspaper were the officers and men of the Red Army and local cadres and people in the Central Soviet Area. The educational level of most of these officers and men, who were predominantly of worker-peasant origin, was very low, and many were even illiterate. There were no other newspapers, books, or periodicals, let alone radio or television. *Red Star* naturally became the source of information and study material for the officers and men of the Red Army and a very good channel for disseminating the Party's ideology and cultural knowledge.

Over seventy issues of *Red Star*, all edited by Father, were published between August 1933 and the eve of the Zunyi Meeting in 1935. A dozen issues were published after the Zunyi Meeting, until August 3, 1935. Publication was stopped because the Red Army had begun to take the arduous Long March.

Father was not a member of the cultural circles, nor was he a journalist. But he was familiar with running a newspaper, a revolutionary newspaper in particular. Ten years before, with Zhou Enlai and others, he had run *Red Light*, the CPC and CCYL journal in France, where he had also won a reputation as a master of mimeography.

The young twenty-year-old Communist in Paris had grown into a mature cadre of the Red Army. Those ten years had been dazzling. He had gone through many experiences, met many people, and had many thoughts. Those ten years, especially the improvement of his political and theoretical level, enabled him to run the newspaper with skill and ease.

When we examine *Red Star*, we can find Father's handwriting everywhere. Although the newspaper was printed in type, Father often handwrote headlines, apparently to attract attention and diversify the format. Father's handwriting then was quite attractive and forceful.

Father told us that he had only a few assistants in editing *Red Star*, and only two for a long time. So he had to select and edit materials, print and write news reports and articles, himself. The headlines he wrote were carved by other comrades on wooden type before being printed in the newspaper. Father said that many unsigned information pieces, news reports, and even many important articles and editorials were written by him. I once showed him the collections of *Red Star* gathered by the Central Archives and asked him to identify the articles he wrote. Waving his hand, he said, "Quite a lot! How can they be identified!"

One editor-in-chief and two assistants had to publish an octavo newspaper of at least four pages an issue every five days on average. That was indeed a heavy workload.

As the mouthpiece of the Revolutionary Military Commission of the Central Committee, *Red Star* carried quite a few resolutions and orders of those two bodies. Leaders of the Party, the government, and the army, including Zhou Enlai, Zhu De, Bo Gu, and Luo Fu (Zhang Wentian), wrote articles and editorials for the newspaper. As one of the principal leaders of the Military Commission, Zhou Enlai was the number-one contributor.

Half of the editorials published in *Red Star* were signed, and the rest were unsigned. I guess most of the unsigned editorials were written by Father. We can see from these articles alone that Father had made great progress in his political and theoretical development and in his practical experience compared with his experience and level of development when he was studying in France. These articles revealed a mature political ideology, great militancy, and no trace of the naiveness he had shown in France.

Apart from reporting news and operations, *Red Star* published many articles about daily life—for instance, articles on ordnance or disease prevention, even interesting questions-and-answers and riddles. We can see from these insignificant pieces that Father really devoted all his energy to editing this newspaper. He tried hard to do his job conscientiously and wholeheartedly by paying attention to every sentence and every word.

Some old comrades told me that Father was always optimistic and could work at either a high or a low level, as required. I think *Red Star* alone reveals this adaptability of his. Such Communists are truly selfless and do not consider their own honor, disgrace, or interests. They can command powerful, massive armies, as well as do their jobs well in ordinary posts.

In our childhood, we often heard: "For the cause of the Party and the people, I am of little account!" Today some people may consider this sentiment a cliché. But the revolutionaries and Red Army fighters of Father's generation truly believed it and acted on it. Thousands upon thousands of revolutionaries besides Father also kept their will to fight in the face of difficulties and hardships and lived tenaciously and optimistically. A paragraph from Li Weihan's *Recollections and Research* gives us a detailed and vivid account of these revolutionary fighters:

> Life in the Central Soviet Area was very hard, and the youths today do not know it very well. The slogan put forth in the Central Soviet Area at that time was "all for the front." Actually, the life of the fighters on the front was not good either. The only difference was that they had a little more salt and could eat their fill. We did not know then to engage in production ourselves. Grain was carried on shoulders to Ruijin. As there was not enough grain, personnel in the rear area could only have two meals a day and could not eat their fill.
>
> Before each meal, everybody put his or her share of rice in a cattail bag with his or her name on it and sent it to the kitchen for steaming. If the rice was not divided, those who ate quickly could eat more, while others who ate slowly would eat less. When rice was not enough, it did not work to eat from the same big pot. At that time, military communism was practiced, and the supply system applied. At each meal, we had very small servings, with no cooking oil. The dishes were small bowls made of iron, and the servings were so small that they did not cover the bottoms of the bowls. We got very hungry between 10:00 A.M. and 12:00 noon every morning. The same thing happened every evening. When we went hungry beyond endurance, we lay on our beds and rested for a while before getting up to work again. . . .
>
> There is a world of difference between the life of the Red Army in those years and the life of the Liberation Army today. As salt could not be transported to the Soviet area then, we had to produce it ourselves. Because of the enemy's blockade, we had difficulties not only in feeding but also in clothing ourselves. The clothes were blue in color. White cloth was dyed to make the clothes, and the color faded easily. Wherever we were, we slept at night without taking off our clothes so as to be ready at all times to march and fight. The bullet supply was more difficult. In operations, the bullet jackets had to be kept for refilling powder.
>
> The political and ideological work was well done then. Under those difficult conditions, nobody had any complaints and everyone devoted themselves wholeheartedly to the revolution. The people shared the fruits of victory of the revolution and gained their benefits, thereby giving their great support to the revolution. I saw with my own eyes how the Red Army expanded. Most of those who joined the Red Army were core militia members. The grand occasion of parents sending their sons and wives their husbands to join the Red Army could be seen everywhere. Those who joined the Red Army carried handkerchiefs and straw sandals on their backs. At the time, the Party and the Red Army maintained very close ties with the masses and enjoyed very high prestige among them.[2]

True, the life at that time was difficult beyond our imagination, and the struggle against the enemy cruel beyond description. Had there not been the fine tradition of unity between the army and the people—the army fighting for the people, and the people supporting the army—and between officers and men, the Red Army would have been defeated by the strong enemy long before. Had there not been a group of so exceptionally heroic and staunch Communists who were always loyal to the people, to the revolution, and to the cause of the Party, no matter what the circumstances, it would have been impossible for the Communist Party of China to achieve the final victory in the Chinese revolution.

CHAPTER 41

THE FAILURE OF THE FIFTH COUNTERCAMPAIGN AGAINST ENCIRCLEMENT AND SUPPRESSION

In the history of both the Communist Party of China and the Chinese Workers' and Peasants' Red Army, 1933 and 1934 were the years in which misfortunes did not come singly.

For over a year, the Party and the Red Army were confronted not only with the enemy's ever more frenzied and cruel campaigns of encirclement and suppression but, at the same time, with Wang Ming's increasingly serious "leftist" mistakes.

By March 1933 the Japanese invaders had marched north and occupied Chengde, the capital of Rehe Province. Inspired by the anti-Japanese upsurge across the country, the defending Chinese troops stationed in areas north and south of the Great Wall rose in resistance. The 29th Corps, commanded by Song Zheyuan, fought a bloody battle against the Japanese troops in the area of Xifengkou and Luowenling and won major victories.

However, in early March Chiang Kai-shek, Generalissimo of the Military Council of the Kuomintang Government, issued an order: "Those who talk glibly about resistance against Japan should be executed without pardon!" He gave an order banning the anti-Japanese organizations, such as the Army of Volunteers and the National Salvation Corps, in all parts of Hebei Province. Because of the concessions made by Chiang Kai-shek, in May the Japanese army easily captured seven counties in Chahar Province, including Duolun and Zhangbei. After capturing the strategic passes along the Great Wall, the massive Japanese army went on to take Yutian, Tongzhou, and other places, encircling Beiping and Tianjin and thus capturing them easily.

On May 31, when the situation in Beiping and Tianjin was critical, the National Government sent its representative to sign the Tanggu Agreement with Yasuji Okamura, the representative of the Japanese Kwantung Army, in Tanggu. The agreement stipulated:

1. The Chinese troops would withdraw to the areas west and south of Yanqing, Changping, Gaoliying, Shunyi, Tongzhou, Xianghe, Baodi, Lintingkou, Ninghe, and Lutai and would "not advance across this line hereafter or commit any acts of provocation and harassment."

2. "To ensure the implementation of paragraph 1, the Japanese army can make inspections by aircraft and other means at any time, and the Chinese side will provide protection and all conveniences."

3. "If the Japanese army has confirmed that the Chinese army observed paragraph 1, it will not cross the line for pursuing and attacking the Chinese army and will return to the line of the Great Wall automatically."

4. "The maintenance of public order in the areas south of the line of the Great Wall and north and east of the line mentioned in paragraph 1 is the responsibility of the Chinese police organs, and no organization of armed forces that may provoke the anti-Japanese feelings should be used in the police organs."

According to this agreement, the Kuomintang government surrendered China's four provinces to the Japanese aggressors and opened wide the passage to northern China, thus paving the way for the avaricious aggressors to escalate the war.

The signing of the Tanggu Agreement was another shameful betrayal of China, like the national betrayal and humiliation effected by previous unequal treaties, including the Sino-British Treaty of Nanking, the Sino-American Treaty of Wangxia, the Sino-Franco Treaty of Whampoa, the treaties of Tientsin between China and Britain, France, the United States, and Russia, the Convention of Peking between China, Britain, and France, the Sino-Russian Ili Treaty, the Sino-British Treaty of Yantai, and the Treaty of Shimonoseki between China and Japan.

According to statistics, between 1840 and 1949 China signed with foreign aggressors a total of more than 1,100 such treaties of national betrayal and humiliation. The feudal governments signed these treaties, as did the National Government. The old warlords signed these treaties, as did the new warlords.

Chinese history of the last century has been like a sad melody or a mourning song so plaintive that one cannot bear to listen to it. Because of the sense of national peril caused by the large-scale Japanese aggression and the righteous indignation over the traitorous policy followed by the Kuomintang government, the people of China rose up in a renewed and militant anti-Japanese and anti–Chiang Kai-shek upsurge. The people of Beiping, Tianjin, Nanjing, Shanghai, and other cities held rallies and sent open telegrams demanding that the National Government declare war against Japan. The Provisional Central Government of the Chinese Soviet Republic and the Revolutionary Military Commission of the Chinese Workers' and Peasants' Red Army published a declaration proposing that all armed forces conclude a truce agreement on joint resistance against Japan. In Zhangjiakou in May 1933, Gen. Feng Yuxiang, an important official of the National Government, sent an open telegram together with Fang Zhenwu and Ji Hongchang and organized the Anti-Japanese Allied Army. By July of the same year, all Japanese and puppet troops had been driven out of Chahar Province. In November the 19th Route Army of the Kuomintang, which had been sent from the anti-Japanese front in Shanghai to Fujian to fight the Red Army, together with the Fujian Provincial Government, started anti-Japanese and anti–Chiang Kai-shek activities. Li Jishen, Chen Mingshu, Jiang Guangnai, Cai Tingkai, and others were elected to form Fujian People's Government, which concluded a treaty with the Soviet Central Government on joint opposition to Chiang Kai-shek and resistance to Japan.

In the face of the increasingly strong resistance to Japan, Chiang Kai-shek continued to adhere to his policy of "putting domestic stability before resistance to foreign aggression," showing not the slightest repentance. He continued to make concessions to Japan, formulated a traitorous foreign policy of "no severance of diplomatic relations, no declaration of war, no making of peace and no conclusion of treaty,"[1] and talked volubly about the "doctrine of friendship and good neighborliness," indicating his determination to "apply sanctions against impulsive anti-Japanese actions so as to show his good faith."[2] Making great efforts to suppress the anti-Japanese actions, he first divided, demoralized, and suppressed Feng Yuxiang's Anti-Japanese Allied Army. He concentrated 150,000 troops to launch campaigns of encirclement and suppression against that army. Feng Yuxiang was attacked in both the front and the rear and, very reluctantly, had to leave his post. The famous anti-Japanese general Ji Hongchang was eventually killed by Chiang Kai-shek. In November Chiang once again concentrated 150,000 troops to attack the Fujian People's Government and the anti-Japanese 19th Route Army. By January the next year, Chiang Kai-shek had put down the two-month-long rebellion in Fujian and forced Li Jishen, Jiang Guangnai, Cai Tingkai, and others to leave Fujian Province.[3]

Meanwhile, although he had failed in his four campaigns of encirclement and suppression against the Central Soviet Area, Chiang Kai-shek did not give up his attempt to kill all Communists. In September 1933 he started his fifth campaign by concentrating one million troops and two hundred aircraft to launch attacks on all Soviet areas simultaneously. Changing his previous strategy of "driving straight in," he employed the new strategy of "advancing gradually and entrenching himself at every step" in an attempt to reduce the size of the Soviet areas step by step, wipe out the effectives of the Red Army, fight a decisive battle with its main forces, and finally eliminate it in the Soviet areas.

Having won victories in its four countercampaigns against the enemy's attacks, and propelled by agrarian reform and economic development, the Red Army had expanded rapidly to 300,000 men before Chiang Kai-shek's fifth campaign began.[4] The peasants in the Soviet areas had obtained land, so their revolutionary, militant enthusiasm was unprecedentedly high. Under these circumstances, it was more than possible to defeat the enemy in its fifth campaign. However, Bo Gu and other leaders of the Central Committee continued to follow Wang Ming's erroneous "leftist" line and excluded Mao Zedong and others from the Central Committee and leading military posts. As a result, after a year of arduous fighting, the Red Army and the local armed forces and the people in the Central Revolutionary Base Area failed to repel the Kuomintang's fifth campaign, and the Red Army was forced to move away in October 1934.

In late September 1933, when the fifth campaign started, someone new arrived in the Central Soviet Area. He was the military adviser sent by the Communist International, a German called Li Teh. His original name was Otto Braun, and he was a native of Munich, Germany. He joined the Communist Party of Germany in his early years and went to Moscow to study in the Frunze Military Academy in 1928. In 1932 he was dispatched by the Communist International to Shanghai to serve as a military adviser to the Central Committee.

In 1933 Li Teh left Beijing, passed through Shanghai, Shantou, and Fujian, and entered the Central Soviet Area in late September. After he arrived in Ruijin, he served as an adviser to the Revolutionary Military Commission of the Central Government of the Chinese Soviet Republic. Thereafter, the Provisional Central Committee of the CPC relied on Li Teh to take charge of military leadership. In this way, Li Teh, a foreigner who had only studied military science in a military academy of the Soviet Union for several years and had no knowledge of China, especially no knowledge of the Central Soviet Area, arrogated to himself all powers to command the Chinese Workers' and Peasants' Red Army and control its destiny.

As soon as he arrived, Li Teh let it be known that he opposed guerrilla-ism, and he improperly required that the Red Army units be "regularized" and conduct positional warfare. "Regular" warfare relied purely on troops, quick strategic decision-making, and tactical protracted warfare, with fixed lines of operation and an absolutely unified command. These military principles were antithetical to the mobile guerrilla warfare that our army had conducted very skillfully in previous battles, and it completely disregarded the enemy's strength and the Red Army's weaknesses. These principles were mechanically copied from military textbooks and were nothing more than military dogmatism! Dogmatism had never worked in China, neither politically nor militarily. When dogmatism prevailed within the Party, it had only brought major setbacks to the revolutionary cause, as well as great defeat in wars.

People wondered when the Communist Party of China, the Chinese Workers' and Peasants' Red Army, and the Chinese people were going to be able to correctly and independently control their destiny.

In September 1933 Chiang Kai-shek concentrated 500,000 troops to start the fifth campaign against the Central Soviet Area. In late September enemy troops captured Lichuan in the northern part of the Central Soviet Area and prepared to drive directly southward.

Li Teh and others first practiced adventurism in the offensive. They directed the Red Army to "launch attacks on all fronts" on the enemy's strong positions and to recover Lichuan, "engaging the enemy outside the Soviet areas." They ordered the Red Army to attack the enemy position at Xiaoshi, but that attack failed. Then they ordered the Red Army to attack Zixiqiao. Once again, the attack failed. By then, the Red Army had landed in a passive position strategically and tactically. At this time, Mao Zedong made a bold proposal on how to conduct the military operation, but it was ignored.

In January 1934 the Provisional Central Committee of the Communist Party, with Bo Gu and others serving as the key leaders, convened the Fifth Plenary Session of the Sixth Central Committee. The convocation of this plenary session was like adding frost to snow: it was an occasion marked by the culmination of the "leftist" mistakes. Leadership committed to the erroneous "leftist" line, which eventually resulted in the failure of the fifth countercampaign, was consolidated.

In the spring of 1934 eleven enemy divisions launched large-scale attacks. Because of the extremely erroneous military leadership of Li Teh and others, the Red Army continued to retreat and suffered heavy casualties. In spite of the Red Army's increasingly passive position in operations and its mounting casualties, the "leftist" leadership advocated dividing the Red Army troops to defend the passages to the base area. Thus constituted, they found themselves in a completely passive position. They intercepted the enemy in the east and attacked it in the west to cope with the pressing military situation. As a result, many Red Army soldiers died and the base area was diminished.

Mao Zedong once again made a proposal that the main forces of the Red Army advance to central Hunan at once so as to draw the enemy troops to Hunan and wipe them out there. But his proposal was rejected by Li Teh and others once again.

After a year of fighting, the fifth countercampaign ended in failure. "The hope for smashing the Fifth Campaign of 'Encirclement and Suppression' vanished at last, and the Red Army had no alternative but to go on the Long March," observed Marshal Liu Bocheng of the situation.[5]

In October 1934 the "leftist" Central Committee represented by Wang Ming suddenly changed the policy of relying on the base area and ordered the Red Army to leave the Central Revolutionary Base Area without conducting thorough ideological mobilization among the cadres and the masses in advance. Li Weihan, who was serving as the director of the Organization Bureau of the Central Committee, recalled,

> After the Central Red Army failed in the battle of defending Guangchang, the enemy troops from all routes began to mount an all-out offensive against the heartland of the Central Soviet Area. The situation was very unfavorable to us. It was impossible for the Red Army to defeat the enemy on interior lines. Bo Gu asked me to go to his office one day between July and August 1934. Pointing to a map, he told me: "The Central Red Army is about to move away to Hongjiang in western Hunan to establish a new base area. You should go to the Jiangxi Provincial Party Committee and the Guangdong-Jiangxi Provincial Party Committee and tell them this decision. They should prepare to move away and submit lists of cadres who are to leave or stay behind to the Organization Bureau." Only on hearing what Bo Gu said did I know that the Central Red Army was about to move away.
>
> The Central Committee and local organizations, both military and nonmilitary, secretly prepared for the Long March. Only a few leaders knew about it. I knew

only a little bit about it, while the masses knew nothing. Although I was the director of the Organization Bureau of the Central Committee, I knew nothing about the specific plan for the Red Army's movement. They also did not tell me the military situation in the fifth countercampaign against the enemy's "encirclement and suppression." To my knowledge, before the Long March, the Political Bureau of the Central Committee had not discussed this major strategic question that had an important bearing on the success of the revolution. Why did the Red Army have to leave the Central Soviet Area? What were the present tasks? Where would the Red Army go? No explanation was ever made to the cadres and the vast numbers of officers and men. Although these questions were military secrets that had to be kept, the necessary propaganda and mobilization also had to be conducted.[6]

Nie Rongzhen, the political commissar of the 1st Corps of the First Front Army of the Red Army, recalled,

Before the Long March, the 1st Corps attacked Wenfang, Fujian Province, and was then ordered back to Ruijin to wait for new orders. I and Lin Biao [commander-in-chief of the 1st Corps] went back to Ruijin one day earlier. Comrade Zhou Enlai had a talk with us, stating that the Central Committee had decided to make a strategic shift. He asked us to prepare secretly without telling the subordinates. He did not tell us where we were moving. At that time, the discipline for keeping secrets was very strict, so we did not ask more questions.[7]

Who was to leave and who was to stay behind were very important questions in a major strategic shift. Li Weihan recalled:

When I returned to Ruijin, I started to organize formations for the Long March.

According to the directive of the Central Committee, the organs of the Central Committee were organized into two columns. The 1st Column was also called the "Red Star" Column. It was the leading organ as well as the general headquarters. Bo Gu, Luo Fu, Zhou Enlai, Mao Zedong, Zhu De, Wang Jiaxiang, Li Teh, and other leading comrades were included in this column.[8] Deng Yingchao, Kang Keqing, the staff of the radio station, and the cadres corps were also organized into this column. . . . The cadres corps had only a few men but strong combat effectiveness. It was actually the security force of the leading organ and played an important role in the Long March. . . .

The 2d Column was also called the "Red Armband Column." It was composed of the organs of the Central Committee, government organs, the Logistics Department, the Health Department, the General Trade Union, the Youth League, and the stretcher teams, approximately 10,000 strong. The Central Committee appointed me commander and concurrently political commissar of the 2d Column, and Deng Fa deputy commander and concurrently deputy political commissar. . . . Li Fuchun was the acting director of the General Political Department and was also in the 2d Column. Four women comrades moved with the headquarters of the 2d Column. They were Cai Chang, Chen Huiqing (the wife of Deng Fa), Liu Qunxian (the wife of Bo Gu), and Ah Jin (Jin Weiying)[9]. . . In addition, there was the transport corps. It had many porters, whose tasks were very heavy. The organs of the Central Committee did not have many documents or materials, but the organs of

> the Central Government had quite a lot. For instance, the Central Bank had large quantities of silver dollars, and the Ministry of Finance had quite a few Soviet bank notes as well as silver dollars. All these had to be carried by porters. . . . The lithographic press for printing bank notes had to be moved, too. The Logistics Department of the Military Commission decided to move the machines for ordnance production, which had to be lifted by seven or eight people. Almost every department had to move its machines. The Health Department also had to move a lot of things. It was a truly large-scale house-moving. . . .
>
> Before the Long March began, the question of which cadres were to leave or stay behind had not been decided by the Organization Bureau. The Provincial Party Committee made decisions on those cadres under its leadership and reported to the Central Committee. The Party, League, and administrative leaders of the organs of the Central Committee, the government, the army units, the Communist Youth League, and the General Trade Union made decisions on the cadres under their leadership and reported to the Central Committee. In the army, the General Political Department made these decisions. For instance, the decision on Deng Xiaoping to go with the army for the Long March was made by the General Political Department.

In October 1934 the army and the people of the Soviet areas were busy fighting the enemy and preparing for the strategic shift. They never anticipated that that moment was the starting point of an unprecedented heroic undertaking that would go down in history.

The autumn wind was blowing, and the mountains and rivers were solemn and respectful. Clouds swept away by the wind flew past at low altitude. Beyond the clouds was the blue and boundless sky. According to an ancient song of China, the Yishui River is cold when the wind is moaning and the warrior leaves never to return. More than 2,000 years later, the autumn was still cold and the warriors had moved away. But the melody was no longer sad: the warriors would be back again! The villagers of the Soviet area saw their kinsfolk off with tears in their eyes, and their tears glistened with the call: Red Army, you shall be back again! The eyes of the Red Army fighters, ready and waiting to set off, filled with tears, and they vowed in their hearts: Ruijin, we will return.

On October 10, 1934, the Central Committee of the Communist Party of China and the general headquarters of the Red Army left the Red Capital of Ruijin. The 1st, 3d, 5th, 8th, and 9th corps of the Central Red Army, totaling more than 80,000 men, set off separately from Changting and Ninghua, Fujian Province, and the areas of Ruijin and Yudu in southern Jiangxi.

The Long March, the strategic shift that would go down in history, had started.

CHAPTER 42

THE PRELUDE TO THE LONG MARCH AND THE ZUNYI MEETING

At the beginning of the 12,500-kilometer Long March of the Chinese Workers' and Peasants' Red Army, it was not called the "Long March" but a shift.

The 12,500-kilometer distance was not predetermined. The Central Red Army intended to move away first to western Hunan to join forces with other units of the Red Army, the 2d and 6th corps, and then decide what to do.[1]

When the Long March began, the leadership of the Central Committee was composed of Bo Gu (Qin Bangxian), the general leader of the work of the Central Committee; Li Weihan, the director of the Organization Department; and Zhang Wentian (Luo Fu), the director of the Propaganda Department. The chairman of the Central Government of the Chinese Soviet Republic was Mao Zedong, and the chairman of the Chinese Soviet Central People's Committee was Zhang Wentian. The chairman of the Revolutionary Military Commission and the commander-in-chief of the Chinese Workers' and Peasants' Red Army was Zhu De. The general political commissar was Zhou Enlai, the general chief-of-staff was Liu Bocheng, and the director of the General Political Department was Wang Jiaxiang. (Li Fuchun acted for him.) Li Teh was the military adviser to the Red Army.

Long before, Mao Zedong had been made a mere figurehead. The Central Committee was led by Bo Gu, who carried out Wang Ming's line and was under the control of Wang Ming. Military power was controlled by the military adviser Li Teh. Wang Ming had gone to Moscow thousands of miles away long before. With the support of the Communist International, Wang Ming exercised his remote control over the Communist Party of China and the army and had taken the destiny of the Chinese revolution into his own hands.

The Central Red Army set off. Nie Rongzhen recalled:

> After October 16, the units of the 1st Corps left Kuantian and Lingbei west of Ruijin, bidding farewell to the masses in the base areas and crossing the Yudu River to begin the Long March. The sun was setting when we crossed the river. Like many other officers and men of the Red Army, I was very excited and repeatedly looked back, staring at the mountains and the river of the Central Base Area and bidding farewell to the comrades-in-arms and villagers on the bank of the river. This was the place where I had fought for two years and ten months. I witnessed how the people in the Central Base Area had made their great sacrifices and contributions to the Chinese revolution. They had given to the Red Army many of their fine sons and daughters. Most of the soldiers of the Red Army came from Jiangxi and Fujian. The people in the base area gave the Red Army maximum material and moral encouragement and support. I recall all this with nostalgia.
>
> As the main forces of the Red Army left, the people in the base areas and the comrades who stayed behind were surely going to suffer cruel suppression and out-

> rages at the hands of the enemy. This made me worry about their future. Parting reluctantly, I slowed my pace, but the low voice ahead saying, "Follow closely! Follow closely!" prompted me to make the new journey quickly.[2]

The Red Army soldiers moving with the main forces would fight new battles and open up new battlefields, while those who stayed behind would be confronted with inconceivably arduous struggles behind enemy lines. Many of those who stayed behind persisted in the struggle and finally broke through the enemy's tight encirclements to wage new revolutionary struggles. However, many more died heroic deaths in the fierce battles and went to their eternal sleep amid the blue mountains and green water of the Central Soviet Area.

Father was in the Red Armband Column, together with the General Political Department, and started the Long March with the main forces of the Red Army. If Father had stayed in the Central Soviet Area, he would certainly have had a very different revolutionary career. Although there was only one destination at the end of each path, the outcome might have been entirely different.

The autumn wind had just begun to blow, and the passes and mountains looked solemn. The large formations of the Central Red Army advanced west quietly and quickly.

This important strategic action, of life-and-death importance to the Red Army, was directed by Li Teh, the military adviser sent by the Communist International. So many battle-seasoned generals of the Chinese Red Army, with rich experience in operations and leadership, were not employed. Instead, the destiny of the entire Workers' and Peasants' Red Army was controlled by an inexperienced foreign commander who knew nothing about China. This is nothing less than a tragic page in the history of the Chinese Workers' and Peasants' Red Army.

Li Weihan's recollection was emotional:

> The entire deployment of the Central Red Army in the early days of the Long March was wrong and could be said to be a farce.
>
> The two columns of the Central Committee were in the middle and the 1st Column was in front of the 2d Column. On both sides of them were the 1st and 3d corps. They were the combat troops, the main forces of operations. The 5th Corps was the rear guard, and its task was to protect these two columns. In addition, there were the 22d Division and the 9th Corps, both of which were composed of new recruits. They were deployed at a greater distance from the two columns, also as rear guard. Their task was to contain the enemy. The several main corps were primarily to protect the columns of the Central Committee, to protect, in effect, the large-scale house-moving. How could mobile warfare even be discussed? How could they hit the enemy flexibly? Such a deployment tied our hands and landed us in a defensive position, and we suffered blows everywhere.[3]

Nie Rongzhen recalled:

On starting off, the Red Star Column acted really like a large-scale house-moving. The column brought with it all possessions, such as machines for printing bank notes and propaganda leaflets, printed leaflets, and paper and machines for ordnance production. Thus it became a very large and cumbersome formation. When it entered the paths in the Wuling Mountains area later, the paths were extremely crowded, and it was very difficult for the column to move. Sometimes it advanced only a few or ten to fifteen kilometers a day.[4]

Liu Bocheng recalled:

When the Long March began, the Red Army continued to suffer heavy losses because of the mistakes of flightism in military actions of the "leftist" line. After leaving the Central Soviet Area, the 5th Corps of the Central Red Army remained to be the rear guard for protecting the whole army for a long time. It protected the mules, horses, and supplies as they moved west along the Guangdong-Guangxi-Hunan border. The whole army, over 80,000 strong, advanced over narrow winding trails in the mountains in great congestion. It often crossed over only one column in a night. We all felt very fatigued. The enemy troops advanced quickly on roads, so there were no means by which we could break away from the pursuing enemy.[5]

Li Weihan wrote in his memoirs:

To evade the enemy, we either marched at night or crossed over big mountains. When we could not evade, we had to fight recklessly. As the commander of the 2d Column, I did not know the military situation and just advanced according to orders. The Military Commission issued orders to the chief-of-staff of the 2d Column, who then relayed them to me. I analyzed the direction of our march according to the contents of the orders. Some days I thought we might climb a mountain, as we could not take the road for fear of an encounter with the enemy. It turned out later to be true we climbed a mountain.

Because of the lack of mobilization and explanation before the Long March, the situation on the march was very bad. The formation was thinly scattered. Sometimes, when the advance troop set off, the rear guard just reached the camp sites. We were tailed by the enemy almost every day. Many of us dropped out. Personnel in the collecting group [in charge of collecting sick and wounded personnel as well as those who dropped out during the march] increased considerably, whereas the number of troops was significantly reduced. Our Red Armband Column was also thinly scattered. Only when it was unable to continue to bring things did the column discard them, burn bundles of bank notes issued by the Soviet Bank, and destroy the machines. I was deeply grieved to see young soldiers lying dead by the roadside. Because of the great reduction of its size, the Red Armband Column was later reorganized into three echelons. Its training division was sent to the front for operations, and I, the former column commander, became an echelon commander.

As we marched every night, we felt very tired and often dozed off while walking. We walked for almost one month and arrived at Wenmingsi near Rucheng, Hunan Province. We walked more than 500 kilometers, over twenty kilometers per day on average.[6]

Although after the failure of the fifth countercampaign the main forces of the Red Army had to move away, Chiang Kai-shek, enormously proud of his success, could not, of course, allow it to slip away. Therefore, as soon as the Red Army began to advance westward, Chiang Kai-shek established three blockade lines in an attempt to eliminate the Red Army completely on its westward advance.

Not long after it left Ruijin, and before it crossed the boundary of Jiangxi Province, the Red Army encountered in the area of Ganzhou, Xinfeng, and Anxi in southern Jiangxi the enemy's first blockade line, extending north to south. On October 21, the 1st Corps had a skirmish with the enemy and fought the first battle during the Long March. After two days of fierce fighting, the enemy troops retreated in defeat; the Red Army had annihilated one enemy regiment and captured more than 300 enemy soldiers. Under the protection of the 1st and 3d corps, the columns of the Central Committee and the follow-up forces passed through safely. The Red Army continued to advance westward.

The enemy's second blockade line was set up along Guidong and Rucheng, Hunan, in the Hunan-Jiangxi border area, and Chengkou, Guangdong, also extending north to south. The enemy expected to annihilate the Red Army in one battle. Around November 2, the Red Army skillfully wiped out enemy troops by making surprise attacks and outflanking operations. By making a detour, the whole formation successfully broke through the second blockade line and continued to advance westward.

The enemy's third blockade line was set up along the area of Chenxian and Yizhang in southern Hunan. This line was strongly defended, and massive numbers of enemy troops pursued the Red Army from Jiangxi and Fujian. The enemy resolved to annihilate the Red Army there completely. In early November, under the effective command of Nie Rongzhen and other commanders, the 1st Corps occupied the positions beside the Jiufeng Mountains on the left flank earlier than the enemy, and the 3d Corps captured Yizhang, Liangtian, and other towns on the right flank, thereby covering the formation's safe passage through the third blockade line. The three blockade lines, which the enemy had boasted would be "steel blockade lines," were thus broken through by the Red Army.

After the main forces of the Red Army broke through the enemy's three blockade lines, Chiang Kai-shek immediately concentrated 400,000 troops to block the advance of the Red Army from three routes, swearing to annihilate the Red Army on the bank of the Xiangjiang River. Confronted with this fourth blockade line,

> the leaders of the "leftist" line were at their wits' end. They only ordered the troops to attack obstinately in a bid to break through the enemy's encirclement, resting their hope on joining forces with the 2d and 6th corps. A fierce battle was fought for one week in the area south of Quanxian, Guangxi Province, and on the eastern bank

> of the Xiangjiang River. They went so far as to employ massive numbers of troops to conduct corridor-type covering operations on both flanks. Although the formation broke through the enemy's fourth blockade line and crossed the Xiangjiang River, it suffered heavy casualties, with a loss of over half of its personnel.[7]

After the Red Army crossed the Xiangjiang River, its troops dropped drastically in number from over 86,000 at the beginning of the Long March to 30,000.[8]

In only one and a half months, from when it set off in mid-October to when it crossed the Xiangjiang River on December 1, the Central Red Army was in a passive position the entire way; it was pursued, encircled, blocked, and intercepted by the enemy; and it suffered heavy losses. This harsh reality not only caused the loss of over half the personnel but also gave rise to suspicion and dissatisfaction among the troops. Flightism could only destroy the whole army! The many officers and men of the Red Army began to make anxious demands for a change in the leadership.

On December 11, the main forces of the Red Army reached Daotong County on the southwestern border of Hunan and were ready to advance northward to western Hunan. The enemy had deployed massive numbers of troops to block the Red Army as it advanced to join forces with the 2d and 6th corps. At the same time, it was using the Guangxi army to pursue the Red Army on its rear flank. At such a critical moment, Bo Gu and others willfully insisted on the Red Army making a northward advance to western Hunan to join forces with the 2d and 6th corps.

In view of this extremely grim situation—the Red Army was facing the possibility of total annihilation—Mao Zedong proposed abandoning the plan to join forces with the 2d and 6th corps and, instead, advancing toward Guizhou, where the enemy forces were weak, so as to gain the initiative. This proposal immediately won the support of Zhou Enlai, Zhu De, Luo Fu, Wang Jiaxiang, and others. Thus, the Red Army halted its northward advance and turned to Guizhou. On December 15, it captured Liping, Guizhou Province.

On December 18, the Political Bureau of the Central Committee held a meeting in Liping, presided over by Zhou Enlai, to discuss the question of where the Red Army should advance. As Bo Gu and Li Teh still insisted on advancing north to join forces with the 2d and 6th corps, a heated debate took place. Finally, because an overwhelming majority of the participants agreed with Mao Zedong's view, they decided to abandon the plan to join forces with the 2d and 6th corps and to march the Red Army to Guizhou, where the enemy forces were fairly weak, and to establish a base in the border area between Sichuan and Guizhou with Zunyi at the center.

One cannot help recalling this situation sixty years later with deep emotion. Had it not been for Mao Zedong, who proposed changing the direction of advance of the Red Army, and had this view not received strong support from vast numbers of the officers and men of the Red Army, the main forces of the

Central Red Army would have been totally annihilated. In the common wisdom, Mao Zedong saved the Red Army.

As early as 1933, many cadres had doubts about the erroneous "leftist" leadership of Wang Ming's Central Committee. The failure of the fifth countercampaign, the defeats suffered since the beginning of the Long March, and the significant reduction in the size of the Red Army—all these distressing losses caused doubts about and discontent with the "leftist" Central Committee and its erroneous leadership among more and more officers and men. The demand increased for a change in the status quo and for getting rid of the erroneous leadership.

On the Long March, Mao Zedong contracted tuberculosis and had to be carried on a stretcher. Wang Jiaxiang and Zhang Wentian were also carried on stretchers because they were seriously ill. Therefore, these three were always together. On the way, Mao Zedong talked often and carefully with his two stretcher companions, analyzing for them the mistakes in military leadership made by the Central Committee during the fifth countercampaign and the Long March. Wang Jiaxiang and Zhang Wentian gradually came to accept Mao Zedong's views.

Meanwhile, the views of some leaders of the Central Committee on the one hand and of Bo Gu, Li Teh, and others on the other diverged more and more as the march continued. "Debates were held all the way from Laoshanjie to Liping and from Liping to Houchang."[9]

After the Liping meeting, the Red Army units underwent reorganization and lightened their packs. It crossed the Wujiang River in January 1935 and captured the ancient city of Zunyi, Guizhou Province, on January 7. During this period, the troops fought triumphantly, and their morale gradually increased. Then they had their rest and reorganization in Zunyi for twelve days.

Between January 15 and 17, 1935, the Central Committee of the Communist Party of China called an enlarged meeting of the Political Bureau in Zunyi, the now-famous Zunyi Meeting. Present at the meeting were: members of the Political Bureau, Bo Gu, Zhang Wentian, Zhou Enlai, Mao Zedong, Zhu De, and Chen Yun; alternate members of the Political Bureau, Wang Jiaxiang, Liu Shaoqi, Deng Fa, and He Kequan; leading comrades of the general headquarters of the Red Army and all corps, Liu Bocheng, Li Fuchun, Lin Biao, Nie Rongzhen, Peng Dehuai, Yang Shangkun, and Li Zhuoran; and the secretary-general of the Central Committee, Deng Xiaoping. The military adviser Li Teh and the interpreter Wu Xiuquan also attended the meeting. The Zunyi Meeting was of great historic importance. Because many works and articles on this meeting have been published, I am not going to describe its contents or process. Two outcomes of the meeting are relevant here.

First, a famous resolution was adopted, the "Resolution on the Summary of Opposition to the Enemy's Five Campaigns of 'Encirclement and Suppression.'" The resolution clearly pointed out that the "purely defensive military line" followed by Bo Gu and Li Teh had caused the Red Army to suffer setbacks in the fifth countercampaign and to practice strategic retreat when making its strategic shift and breakthrough of the enemy encirclement. The resolution completely ended Wang Ming's erroneous "leftist" leadership as it had been applied militarily.

Second, the leading body of the Central Committee was reorganized. Mao Zedong was elected a member of the Standing Committee of the Political Bureau of the Central Committee, and Zhou Enlai and Zhu De took charge of military command.

The Zunyi Meeting was held at the most critical moment in the history of the Chinese revolution. It was of extraordinary importance to the Communist Party of China, to the Chinese Workers' and Peasants' Red Army, and even to the future and destiny of the Chinese revolution. It put an end to the erroneous "leftist" military leadership. It put an end to the organizational rule of "leftist" dogmatism in the Communist Party of China. At the Zunyi Meeting, the CPC handled its own affairs in light of the actual conditions in China, the Party, and the Chinese army, without external intervention. And the most important outcome was that a nucleus of leadership was formed within the CPC, with Mao Zedong serving as the main pillar.

Father often said that the Party's nucleus of leadership was not truly formed until the Zunyi Meeting. It was the first-generation nucleus of leadership of the Communist Party. No genuine nucleus of leadership had been formed earlier. Father highly endorsed the establishment of this nucleus of leadership.

Mao Zedong did not proclaim his own position in the nucleus of leadership, nor was it granted by foreigners. It emerged in the Chinese revolution after nearly fourteen years of practice. He was chosen by the Chinese Communists after years of trials, difficulties, and perplexities. To the Chinese revolution and the Communist Party, the indispensable process of the emergence of the people's leader had not been easy.

The definition of *leader* given in the *Chinese Collection of Words* is, "the top leading person of the state, the political organization, and the mass organization." This definition is both accurate and inaccurate when applied to Mao Zedong, who brought much more profound meaning to the word.

Mao Zedong was a leader and a great man. Once his position in the nucleus of leadership was established, he remained there for a full forty-one years. He was a statesman, a military strategist, a thinker, and a poet—a man of many talents. He was well versed in both belles lettres and military science. He had the abundant knowledge and lofty ideals of a great man and a leader. At the same time, he had a free and easy style and the romanticism of a man of letters. His life was legendary. He both achieved successes and made mistakes.

There have been too many commentaries on the life of Mao Zedong and his achievements and failures. Some people have considered him the ideal Communist; others have regarded him as a rare and unique talent and a great man of politics. Some have made a fetish of him, while others have denounced him as an Oriental-style monarch. In my opinion, Mao Zedong encompassed Marxism, idealism, communism, nationalism, and feudalism; he was a great revolutionary leader strongly marked by the times and the Chinese national character.

Whatever one's appraisal of Mao Zedong, it is undeniable that only after he steered the course of the Chinese revolution at the Zunyi Meeting did it regain initiative and advance to victory! Therefore, the Zunyi Meeting can truly be called "a turning point of vital importance" in the history of the Communist Party of China."[10]

The Zunyi Meeting enabled the CPC to extricate itself from four years of the damage done by Wang Ming's "left"-adventurism, "leftist" dogmatism, and "leftist" factionalism and from the defensive position of subjection to external intervention.

Wang Ming, the leading figure responsible for the harmful, long-term "leftist" mistakes, was then in Moscow. Since he did not take part in the Long March, he did not share weal and woe with his comrades, nor did he accept the criticism of his mistakes by the Communist Party of China. After 1956 Wang Ming stayed in the Soviet Union and wrote articles opposing the CPC and his own motherland under the aegis of the Soviet Union. He died there in 1974. The overwhelming majority of the one billion Chinese people and the tens of millions of Chinese Communists neither knew nor cared about his death. This man, who tried to control the destiny of China with the aid of foreign forces, was spurned by the Chinese people in the end.

As to Li Teh, the foreign military adviser who had been dispatched to China by the Communist International and was once regarded as the "overlord," he sat dejectedly at the door of the room where the Zunyi Meeting was held. Afterwards, he followed the Red Army during the Long March and reached northern Shaanxi. He went back to the Soviet Union in 1939. He took to heart the CPC's criticism of him. He wrote a series of anti-Chinese articles and, in the 1970s, wrote a book attacking China and defending the ignominious role he had played in the history of the Chinese revolution.

All foreigners who came to China to take part in and give support to the cause of the Chinese revolution were sincerely welcomed and later remembered by the Chinese people. Among them were Dr. Norman Bethune from Canada; Dr. D. S. Kotnis from India; Dr. George Hatem from the United States; Dr. Hans Miller from Germany; and the American journalists Agnes Smedley and Anna Louise Strong. They sympathized with and supported the Chinese revolution, and some even gave their lives to the Chinese people and the cause of the Chinese revolution. The Chinese people love them ardently and cherish their memories forever. But probably few young Chinese today know the name of a person like Li Teh. He was ruthlessly drowned in the tide of the Chinese revolution.

After the Zunyi Meeting, the main forces of the Red Army, having banished military dogmatism, conducted flexible mobile warfare and made great strides with no trace of regret.

At this time, Chiang Kai-shek had concentrated hundreds of thousands of troops to press on Zunyi from all directions in an attempt to annihilate the Red Army in western Guizhou. Under the leadership of Mao Zedong and others, the main forces of the Red Army crossed the Chishui River in the border area between Sichuan and Guizhou four times between January and March 1935. They advanced first to Zhaxi, in southern Sichuan, from Guizhou and then turned back to Guizhou. After they entered southern Sichuan the second time, they turned back to Guizhou again and then crossed the Wujiang River on the southern bank. They made a feint to advance to Guiyang, the capital of Guizhou Province, so as to lure the enemy troops into Guizhou. When the Yunnan enemy troops arrived in Guizhou, the main forces of the Red Army advanced suddenly and rapidly toward Yunnan Province and then turned just as rapidly northwestward after making a feint. In early May the main forces of the Red Army crossed the torrential Jinsha River at the juncture of Sichuan and Yunnan. Thus, the main forces of the Red Army jumped out of the ring of encirclement, pursuit, blocking, and interception by the enemy troops and left them behind on the southern bank of the Jinsha River, thereby eventually gaining the strategic initiative.

Such swift and massive mobile warfare movements made this a military operation of a brilliance rarely seen in either Chinese or foreign military history. This method of fighting smashed the wishful thinking of Chiang Kai-shek—who boasted of being a brilliant military strategist himself—about encircling and annihilating the main forces of the Red Army. It was very easy for Chiang to control, in the name of the orthodox Central Government, all the feudal warlords who had established their separate regimes with their own troops by relying on his one million crack troops and his political trickery. However, these assets were not enough to cope with the Communist Party and Mao Zedong.

This was merely the beginning of the contest of strength between the Communist Party and the Kuomintang, and between Mao Zedong and Chiang Kai-shek. The major part of the contest lay ahead!

CHAPTER 43

THE RED ARMY BRAVES ALL DIFFICULTIES ON THE LONG MARCH

Once I asked Father, "What work did you do in the Long March?"

"Just followed!" he replied with his usual simplicity of expression. Every old Red Army man who took part in the world-renowned 12,500-kilometer Long March has many recollections of it and can tell endless stories about it. But this was all Father had to say.

However, his words are true. At the beginning of the Long March, the "right-deviationist" label that had been put on him had not yet been cast off, and he did not hold an important military post until later. Since Father said no more about his involvement in the Long March, I had to make inquiries elsewhere. As a result, I gained a general picture of him during the Long March.

In October 1934 Father followed the organs of the General Political Department to start the Long March. (The *Red Star* newspaper he edited had to be mimeographed with cut stencils because of the march.) Father followed the Central Column to pass four blockade lines, cross the Xiangjiang River, and reach Liping, Guizhou Province, and Houchang on the south bank of the Wujiang River, and then in early January 1935 he followed the organs of the Central Committee to enter Zunyi, Guizhou Province.

Father really "just followed" during this period of more than two months. While following, he overcame all kinds of difficulties to edit and mimeograph seven or eight issues of *Red Star* in intense marching and fighting between October 20 and January 7, when Zunyi was captured.[1]

Father was appointed secretary-general of the Central Committee in early January 1935 and in this capacity took part in the Zunyi Meeting. Father was appointed secretary-general because the great majority of the senior cadres, resenting the erroneous "leftist" leadership before they arrived at Zunyi, had initiated a change in leadership. Quite a few senior cadres went frequently to inform Mao Zedong of the situation and to exchange views with him. The influence of Mao Zedong within the Party and the army was increasing day by day. Thus, under Mao Zedong's influence, the Central Committee appointed Deng Xiaoping its secretary-general.

This was his second time in the post. His first appointment was made during the dangerous situation after the failure of the Great Revolution. His second appointment occurred before the critical turning point of the Long March.

Soon after Father took up his new post, he participated in the Zunyi Meeting. He did not speak at the meeting, but undoubtedly he was a firm supporter of Mao Zedong.

After the Zunyi Meeting, Father followed the army to cross the Chishui River four times and the Wujiang River once again. Father told me once that this type of mobile warfare, going around in circles and launching surprise attacks on enemy troops, was like "the cat catching the mouse and the mouse catching the cat"! What he meant was that the strong enemy wanted to "catch" the Red Army but was unexpectedly led and turned around, in great befuddlement, by the Red Army, suffering heavy losses time and again. A big cat wanted to catch a small mouse, but the mouse played a trick on the cat instead.

Hearing father's analogy, I burst into laughter, for two reasons. First, it was lively and vivid. Second, he had been inspired by his grandchildren: my daughter Yang Yang and my younger brother's son Xiao Di liked to watch the animated cartoon "Cat and Mouse" in their grandpa's room.

In early May 1935 the main forces of the Red Army were in Yunnan Province and made plans to cross the Jinsha River to the north and enter Sichuan Province. "The Jinsha River flows between remote mountains and narrow valleys in the border area between Sichuan and Yunnan provinces, is very wide, has a rushing current, and is strategically located and difficult of access. If our army cannot cross the river to the north, we will be in danger of being extruded and annihilated in the deep valleys."[2] With the dangerous river in front of it and the enemy troops pursuing it from behind, the Red Army used every means to find seven small boats from the Jiaoping Ferry. The 35,000 men all crossed the river in nine days and nights.

The commander of the 1st Division of the 1st Corps, Li Jukui, led his troops to cross the river at the Longjie Ferry but failed several times. Then he was ordered to lead his troops to cross the river at the Jiaoping Ferry sixty kilometers away. He recalled:

> I marched in front of the column, and on arriving at the Jiaoping Ferry, I met Comrade Deng Xiaoping first. At the first sight of me, he asked: "Are your troops here?"
>
> "Yes, they are, they come in a scattered way," I answered.
>
> Comrade Deng Xiaoping said, "Send somebody at once to urge the troops to come more quickly and cross the river immediately." Then he added, "The troops are to be commanded by Comrade Liu Bocheng, and I will take care of the mules and horses as well as the carrying poles with their loads."
>
> As we arrived on the other side of the river, we saw Mao Zedong, Zhou Enlai, Zhu De, and other leading comrades of the Central Committee in a cave at the ferry, watching the troops cross the river. I heard that they stayed there to watch the troops for several days and left only after all the Red Army units had crossed the river.[3]

When I went to the residence of Old General Li in the western suburbs of Beijing for a visit in the early winter of 1991, he held my hand and told me cheerfully, "Before that time, I had only heard about your father and the 7th Corps of the Red Army. This was the first time, in crossing the Jingsha River, that I met your father. Since then, we have been familiar with each other. I fought years of battles under his leadership later!" The eighty-seven-year-old General Li, with gray hair and eyebrows, smiled very, very kindly.

After crossing the Jingsha River, the Red Army continued to advance northward. It passed through the area inhabited by the Yi people first and then crossed the natural barrier, the Dadu River. The Dadu River is where Shi Dakai, the famous general of the Taiping Heavenly Kingdom, toward the end of the Qing Dynasty, failed to cross, was defeated, and had to surrender. Seventy-two full years later, the heroic Red Army first sent seventeen brave fighters to cross the river with a small boat, then another twenty-two to seize the Luding Bridge rapidly. After that, more than 30,000 Red Army troops swiftly crossed the river across the chain bridge, like divine troops, thus crushing Chiang Kai-shek's dream of sending the Red Army to the same fate as Shi Dakai.

Father said he attended many important meetings of the Political Bureau in the capacity of secretary-general of the Central Committee after the Zunyi Meeting. The military chiefs of all the corps were present at many of the meetings. The meeting he remembered most clearly was held in Huili after the Red Army crossed the Jingsha River in May. That meeting lasted two days and focused on criticizing Lin Biao.

Father told me that he had lived with Chairman Mao during the Zunyi Meeting. After the meeting, he continued on the Long March with Chairman Mao and Zhang Wentian. During that period, they marched during the day and got very tired. They stopped at night and hurriedly found a place to sleep. They marched and lived together the whole way.

On the Long March, the Red Army met with and survived endless dangers. Of course, the Long March would not have come to be seen as an epic event and have gone down in history if these inconceivable difficulties and hardships had not happened.

Soon after crossing the Dadu River, the Red Army encountered a mountain covered with snow that had been accumulating for years. Called the Jiajing Mountain, it was the first big snow-capped mountain on the Long March. Steep, pathless, and perennially covered with snow, it is difficult for people and horses to climb. The air is so thin on the top of the mountain that climbers cannot sit down to rest there because they probably will not be able to stand up again. The Red Army soldiers had been marching for a long time with inadequate food and clothing, and they did not have the physical strength necessary to withstand the cold and climb this lofty mountain. Some of them did not get up again after sitting there and went to their eternal sleep amid the thousand-year-old snow of the Jiajing Mountain.

Father said his horse died before reaching the top of the snow-capped mountain, so while others climbed by holding onto their horses' tails, he climbed the mountain step by step.

In memory of the Red Army's magnificent feat, a golden monument has been erected on the Jiajing Mountain. With the blue sky and white clouds above and the age-old snow beneath, this monument reflects under sunshine the golden light that can be seen from many kilometers away. It is a marvelous spectacle. Today, when people see this golden monument, it seems they can see the tortuous and endless path traversed over the white snow by the Red Army soldiers.

After climbing the snow-capped mountains, the Red Army arrived at Maogong in the westernmost part of Sichuan. On June 14, the First Front Army of the Red Army joined forces with the Fourth Front Army from the Sichuan-Shaanxi Base Area. The Fourth Front Army had established the Hubei-Henan-Anhui Revolutionary Base Area along the borders of Hubei, Henan, and Anhui provinces. As a result of the encirclement and suppression campaign by the massive troops of Chiang Kai-shek, the Fourth Front Army had been forced to leave the Hubei-Henan-Anhui Base Area. After innumerable hardships and bloody battles, the Fourth Front Army marched over 1,500 kilometers, in just over two months, and established the Sichuan-Shaanxi Base Area along the borders of southern Shaanxi and northern Sichuan in 1932. After fierce fighting with the enemy, a vast new area extending 100 to 150 kilometers was formed, and the strength of the troops increased to more than 80,000. At this juncture, the Central Red Army had failed in its fifth countercampaign and departed from the Central Soviet Area in Jiangxi Province to start its western expedition. The Fourth Front Army had also left the Sichuan-Shaanxi Base Area to start moving away after starting to cross the Jialing River in March 1935 before it joined forces with the First Front Army in the Maogong area in June.

After the First and Fourth front armies joined forces at Maogong, Father met Fu Zhong, who had studied in France and with whom he had conducted revolutionary activities. They had also studied in the Soviet Union together, so they maintained friendly relations.

Coming out of the Sichuan-Shaanxi Base Area, the Fourth Front Army had strong soldiers and sturdy horses. Fu Zhong then served as the director of the Political Department of the Fourth Front Army and had some power. Upon learning that his old comrade-in-arms had lost his horse, Fu Zhong generously helped Father at once. Father said, "After climbing the snow-capped mountains, Fu Zhong offered me three treasures, namely, a horse, a fox-fur coat, and a pack of dried beef. These three things were of great use indeed!"

After the First and Fourth front armies joined forces, the Communist Party of China and the Red Army were confronted with an urgent task—deciding on a strategy for developing the Red Army.

The area of northwestern Sichuan where the Red Army was stationed was an area where minority nationalities lived in compact communities. It was far away from the central hinterland under enemy control, but it was sparsely populated, its soil was impoverished, its inhabitants were poor, and communications and provisions were inadequate. It was impossible for a large army over 100,000 to stay there for long. In the meantime, the movements to resist Japan and strive for democracy were surging across the whole of China, and northern China had become the front of the anti-Japanese struggle.

After analyzing the nationwide situation, the Central Committee confirmed that the Red Army should continue its northward advance to establish base areas in Shaanxi and Gansu provinces. Shaanxi and Gansu encompassed a vast expanse, had plentiful resources, and were areas where enemy control was weak, making it easy to survive and develop. Meanwhile, the establishment of base areas there could lead to the establishment of an anti-Japanese forward base in the north and enable the Red Army to go to the outpost of the anti-Japanese democratic movement.

This proposition of the Central Committee met with the opposition of Zhang Guotao. On June 26, the Central Committee convened a meeting of its Political Bureau at Lianghekou to the north of Maogong. After discussion, Zhang Guotao agreed reluctantly to the views of the Central Committee on advancing northward. After the meeting, the Central Committee led the First Front Army in the continued northward advance.

On July 10, the advance troop reached Mao'ergai near Songpan. At this time, Zhang Guotao coerced the Central Committee by taking advantage of having 80,000 troops under his command—twice the number of troops in the First Front Army—and proposing to reorganize the Military Commission of the Central Committee and the general headquarters and to appoint himself chairman of the Military Commission, with the power to "act arbitrarily." This action was actually designed to usurp the military power of the Central Committee.

While flatly rejecting Zhang Guotao's unreasonable demand and criticizing his mistake, the Central Committee appointed him general political commissar of the Red Army to avoid a split.

On August 3, the general headquarters of the Red Army decided to combine and reorganize the First and Fourth front armies into the Left Route Army and the Right Route Army. The Left Route Army was led by Zhu De, Zhang Guotao, and Liu Bocheng, and the Right Route Army by Mao Zedong and Zhou Enlai.

On August 4, the Political Bureau of the Central Committee held a meeting at Shawo near Mao'ergai to discuss the current situation and tasks. The meeting reiterated the strategy of advancing northward and establishing the Sichuan-Shaanxi-Gansu Base Area and pointed out the need to further strengthen the Party's absolute leadership over the Red Army, to maintain unity, and to correct the right-wing mistake of being pessimistic about the future of the revolution.

On August 20, the Political Bureau of the Central Committee held a meeting at Mao'ergai, reconfirming the strategy of advancing northward and criticizing Zhang Guotao's mistake of attempting to lead the Red Army to cross the Yellow River to the west. After the meeting, the Left and Right route armies advanced northward separately and started to enter the uninhabited, boundless grasslands.

Climbing snow-capped mountains and plodding through grasslands has traditionally been the symbol of the difficulties encountered on the Long March. Perhaps plodding through grasslands was more difficult than climbing snow-capped mountains. On the vast grasslands in northwestern Sichuan, there was a limitless expanse of weeds above and black water below. There were neither villages nor signs of human habitation, nor anything to eat. The weather was fast-changing, and it drizzled sometimes before suddenly clearing up.

The Red Army soldiers had nothing to eat. They ate the meat of their horses when they died. The follow-up troops had no horse meat to eat and had to gnaw the horse bones, and when there were no horse bones to gnaw, they had to eat grass roots and bark and even belt leather. They dug pits in the ground and propped up canopies to take shelter from the wind and the rain.

It took seven full days and nights to plod through the grasslands. Many of my seniors who took part in the Long March told me that the most difficult time during the Long March was plodding through the grasslands and that the greatest number of their comrades died at that time. They died of starvation, illness, eating poisonous weeds, and sinking into swamps.

Father said that when plodding through the grasslands Zhou Enlai was critically ill and and that he passed through the grasslands on a stretcher. Father did not accompany Zhou Enlai through the grasslands because he had left the Central Committee and the Central Column.

Between June and July 1935, Father had been transferred from the post of secretary-general of the Central Committee to the Political Department of the 1st Corps of the Red Army to serve as the director of the Propaganda Department. An old Red Army man, Liu Daosheng, recalled that not long after the Lianghekou meeting held on June 26, he was transferred to work with the Organization Department of the 1st Corps and Deng Xiaoping went with him to the 1st Corps. Liang Biye, who was then the political instructor in the Political Department of the 1st Corps, told me, "Your father came to the Political Department of the 1st Corps to serve as the director of the Propaganda Department in Mao'ergai in July, and he worked in the 1st Corps till the Long March ended."

I asked Father why he was transferred from the post of secretary-general of the Central Committee to that of director of the Propaganda Department of the 1st Corps. He answered that in those days they were marching every day without anything to do. Mom Liu Ying, who was then in the Central Column, told me,

> When I was transferred to work in the Central Column, your father had already left. I straightened out the things in an iron-sheet box he had left behind, and there were some books and documents. I had been with the rear echelon previously. It was Chairman Mao who had me transferred to work in the Central Column. He said the life in the rear echelon was hard and there was nothing to eat. Female comrades would wear themselves down there. At that time, the organs of the Central Committee had a small staff. All capable comrades were sent to the front to

This photo was taken of Father while he studied in France on a work-study program.

Delegates to the Third Congress of the European Branch of the Chinese Socialist Yout
League in Paris. In the front row, first from the left is Nie Rongzhen, fourth from the left
Zhou Enlai, and sixth from the left is Li Fuchun. Father is third from the right in the last ro

Father's file card at the Hutchinson Rubber Plant. The personnel department marked it: resigned and employed no more.

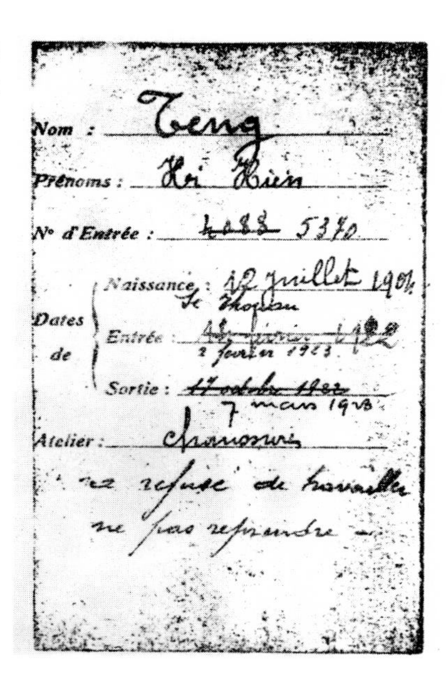
Nom : Teng

Prénoms : Hi Hien

N° d'Entrée : ~~4088~~ 5370

Dates de
Naissance : 12 juillet 1904
Se Tchouan
Entrée : ~~[illegible] 1922~~
2 février 1923
Sortie : ~~17 octobre 1922~~
7 mars 1923

Atelier : Chaussures

s'est refusé de travailler
ne pas reprendre

RÉFÉRENCES

du au chez qualité
du au chez qualité
du au chez qualité
du au chez qualité
du au chez qualité

Études
Diplômes universitaires :
Langues étrangères :
Aptitudes spéciales :

SALAIRE DÉCLARÉ

Année 19	Année 19	Année 19
Année 19	Année 19	Année 19
Année 19	Année 19	Année 19
Année 19	Année 19	Année 19

APPOINTEMENTS

Dates	Appointements	Total	Dates	Appointements	Total	Nombre d'Enfants	Allocations

Motif de départ : Retourne en Chine

Valeur professionnelle : A. B.

Autres [illegible]

824094

Father's file card at the Renault Automobile Factory. His reason for departure: to return to China.

Together with Zhang Zunyi and Wei Baqun, Father launched and led the Bose Uprising and established the 7th Corps of the Chinese Workers' and Peasants' Red Army and the Youjiang Revolutionary Base Area on December 11, 1929. Father served as the Secretary of the CPC Front Committee and as Political Commissar of the 7th Corps. This picture of Father was taken at that time.

Some leading cadres of the 1st and 15th Corps of the Red Army at Chunhua County, Shanxi, in 1936. Pictured from the left are Wang Shoudao, Luo Ruiqing, Yang Shangkun, Cheng Zihua, Nie Rongzhen, Chen Guang, and Xu Haidong. Father is at the right.

Father served as the Political Commissar of the 129th Division of the Eighth Route Army in January 1938. This photo was taken of the leaders of the 129th Division in Tongyu Town, Liaoxian County (now Zuoquan County), Shanxi. From the left, they are Li Da, Father, Liu Bocheng, and Cai Shufan.

This photo of Father (first from the right) was taken with (from left) Zuo Quan, Peng Dehuai, Zhu De, Peng Xuefeng, and Xiao Ke at the General Headquarters of the Eighth Route Army at Mamu Village, Hongdong County, Shanxi.

During the War of Liberation, the Shanxi-Heibei-Shandong-Henan Field Army carried out the movement of Party consolidation and the new ideological education movement in the army while also conducting operations. This photo shows Father making the mobilization report.

In November 1948, the Military Commission of the CPC Central Committee decided to establish the five-member General Front Committee to exercise unified leadership and command over the Central Plains and East China Field Armies. Father served as the Secretary of the General Front Committee. Pictured from the left are Su Yu, Father, Liu Bocheng, Chen Yi, and Tan Zhenlin.

Uncle Liu Bocheng and Father worked together for thirteen years, and the army they led was called the "massive Liu-Deng troops." This picture of them was taken in the early days after the founding of the People's Republic.

The simple wedding ceremony of Father and Mother and Uncle Kong Yuan and Aunt Xu Ming was held at Yan'an in September 1938. This picture of the two couples was taken at the ceremony.

Mother Zhang Xiyuan.

Father and Mother in the Taihang Mountains.

Father and Mother are holding my elder brother and sister, with Uncle Liu Bocheng, Aunt Wang Ronghua, and their daughter in 1945.

This photo of Father and Mother with Uncle Chen Yi and Aunt Zhang Qian and their children was taken after the liberation of Shanghai.

Father and Mother with me in the early 1950s.

Father and Mother with my family in the spring of 1991. First from the left is my husband, He Ping, and second from the left is my daughter, Yang Yang.

I accompanied Father to Hangzhou in early 1993.

The whole family. Top row (from left): Deng Zhifang, Deng Lin, Meng Meng, He Ping. Middle row: Deng Pufang, Yang Yang, Mao Mao, Mian Mian, Deng Nan. Front row: Deng Xiaoping (holding Xiao Di), Zhuo Lin.

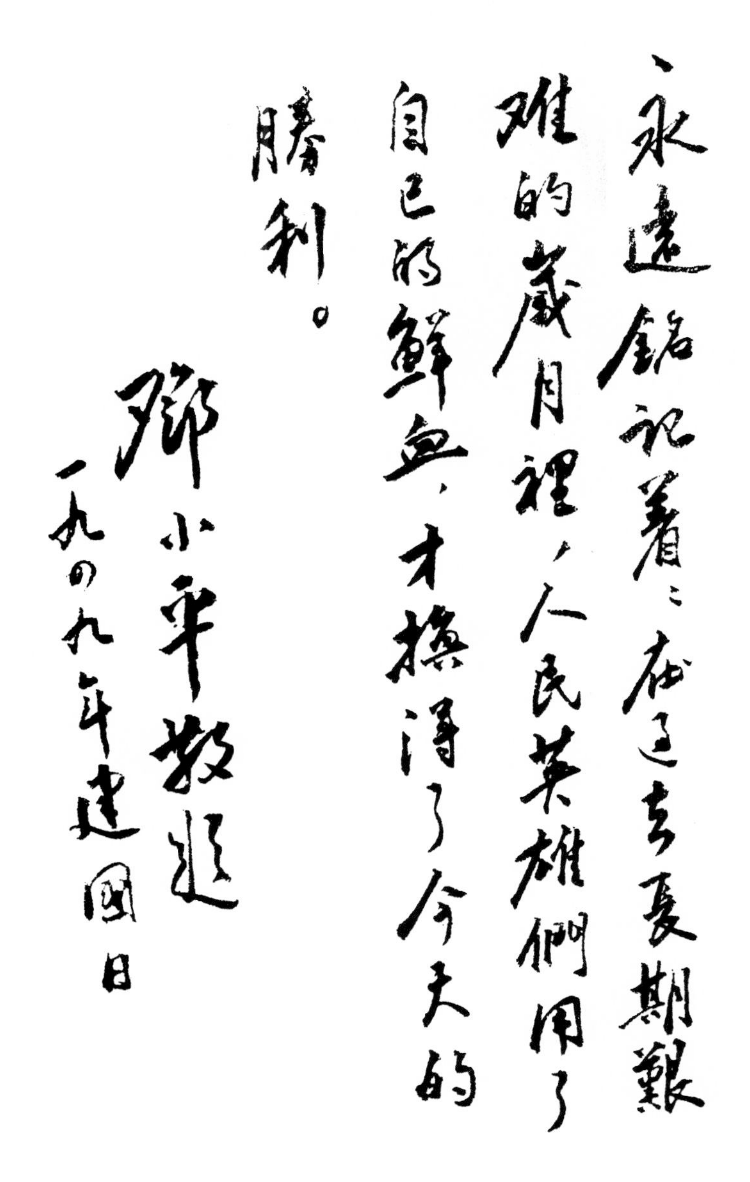

Remember forever that the victory of today was won with the blood of the people's heroes in the long and difficult years of the past. Deng Xiaoping. Written on the day of the founding of the People's Republic of China, 1949.

> strengthen the combat forces. Wang Jiaxiang told me that as the Central Committee didn't have much work to do then, Comrade Xiaoping was sent to the front. I also asked Chairman Mao why Comrade Xiaoping was transferred. Chairman Mao said it was due to the need on the front. When your father served as the secretary-general of the Central Committee, he took charge of the life of the leaders of the Central Committee, kept the minutes of the meetings, and was responsible for the security work as well.

When I went to see Mom Liu Ying, she had put on neatly pressed, dark blue Western-style clothes, with a pretty, colorful gauze handkerchief. She was short and slight of stature but full of vigor. She had much to say about the Long March. She told me,

> At the beginning of the Long March, I was always together with your father and others. So long as there was half a day for rest, we often gathered together, and as we had nothing to do, we chatted about anything. Those present established for fun a talking company, with Chen Yun as general manager and your father deputy general manager. As we had nothing to eat, we talked about food and had a dinner party in our minds. Your father often boasted about delicious Sichuan dishes. When we came to the border area of Sichuan, we found people there in dire poverty. I told him, "What is there in Sichuan, nothing but fermented glutinous rice!" Then he said, "Here is the border area!" and the Sichuan dishes are delicious anyway. Your father was always cheerful and humorous. All of us were young and optimistic then.

After they arrived at Mao'ergai in early August, the First and Fourth front armies held a get-together. At the party, Li Jukui met with Deng Xiaoping once again. Old General Li said,

> A stage was set on the ground for the get-together, and Zhang Guotao was asked to make a speech. Several of us cracked jokes below the stage, and Comrade Xiaoping was among them. At that time, our 1st Division had just got some cut tobacco. Comrade Xiaoping told me, "If you give me some tobacco, I'll tell you some good news." "It's easy!" said I. Then I took an iron case out of my pocket, handed it over to him, and said, "Smoke!" Comrade Xiaoping said with a smile, "Let me tell you the good news. You are promoted!" He told me, "The Military Commission has decided to transfer you to the Fourth Front Army to serve as the chief of staff of the 31st Corps, and the order has been issued." At that time, the First Front Army had a large number of cadres, while the Fourth Front Army had more soldiers but lacked cadres. So the Fourth Front Army had asked the First Front Army for cadres. After hearing what your father told me, I went to ask Senior Commander Nie, who confirmed this news.

I went to see Mom Lin Yueqin, the wife of Marshal Luo Ronghuan, in Yu'er Lane in Beijing. Members of our family and the Luo family have been acquainted for a long time. The elders are longtime comrades-in-arms and have maintained close relations, and the children of the two families have been good friends, too. When Mom Lin learned of my search for information about the

Long March, she told me straightforwardly, "Your father was with Uncle Luo when they were plodding through the grasslands."

Luo Ronghuan was born in Hengshan, Hunan Province, in 1902. He took part in the patriotic student movements in his youth, joined the Communist Party of China in 1927, and participated in the rebellion in southern Hubei and the Autumn Harvest Uprising. He held important posts in the 4th Corps of the Red Army led by Mao Zedong and Zhu De and served later as the political commissar of the 4th Corps and the director of the Political Department of the 1st Corps. During the Long March, he suffered attacks from those who took the erroneous "leftist" line, was removed from office, and took up the new post of inspector until he was reappointed deputy director of the Political Department of the 1st Corps in September 1935.

Mom Lin Yueqin has a round face and squints her eyes when smiling. Drinking tea, she told me,

> Uncle Luo and your father had one horse each. They were together throughout the Long March in 1935. At that time, they marched every day. Uncle Luo did not like to talk, but your father often chatted and laughed heartily. When these people got together, they often talked at random. About what? About delicious food. They said pepper was delicious. As soon as pepper was mentioned, they slobbered. They said the twice-cooked pork was delicious. One said the Sichuan-style twice-cooked pork was delicious, whereas the other insisted this dish in the Hunan style was delicious. Anyway, they had nothing to eat and could only have their dinner party in their minds! They had no tobacco to smoke then. They had to find some used paper and dry leaves and then wrap the leaves in paper as cigarettes. Your father said jokingly, "I am a cigarette-maker from a tobacco factory." In the course of plodding through the grasslands, both of them had a bath in a brook. The natives of Sichuan and Hunan are fond of cleaning themselves.

In the course of plodding through the grasslands, reaching Ejie, climbing the Minshan Mountain, coming to Hadapu, crossing the Weihe River, climbing the Liupan Mountain, and arriving at last at Wuqi Town in northern Shaanxi, Father was with Luo Ronghuan the whole time. They marched, worked, played chess, and looked for tobacco together. One of them was the deputy director of the Political Department of the 1st Corps and director of the Local Work Department, and the other was the director of the Propaganda Department of the 1st Corps. The age difference between them was only two years, and their prior experiences were quite similar. On the Long March, they "rode their horses side by side in marching, had heart-to-heart talks while taking a rest, and shared the same bed in camping. They discussed frequently the harms done to the revolutionary cause by "left"-adventurism."[4] Father said once, "We kept no secrets from each other."[5] Mom Lin also said, "Your father and Uncle Luo had different dispositions, one active and the other passive. During the Long March, they marched and camped together. They got along very well!"

Yes, indeed, Father and Marshal Luo got along very well. After liberation, they had frequent contacts. When Uncle Luo was ill, my parents often went to visit him. Four houses were built in Dongjiaominxiang in Beijing in the 1950s, and one of them was distributed to Father. But Father said, "I'll not move there. Let Marshal Luo live there!" When Uncle Luo was reluctant to do so, Father set a deadline for him to move in. Once my parents went to see Uncle Luo and gave him a shower nozzle brought back from the Soviet Union. On the occasion of being granted the military rank of marshal in 1955, Uncle Luo gave a dinner party. My parents as well as Uncle Nie Rongzhen and Mom Zhang Ruihua were all present. Father was on very intimate terms with these old comrades-in-arms! Although Lin Biao and Marshal Luo had worked together for a long time, they did not have a close friendship. Father once said, "Lin Biao cannot unite with even such a person as Marshal Luo!"

Father considered Marshal Luo to be plain, sincere, honest, and kind, a comrade who enjoyed high prestige among the cadres. Father had a high respect for him. When Uncle Luo passed away in 1963, our whole family was deeply grieved, and Mother asked me to go to Luo's home to be with Qiao Qiao, Uncle Luo's daughter, for several days. Today, our two families still have very close contact, and the two old women, my mother and Mom Lin Yueqin, pay occasional visits to each other.

Well, I have wandered from the subject! Where have I stopped?

I have written that after the difficult journey of seven days and nights, the Red Army eventually passed through the boundless large grasslands. After plodding through the grasslands, the Left and Right route armies, over 100,000 strong, should have had high morale and energy to carry out their great plans. However, too numerous are the tribulations in the land of the living!

The critical split in the Party and the Red Army brought on by Zhang Guotao occurred right after the Red Army had plodded through the grasslands. Zhang Guotao had delayed time and again and refused to join the Central Committee and the Right Route Army. Ignoring the repeated persuasion of the Central Committee, he sent a secret telegram to Chen Changhao, who then served as the political commissar of the Right Route Army, asking him to lead the Right Route Army southward. Zhang Guotao was plotting to split and endanger the Central Committee. Fortunately, Ye Jianying, chief-of-staff of the Right Route Army, saw this secret telegram and reported it to Mao Zedong immediately.

After urgent consultations, Mao Zedong, Zhou Enlai, Bo Gu, and others decided that, under the dangerous conditions, the 1st and 3d corps and the column of the Military Commission from the Right Route Army should first be rapidly pulled out and moved northward. Thus, they got out of danger.

At that time, the Central Committee and the Right Route Army had only seven or eight thousand men left.[6] They started off on September 10[7] and reached Ejie, Diebu County, Gansu Province.

On September 12, the Political Bureau of the Central Committee held an enlarged meeting at Ejie that made "The Decision on the Mistakes of Zhang Guotao." The decision pointed out that the essence of Zhang Guotao's mistake in opposing the strategy of northward advance lay in a principled divergence of views on the political situation and on the relative strength of the Red Army troops and those of the enemy. It called on the comrades of the Fourth Front Army to rally around the Central Committee, to fight against Zhang Guotao's mistakes, and to urge him to carry out the strategy of northward advance.

After the Ejie meeting, the Central Committee led the 1st and 3d corps and the column of the Military Commission in the northward advance. They seized Lazikou, the natural barrier where "one person can block the advance of ten thousand men," climbed the Minshan Mountain across Sichuan and Shaanxi provinces, and arrived at Hadapu to the south of Minxian County, Gansu Province. Here the troops were renamed the Shaanxi-Gansu Detachment of the Chinese Workers' and Peasants' Red Army, with Peng Dehuai as commander and Mao Zedong as political commissar.

After learning that the Northern Shaanxi Red Army and that the Northern Shaanxi Base Area still existed, the Central Committee and the Shaanxi-Gansu Detachment continued to advance northward, climbed the Liupan Mountain in northwestern Gansu, and arrived at Wuqi Town in the Shaanxi-Gansu Base Area on October 19.

On October 22, the Political Bureau of the Central Committee held an enlarged meeting. The meeting pointed out that the yearlong Long March had been completed, and the future tasks were to establish revolutionary base areas in northwestern China and to exercise further leadership over the nationwide revolution.

During the year from October 1934 to October 1935, the Central Red Army started off from the Central Soviet Area in Jiangxi Province, passed through eleven provinces, and at last arrived at northern Shaanxi in northwestern China. Along the way, the Central Red Army overcame numerous hardships and dangers, defeated the encirclements, pursuits, and interceptions by hundreds of thousands of enemy troops, achieved an unprecedentedly great historic move, and triumphantly completed a magnificent feat. The Long March shocked the country and the whole world.

Mao Zedong said that the Long March was the first one in history and that it was a manifesto, a work of propaganda, sowing the seeds of revolution. The Long March ended in our victory and the defeat of the enemy.[8]

The Long March became world-famous not because of the long distance covered, but because Chinese Communists and the Chinese Workers' and Peasants' Red Army conquered, with an unrivaled heroic spirit, a strong enemy, natural barriers, and their own mistakes; completed the process of renewal from being in a passive position to taking the initiative, from failure to victory; and created on this basis an entirely new revolutionary situation that endured until victory was won throughout the country.

The romanticism of the Long March has aroused the interest of many people, both Chinese and foreigners. Some have written books about the Long March, while others have retraced the footsteps of the Red Army in order to have a firsthand experience of the Long March. The romanticism, the experience, and the illustrious name recorded in history are precisely the glorious portrayal of the Chinese Communists.

The Long March in the 1930s was the first such effort in China's five-thousand-year history, but not the last. Today, over four decades later, China has started a new long march. The first Long March was a revolutionary struggle for political power waged by the people. This second long march is an entirely new effort at socialist construction of a more brilliant and magnificent future for the Chinese nation.

It took only one year to complete the first Long March, but this new long march will take a dozen years, decades, perhaps even centuries. However, after experiencing the dauntless heroism of the first Long March, the Chinese people and Chinese Communists are convinced that they can fulfill this new historic mission so that their country will be more powerful and prosperous, the people will be happier, and later generations will continue their undertakings throughout the ages!

CHAPTER 44

ON THE LOESS PLATEAU OF NORTHWESTERN CHINA

In October 1935 the Shaanxi-Gansu Detachment of the Chinese Workers' and Peasants' Red Army led by Mao Zedong arrived in northern Shaanxi.

In November the Central Committee decided to restore the designation of the First Front Army of the Chinese Workers' and Peasants' Red Army, which had under it the 1st and 15th corps and other units and totaled more than 11,000 men. On November 7, the Central Committee organs reached Wayaobao, Anding County, in the Shaanxi-Gansu Base Area.

The Kuomintang was worried about the movement of the First Front Army of the Red Army in northern Shaanxi. It dispatched five divisions of the Northeast Army to launch a campaign of encirclement and suppression against the First Front Army and attack it from two directions. From November 21 to 24, the First Front Army fought a countercampaign, the Zhiluo Town Campaign, and annihilated one enemy division plus a regiment, totaling 8,300 men. That great victory laid the foundation for basing the supreme headquarters of the national revolution in northwestern China.

Father said that after the campaign started, he and Luo Ronghuan "watched the battle" from a hilltop. Suddenly, a unit of the Kuomintang army attacked them with heavy fire, and they were in great danger. His fox-fur overcoat, which Fu Zhong had given him, received several bullet holes. Fortunately, he was not wounded. At the critical moment, a company of the former 7th Corps came to their rescue. Father often said it was not easy for him to avoid, for decades, being arrested while doing underground work or being wounded in battle.

After the Zhiluo Town Campaign, the First Front Army got a good chance for rest and reorganization. It did not march very much, nor did it fight many battles.

Liang Biye, who worked in the Political Department of the 1st Corps, knew a lot about Deng Xiaoping, who served as the director of the Propaganda Department of the Corps Political Department. Liang said:

> The task of our Propaganda Department was to ensure that during the march and operations, the troops had enough good food, kept fit, and did not drop out, and that the number of combatants was not reduced. We mainly did propaganda work. Even in the most difficult times, we also propagated the firm ideal that we were sure to win the victory in the revolution. We explained why the army advanced northward to fight the Japanese invaders. The Propaganda Department also edited a mimeographed tabloid, *Fighters' News*, all along the way and after the Long March. The cadres of the Propaganda Department and other cadres of the Political Department often went down to the divisions and regiments to relay important instructions and review work. After the Red Army passed through the grasslands,

> cadres often went down to the grass-roots units. As there were not many units, they went there in the morning and returned in the evening. At that time, the propaganda work escalated and included telling cadres and soldiers about the situation and the heroic deeds. After we arrived at Wuqi Town, the 1st Corps organized a visiting group headed by Li Fuchun and Huang Kecheng to visit the 15th Corps, which was composed of the 25th Corps of the Red Army from the Hubei-Henan-Anhui area and the Red Army in Shaanxi and Gansu. We also organized a theater troupe to give them a special performance as an expression of appreciation. Comrade Xiaoping did not attend, but he gave us some points to cover. He especially invited a performing artist to teach our theater troupe.

Liang Biye was born in Pitou County, Jiangxi Province. He joined the revolutionary ranks at the age of fourteen in 1930, when Mao Zedong and Zhu De led the 4th Corps to his hometown. Then he became the head of a children's corps and later joined the Communist Youth League. In the same year he joined the Red Army. At that time, "child soldiers" like Liang Biye were common in the Red Army. Hu Yaobang, who once served as the general secretary of the Central Committee after liberation, and Chen Pixian, who was a member of the Political Bureau of the Central Committee for some time, were among them.

In January 1936 the twenty-year-old Liang Biye returned to his unit after recovering from an illness. At Linzhen Town he saw many familiar faces while watching the performance given by the propaganda team. There was a dinner party after the performance, and among those present were Zhu Rui and Luo Ronghuan, director and deputy director of the Political Department of the 1st Corps, and Deng Xiaoping, director of the Propaganda Department. It was decided then to appoint Liang Biye leader of the propaganda team. Liang said:

> After that, I worked under the direct leadership of Comrade Xiaoping. Comrade Xiaoping attached importance to the propaganda team. He said, "The propaganda team should not only do propaganda work but also become a place for training cadres. The propaganda team should do work among the people, among the soldiers, and even among the enemy troops." Deng always said, "The propaganda team is the best place for training cadres." At that time, most of the members of the propaganda team were cadres. In one performance, three regiment cadres sang songs. They were Zhang Guohua, Tan Guansan, and Chen Xiong. They sang "The Song of the Road," Soviet songs, and the French song "La Marseillaise." On the propaganda team, we often studied politics and gave tests. There was more political education after the Eastern Expedition. We were short of hands. Once when a new recruit came, he was educated somewhat, but he was already thirty years old. Deng told his bodyguard, "Assign him to the propaganda team." When I took a look at him, oh, he was like an old man. I refused to take him. The bodyguard reported to Deng on my refusal. Deng said, "Take him, willing or not." That was the way Deng handled affairs, brief and to the point. It turned out that the man was very good. He was good at playing the role of an old woman. He worked diligently and took care of the children on the team. Our propaganda team presented programs propagating

resistance to Japan and composed a song, "Song of the Long March of the Central Red Army."

Then Gen. Liang Biye began to sing the song with his eyes glistening and a fist waving. The song went like this:

The Central Red Army set off from Jiangxi.
It marched 12,500 kilometers in twelve months,
climbing high mountains and crossing swift currents.
It fought more than 500 battles, big and small,
defeating the White army and local bandits, totaling 410 regiments.
Brave are the red heroes, who are all-conquering.
They finally arrived in northern Shaanxi to join forces with the 15th Corps of the Red Army.
They smashed the enemy's campaign of "encirclement and suppression" and continued to advance triumphantly!

I interviewed Gen. Liang Biye between the autumn and winter of 1991 in his house. It had already been fifty-six years since he and his comrades-in-arms composed this song. With profound respect, I noticed that he had not forgotten one word of it and still sang it imposingly.

Before the Eastern Expedition by the Red Army, the Political Department of the 1st Corps had been in the Daotong area of northern Shaanxi. In 1935 the situation was changing constantly. The Japanese imperialists occupied all the three provinces in northeastern China and had stepped up their aggression against northern China, their attacks being mainly directed at Hebei, Shandong, Shanxi, Chahar, and Suiyuan provinces.

Under Japanese pressure, the KMT government in Nanjing signed with Japan the Qin (Dechun)-Doihara Kenji Agreement and the He Yingqin-Umezu Yoshijiro Agreement in June and July, respectively, of 1935. According to these two agreements, the Kuomintang government, yielding to Japan's demands for expansion, agreed to withdraw Chinese troops from Hebei and ban all anti-Japanese activities across the country. The Kuomintang government, in effect, surrendered to Japan most of China's sovereign rights in Hebei and Chahar, including Beiping and Tianjin.

While the Japanese invaders were intensifying their economic control and plunder of northern China, they were also engineering the "autonomy of northern China." This aggression against northern China and the Kuomintang government's policy of national betrayal and humiliation made the Chinese people feel ever more strongly the seriousness of the national crisis. The voice of the people from all strata demanding that the Kuomintang government change its policy toward Japan of submitting to national humiliation became louder and louder.

At this critical moment, the Communist Party of China published "An Appeal to All Fellow Countrymen to Resist Japan and Save the Nation" (in

August), "The Declaration on the Annexation of Northern China by Japanese Imperialists and the Selling-out of Northern China and China by Chiang Kai-shek" (in November), and "The Declaration on Resisting Japan and Saving the Nation" (also in November). The three documents appealed to all parties, armies, and compatriots from all strata to stop civil war and concentrate all the national strength on resisting Japan and making concerted efforts to save the nation, irrespective of differences in political views and interests or of any hostile actions in the past or at present. The documents proposed that all parties, organizations, celebrities, scholars, politicians, and local army and government organizations that were willing to participate in the resistance to Japan and national salvation hold talks, establish a joint national defense government, and organize a unified anti-Japanese allied army. The documents declared that Chinese Communists would be the first to join an anti-Japanese allied army to perform their duties of resisting Japan and saving the nation.

The CPC appeal caused strong repercussions among people throughout the country. The long pent-up fury and patriotic zeal of the people erupted like a volcano. On December 9, 1935, a student movement broke out in Beiping. Thousands of young students who felt deeply the danger of national subjugation unfurled banners and marched in the streets. They held an exciting mammoth demonstration for resisting Japan and saving the nation and presented their petition to the Kuomintang authorities in Beiping. The marchers were ruthlessly suppressed by the Kuomintang army and armed police. More than thirty students were arrested, and hundreds were wounded.

The following day the Beiping students waged a general strike. On December 16, the patriotic students in Beiping marched into the streets again, with some 10,000 patriotic residents joining in the even larger demonstration. Again they were suppressed by the National Government, with dozens of students arrested and more than 300 wounded.

Under the influence of the patriotic struggle waged by the students in Beiping, students in Tianjin, Baoding, Taiyuan, Hangzhou, Shanghai, Wuchang, Chengdu, Chongqing, Guangzhou, and Nanning followed suit and held anti-Japanese rallies and demonstrations. The workers in Guangzhou and Shanghai also held rallies and sent open telegrams demanding resistance to Japan. Shen Junru, a famous man of culture, and others organized the National Salvation Federation. For a time, the angry patriotic voices resisting Japan resounded throughout the country, forming an irresistible upsurge in the mass movement.

On December 17, 1935, the Central Committee called a meeting of its Political Bureau at Wayaobao in northern Shaanxi. According to the resolution adopted by the meeting, the basic features of the political situation in China were that the Japanese imperialists "were preparing to annex the whole of China and turn China, a semi-colony dominated by all imperialists, into a colony of Japan," and that the only way out for all those Chinese who refused to be slaves of a foreign power and traitors was to "wage a sacred national revolutionary war against Japanese imperialism and its running dogs and traitors." The resolution called for the establishment of the broadest possible national united front against Japanese aggression.

The Communist Party of China and the Red Army in northern Shaanxi strengthened their work in the Northeast Army stationed in Shaanxi for the purpose of establishing a united front. Mao Zedong and Zhou Enlai worked particularly hard on Zhang Xueliang, the commander of the Northeast Army. In February 1936 the Red Army and the Northeast Army reached an oral agreement on mutual nonaggression. Then Zhou Enlai and Zhang Xueliang held secret talks and reached an agreement on mutual nonaggression and exchange of representatives. Zhang Xueliang further proposed to try to persuade Chiang Kai-shek to resist Japan.

At the same time, the Communist Party of China strengthened its contact with Gen. Yang Hucheng, commander of the Northwest Army stationed in Shaanxi. Both sides stopped their hostilities and exchanged representatives in an effort to resist Japan jointly. The establishment of cooperative relations with Zhang Xueliang and Yang Hucheng provided favorable external conditions for the Central Committee and the Red Army to set up stable base areas in northern Shaanxi.

Facing the increasing nationwide demand to resist Japan, Chiang Kai-shek also began to hold secret talks with the Communist Party of China.

To show the determination of the Red Army to resist Japan with actual deeds, the First Front Army, led by Mao Zedong, Peng Dehuai, and others, started the Eastern Expedition on February 20, 1936, in the name of the Anti-Japanese Vanguard Army of the Chinese People. The Vanguard Army broke through the defense line established by the Shanxi warlord Yan Xishan and successfully crossed the Yellow River.

Chiang Kai-shek's talks with the Communist Party were just a cover. His real purpose was to dissolve the Communists through the talks. When he learned that the Communist-led troops had crossed the Yellow River to resist the Japanese invaders, he immediately concentrated 200,000 troops to reinforce Yan Xishan in Shanxi in an attempt to wipe out the Red Army in areas east of the Yellow River.

After crossing the Yellow River, the Eastern Expeditionary Red Army units controlled the vast area of more than fifty kilometers from north to south and thirty-five kilometers from east to west on the eastern bank of the river in just three days and annihilated a regiment of Yan Xishan's army during a battle at Guanshang Village. By the end of March the Red Army had fought separately in three directions and swiftly won more victories. Gen. Liang Biye said: "On the Eastern Expedition, the Propaganda Department led by Comrade Xiaoping disseminated the propositions of the Communist Party and the idea of resisting Japan. We also had to do our work among the enemy troops and prisoners of war. Comrade Xiaoping wrote outlines for the propaganda work and teaching materials himself during the Eastern Expedition."

In May 1936, to avert a civil war, the Central Committee decided to withdraw the Red Army to northern Shaanxi west of the Yellow River, thus ending the two-month Eastern Expedition.

Not long after the return from the Eastern Expedition, Luo Ronghuan, deputy director of the Political Department of the 1st Corps, was instructed to study at the Red Army College, and Father succeeded him as deputy director. The Political Department of the 1st Corps was then located at Yuzhu in northern Shaanxi.

During this period, the Central Committee called a meeting at Daxiang Temple to sum up the work and rectify its style. Father and other senior cadres of the 1st and 15th corps attended the meeting. Father and Luo Ruiqing were also dispatched directly by the Central Committee to some units of the 1st Corps to make investigations and study and assess cadres. This was the first time the Central Committee had dispatched Deng Xiaoping to make investigations, as well as the first time that Gen. Wang Ping met Father. He was then a regiment political commissar of the 4th Division. He told me:

> The Central Committee sent Luo Ruiqing, director of its Security Bureau, and your father to investigate the situation. Your father talked with most of our cadres. I told him everything I knew, and your father appreciated my frankness. Later he and Luo Ruiqing delivered speeches to us. My impression was that your father was cool and serious. He did not talk much, but his speech was simple and to the point. He spoke in short sentences, and it was easy to take notes. His viewpoints were clear, and what he said was useful.

After the Red Army withdrew to northern Shaanxi, Chiang Kai-shek concentrated sixteen divisions and three brigades to launch a new campaign against the Shaanxi-Gansu Base Area of the Red Army. Under these circumstances, the Central Committee decided that the Red Army would launch the Western Expedition against the border areas of Shaanxi, Gansu, and Ningxia, where the Kuomintang forces were weak.

From May to the end of July 1935, the Red Army quickly established a new base area on the borders of Shaanxi, Gansu, and Ningxia. Covering more than 400 kilometers in circumference, this area was linked up with the old Shaanxi-Gansu Base Area. The Red Army and the local armed forces expanded.

For some time thereafter, the situation was relatively stable, and no significant battles were fought on the front. The army took this opportunity to conduct training and education. Gen. Liang Biye recalled:

> Returning from the Eastern Expedition, we raised money and grain and brought back quite a lot of mules. After the Western Expedition, we fought fewer battles, and Zhang Xueliang and Yang Hucheng began to form a united front with us. At this time, Comrade Xiaoping was the deputy director of the Political Department

> of the 1st Corps, and Zhu Rui was the director. Deng was responsible for the Party's organizational, propaganda, and educational work, and for the cadres' education in particular. It was at that time that we, joining the army at an early age, acquired common political knowledge in a relatively systematic way. The study class started with courses on political parties, leaders, and the masses, including the history of social development. We listened to lectures, held discussions, and had to take graded examinations. Comrades Zhu Rui and Xiaoping gave lectures personally. Many comrades from the combat units began to heighten their formerly simple class awareness to the theoretical level. Our class was located in the area of Qiyingchuan in Ningxia.

Gen. Su Jing, who did reconnaissance work in the 1st Corps at the time, recalled:

> In 1936 Comrade Xiaoping organized us into a study class, which lasted for more than a month. We learned about the world, the history of social development, and Marxism-Leninism. Comrade Deng Xiaoping gave us lectures, issued study materials to us, and gave us tests and marks. Sometimes we held discussions. We raised questions, and he gave us answers. Before that, we had spent most of our time fighting battles and marching. This time, we learned a lot from the class organized by Comrade Xiaoping.

Gen. Liang Biye told me that apart from educational work, the Political Department was also in charge of the work among the enemy troops and the united front work in the Northeast Army. At the same time, Ningxia was an area where the people of the Hui nationality lived in compact communities. So political work had to involve nationality issues. Besides, the secret society in Sichuan, the Society of Brothers, was quite influential there, and attention had to be paid to the work among its members. It was Father who took charge of most of the work.

General Liang also recalled that between August and September 1936, the Political Department was located at Wulidong in the Yuwang area in Ningxia. At that time, the Central Committee designated Deng Xiaoping to lead an inspection team to examine the work of the 15th Corps. General Liang said:

> Deng took me, Tang Liang, and Cai Yuanxing along and went to the 81st and 75th divisions of the 15th Corps stationed in northern Shaanxi under the cover of a twelve-member guard squad. Deng mainly talked with cadres at the division and regiment levels, while we talked with cadres and soldiers at lower levels. Deng was not sent by the 1st Corps but by the Central Committee and its Military Commission. It was a very important task. After returning, Deng reported to the Central Committee.

This was the second time that the Central Committee sent Deng Xiaoping to make investigations. The leaders of the Central Committee and its Military Commission, including Mao Zedong and Zhou Enlai, trusted Deng Xiaoping very much.

Liang Biye and Su Jing said that during this period, many important materials of the 1st Corps and the editorials of the *Fighters' News* were written by Father personally. "Comrade Xiaoping was a generally acknowledged quick hand at writing. Once Director Zhu Rui urged him to write a speech to be delivered to combat units. He said, 'That is okay.' He immediately took up a piece of paper and a pencil and wrote by using his knees as a desk. In no time, he had finished."

When the First Front Army was stationed in the Shaanxi-Gansu-Ningxia Base Area, a major event in the history of the Red Army took place. In October 1936 the Second and Fourth front armies arrived at last in the Shaanxi-Gansu-Ningxia Base Area, after overcoming untold hardships and Zhang Guotao's separatism, and joined forces with the First Front Army. After separating from the Central Committee during the Long March, Zhang Guotao had taken the Fourth Front Army to western Sichuan and the Xikang area. There, counting on the great number of troops and arms under his command, he established a new Party "central committee" and called himself its chairman. But because of the massive campaign launched by the Kuomintang troops, he was unable to gain a foothold in western Sichuan. The Fourth Front Army was continuously forced to retreat to the west until it reached the Ganzi area. At that time, there were only some 40,000 men left in the army. The reality of the situation forced Zhang Guotao to announce the abolition of his "central committee." In July 1935 the Fourth Front Army joined forces with the Second Front Army led by He Long. Pressed hard by Zhu De, Liu Bocheng, Ren Bishi, He Long, and Guan Xiangying, Zhang Guotao agreed to move the army northward to join forces with the Central Committee. After overcoming untold hardships in the expedition for three months, the Second and Fourth front armies advanced northward and at last reached the Shaanxi-Gansu-Ningxia Base Area, where they joined forces with the First Front Army. Then, as discussed earlier, the Central Committee called an enlarged meeting in March 1937 and made "The Decision on the Mistakes of Comrade Zhang Guotao."

Zhang Guotao admitted his mistakes at first, but in reality, he refused the Party's efforts to redeem him. In April 1938 he sneaked out of the Shaanxi-Gansu-Ningxia Base Area and threw in his lot with the Kuomintang to serve as an anti-Communist lackey despised by the people. On April 18 of the same year, the Central Committee decided to expel Zhang Guotao from the Party.

In the course of its development, the Communist Party of China corrected the right-capitulationism of Chen Duxiu, the "leftist" putschism of Qu Qiubai, the "left"-adventurism of Li Lisan, and Wang Ming's "leftist" dogmatism and factionalism, and finally it overcame the "splittism" of Zhang Guotao as well. At this time, it had been only sixteen years since the Communist Party of China was founded. Sixteen years seemed like a very short time, but also a very long time. During those sixteen years, the CPC and Chinese Communists went through untold hardships, and it was these hardships that enabled Chinese

Communists to become more mature and the Communist Party, representing the Chinese proletariat and the laboring people, to become stauncher and stronger.

The struggle waged by the Communist Party against the hostile classes was inevitable. So were the struggles between right and wrong and between truth and falsehood within the Party. The Communist Party is a great Party in that it dares to face up to its own shortcomings and to correct its own mistakes. This was true during the CPC's first sixteen years and has been true throughout the long years of the revolution and economic development. Otherwise, it would have long died in the torrents of the Chinese revolution. Father has often said that in the history of the Party, both "left"- and right-wing ideas have influenced us at any given time. But apart from the right-capitulationism of Chen Duxiu, "leftist" ideas have had a more deep-rooted effect on the Party. He said, "The 'leftist' influence has been terrible in the history of the Party! It could bury a good thing overnight." He added, "Right ideas can damage our cause, and so can 'leftist' ideas." He said, "We must be sober-minded on this problem, so that we will not make serious mistakes and will easily solve problems when they arise." This is the idea and philosophy of dialectical materialism.

The Communist Party and Chinese Communists have overcome setbacks and corrected mistakes one after another and have advanced along the road to truth step by step. It is precisely because they can face their own problems and mistakes soberly and honestly, and correct them before continuing to advance toward truth, that they have been so vital and have been able to achieve a final great victory.

A party, a cause, and even a person can make progress only through constant efforts at self-perfection. Dogmatism and ossification of thinking can lead only to self-suffocation and even self-destruction. On the other hand, all rashness or pessimism that is divorced from reality will make a cause suffer defeat when victory is at hand. How difficult it is for the leader of an organization, a political party, or a state to stay sober-minded, persist in seeking truth, and be ready to prevent and correct both "left"- and right-wing mistakes at all times! What a great responsibility!

The history of those first sixteen years of the Party, and even of the whole seventy-year history, may well teach something to present and future generations.

After joining forces in the mid-1930s, the First, Second, and Fourth front armies overcame the splittism of Zhang Guotao and defeated the large-scale attacks mounted by the Kuomintang troops, thus gaining a firm foothold in the loess plateau of northwestern China, the Shaanxi-Gansu-Ningxia Base Area. A new historical period had been entered.

Chinese Communists then held even higher the banner of resistance to Japan and of national liberation to create an entirely new situation. Likewise, Father and his comrades-in-arms began to march to the front of resistance to Japan.

THE XI'AN INCIDENT

In 1936 the political situation in China was extremely volatile.

Because Japan had intensified its aggression against China, the Chinese people's sense of national peril was growing steadily, and the voices of resistance to Japan were becoming louder and louder.

The movement against Japan and for national salvation was surging ahead. In northwestern China the united front was formed by the Northeast Army under Zhang Xueliang and the Northwest Army under Yang Hucheng, together with the Communist-led Red Army. In northeastern China the seven corps of the Anti-Japanese Allied Army founded by the Communist Party of China and other anti-Japanese forces resisted the invading Japanese troops heroically. In southern China Chen Jitang in Guangdong and Li Zongren and Bai Chongxi in Guangxi announced their intention to move their troops northward together to put up resistance to Japan. In eastern China the Shanghai Federation for the National Salvation of People from All Walks of Life was set up. In eastern and northern Suiyuan Gen. Fu Zuoyi led his army to smash Japanese attacks, thus greatly promoting the movement of resistance to Japan. At the same time, the rapid growth of the Japanese forces in China was causing uneasiness among the British and American forces, and the conflicts between the pro-British and pro-American forces and the pro-Japanese forces were aggravated.

Chiang Kai-shek still cherished the illusion that he could make concessions to Japan in order to retain his sovereignty over a part of the country, stating that "peace will never be given up until it is completely hopeless, and neither will sacrifice be talked about readily until the last moment." Internally, he made vigorous efforts to suppress the anti-Japanese actions, plotted to foil the anti-Japanese alliance between Guangdong and Guangxi, rejected the demand of Zhang Xueliang and Yang Hucheng in northwestern China that civil war there be stopped, and arrested seven patriotic leaders, namely, Shen Junru, Zou Taofen, Li Gongpu, Sha Qianli, Shi Liang, Zhang Naiqi, and Wang Zhaoshi. He persisted in perceiving the Communist Party and the Red Army as his "serious hidden danger," vowing to "eliminate the remnants of the red bandits."[1]

At this critical moment of national crisis and domestic trouble, the Xi'an Incident broke out, shocking the country and the world. Stationing their troops in northwestern China, Zhang Xueliang and Yang Hucheng, in compliance with the overall national interest, advocated alliance with the Communist Party in resisting Japan. This proposal caused Chiang Kai-shek to panic. To prevent the situation from "developing into rebellion,"[2] on December 4, 1936, Chiang Kai-shek went to Xi'an personally, ordered Zhang Xueliang and Yang Hucheng to march all their troops immediately to northern Shaanxi to "suppress the Communists," and threatened them by saying that if they disobeyed, he would move their troops away from Shaanxi. Zhang and Yang repeatedly failed to persuade Chiang to consider the overall interests of the nation. Instead, Chiang reprimanded them.

On December 9, 10,000 patriotic students in Xi'an took to the street, demanding the cessation of civil war and concerted efforts to resist Japan. As Kuomintang special agents opened fire and wounded some students, the indignant students rushed out of Xi'an and planned to go to the Huaqing Pool in Lintong to petition Chiang Kai-shek. Moved by the patriotism of the students, Gen. Zhang Xueliang promised to reply to their demand with actual deeds.

On December 10 and 11, Generals Zhang and Yang called on and remonstrated with Chiang Kai-shek again, but Chiang rebuked them for going against their superior and creating trouble. By then Zhang and Yang had come to realize that to avert civil war and resist Japan, they had no alternative but to make a "military remonstration"!

On December 12, 1936, Zhang and Yang launched what was known as the Xi'an Incident. They detained Chiang Kai-shek at the Huaqing Pool. Then they sent an open telegram to the nation, stating that in the current situation of national crisis, they were compelled to launch the incident for the purpose of urging Chiang Kai-shek to resist Japan and putting forward an eight-point proposal, including ending the civil war, releasing the patriotic leaders in Shanghai, setting free all political prisoners, encouraging the people's patriotic movement, and immediately calling a meeting for national salvation. The Xi'an Incident immediately sent shock waves through the land of China.

The Kuomintang government in Nanjing was in chaos, and there was a public outcry. The followers of Chiang Kai-shek demanded a peaceful settlement so as to save his life. Some government officials with ulterior motives called for bombing Xi'an and sent a telegram to Wang Jingwei, head of the pro-Japanese forces, asking him to return to China. To maintain the rule of Chiang Kai-shek, Britain and the United States held that Chiang might as well have some form of cooperation with the Communist Party in a joint resistance to Japan. The Soviet Union hoped for a peaceful solution to the incident but mistook Generals Zhang and Yang for having close relations with the pro-Japanese forces. Japan claimed clearly that Zhang and Yang had become "red" in an attempt to provoke the outbreak of a civil war in China.

How the Xi'an Incident would develop and what Generals Zhang and Yang would do became the focus of attention of the whole country, even the whole world.

After repeated analysis and study, the Central Committee in northern Shaanxi held that the Xi'an Incident stemmed from the desire to resist Japan and seek national salvation and was, therefore, entirely justifiable. However, the way Zhang and Yang acted landed Nanjing in a position of hostility toward Xi'an, so there was the possibility of a new large-scale civil war. This was what Japan desired. So the Central Committee established the basic principle that the Xi'an Incident should be settled peacefully and then acted accordingly.

On December 15, the Communist Party sent an open telegram to Zhang and Yang, giving its support to their eight-point proposal and opposing the attempt of the pro-Japanese forces to take the opportunity to launch a civil war. On December 17, Zhou Enlai, the representative of the Communist Party, flew to Xi'an to explain to Zhang and Yang the propositions of the Party on encouraging resistance to Japan and averting a larger civil war. He won the approval of Zhang and Yang. On December 19, the Political Bureau of the Central Committee officially put forward the principle of peaceful settlement of the incident. On December 22, the Nanjing government sent Chiang's wife, Soong Mei Ling, and her elder brother, T. V. Soong, to Xi'an. Zhang Xueliang, Yang Hucheng, and Zhou Enlai held talks with Soong Mei Ling and T. V. Soong for two days and agreed to six conditions:

1. to reorganize the Kuomintang, expel the pro-Japanese elements, and admit anti-Japanese elements;
2. to release patriotic leaders in Shanghai and all political prisoners and ensure the right of freedom of the people;
3. to stop the policy of "suppressing Communists" and cooperate with the Red Army to resist Japan;
4. to call a meeting for national salvation to be attended by all parties, people from all walks of life, and representatives of all armies to decide on the policy of resistance to Japan and national salvation;
5. to establish cooperative relations with all countries that sympathized with China's resistance to Japan; and
6. to implement other specific measures for national salvation.

On the evening of December 24, Zhou Enlai met with Chiang Kai-shek and explained to him the propositions of the Communist Party. In the end, Chiang Kai-shek indicated his willingness to accept the conditions, including putting an end to his efforts to "suppress Communists," and to resist Japan jointly. On December 25, Chiang Kai-shek flew back to Nanjing. Zhang Xueliang suddenly decided to accompany him to Nanjing as a gesture of thanks to the people of the whole country.

Thus, the thirteen-day Xi'an Incident ended. The peaceful settlement of the Xi'an Incident foiled the schemes of Japan. Chiang Kai-shek was forced to accept the policy of stopping the civil war and resisting Japan jointly. The cooperation between the Kuomintang and the Communist Party for the second time since the national revolution led by Dr. Sun Yat-sen was then realized under the important precondition of uniting to resist Japan. But Gen. Zhang Xueliang lost his freedom.

Burning with righteous indignation and resolved not to fight a civil war,

Zhang Xueliang resolutely detained Chiang, together with Yang Hucheng, and forced him to resist Japan. When the incident came to an end, the honorable, sincere, and uniquely chivalrous young marshal boarded Chiang Kai-shek's plane with no hesitation. It had probably never occurred to Zhang Xueliang that his benevolence was not reciprocated by Chiang Kai-shek.

But Chiang Kai-shek bore grudges for life. Gen. Zhang Xueliang never came back. He had detained Chiang for thirteen days. Chiang detained him for more than fifty years. When Chiang Kai-shek died of illness in Taiwan in 1975, Zhang Xueliang was still detained there. Now Zhang Xueliang is advanced in age and no longer the bright and brave young marshal. Does he know that his one billion compatriots on the mainland have always missed him? His homeland has been expecting his return for decades. The murmuring stream of the scenic Huaqing Pool at Lintong has been telling tourists the moving and tragic story of the earthshaking incident.

Zhang Xueliang was not the only player in the Xi'an Incident toward whom Chiang Kai-shek bore a grudge. Gen. Yang Hucheng was arrested and imprisoned for twelve years by Chiang Kai-shek, who killed Yang on the eve of his flight to Taiwan. Chiang killed not only General Yang but also his seventeen-year-old son, his nine-year-old daughter, his secretary and his secretary's wife, and another son less than ten years old. A total of nine people, including General Yang's wife—who had died of illness in prison—and two of his aides-de-camp died at the hands of Chiang Kai-shek. Chiang Kai-shek thus vented his spleen for his arrest by Zhang and Yang in the Xi'an Incident.

Although General Zhang was imprisoned and General Yang was killed, their heroic deeds brought about the second cooperation between the Kuomintang and the Communist Party and the resistance to Japanese invaders by the whole nation. This new situation constituted a key to the change in the overall situation in China. Their launching of the Xi'an Incident was of immortal importance.

Father heard of the Xi'an Incident, which broke out on December 12, when he was suffering from a serious illness. Toward the end of 1936, he contracted a very serious paratyphoid. One day in 1991 I went to see Pa Yang Shangkun. I have been his half-daughter since my childhood, and we can talk informally. When I asked whether he knew about this illness of Father's, Pa Yang said, "How couldn't I know!" He told me that Father

> suffered from a very serious paratyphoid in the area of Qingyang in Gansu. He was in a coma. He could not eat anything. If he ate something, his intestines would be broken. So he had to be fed with some rice water. It happened that we formed a united front with Zhang Xueliang at that time, and he sent his aide-de-camp to express his regard and concerns for our Red Army with two trucks of food, cigarettes, and some other materials, including canned condensed milk. Nie Rongzhen decided to give all the condensed milk to Xiaoping. It was the milk that saved the

life of your father. We, several smokers, shared one good cigarette. We made marks on it and smoked in turn. That was before the December 12th Incident.

Old Gen. Xiao Ke told me in my interview with him:

I came to know your father after the Party congress held in the Central Soviet Area in 1931. Both of us knew how to cut mimeograph stencils, so we were soon familiar with each other. We often cracked jokes. Your father mocked me, saying that I had not even been to Shanghai. Between November and December 1936 our two armies advanced on the same route. I heard that your father was ill, so I went to see him. He was carried on a stretcher, in a very serious condition, in a coma, a critical state.

Father also said that he almost died that illness. He dimly heard somebody talk about the Xi'an Incident but was then in the coma again. He said that in his life typhoid fever struck him twice. The other time was in France. He was in very critical condition on both occasions.

After the Xi'an Incident broke out, the main forces of the Red Army advanced southward, first to the area of Qingyang toward the end of December, and then to the area of Sanyuan north of Xi'an. Gen. Liang Biye remembered that the 1st Corps arrived at Donglipu on January 8, 1937, and in the area of Gonghe Town in Gansu on February 22. The Political Department of the corps was located at Wangjialou.

To prepare for operations against the Japanese invaders, the Red Army concentrated on military and political training. In January 1937 Deng Xiaoping became director of the corps Political Department, succeeding Zhu Rui, who was transferred to the Second Front Army to the post of director of the Political Department there. As director of the Political Department of the 1st Corps, Father was in charge of political training.

The Political Department ran a political training class. The cadres of the organs under the direct leadership of the corps studied Marxist philosophy, political economy, and the history of social development. General Liang vividly recalled life then:

We received both military and political training. We studied the resolution of the Wayaobao Meeting of the Central Committee and learned the principles and policies on the united front. Every morning we went out to have drill and running and learn military skills. We held demonstrations of, and competitions in, military skills. Comrade Xiaoping gave us political lectures. He read books and prepared lessons every morning and did not allow us to make noise. He lectured on political economy, starting from the dual characters of commodity. He told us what labor was, how labor created value, and why it was inevitable that socialism would replace capitalism. We had one class a week. We put up our own classroom. It was a mat shed. We had a blackboard and borrowed over twenty benches from local peasants.

Deng was very punctual. Once cadres under Kuang Rennong, director of the Supply Department, were late for class. When starting his lecture, Deng wrote on the blackboard: "The Supply Department was late for class." When cadres from the Supply Department came, they sat down quietly when they saw what was written on the blackboard. Deng did not criticize them, but no one was late for class thereafter. What Comrade Xiaoping taught us were all basic principles, very simple ones. It was the first time many cadres of peasant or worker origin had received such a systematic education. He also taught us to sing "The International." As it was a foreign song, many of us found it difficult to sing correctly. I learned the tune of the song from Deng.

At Wangjialou, we lived in the same small courtyard with two cave dwellings. Comrade Xiaoping and I lived in the one facing south, and the guard squad in the other facing north. The dwellings were surrounded by a wall. To the east was a small vegetable garden. We leased it and turned it into a sports ground. Each of us got a monthly allowance of five silver dollars. I took charge of Deng's money. He liked to drink cocoa, so whenever I went to Sanyuan, I bought some cocoa for him. We from the Political Department ate from the same pot. Food was very simple, sometimes it was meat. In the corps Political Department there was a cook from Jiangxi who could prepare pork braised in brown sauce very well. All cadres who attended meetings of the corps liked to eat the pork cooked in the Political Department.

Deng lived a simple and regular life. After supper, he would go for a walk and then sat down to read. When tired, he engaged in sports activities or watched soldiers playing basketball. Deng kept in close contact with general headquarters, especially with Yang Shangkun, deputy director of the General Political Department. The long letters from Yang were put in big envelopes. Almost every day, Deng went to the headquarters of the corps in Gonghe Town to read telegrams or to discuss matters with Comrades Nie Rongzhen and Zuo Quan. Deng was very strict with the cadres. He said, "As the director, I would manage division commanders!" All division commanders and political commissars of the 1st Corps went to Deng for instructions whenever they came to the headquarters.

At the time, I was the chief of the General Affairs Division. Some comrades in the office wanted to buy something of good quality, like envelopes and writing paper and even fragrant pastes, instead of making them themselves. Deng criticized them, and no one dared to buy thereafter. In the first half of 1937, when the Aiding-West Army led by Liu Bocheng and Xiao Ke passed by Gonghe Town, they all came to Wangjialou to see Comrade Xiaoping. Deng told them, "Your tasks are very arduous!" After the Western Route Army was defeated, the Aiding-West Army stopped in the area of Qingyang and later withdrew to northern Shaanxi. The Central Committee called a meeting attended by cadres at and above the regiment level from the First and Fourth front armies to criticize Zhang Guotao for his mistakes. The Central Committee entrusted Comrades Yang Shangkun, Luo Ruiqing, and Xiaoping to preside over the meeting. The meeting place was at Wangjialou. Comrades Shangkun and Xiaoping and I lived in one room. Luo Ruiqing was tall, and he shared the room of the guard squad. Our Political Department was responsible for the organizational work of the meeting, including the meeting work and the material welfare and safety of the participants. It was a very important meeting.

There were altogether five terms of the director of the Political Department of our 1st Corps. Luo Ronghuan served for two terms, the longest. Comrade Xiaoping served for two years. In that period, the director of the Political Department was mainly responsible for the political work of the corps. Zhu Rui served for some time. Li Zhuoran's term of office was the shortest of all. I learned how to do political work, first from Marshal Luo Ronghuan and then from Comrade Xiaoping. Comrade Xiaoping had a strong grasp of theory and was good at writing. He seemed to have an inexhaustible energy. He could keep a firm hold on important aspects of questions and lay aside minor ones. He could grasp principled issues while leaving minor ones.

General Liang said there were more than one hundred people in the Political Department of the 1st Corps, including seventy to eighty cadres. The Political Department had departments that handled organization, propaganda, security, mass movements, and sabotage, as well as a general affairs division and a cadre inspection corps for would-be cadres. The cadres of the inspection corps could hold higher or lower posts, and the size of the corps could be large or small.

One day between late June and early July 1937, Father told Liang Biye, "I am going to be transferred."

Liang asked: "Where to?"
Deng replied: "To the headquarters."
Liang: "Who is to succeed you?"
Deng: "Luo Ronghuan."
Liang: "When do you leave?"
Deng: "Very soon."
Liang: "What will you do about your remaining food allowances of a few silver dollars?"
Deng: "Why are you so serious!"

Later, Liang Biye used the money to buy a few pieces of ham for Deng. Liang said that Deng and Luo did not meet. Luo arrived only after Deng left. According to the *Biography of Luo Ronghuan*, Luo was appointed director of the Political Department of the 1st Corps on the third day after the July 7th Incident of 1937. Later the 1st Corps moved to Anwupu. Before he left on August 22, Deng made a special trip from Sanyuan to see Luo Ronghuan.

Between June and July 1937 Deng Xiaoping succeeded Fu Zhong as the deputy director of the General Political Department of the front headquarters of the Chinese Workers' and Peasants' Red Army and concurrently as deputy director of the General Political Department. The general headquarters staff included Commander-in-Chief Peng Dehuai, the general political commissar Ren Bishi, who was also the director of the Political Department, Chief-of-staff Zuo Quan, and Deputy Director Deng Xiaoping.

After the Xi'an Incident, the principal task of the CPC was to mobilize the whole Party and the people throughout the country to consolidate peace, strive for democracy, and put up effective resistance to Japan as soon as possible.

On January 13, 1937, the organs of the Central Committee were moved from Bao'an to Yan'an. Yan'an, the ancient city in northern Shaanxi, became the sacred heart of the Chinese revolution that all patriotic progressive youths yearned to go.

In the first half of 1937 the second cooperation between the Kuomintang and the Communist Party progressed. The two parties held three rounds of talks, with the CPC represented by Zhou Enlai and the Kuomintang by Gu Zhutong. The Kuomintang sent the eighteen-member Central Inspection Delegation to Yan'an for an inspection tour. In the first twenty days of June, Chiang Kai-shek and Zhou Enlai held talks on the Lushan Mountain. Although the Kuomintang did not show very much sincerity in the talks and created difficulties in every possible way—it even put forward the absurd proposal that Mao Zedong and Zhu De "go abroad"—with the result that no substantial breakthroughs were made, the trend of cooperation between the two parties signaled by the talks became clear and irreversible.

During this period, the Red Army and other revolutionary armed forces across China swelled to about 100,000, and the CPC had a membership of more than 40,000.[3] The Shaanxi-Gansu-Ningxia Base Area had expanded to the bank of the Yellow River in the east, to the Great Wall in the north, to Guyuan in the west, and to Cunhua in the south, covering thirty-six counties, with a total area of 130,000 square kilometers and a population of two million.[4]

With the consolidation and expansion of the revolutionary base area, the continued growth of the revolutionary military forces, the trend of cooperation between the Kuomintang and the CPC, and particularly the establishment of the staunch and correct nucleus of leadership within the Party, Party members and the officers and men of the Red Army were in high spirits and were ready to march to the battlefields to fight the Japanese invaders.

MARCHING TO BATTLE AGAINST THE JAPANESE

The Lugouqiao (Marco Polo Bridge) Incident plotted by the Japanese invaders occurred on July 7, 1937. The incident was staged at Lugouqiao, near Beiping, by the Japanese in their attempt to annex the whole of China. Afterwards, the Japanese aggressors built up their forces in northern China and launched an all-out war of aggression against China.

In late July the Japanese army mounted large-scale attacks on Beiping and Tianjin and occupied them. Then it launched large-scale attacks on the heartland of northern China. On August 13, the Japanese troops were so confident that they spread the flames of war to eastern China. They attacked Shanghai—the most important city in eastern China—and directly threatened the KMT government in Nanjing.

At this critical moment, when the enemy troops were launching an all-out war of aggression and a national crisis was at hand, Chiang Kai-shek was at last forced to resolve to fight the war of resistance.

The day after the Lugouqiao Incident, the Central Committee sent an open telegram declaring that a nationwide war of resistance was the only way out for China. In the same month, the CPC again called for a general mobilization of the navy, air, and ground forces and a general mobilization of all the Chinese people to conduct a unified and active war of resistance, immediately concentrate the military leadership in charge of the anti-Japanese war, coordinate efforts to command troops at all battlefronts, adopt the strategy of offensive defense, and launch large-scale guerrilla warfare in areas occupied by the Japanese invaders in cooperation with the operations of the main forces of the army.

Under the pressure of this pressing situation in northern and eastern China, the KMT government began to take seriously the major question of KMT-CPC cooperation. In August the KMT held a national defense meeting in Nanjing. In the meantime, the CPC sent, upon invitation, Zhou Enlai, Zhu De, and Ye Jianying to lead a delegation to Nanjing to attend an informal meeting organized by the Ministry of Military Affairs and to hold negotiations with the KMT.

Father accompanied this delegation to Nanjing. He said that they worked behind the scenes and that Zhou Enlai, Zhu De, and Ye Jianying actually negotiated at the table. Gen. Liang Biye said he heard that the meeting needed to draft a document on political work in the war of resistance, that the KMT side did not have the right person to prepare it, and that Deng was then made responsible for writing the draft, which was accepted by the KMT. General Liang also heard that when the CPC delegation was in Nanjing, the Japanese aircraft often came to the city to drop bombs. As soon as the aircraft flew overhead, the KMT officials went into hiding; only the Communists were unafraid, because the Red Army had long been used to bombardment by the KMT aircraft.

Through negotiation, the KMT and the CPC eventually reached an agreement. The CPC agreed to reorganize the main forces of the Red Army stationed in the Shaanxi-Gansu-Ningxia region. On August 22, the Military Council of the KMT government in Nanjing formally issued an order to reorganize the Red Army into the Eighth Route Army of the National Revolutionary Army and appointed Zhu De and Peng Dehuai commander-in-chief and deputy commander-in-chief, respectively.

Between August 22 and 25, the CPC Central Committee held an enlarged meeting of its Political Bureau in Luochuan in northern Shaanxi. The meeting pointed out that the political situation in China had begun a new phase—that of a nationwide war of resistance. This phase's central task was mobilizing all forces to win victory in the war of resistance and achieving democracy in the process of winning such a victory.

On August 25, the Military Commission of the Central Committee issued the order reorganizing the First, Second, and Fourth front armies of the Chinese Workers' and Peasants' Red Army and the Red Army in northern Shaanxi into the Eighth Route Army of the National Revolutionary Army. It renamed the front headquarters of the Red Army the general headquarters of the Eighth Route Army and appointed Zhu De commander-in-chief, Peng Dehuai deputy commander-in-chief, Ye Jianying chief-of-staff, Zuo Quan deputy chief-of-staff, Ren Bishi director of the Political Department, and Deng Xiaoping deputy director. The Eighth Route Army had three divisions—the 115th, the 120th, and the 129th. In December 1937 the guerrilla forces of the Red Army in the south were reorganized into the New Fourth Army.

The second cooperation between the KMT and the CPC was based on the sacred war of resistance. This cooperation received the warm support of all the Chinese people and of patriotic persons from all circles and strata. On hearing this news, Madam Soong Qing Ling, the leader of the left wing of the KMT and the widow of Dr. Sun Yat-sen, said excitedly, "I was almost moved to tears."[1]

On July 31, the seven patriotic leaders in Shanghai were set free. Immediately on their release, they gave their support to the unity necessary for the nationwide war of resistance, based on the cooperation between the KMT and the CPC.

All political parties and mass organizations in China expressed support for KMT-CPC cooperation and the government's policy of resistance to Japan and showed great enthusiasm for the movement to resist Japan and save the nation. The national industrial and commercial circles enthusiastically bought government bonds for national salvation, donated materials to the officers and men at the battlefronts, and took concrete actions to support the protracted war of resistance. Tens of millions of patriotic Chinese residing abroad donated money or did something, through various means, to support the motherland in its war of resistance.

The Red Army men in northwestern China took off their octagonal caps with red stars and put on the new uniforms of the Eighth Route Army. Tens of thousands of men were full of energy and ready for orders to instantly march to the anti-Japanese battlefield.

At the time, Mao Zedong, Zhu De, Zhou Enlai, and the Central Committee were in Yan'an—the heart of the Chinese revolution. The general headquarters of the Eighth Route Army was located in Yunyang Town, Sanyuan County, Shaanxi.

After the Red Army was reorganized into the Eighth Route Army, the whole army conscientiously studied the documents adopted at the Luochuan meeting and prepared to march to the anti-Japanese battlefield. But some officers and men were rather depressed by the reorganization and the change in military uniforms. How could Red Army men set their hearts at ease when they were to take off their treasured red-starred caps, which they had worn for ten years, and put on the military uniforms of the KMT troops who had fought them so ferociously before?

At that time, an important component of the political work in the Eighth Route Army was helping the officers and men study and understand the guiding principles and policies of the Central Committee on KMT-CPC cooperation and the anti-Japanese united front. Wang Ping, an old general, served then as the director of the Organization Department of the Political Department of the general headquarters of the Eighth Route Army. He had many contacts with Deng Xiaoping. To get an idea of how Father worked in the general headquarters of the Eighth Route Army, I paid a special visit to him.

Old Gen. Wang Ping is a distant relative of ours, so he was very kind to me. This old man likes drinking wine and has been known as the "celestial of wine." When I visited him, I brought a special bottle of high-quality wine to give him. When he saw me, he smiled happily. When he saw the wine, he laughed heartily. So he told me many stories:

> At the time, the Central Committee was in Yan'an, and the general front headquarters was located in Yunyang Town, Sanyuan County. The members of both the headquarters and the Political Department lived together and ate from the same big pot. The Political Department had a staff of more than 100 and had under it the Organization Department, the Department of Mass Work, the Department of Enemy Work, and the General Affairs Division. Since Comrade Ren Bishi was in Yan'an, Deng Xiaoping was mainly responsible for the work of the Political Department.
>
> Once, that is, half a month before we set off for the battlefront, your father summoned me into his presence, saying, "There is still half a month left before we

set out. You go down to the army units to find out what our personnel are thinking about after the reorganization and see whether there are any problems." I went to the 30th Corps and met Xiao Ke and Li Jukui. From them I knew that our troops were dissatisfied with the reorganization into the Eighth Route Army and the change of the Red Army's caps into those of the KMT troops, and they were still in low spirits. Many of them took off their octagonal caps but secretly hid them. They said that outwardly they were white, but inwardly they were red forever! After I returned to the general headquarters, I reported all this to your father. He said what I reported was very true and good.

Yan'an was the rear area, but the place where we lived was at the front. At the time, the anti-Japanese war had just begun, so we were very busy with our work. We had to transfer and assign personnel, administer cadres, conduct ideological education among army units, undertake the united front work, and take charge of doing work among the Japanese invaders to persuade them to oppose war. So we were extremely busy with our work. Under the influence of the war of resistance and in response to the call of the CPC for concerted efforts to fight the war of resistance, a great many young students from all parts of the country came to northern Shaanxi and asked to join our army. So many students arrived, and our Organization Department was especially busy. In our Political Department, personnel from the organizations at the higher level came here to handle affairs, and people from other parts of the country also came here. So people came and went day and night, and we were busy all the time. The Eighth Route Army also ran a newspaper, and all its editorials had to be approved by your father, who served as the deputy director. We just worked around the clock and even had no time to play cards. Our objective was to prepare our troops to march to the anti-Japanese battlefield ideologically, organizationally, and in all other aspects of work!

Hearing Old General Wang make these remarks, and looking at his slightly red face, I was moved by his great enthusiasm for resisting Japan. After a short pause, he went on:

Once, our troops and the KMT troops held a get-together. The KMT soldiers were neatly dressed, but our soldiers' uniforms were worn out. Our soldiers were singing songs and shouting slogans, showing a high morale. In comparing the two troops, the civilians who were watching them said, "The KMT troops looked good outwardly but bad inwardly, whereas the CPC troops looked poor outwardly but good inwardly!" At the time, ordinary people were afraid of the KMT troops. Because they were arrogant to the people, looting things and taking women away by force, people ran away at the sight of them. After the Red Army was reorganized into the Eighth Route Army, although our soldiers wore the KMT uniforms, people didn't take to their feet at the sight of us because they knew we were the Red Army men.

The CPC troops were utterly different from the KMT troops in their behavior, their work style, and especially their attitude toward the people. So people were able to distinguish KMT from CPC.

Since the CPC and KMT troops then wore the same caps and marched to the same anti-Japanese battlefronts, they should have fought bloody battles together against the strong enemy troops. But these two armies finally took utterly different roads. Why? Because they had different viewpoints, different ultimate purposes, and different political parties leading them.

At the time, who would have expected that the Red Army, a workers' and peasants' army whose soldiers came from poor families, put on worn-out uniforms and caps, and were poorly equipped, would defeat the Japanese invaders as well as the KMT troops, and with the support of the people would finally win political power in China?

In late August 1937 the Eighth Route Army, its morale high, started off. It began advancing toward the east, the Yellow River, and the land of northern China that had been trampled on by the Japanese invaders.

In early September the general headquarters of the Eighth Route Army set off to advance eastward. Father said that, under the command of Commander-in-Chief Zhu De, they started off from Sanyuan, first rode on horseback, ferried across the Yellow River at Fengling Ferry to the east, and then got to Houma, Shanxi Province.

General Wang Ping said that in Houma, Deng once sent him to find out the situation in the army units and that both Deng and he spoke at a meeting of activists from a special service regiment. Deng talked about the current situation, the concept of the united front, the reasons the Red Army was reorganized into the Eighth Route Army, and the spirit of the Luochuan meeting held by the Central Committee.

On September 21, Zhu De, Peng Dehuai, Ren Bishi, Zuo Quan, Deng Xiaoping, and others took a train to Taiyuan. The Office of the Eighth Route Army in Taiyuan was located in the Chengcheng Middle School. At the time, the senior leaders of the Eighth Route Army and the Northern Bureau of the Central Committee, like Zhu De, Peng Dehuai, Zhou Enlai, Ren Bishi, Liu Shaoqi, Deng Xiaoping, and Xu Xiangqian, all arrived in Taiyuan one after another. So many outstanding people gathered there.[2]

In Taiyuan, key leaders at the general headquarters of the Eighth Route Army and the Northern Bureau of the Central Committee called a meeting to discuss the situation of resistance to Japan in northern China and the course of action of the Eighth Route Army. The meeting pointed out that a danger was that all of northern China could fall. So the Party and the army had to be prepared to wage extensive guerrilla warfare, expand the Eighth Route Army into a powerful army with several hundred thousand men and guns, and establish many base areas so as to undertake the important task of independently persisting in resisting Japan in northern China.

On September 23, Commander-in-Chief Zhu De led the general headquarters of the Eighth Route Army to Nanru Village, Wutai County, Shanxi Province. From then on, the general headquarters of the Eighth Route Army was located in Wutai County. The Political Department led by Father was located in nearby Dongru Village. From then on, the CPC-led Eighth Route

Army formally marched to the forefront in the resistance to the Japanese invaders and threw itself into the flames of war on the anti-Japanese battlefield.

In July, after occupying the two big cities in northern China—Beiping and Tianjin—the militarily superior Japanese invaders began to launch a large-scale war of aggression in China. To resist the aggression of the Japanese army, the Chinese army launched an all-out war of resistance on all fronts. The flames of war spread every day. The war became increasingly cruel and intense. The anti-Japanese war, which would last eight years, was under way.

In September the Japanese troops advanced southward along several routes: along the Tianjin-Pukou Railway they occupied the ancient city of Cangzhou in Hebei Province; along the Beiping-Hankou Railway they captured a city of strategic importance, Baoding, Hebei; along the Beiping-Suiyuan Railway they advanced westward to Shanxi, occupied the famous city of Datong, Shanxi, and invaded Pinxingguan, southeast of Datong.

On September 23, the Japanese troops fought a fierce battle with the defending Chinese troops at Pingxingguan. On September 25, the main forces of the 115th Division of the Eighth Route Army, which had come as reinforcement, took advantage of favorable terrain, launched a fierce surprise attack on the enemy, used the close combat and mountain warfare that the Chinese army excelled at, dealt a heavy blow to the enemy troops, and finally wiped out over 1,000 enemy troops, destroyed more than 100 vehicles, and seized large quantities of military supplies and weapons.

The great victory at Pingxingguan was a major one for the Chinese troops, who took the initiative and wiped out the enemy on the battlefield in northern China. The victory exploded the myth that the Japanese army was invincible and greatly increased the prestige of the Chinese army, especially the Eighth Route Army.

The Great Wall was a military barrier that had been used by Chinese warriors to resist strong enemy troops in ancient times. Now, at the foot of the Great Wall, the Chinese armymen, the brave officers and men of the Eighth Route Army, dealt a heavy blow to the invaders once again, rendering meritorious protective service, once again, to the Chinese nation.

On October 2, the Chinese army lost a fort of strategic importance—the Yanmen Pass in northern Shanxi.

To prevent the enemy troops from advancing southward, the Chinese army decided to defend Taiyuan by conducting a campaign at Xinkou, north of Taiyuan. The Xinkou Campaign lasted twenty-one days and was the largest, fiercest battle of the war of anti-Japanese resistance fought in northern China.

In northeastern and northwestern Shanxi and at Niangziguan, the KMT and CPC troops supported each other in their operations against the enemy. The KMT troops, under the command of Wei Lihuang, inflicted heavy casualties on the Itagaki Division of the Japanese army. The 115th and 120th divisions of the Eighth Route Army frequently mounted attacks on the enemy from the flanks and in the rear and cut off the enemy's communication lines in the rear many times, destroyed large numbers of military vehicles, and launched a surprise attack on the enemy troops. The 129th Division of the Eighth Route Army made a night raid on the Yangmingpu Airfield and destroyed more than twenty enemy aircraft. The Eighth Route Army ambushed the Japanese troops in Pingding, Xiyang, and Yuci many times. But since the Japanese troops mounted powerful attacks, they captured Taiyuan on November 8. The Xinkou Campaign ended.

By then, the Japanese troops had occupied Guisui and Baotou in the north, they had approached Jinan and Qingdao in the east, and they had invaded Taiyuan, Shijiazhuang, and Dezhou in the south. Vast areas of northern China were occupied by the Japanese troops, and the battlefield gradually shifted to central China.

Faced with the critical military situation of needing to slow the southward advance of the Japanese troops, Chiang Kai-shek decided to concentrate 700,000 troops to launch the Wusong-Shanghai Campaign, with Shanghai proper as the main battlefield, and he personally served as the commander-in-chief of the operation. During the three months of the Wusong-Shanghai Campaign, the Chinese army fought bloody battles and doggedly resisted the enemy. But on November 12, the Japanese troops captured Shanghai, and the Wusong-Shanghai Campaign ended. The Japanese army had suffered over 40,000 casualties, and the Chinese army over 200,000.

On December 13, the Japanese army captured Nanjing—the capital of the National Government. The brutal Japanese invaders massacred the Chinese for six weeks in this ancient city (formerly known as Jinling), and more than 300,000 innocent ordinary Chinese people were killed, burnt, or buried alive.

The blood of the Chinese victims, both civilian and military, reddened the Huangpu River, the surging Yangtze River, and the land of eastern China.

After it arrived in Wutai County, Shanxi, the Eighth Route Army, while waging face-to-face armed struggles against the Japanese army, carried out all kinds of anti-Japanese activities. On September 20, Zhou Enlai and Peng Dehuai made a proposal to Yan Xishan, commander of the Second War Zone, that a united front organization be established—the Field General Mobilization Committee for National Revolutionary War of the Second War Zone. This committee functioned like a local wartime government, and it conducted mobilization in Suiyuan, Chahar, the Yanbei area, and northwestern and northeastern Shanxi. Xu Fanting, a patriotic KMT general, served as its chairman, and Deng Xiaoping was its Eighth Route Army representative. The field mobilization committees, at all levels, were the leading bodies for mobilizing the people to support and join the war of resistance.

Father gave detailed instructions at a meeting attended by the leading cadres of the political departments of the 115th and 120th divisions that they should rapidly establish field mobilization committees in light of the war situation and that they should boldly and resolutely carry out work among the masses. Father also gave concrete instructions to Wang Ping and others whom the Party sent to work on the Hebei Provisional Provincial Committee and assigned tasks to them. The Field General Mobilization Committee did a great deal to mobilize resistance to Japan, organize and arm the masses, train cadres, wage guerrilla warfare, and establish base areas.

Fu Zhong, Father's old comrade-in-arms, recalled that it was Deng Xiaoping who took charge of the routine work of the Political Department of the Eighth Route Army. He wrote:

> Xiaoping was just over thirty and in his prime, and he had rich experience in going about the work in army units and the localities. He was capable and experienced, handled affairs reliably, treated others sincerely, showed concern for his subordinates, and won the trust of all comrades. . . . In solving the problems concerning the Party's policies and tactics, Comrade Xiaoping was consistently strict with us and instructed us to pay particular attention to them in our practical work.

Relying on the masses, organizing them, and mobilizing them had always been the fine tradition of the Red Army. Yan Xishan had anticipated that it would take at least three months to mobilize the people. But unexpectedly, the Eighth Route Army spent only twenty days organizing and arming the people. Yan Xishan and his subordinates regretfully admitted that they could not do that so quickly, saying, "The Eighth Route Army did things too quickly!"

In Dongru Village, Father received many visiting correspondents, patriotic personages, and ardent youths who came from far away to join the war of resistance and, at the same time, accomplished a great many tasks assigned by the Central Committee and its Northern Bureau.

Before the great victory at Pingxingguan, the Communist-led army had found it hard to gain a foothold in the area of Wutai, Shanxi Province. Mao Zedong pointed out that the troops needed to create a new situation in the four areas of northwestern, northeastern, southeastern, and southwestern Shanxi so that they could support and closely cooperate with each other and gain a firm foothold.

On October 12, Father, sent by the general headquarters of the Eighth Route Army, led Fu Zhong, Lu Dingyi, Huang Zhen, and other cadres totaling 500 to 600 to move far away from the main forces to start their work in southwestern Shanxi. Fu Zhong recalled:

> We went south from Taiyuan. Wherever we went, we conducted propaganda, and we were warmly received by the masses all the way. Although the size of our contingent was small, we dramatically spread the influence of the Red Army. As soon as we put up the notice for enrolling students for the School Accompanying the Army in Taiyuan and other places, young intellectuals immediately entered their names to join our ranks. Our contingent eventually swelled from three groups to four. After we came to Sanquan Town, Fenyang County, the leaders of the local branch

of the League of Self-Sacrifice for National Salvation and the comrades from the Shanxi Provincial Committee kept reporting their work to Comrade Xiaoping and asking for instructions. Several days later, there were many KMT troops retreating over the mountains and the plains from the Xinkou battlefield. Yan Xishan also left Taiyuan and ran into the Lüliang Mountains.

On November 8, we lost Taiyuan. The KMT troops retreating along roads both big and small in all directions from the Taiyuan area were in an unrelenting state of fear. When Zhou Enlai arrived at Fenyang, he told the comrades working on the General Field Mobilization Committee that they should try every means to prevent the defeated KMT troops from disturbing ordinary people, that even if Yan Xishan withdrew his cadres from the committee, they must keep to their posts, and that even if he didn't cooperate with us, we should work independently and live or die together with the people in northern China. Xiaoping calmly encouraged all of us, saying that the KMT abandoned the territory and the ordinary people so we had to undertake the responsibility for the war of resistance, that we should not decline such a responsibility but should have the greatest determination and courage to stand in the forefront, fight the Japanese invaders in defiance of death, and carry on the war of resistance to the end with the people of Shanxi! After tens of thousands of the KMT troops under Yan Xishan left, only this contingent of cadres from the Eighth Route Army and the General Field Mobilization Committee led by Comrade Xiaoping stood firm on the main road southwest of Taiyuan.

After Comrade Xiaoping and we left Sanquan Town, we came to Xiapu Town, Xiaoyi County. It was located at the foot of the Lüliang Mountains and a passage to southwestern Shanxi. At the time, a unit of the Japanese army had rushed to Pingyao County and the county magistrate under Yan Xishan was so frightened that he had fled from the county town with his military and civil officials and materials. When informed of this, Comrade Xiaoping told our personnel that at this critical moment of national crisis, we should resolutely oppose flight, with perfect assurance, and dare to lead the masses to wage the struggle against flight. At the same time, he summoned the county magistrate, informed him about the overall national interests, and persuaded him to return to Pingyao to wage guerrilla warfare; otherwise, he would be held responsible for not "defending the territory and resisting Japan." Comrade Xiaoping said sternly, "When the Japanese troops come, if we abandon the ordinary people and flee, they will never agree! Every gun, every bullet, and every coin is the blood and sweat of the ordinary people. If we don't use them for the war of resistance, the people will have their say!" That county magistrate did not heed this persuasion and insisted on fleeing, but under the pressure of our side, he finally handed over guns, bullets, materials, and personnel. Afterwards, Comrade Xiaoping sent several cadres from the Eighth Route Army to the unit remaining there to tighten discipline and train it. This unit continued to fight in Pingyao and rapidly swelled to an anti-Japanese guerrilla force of 500 to 600 strong, thus achieving victory in the struggle against fleeing.

After that, Comrade Xiaoping enforced martial law at Xiapu, Xiaoyi County, and confiscated the guns and ammunition of the county magistrate and police who wanted to flee. He also sent the cadres of the Eighth Route Army to organize an anti-Japanese guerrilla force in Xiaoyi. Subsequently, Comrade Xiaoping personally

> drafted a telegram to Yan Xishan to expose the acts of the county magistrates of Pingyao and Xiaoyi who had left their posts without permission and abandoned the territory to flee, stating that when the Japanese invaders came, we "are determined to fight to death but have no excuse for fleeing." The resolute actions of the Eighth Route Army and its determination to resist Japan were soon widely known among the masses. People praised us highly, saying, "The cadres of the Eighth Route Army are made of iron and steel, but the officials under Yan Xishan are made of clay and wood."

In southwestern Shanxi, Father further instructed our cadres to unite with the patriotic personages from all circles and strata, broaden the united front, mobilize the young people to join the army, recruit new Party members, and establish Party organizations. He set rigorous standards for the military's bearing and discipline, especially in its relations with the people. He pointed out that the more ardently the people loved the Eighth Route Army, the stricter its members had to be with themselves. He personally went to a few villages to make inspections and mobilize new recruits. Since the work was done in a thoroughgoing way, the local people showed great enthusiasm for the resistance to Japan, and the army was successfully expanded. In Xiaoyi County alone, in less than two months, more than 3,000 young men joined the ranks of the Eighth Route Army.

Father did not leave southwestern Shanxi and return to the general headquarters of the Eighth Route Army until the troops of the 115th Division had arrived in Xiaoyi County, toward the end of 1937.[3] The general headquarters had already moved south to Heshun County in the eastern part of central Shanxi.

During the operations from late October to early November, the Eighth Route Army attacked the enemy on its own initiative and fought the war of resistance bravely on the various battlefields in Shanxi, wiping out over 2,000 enemy troops and capturing large quantities of weapons, horses, and matériel.

After the fall of Taiyuan on November 8, in accordance with the Central Committee directive on creating a new situation in northwestern, northeastern, southeastern, and southwestern Shanxi, the Eighth Route Army took immediate action. A unit of the 115th Division, led by Commander Nie Rongzhen, established the Shanxi-Chahar-Hebei Military Area centering on Fuping and Wutai, smashed the converging attack mounted by 20,000 Japanese troops, and killed or wounded more than 2,000 Japanese and puppet troops. The 120th Division, led by Division Commander He Long and Political Commissar Guan Xiangying, moved to the vast mountain areas and villages in northwestern Shanxi to wage guerrilla warfare and begin to establish the Northwestern Shanxi Base Area. The 129th Division, led by Division Commander Liu Bocheng and Political Commissar Zhang Hao, worked to establish the Taihang and Taiyue mountains area of the Shanxi-Hebei-Henan Base Area by waging extensively guerrilla warfare in eastern Shanxi and setting up the anti-Japanese democratic governments there. In late December it repulsed the six-pronged converging attack mounted by more than 5,000 Japanese calvary troops in the area of Shouyang and Xiyang. A unit of the 115th Division, under Commander Lin Biao, moved south to the Lüliang Mountains and prepared to establish the Southwestern Shanxi Anti-Japanese Base Area.

By the end of 1937, under the unified command of the Central Committee, the Eighth Route Army had marched to the anti-Japanese battlefield, created a new situation of resistance to Japan, and become a staunch armed force of the Chinese nation against the Japanese invaders.

CHAPTER 47

THE POLITICAL COMMISSAR OF THE 129TH DIVISION

In January 1938 the general headquarters of the Eighth Route Army appointed Deng Xiaoping political commissar of the 129th Division to succeed Zhang Hao. He was less than thirty-four years old.

On January 18, Father arrived at the headquarters of the 129th Division at Xihetou Village, Liaoxian County, Shanxi Province. Liaoxian County is located at the southeastern sector of the Taihang Mountains in southeastern Shanxi. The small village of Xihetou is not well known and is even not marked on the map. This was where the headquarters of the 129th Division was located. At that time, midwinter was not over yet. The weather was very cold, the ground was frozen, and icicles abounded.

The day before, Liu Bocheng, commander of the 129th Division, had gone to Luoyang to attend a meeting, convened by Chiang Kai-shek, of senior generals at and above the level of division commander in the Second War Zone. In the diary kept by Yang Guoyu, who worked in the Confidential Section, that day's entry read as follows:

> January 18. Sunny. Xihetou.
>
> The director of the General Political Department of the 18th Army [the Eighth Route Army], Deng Xiaoping, arrived at headquarters. He is not tall, and when he saw us, he often smiled. We said he came as soon as Division Commander Liu Bocheng had just left. He did not live in the Political Department but in the headquarters with Liu. Probably he was to succeed Liu. It was surprising that no one knew when our Political Commissar Zhang Hao had left.

Two days later, Yang Guoyu continued:

> On January 20, a meeting on political work was held in Liaoxian County, and the cadres at and above the battalion level attended the meeting. Because I was busy, I was not present. Later I heard that Director Deng delivered a report.
>
> On January 21, our army was fighting fierce battles at the front, and the meeting on political work continued.
>
> On January 24, the meeting on political work concluded, and the participants on the bank of the river at Xihetou Village returned one after another to their own units. It seems that Director Deng is in overall charge of all work when the division commander is away.[1]

On January 27, Division Commander Liu Bocheng returned to Xihetou Village and met with the newly arrived political commissar. Liu Bocheng and Deng Xiaoping already knew each other. From that day on, they formally worked together for thirteen years. One was division commander, and the other political commissar; one was the chief military officer, and the other the chief political officer.

Uncle Liu was born in 1892; he was twelve years Father's senior. Both were natives of Sichuan and born in the year of the dragon. There was a long story about Uncle Liu.

In 1911 Uncle Liu joined the student army during the Revolution of 1911 in Wanxian County, Sichuan Province. In 1912 he was admitted to the military academy of the Chongqing Military Government. Later he participated in the campaign against Yuan Shikai and for the protection of the republic, and in the early 1920s he became a famous general in central Sichuan. In 1926 Uncle Liu joined the Communist Party of China and took part in the Lushun Uprising in Sichuan. In 1927 he participated in the Nanchang Uprising. He went to the Soviet Union and graduated from the Frunze Military Academy. After he returned home, he worked in the Military Commission of the Central Committee, and in 1932 he served as chief of the general staff of the Central Revolutionary Military Commission. During the Long March, he courageously directed his troops to cross the Wujiang River, take Zunyi by stratagem, seize the Jiaoping Ferry, and cross the Daduhe River. After the First and Fourth front armies joined forces, he was sent to the Left Route Army and waged a resolute struggle against Zhang Guotao's splittism. After the War of Resistance Against Japan broke out, this famous general of the Red Army was appointed commander of the 129th Division of the Eighth Route Army and led an army unit to fight bravely on the anti-Japanese battlefields. At the beginning of the anti-Japanese war, he applied flexible strategy and tactics to lead his troops to make a night raid on the Yangmingpu Airfield, ambush the Japanese troops at Qigen Village, and wipe out over 1,000 enemy troops in the area south of the Zhengding-Taiyuan Railway. In October 1937, on the instructions of Mao Zedong and the Central Committee, Liu Bocheng led the 129th Division to the Taihang Mountains to establish the Shanxi-Hebei-Henan Border Area Anti-Japanese Base Area.

In his early years, Uncle Liu lost his right eye in battle, he was thus known as "the one-eyed general." Because he directed military operations with miraculous skill, he was also known as "the ever-victorious general." His resourcefulness drew comparisons with Liu Bowen, the famous Ming Dynasty official. He was an illustrious and important general of the Red Army and the Eighth Route Army and an unusually great strategist of the people's army.

Father got acquainted with Uncle Liu in the Central Soviet Area in 1931. Father said, "When we met for the first time, I got a deep impression that he was honest, tolerant, sincere, and affable." After they began working together in the anti-Japanese war, Father said that they "were on very friendly terms and coordinated their work very well." Afterwards, Father also said, "I am more than ten years younger than he, and we have different dispositions and hobbies, but we can cooperate closely. People are accustomed to linking Liu with Deng. In our hearts, we feel inseparable. I feel very happy to work with Bocheng and fight together with him."[2]

The profound friendship between Father and Marshal Liu lasted for decades. In 1986 Liu Bocheng died. In his memorial article, Father said that he had worked with Bocheng for a long time and they had known each other very well, that Bocheng was a great intellectual and military strategist, that in his childhood he was determined to "save the people from a hell on earth," and that he eventually became truly selfless. Bocheng's passing deeply grieved Father.

Father is usually rather serious, even stern, and reticent, rarely revealing his feelings. His article mourning Marshal Liu is unusual for the profound feeling conveyed in his remarks and reminiscences. We can see from this how profound the revolutionary friendship between Father and Marshal Liu was.

In January 1938, after Father had arrived at the 129th Division, he and Division Commander Liu Bocheng immediately threw themselves into the intense and bitter fighting of the anti-Japanese war. On January 28, the army and people in Liaoxian County held a rally to commemorate the sixth anniversary of the resistance to Japan at Wusongkou and Shanghai. People delivered speeches and paraded. On February 3, the 129th Division held a meeting of cadres. "The meeting was held in the rear to discuss important matters while small battles were fought at the front."[3] On February 5 and 6, the 129th Division held a meeting of high-ranking cadres presided over by Political Commissar Deng Xiaoping. Division Commander Liu Bocheng talked about tactics, Deputy Division Commander Xu Xiangqian spoke about operations, and Political Commissar Deng Xiaoping communicated the gist of the meeting of the Political Bureau of the Central Committee to those assembled. Participants reviewed the work after the fall of Taiyuan and made a plan for carrying out strategic deployment, waging guerrilla warfare, and establishing base areas.

On February 15, the headquarters of the 129th Division started to move north. In the first half of 1938 the Communist-led army began to fight a series of fierce and bitter battles with the Japanese troops. The Shanxi-Hebei-Henan area was based on the Taihang Mountains and at their southern end, and the key area was in that part of the mountains in Shanxi Province. On the northeast it bordered on Hebei and was close to Xingtai and Handan, and on the southeast it bordered on Henan and was close to Anyang and Linxian. The area was surrounded by Japanese troops on three sides. Pingyao and Fenyang west of it, Anyang and Xinxiang east of it, and its northern area were already occupied by the Japanese army. In mid-February 1938 over 30,000 Japanese troops launched an attack on southern and western Shanxi and concurrently on Tongguan, Xi'an, and northern Shaanxi.

Chiang Kai-shek ordered his troops to launch a counterattack on Taiyuan, and the task of the Eighth Route Army was to cut off the enemy's communication lines in the rear so as to coordinate the war effort of friendly troops. The 129th Division was instructed to concentrate its main forces so as to attack the enemy in the area between Yangguan and Jingxing along the Zhengding-Taiyuan Railway in coordination with some units of the 115th Division.

Under the leadership of Liu and Deng, the members of the 129th Division headquarters moved north day and night in the cold winter weather and crossed over mountain after mountain. They sometimes marched on narrow winding trails and sometimes passed through stone dams and desolate places. On the way they saw the villages burnt by the Japanese troops, and the villagers, men and women, old and young, who had become homeless. The miserable scenes were too horrible to look at.

On February 21, the Communist-led army fought the battle at Changshengkou. The Communist-led troops first launched a surprise attack on Jiuguan to attract the enemy troops at Jingxing to come for reinforcement and then laid an ambush at Changshengkou. When the eight trucks carrying 200 enemy reinforcements came, the Chinese troops waiting in ambush suddenly attacked them. After five hours of fierce fighting, the Chinese troops wiped out over 130 enemy troops, destroyed five trucks, and captured many weapons.

In the battle at Changshengkou, the Communist-led army had achieved splendid results. On February 27, the headquarters of the 129th Division triumphantly moved back to Xihetou, Liaoxian County.

On March 4, the 129th Division set out again to march south, looking for opportunities to wipe out enemy troops in the area north of the Handan-Changzhi Highway. The highway from Handan in Hebei to Changzhi in Shanxi was an important communication line for the Japanese army, and it defended all the county towns along the highway. Licheng County to the south of Liaoxian County was an important depot base for the Japanese army where over 1,000 infantry and cavalry troops were stationed. There were 2,000 enemy troops in Lucheng County to its south. The commanders of the 129th Division resolved to fight the enemy there. They planned to attack Licheng first so as to attract the enemy troops stationed in Lucheng to come for reinforcement; they would then take advantage of the complex terrain at Shentouling to ambush the reinforcements.

Before dawn on March 16, at 4:00 A.M., the battle began. It progressed as planned. The 129th Division first launched a surprise attack on Licheng, and 1,500 enemy troops stationed in Lucheng immediately moved to reinforce their compatriots. The 129th Division had cut off the communication line between Shentouling and Licheng, and when the enemy came, it demolished the bridge and cut off the road. After the enemy came into the pocket where the 129th Division had laid an ambush on three sides, they were attacked, in hand-to-hand combat, from those three directions. After two hours of fierce fighting, the 129th Division had annihilated over 1,500 enemy troops and captured several hundred guns, mules, and horses.

The ambush operation at Shentouling was crowned with brilliant results. After the battle, Liu Bocheng, Deng Xiaoping, and Xu Xiangqian decided to fight another battle along the Handan-Changzhi Highway. On March 31, the battle at Xiangtangpu began.

To support the troops in southern and western Shanxi in their attacks on all the ferries along the Yellow River, the Japanese army used the Handan-Changzhi Highway to speed up the transportation of matériel; its trucks went along the road every day. The 129th Division decided to lay an ambush at Xiangtangpu, Shexian County, and take advantage of the mountains to mount a surprise attack on the Japanese transportation troops. On the night of March 30, the ambush ring was formed, and on the morning of March 31, two enemy squadrons of 180 trucks, under the cover of Japanese troops, came from Licheng. When they entered the ambush ring, the battle began. The 129th Division neutralized the enemy's fire with strong firepower and then conducted hand-to-hand combat against the enemy. After two hours, the battle ended. The 129th Division had wiped out over 400 enemy troops, burnt 180 military vehicles, and captured large numbers of weapons and equipment. At the same time, it had repulsed the attack mounted by over 1,000 enemy reinforcements from Licheng and Shexian. Liu Bocheng considered the battle at Xiangtangpu a brilliant example of ambush operation.

Within one and a half months, the 129th Division achieved victory in the three battles at Changshengkou, Shentouling, and Xiangtangpu. It flexibly applied the strategy and tactics of guerrilla and mobile warfare and concentrated a superior force to attack the enemy. It won brilliant victories, and its soldiers fought bravely.

The battles fought by the 129th Division in the Shanxi-Hebei-Henan Liberated Area not only struck heavy blows to the Japanese aggressors but also established and consolidated anti-Japanese base areas behind enemy lines. At the same time, the victories won by the Eighth Route Army in southeastern Shanxi frightened the Japanese army. To remove the threat to its rear, the Japanese army decided to launch a converging attack on southeastern Shanxi in early April.

At the end of March, the Eighth Route Army's Commander-in-Chief Zhu De and Deputy Commander-in-Chief Peng Dehuai called a meeting to study and formulate the strategy of smashing the enemy's converging attack.

On April 4, a dozen Japanese regiments, over 30,000 strong, mounted a large-scale converging attack on the Eighth Route Army and the KMT army in southeastern Shanxi from nine directions. Around April 10, the Japanese troops invaded the base areas from the east, west, and north. Under the leadership of the general headquarters of the Eighth Route Army, the Chinese troops pinned down the Japanese invaders from six of those directions. The enemy troops from the other three directions entered the heartland of the base areas and occupied Qinxian, Wuxiang, and Liaoxian, but they were isolated and blocked by our army and guerrilla forces, thus finding themselves subject to hunger, fatigue, and panic. Under these circumstances, the 129th Division decided to seize the opportunity to fight a battle of annihilation.

On April 16, the Communist-led troops launched a pincer attack on the enemy at Changle Village east of Wuxiang. After repeated, intense, and bitter fighting, the Chinese army wiped out over 2,200 enemy troops but also suffered over 800 casualties of its own.

The battle at Changle Village was of decisive significance for smashing the converging attack launched by the Japanese army from nine directions. Afterwards, the enemy's momentum was lost completely. Taking advantage of the enemy's adjustments to its deployments, our army recovered Liaoxian, Licheng, Lucheng, Xiangyuan, Tunliu, Qinxian, Gaoping, Jincheng, Shexian, and Changzhi.

After twenty-three days of operations against the enemy's converging attack, the Communist-led army frustrated the Japanese army in its attempt to wipe out the defending forces in southeastern Shanxi, annihilating over four thousand Japanese troops, recovering eighteen county towns, and driving all Japanese troops out of the area. The northern part of the Shanxi-Hebei-Henan area (north of the Taihang Mountains) came under the control of the Communist-led army, and its prestige rose to an unprecedentedly high level. After suffering outrages at the hands of the Japanese army, even massacre, the Chinese people had greater confidence in their efforts at anti-Japanese resistance, an attitude that created favorable conditions for the establishment of the Shanxi-Hebei-Henan Anti-Japanese Base Area.

On April 22, the headquarters of the 129th Division moved back to its "capital," Xihekou, Liaoxian. The excited people of Liaoxian, people from all walks of life, went to the headquarters to pay their respects to Liu Bocheng, Deng Xiaoping, and Xu Xiangqian.

Three months had passed in a flash since Father came to the 129th Division in January. During those three months, he was burdened with pressing military duties. Because of the intense fighting, he had had no time to spare.

On April 25, Political Commissar Deng Xiaoping called a meeting of the Military and Administrative Commission of the division. It was decided at the meeting to establish the Shanxi-Hebei-Henan Military Area and divide the main forces of the 129th Division into two columns, the eastern and western columns of the Beiping-Hankou Railway. The eastern column, commanded by Deputy Division Commander Xu Xiangqian, marched to southern Hebei, and the western column, commanded by Brigade Commander Chen Geng, marched to western Hebei. Liu Bocheng and Deng Xiaoping directed the headquarters of the forward echelon and the 386th Brigade to advance to the area west of Xingtai, Hebei Province, and to organize and direct the operations against the enemy in the mountains and on the plains.

In late April, under the command of the division headquarters, the 386th Brigade advanced to Xingtai in western Hebei west of the Beiping-Hankou Railway. From there southward, it attacked the puppet troops stationed west of

Xingtai, Shahe, Wu'an, and Cixian. By the end of May these troops had largely put an end to the chaos that arose after the Japanese army occupied the Wu'an-Shexian Highway.

The 129th Division originally had two brigades, the 385th and 386th, both comprising over 13,000 men. The 4th Corps of the Fourth Front Army of the Red Army had been reorganized into the 385th Brigade, with Wang Hongkun serving as brigade commander, Wang Weizhou as deputy brigade commander, Geng Biao as chief-of-staff, and Su Jingcheng as political commissar. On June 12, 1938, a new 385th Brigade of the 129th Division was organized, with Chen Xilian serving as brigade commander and Xie Fuzhi as political commissar.

The 31st Corps had been reorganized into the 386th Brigade, with Chen Geng serving as brigade commander, Chen Zaidao as deputy brigade commander, Li Jukui as chief-of-staff, and Wang Xinting as political commissar.

In April the reorganized main forces formed the eastern and western columns. After its reestablishment in June, the 385th Brigade operated in western Hebei, wiping out large numbers of puppet troops and repulsing many enemy attacks. By then, after almost one year of expansion, the 129th Division had under it the 385th Brigade, the 386th Brigade, the Shanxi-Hebei-Henan Military Area, the southern Hebei guerrilla zone (later called the Southern Hebei Military Area), the Eastward Advance Column, and the Youth Anti-Japanese Column, and it also commanded the 344th Brigade of the 115th Division and the 5th Detachment of the Eighth Route Army.

Large numbers of high-ranking famous generals of the Eighth Route Army gathered on the anti-Japanese battlefields of the Taihang Mountains and the Shanxi-Hebei-Henan area. Almost all these high-ranking generals of the 129th Division and other army units were battle-seasoned veterans of the Red Army, and most of them were valiant, heroic, and under thirty. Under the leadership of Mao Zedong, the Central Committee, and the general headquarters of the Eighth Route Army, and under the direct command of Liu Bocheng and Deng Xiaoping, they fought bravely on the battlefield. The anti-Japanese battlefields in north China became the ideal places for displaying the brilliant military skills many of them had acquired so young. In the later operations, in the march toward the liberation of the whole of China and in the building of the people's army in New China, they would make many more heroic appearances and their famous names would be heard many more times.

In June 1938 Father left Liaoxian for western Hebei in the northeast on an inspection tour of the Western Hebei Military Subarea. The western Hebei area was between Shijiazhuang and Xingtai, including Yuanshi, Zanhuang, Gaoyi, Lincheng, and Neiqiu. In March 1938 Liu Bocheng and Deng Xiaoping had sent Zhang Yixiang and others to western Hebei to establish the anti-Japanese base area, and now two months later Father personally went there to guide the work.

After coming to the Western Hebei Military Subarea, Father first heard reports from Zhang Yixiang and others. He instructed them to further organize guerrilla forces and mobilize the masses to defeat the enemy. Afterwards, he went on an inspection tour of the 385th Brigade in the mountains of Nanjian, Shijiazhuang. Father spent about one week in western Hebei and hurried back to the headquarters of the 129th Division in Liaoxian, Shanxi Province.[4]

To destroy the enemy's traffic arteries, all army units of the 129th Division, under the unified command of the division headquarters, conducted ten large-scale sabotage operations and numerous small ones against the Beiping-Hankou, Zhengding-Taiyuan, and Daokou-Qinghua railway lines. Along the 500-kilometer railway lines, sabotage operations were conducted one after another, and the people participated enthusiastically. As a result, the Japanese army's ability to transport its troops became sporadic, then came almost to a standstill. The main forces of the 129th Division continued to operate in many places and strike effective blows at the Japanese army. At the same time, with the help of local Party organizations, they mobilized the masses extensively and easily recruited more soldiers. By September the size of the 385th and 386th brigades had increased to some 7,000 men, and their military and political quality had improved considerably.

On July 5, 1938, Father went southward from the Taihang Mountains on an inspection tour of the Southern Hebei Anti-Japanese Base Area. The Southern Hebei Anti-Japanese Base Area was established when the advance detachment of the 129th Division led by Chen Zaidao marched to southern Hebei in 1937. In May 1938 the deputy commander of the 129th Division, Xu Xiangqian, personally went to southern Hebei to lead the struggle against Japan. The Southern Hebei Base Area, with Nangong County in the Xingtai area as its center, expanded rapidly. In a few months over twenty anti-Japanese people's governments were established, the troops swelled from 500 men to over 10,000, and the Eastward Advance Column grew from five companies to three regiments nearly 7,000 strong.[5]

In the fall of 1991 I went to the home of Old Gen. Chen Zaidao to interview him. In the golden season of the fall, chrysanthemum was in full bloom. The general grasped my hands and shook them happily, asking "How is your father?"

I had known of the "great general Chen," the famous and valiant general of the Red Army, for a long time, but this was the first time I had seen him. His hair and eyebrows were white, and his dark face was all wrinkles when he laughed. I could not see the pockmarks that everyone knew about.

The old general sat with his legs crossed and rested them on the chair, saying, "I was acquainted with your father when crossing the Yellow River at the beginning of the anti-Japanese war." He fixed his eyes on the ceiling, speaking loudly and clearly.

> In July 1938 your father went on an inspection tour of southern Hebei. We held a meeting attended by members of the Special Party Committee and army cadres at and above the regiment level. Political Commissar Deng made a report in which he analyzed the situation. He pointed out that Chiang Kai-shek's resistance to Japan might shift to compromise and surrender and that the greatest danger was partial resistance coupled with such compromise and surrender. At present, the Japanese army was busy attacking Wuhan, and its forces in northern China decreased. This gave us an excellent opportunity to wage guerrilla warfare behind enemy lines. He also said that in handling relations with Lu Zhonglin, governor of the Hebei Provincial Government, we should unite with him in a concerted effort to resist Japan but keep on guard to uphold the principle of independence in the united front and to expand our forces.
>
> After the meeting concluded, Political Commissar Deng dined with us, and the food was very simple. He spoke eloquently. Later on, Liu and Deng came to southern Hebei once again, and they wanted me to put my four regiments under their command. We in southern Hebei also supported the Taihang Mountains with many things, including clothes, cloth, quilts, and bedding. We were then in great difficulties ourselves! We lived on the plains, with a strong wind and a lot of dust. When the wind blew, the sand on the clothes made of handwoven cloth weighed half a kilogram. But we did our utmost to support the Taihang Mountains because people there lived a harder life than we did.

His wife was hospitalized at the time of my interview, and he was home alone. So my visit made him very happy. He was known for drinking much wine. He drew me beside him, telling me in a low voice, "I have good wine everywhere!" True, there was even a wine jug by his bed. The old general pointed to a glass jar, saying, "There are three very poisonous snakes inside it. This snake wine is very good. You dine with me here, and I'll treat you to wine!"

Because the old general was so kind to me, I, as his junior by one generation, should not have left. But at the sight of the wine jug with three poisonous snakes inside, I apologized and said good-bye to him repeatedly. In the end, he saw me out of the room. In the courtyard he asked me again and again to revisit him in the future. I knew he had enjoyed my visit not because he met me but because I am the daughter of Deng Xiaoping, with whom he had forged a revolutionary friendship decades before.

By the middle of 1938, the 129th Division of the Eighth Route Army had conducted a series of successful operations and achieved victories in southeastern Shanxi and in western and southern Hebei. It also wiped out several thousand enemy troops and established anti-Japanese democratic governments over these vast areas so that the Shanxi-Hebei-Henan Anti-Japanese Base Area, based in the Taihang Mountains, was further consolidated and expanded.

Between late August and early September, to contain the Japanese army in its attack on Tongguan and Luoyang, the units of the 129th Division commanded by Chen Zaidao launched the Zhangnan Campaign, wiping out over 4,000 puppet troops and capturing over 1,500 enemy troops. It also established local anti-Japanese democratic governments in the counties of Anyang, Neihuang, and Tangyin, built new base areas nearly fifty kilometers in length from north to south in northern Henan, west of the Weihe River, established anti-Japanese democratic governments in more than thirty counties in the border area between southern Hebei and northern Henan, and strengthened the ties between the Southern Hebei Anti-Japanese Base Area and the combined areas of Hebei-Shandong-Henan and the Taihang Mountains.

While the Shanxi-Hebei-Henan Anti-Japanese Base Area expanded, other army units of the Eighth Route Army seized opportunities to fight the enemy and created new pockets of resistance to Japan in different areas. The 120th Division waged guerrilla warfare in the vast areas of northwestern Shanxi behind enemy lines, smashed a converging attack mounted by over 10,000 Japanese and puppet troops, recovered seven county towns, and wiped out over 1,500 enemy troops. Its forces swelled from 8,000 men to over 25,000, and it established anti-Japanese base areas in northwestern Shanxi and the Yanbei area.

The 115th Division advanced to the Lüliang area in southwestern Shanxi, mobilized the masses energetically, waged guerrilla warfare, and wiped out over 1,000 enemy troops. It repeatedly ambushed the enemy's transportation troops and created conditions for establishing the Southwestern Shanxi Base Area.

The flames of the anti-Japanese war were spreading over the lands of Shanxi, Henan, and Hebei and could not be extinguished. At the same time, the Eastward Advance Anti-Japanese Column of the Eighth Route Army was being formed in Shandong to establish a guerrilla base area on the plains. In central China, the New Fourth Army, led by Ye Ting, Xiang Ying, and Chen Yi, smashed many of the Japanese army's "mopping-up" operations, attacked enemy troops on its own initiative, and dealt them heavy blows, thus laying the groundwork for continued Chinese gains in the anti-Japanese war.

When the united anti-Japanese front was further consolidated, the Sixth Central Committee held its Enlarged Sixth Plenary Session in Yan'an from September 29 to November 6, 1938. Seventeen members and alternate members of the Central Committee and over thirty leading cadres from various departments of the Central Committee and different areas attended the meeting.

On August 25, Father set off from the Taihang Mountains to Yan'an to attend the Sixth Plenary Session. Mao Zedong delivered a political report entitled "On the New Stage" in which he pointed out that the war of resistance was in the transition period from defense to stalemate. The Japanese troops had occupied Wuhan, Guangzhou, and other places, but in doing so their fundamental weaknesses, such as too few troops too widely dispersed, were exposed all the more, all of Japan's conflicts at home and abroad intensified, and the

Japanese strategic offensive was inevitably coming to a climax. Mao advocated that the Chinese army and people plan their defense and resistance on the frontal battlefields, wage extensive guerrilla warfare behind enemy lines, and take advantage of the enemy's weaknesses to drain its human and material resources so as to shift the war to the new stage of stalemate between Japan and China. These were the pressing tasks for the whole country, which had to prepare for bitter fighting. At the same time, it was important to continue consolidating and expanding the anti-Japanese national united front to support the protracted war through long-term cooperation.

At the session, Peng Dehuai, Qin Bangxian (Bo Gu), He Long, Yang Shangkun, Guan Xiangying, Deng Xiaoping, Luo Ronghuan, and Peng Zhen made speeches to review the experience gained over the previous fifteen months. The plenary session adopted the political resolution, and approved the line, of the Political Bureau of the Central Committee led by Mao Zedong.

In mid-December, Division Commander Liu Bocheng led the headquarters of the 129th Division to southern Hebei to direct the struggle in southern Hebei and northwestern Shandong. Toward the end of December, Political Commissar Deng Xiaoping returned from Yan'an to southern Hebei.

On December 30, the 129th Division held a meeting attended by army and government cadres at Luohuzhang Village, Nangong County, in southern Hebei. Political Commissar Deng Xiaoping communicated to them the resolution adopted at the Sixth Plenary Session of the Central Committee. To address the struggle in southern Hebei, the meeting established guiding principles on struggle—to rely on the masses of workers and peasants, to be based in the vast rural area, to persist in guerrilla war on the plains in southern Hebei, and to consolidate the anti-Japanese democratic positions.

From November 1937 to the end of 1938, the 129th Division made great progress in the vast areas of northern China where the four provinces of Shanxi, Hebei, Henan, and Shandong crisscross. Over the areas that stretched east as far as the Tianjin-Pukou Railway, west as far as the Datong-Puzhou Railway, north as far as the Zhengding-Taiyuan and Cangzhou-Shijiazhuang railway lines, and south as far as the Yellow River, it established the Shanxi-Hebei-Shandong-Henan Anti-Japanese Base Area with a population of twenty-three million. Its troops, numbering nearly 30,000, swelled to thirteen regiments.

Beginning in October 1938, because the Chinese army and people doggedly rose in resistance, the Japanese invaders suffered over 400,000 casualties, large quantities of their military matériel were consumed, and their troops were dispersed ever more widely. The Japanese aggressors once arrogantly said they could "end the war in two months" and "conquer China in three months," but their fond dream came to nothing and they had to stop their strategic offensive on the frontal battlefields. As predicted by Mao Zedong, the Chinese War of Resistance Against Japan switched from one of strategic defense to one of strategic stalemate.

The year 1939 came. It was the third year of the anti-Japanese war.

The headquarters of the 129th Division was located at Zhangjiazhuang, Qiji, Weixian County, in southern Hebei. On January 1, snow fell, and it was piercingly cold. Liu Bocheng and Deng Xiaoping led the headquarters of the 129th Division to provide guidance on the work in southern Hebei and ordered the troops to strike heavy blows to the enemy on New Year's Day.[6]

To carry out the principles and policies of the united front, to make all possible alliances in the effort to resist Japan, to avoid frictions that could lead to civil war, and to win them over so as to make as concerted an effort as possible in resisting Japan, Liu Bocheng and Deng Xiaoping held talks in southern Hebei with Lu Zhonglin, governor of the Hebei Provincial Government, and Shi Yousan, commander of the 10th Corps of the KMT army. While Liu Bocheng held talks with Lu Zhonglin many times, Deng Xiaoping met with Shi Yousan twice, on January 16 and 25, to let him know the righteousness of the cause, to indicate the willingness of the Eighth Route Army to persist in uniting with friendly forces to resist the Japanese army, and to make clear the solemn and just stand that the Eighth Route Army would never withdraw from the anti-Japanese base areas. Deng Xiaoping's persuasion resulted in Shi Yousan's temporary neutrality and the isolation of anti-Communist activities by Lu Zhonglin and other diehards.

As New Year's Day 1939 passed, the Spring Festival was yet to come. The Japanese army concentrated its forces to conduct large-scale "mopping-up" operations against the anti-Japanese base areas. They conducted these operations more and more frenziedly and cruelly.

Throughout 1939 the Chinese army and people in the anti-Japanese base areas doggedly fought the strong Japanese troops. On January 7, over 30,000 Japanese troops conducted large-scale operations against southern Hebei from eight directions. On January 21, 6,000 Japanese troops conducted operations against Hesun and Liaoxian in the heartland of the Taihang Mountains. On February 12, 2,000 Japanese troops conducted operations against the Xiangchenggu area of Weixian County in southern Hebei. On February 21, Japanese mobile forces conducted operations against the areas of Nangong, Weixian, and Qinghe in southern Hebei. On March 10, the Japanese army conducted operations against the Juye area in southwestern Shandong. On April 1, 2,000 Japanese troops conducted operations against the areas south of Nanpingyao and north of Qinyuan in central Shanxi. On April 10, 3,000 Japanese troops conducted operations against the area south of the Baigui-Jincheng Highway. On April 20, the headquarters of the Japanese Expeditionary Forces in northern China issued a "public security and suppression" plan and stepped up its efforts to conduct mopping-up operations. On April 23, 1,000 Japanese troops launched a converging attack on the areas of Gaotang and Yucheng in northwestern Shandong from four directions. On May 2, 1,000 Japanese troops conducted operations against the area of Nangong in southern Hebei. On June

1, 3,000 Japanese and puppet troops conducted operations against the area of Luluo in southern Hebei. On July 1, over 10,000 Japanese troops conducted operations against southwestern Shandong. On July 3, 50,000 Japanese troops conducted large-scale operations against the Shanxi-Hebei-Henan area, occupied most of the county towns in the base area, and controlled the northern section of the Baigui-Jincheng Railway and the Handan-Changzhi and Gaoping-Liaoxian highways.

In the face of the intense and frenzied mopping-up operations by the enemy troops, the Shanxi-Hebei-Shandong-Hebei Base Area and the 129th Division strengthened the deployment of forces, flexibly fought the enemy troops, and seized opportunities to strike resolute blows.

On January 12, when 30,000 enemy troops conducted mopping-up operations against southern Hebei, Liu Bocheng and Deng Xiaoping called a meeting of cadres to assign tasks of counter–mopping up operations and then issued operational orders to launch the operations. From January to March, under the direct command of Liu Bocheng and Deng Xiaoping, the Chinese army and people in southern Hebei fought over 100 major battles, killing or wounding 3,000 Japanese and puppet troops, and foiling the Japanese army's plan to control the southern Hebei plains.

After March the focus of the struggle gradually shifted to the mountains. On March 7, Liu Bocheng and Deng Xiaoping led the main forces of the 129th Division to return to the Taihang Mountains. On March 18, the 129th Division carried out the army rectification campaign to improve the troops' thinking and work style in an effort to consolidate the army. On April 3, the troops directly under the headquarters of the 129th Division held a parade at Shangzhaozhan Village, Licheng County. Commander-in-Chief Zhu De was present and reviewed the parade.

In July, in view of the large-scale mopping-up operations of the Japanese army against the Shanxi-Hebei-Henan Base Area, Liu and Deng decided to organize large numbers of local armed forces, militiamen, and guerrilla forces to wear down the enemy forces by means of dispersed and protracted guerrilla warfare and to concentrate the main forces so as to seize opportunities to wipe out the enemy. Before the enemy invaded the base areas, the Chinese army and people moved away from the area of military operations, vacated houses, and concealed their property. After the enemy troops invaded the base areas, the militiamen and guerrilla forces continuously ambushed and intercepted them so as to fatigue them. At an opportune time, the main forces ambushed and attacked the enemy troops, launching surprise attacks and besieging them along the enemy-occupied communication lines, striking them hard. After the Japanese troops entered the base areas, they failed to find the main Chinese forces anywhere. Instead, suffering repeated attacks, they were forced to withdraw in late August. The enemy's mopping-up operations was thus concluded.

During this period, the 129th Division took the offensive to attack the enemy, fought seventy-eight battles, both small and large, wiped out over 2,000 enemy troops, recovered many important county towns, and smashed the Japanese army's fierce and large-scale mopping-up operations. At the same time, between January and August, the 129th Division conducted constant sabotage operations against the vital, enemy-occupied communication lines along railways and highways, paralyzing the enemy's transportation capacity.

For over a year after he became political commissar of the 129th Division in January 1938, Father either commanded troops in operations or marched. He was busy with military affairs. Military life at the front was an active life and made one feel busy all the time.

Sometimes Father and his comrades-in-arms strode the plains, sometimes they hurried along the mountain roads amid high mountain ridges. However burdened with pressing military duties, they were never afraid of the strong enemy, had unusual courage and resourcefulness, and were determined to win the victory. They knew it was no easy thing to defeat the strong Japanese troops in a short time. However, Chinese Communists were endowed with the heroic spirit and were resolved to save the Chinese people from subjugation by the strong Japanese invaders. They never wavered in this goal.

Around August 1939 Father temporarily said good-bye once again to the Taihang Mountains and his intimate, Division Commander Liu Bocheng, and went to Yan'an to attend the enlarged meeting of the Political Bureau. After arriving at Yan'an, Father shared a cave dwelling with his old comrade-in-arms Deng Fa. Deng Fa was very active. He had an intimate personal friendship with Deng Xiaoping, so when they were free after the meeting, he warmheartedly insisted on helping Deng Xiaoping find a wife.

Mom Liu Ying told me, "At that time, Deng Fa and your father strolled happily everywhere every day in Yan'an, and people were saying they looked like two roving gods!"

In early September 1939, with the enthusiastic help of many of his friends and comrades-in-arms, Father got married. The bride was named Zhuo Lin. She is my dearest mother.

CHAPTER 48

MY GRANDFATHER PU ZAITING

My mother's name is Zhuo Lin, but her original name was Pu Qiongying. She was born into the family of a noted industrialist in Yunnan.

To talk about my mother and her life, I must start with her family and her father, Pu Zaiting. Nowadays few people know of Pu Zaiting. But many people know about the canned ham made in Yunnan, especially overseas Chinese residing in Southeast Asia. People in Yunnan began to make ham in ancient times. But it was Pu Zaiting who canned it and turned ham-making into an industry, manufacturing products that were sold both at home and abroad.

I have never seen my grandfather. My mother, who left home at an early age, does not know much about her family either. I have gleaned some general information about her family and my grandfather.

Pu Zaiting was born around 1870 in Xuanwei County, Yunnan Province. He belonged to the Han nationality. (Yunnan was a multinational province.) It was said that my grandfather's ancestors' home was in Changshu, Jiangsu Province. During the reign of Emperor Hong Wu of the Ming Dynasty, a member of his family was granted the title "General Wulue" and was ordered by Zhu Yuanzhang, the first emperor of the Ming Dynasty, to conquer Yunnan. After coming to Xuanwei, he settled down there. During the ensuing several hundred years, this branch of the Pu family multiplied, now totaling about several thousand people.

The descendants of the Pu Family said the father of Pu Zaiting passed the imperial examination at the provincial level and later, as a teacher, ran an old-fashioned private school in his hometown. He had four sons. Three of them followed in their father's footsteps by studying hard and engrossing themselves in stereotyped writing. Only one of the sons, Pu Zaiting, was born unruly. From an early age, he disliked reading and writing. At the age of fourteen, he sneaked out of the family to join one of his friends in a horse caravan to learn trading. For generations, members of the Pu family had been scholars. Pu Zaiting's behavior was certainly a humiliation to the family. His father sent people to get him back and gave him a good dressing-down. But Pu Zaiting was resolved to be a trader, so no pressure could bend him. Not long afterwards, he fled again to join the caravan.

Yunnan is a border province in southwestern China. It is noted for its beautiful landscape and the colorful life of its variety of nationalities. Bordering on Vietnam, Laos, and Burma, it had exchanges with businesspeople from these countries. As it was far away from the Central Plains of China, it was an economically and culturally backward province, still strongly influenced by feudalism. Toward the end of the Qing Dynasty, with no highways in the province, trading and the flow of goods were dependent on horses. One horse train after another carrying goods, their bells ringing monotonously but pleasantly, could be seen on the roads and paths in the mountainous and hilly areas of Yunnan.

At first, Pu Zaiting peddled goods locally, using the horse trains. After earning some money, he bought his own horses and, with others, began to carry goods to Southeast Asia. It was not an easy journey from Xuanwei in northwestern Yunnan to Indochina. He had to go through boundless primitive, subtropical forests in Xishuangbanna and travel up to 500 kilometers. On the way, they contended with venomous snakes and beasts of prey as well as bandits. Pu Zaiting was young, ambitious, and brave, with business acumen. As his caravan expanded, he became more and more successful. His reputation grew in Xuanwei, and he was elected president of the Xuanwei County Chamber of Commerce for two successive terms.

After traveling extensively, Pu Zaiting grew bolder and more knowledgeable. He found that Xuanwei produced a lot of delicious and fragrant ham, but the processing was very primitive. It was hard to preserve a whole pig's leg and sell it. It struck him that he could can ham, making it easier to preserve and sell.

Pu Zaiting invited some people in Xuanwei to discuss the matter and raised funds to run a ham-canning factory. With their support, he sent people to Guangzhou to learn the skills needed to make canned products. After raising enough funds, he bought from Hong Kong a set of machinery for producing canned food.

In 1920 the Xuanhe Canned Ham Company was formally established. Pu Zaiting was the chairman of the board of directors and concurrently the general manager. In this way, Pu Zaiting grew from a merchant into an industrialist. Poor and backward Xuanwei County had the first factory that used machines for production.

Yunnan is in southwestern China and borders on Thailand and Burma. It was the area where poppy production and drug trafficking were very popular. At the time, opium was brought from Southeast Asia and sold in Yunnan. Drug trafficking was the quickest way for merchants to become rich. After the Xuanhe Canned Ham Company was founded, it should have been dealing exclusively in ham. But these people were so greedy that they began to can opium and sell it in Southeast Asia disguised as canned ham. Things never go as planned. These fake canned hams of the Xuanhe Company were discovered and seized by French police. (Southeast Asia was a French colony, and Yunnan was also in the French sphere of influence.) The opium was confiscated. The company thus could not operate and declared bankruptcy. The shareholders suffered heavy losses. The machines shipped from Guangdong lay idle like a heap of scrap iron.

Pu Zaiting was stubborn. Although the bankruptcy of the company was a great blow to him, he did not lose heart. He still wanted to run a factory. He told the shareholders of the former company that since the company was closed down and the machines had become useless, it would be better for him to have all the machines and open a new canned ham factory alone. If he failed to make money, the shareholders could just ignore it. If he succeeded, he would return to them all the money they had invested in the Xuanhe Company.

Thinking that the machines were useless, the shareholders agreed to let Pu Zaiting take them and go it alone. Pu Zaiting started again in Xuanwei. The new company was called the Xuanwei Pu Zaiting Brothers Canned Food Company. It was also known as the Dayouheng Double Pig Canned Ham Company. Pu Zaiting was then over fifty. As the new company was his own and what he said counted, it turned out to be a success. From then on, the primitive but canned Xuanwei ham proved to be very popular.

With his abundant knowledge of business, Pu Zaiting knew that to expand his business, he had to promote it in the markets of Southeast Asia and in the international market as well. So he went to Guangzhou, where he proceeded to expand. Meanwhile, his company's products were soon exported, gaining the company access to both domestic and world markets. Since then, the Xuanwei canned ham has been sold in Hong Kong, Macao, Singapore, Burma, Haiphong, Panama, Japan, Germany, and France. The company expanded so fast that it had twenty-six subsidiary companies both at home and abroad, from Southeast Asia to Paris. In 1923 the Xuanwei canned ham was highly rated at the national foodstuff fair held in Guangzhou.

Inspired by Pu Zaiting, a number of other canned ham factories were opened in Xuanwei. The expansion of the canned ham industry promoted ham-making, coal mining, and brewing in the county. National industry in Xuanwei County increased, and by 1939 inhabitants engaged in industry or commerce accounted for 45 percent of the county's population.

I knew about Xuanwei canned ham when I was young because Mother often mentioned it. The ham is indeed very tasty, especially when cooked with a white gourd. The Xuanwei ham and the ham made in Jinhua, Zhejiang Province, are the two most famous varieties of ham in China. In the early 1980s, when I was working in the Chinese Embassy in the United States, I came to know many overseas Chinese, and from them I learned that "Yunnan ham" was the Xuanwei ham and was popular with them, especially the older generations of overseas Chinese and foreign nationals of Chinese origin.

His business experience made Pu Zaiting very experienced, knowledgeable, and open-minded. As a member of the national bourgeoisie, he readily accepted the political ideas of the bourgeois democratic revolution. As soon as Dr. Sun Yat-sen put forward his bourgeois revolutionary ideas, he supported them. It was said that Pu Zaiting was a member of the Tong Meng Hui organized by Dr. Sun Yat-sen. Since Pu died long ago, I cannot find evidence of his membership. But it was true that he supported the Revolution of 1911 and the national revolution led by Dr. Sun Yat-sen.

Pu Zaiting supported Gen. Cai E, a member of the vanguard of the Chinese bourgeois democratic revolution who staged an uprising in Kunming and launched the campaign for protecting the republic and against the plot by Yuan

Shikai, the arch usurper of state power, to restore imperial rule. He took the lead in making donations and established a depot in the building of the Xuanwei County Chamber of Commerce to raise money and provisions for the Protecting-the-Republic Army under Gen. Cai E.

When the Protecting-the-Republic Army returned victorious from Chongqing, Gen. Tang Jirao of the Yunnan Army of the Protecting-the-Republic Army went to Xuanwei and gave Pu Zaiting a special award, a silver medal commending the progressive actions of open-minded businessmen. Tang also presented him with a plate on which he had written the four characters "Ji Gong Hao Yi" (zealous for the common weal).

In supporting the movement to protect the republic and take punitive actions against Yuan Shikai, Pu Zaiting came to know many personages in the army and government and maintained especially close relations with Fan Shisheng, a general of the Yunnan Army. He began to participate in military and governmental activities.

In 1917 Dr. Sun Yat-sen launched the campaign for upholding the Provisional Constitution of the Republic of China against the Northern Warlord government. The Yunnan Army split. The Yunnan provincial military governor, Tang Jirao, supported Dr. Sun Yat-sen overtly but turned to the side of the Northern Warlord government covertly. In 1921 Gu Pinzhen returned to Yunnan and cooperated with Fan Shisheng to overthrow Tang Jirao. After Tang Jirao was driven out of Yunnan, Gu Pinzhen became the commander-in-chief of the Yunnan Army and controlled the military and administrative power.

Not long after that, Dr. Sun Yat-sen called for the Northern Expedition, aimed at overthrowing the Northern Warlord government. Gu Pinzhen responded and appointed Fan Shisheng commander of the Advance Forces of the Northern Expeditionary Army of the Yunnan Army. Dr. Sun Yat-sen then appointed Gu Pinzhen commander-in-chief of the Northern Expeditionary Army. At this moment, Tang Jirao unexpectedly returned to Yunnan and restored his rule. Gu Pinzhen died in battle.

The sudden change in the situation forced the Yunnan Northern Expeditionary Army to leave Yunnan. Deputy Commander Zhang Kairu led some troops in retreat to the area of Xuanwei, from where they marched to the area of Panxian in Guizhou. While reorganizing his troops, Zhang Kairu sent a telegram to Dr. Sun Yat-sen, expressing his desire to serve as the vanguard of the Northern Expedition.

After Gu Pinzhen died, Pu Zaiting went with Fan Shisheng to Guizhou and joined this Northern Expeditionary Army unit of the Yunnan Army. Zhang Kairu appointed Fan Shisheng commander of the 8th Brigade and Pu Zaiting director of both the General Logistics Bureau and the Bureau of the Monopoly of Sales of Tobacco and Alcohol of the Yunnan Army in Guangdong. Pu Zaiting served in the army from then on.

On learning that the Guangdong warlord Chen Jiongming betrayed the revolution, the Northern Expeditionary Army of the Yunnan Army hurried to Guangdong to fight Chen Jiongming. Noted for its fighting spirit and skills, the Yunnan Army marched directly to Guangzhou with irresistible force, and Chen Jiongming fled in panic.

In 1923 Dr. Sun Yat-sen came to Guangzhou and established the Palace of the Generalissimo there. He issued orders that the armies of Yunnan, Guangxi, and Guangdong and the navy be cited for their meritorious services in their joint punitive actions against Chen Jiongming. He also officially conferred military ranks to meritorious high-ranking officers, and Pu Zaiting was granted the rank of major general.

After Dr. Sun Yat-sen entered Guangzhou, the political situation was volatile and unstable. Dr. Sun Yat-sen led the army to defeat the rebellious troops of Guangdong and repulsed the invasion by the troops of the northern warlord Wu Peifu and those of Chen Jiongming, successfully defending Guangzhou. Since the Yunnan Army performed very well in the battles, Dr. Sun Yat-sen relied heavily on its services.

In January 1924 the Kuomintang held its congress in Guangzhou. Dr. Sun Yat-sen put forward the Three Great Policies, reorganized the Kuomintang, and founded the Whampoa Military Academy in Guangzhou. While in Guangzhou with the Yunnan Army, Pu Zaiting called his eldest son to Guangzhou to help him expand his company's business as he himself was engaged in military work. He sent his second son to the newly founded Whampoa Military Academy to study as a member of the first group of trainees.

At this time, Dr. Sun Yat-sen wrote an inscription for Pu Zaiting to commend his achievements in developing national industry. The inscription read: "Yin He Shi De" (bringing prosperity to consumers). A member of mother's family said my grandfather and his family treasured the inscription very much. They put the four large characters in a big plate and hung it in the main hall.

From a businessman to an industrialist, a supporter of Dr. Sun Yat-sen, and a member of the Northern Expeditionary Army, my grandfather had reached the zenith of his career. Thereafter, he suffered great setbacks and lost his vitality.

On March 12, 1925, Dr. Sun Yat-sen died in Beiping. In May the Yunnan warlord Tang Jirao plotted to overthrow the Guangzhou revolutionary government. Collaborating with him, the commander-in-chief of the Yunnan Army in Guangdong, Yang Ximin, betrayed the revolution.

Pu Zaiting was put under house arrest at that time. There were different explanations of his arrest. Some said that he was arrested because he supported the Northern Expeditionary Army and thus became intolerable to Yang Ximin, who framed him. Others said that he was charged with embezzlement when he served as the director of the General Logistics Bureau and as director of the Bureau of the Monopoly of Sales of Tobacco and Alcohol. In the old army, these two posts were lucrative. The incumbent enjoyed authority commensurate with the post, and he could make money by taking advantage of authority.

I am not surprised at all that Pu Zaiting was said by some to have been guilty of corruption, for although he devoted himself to industrial development and supported and participated in the national revolution, he was, after all, an old-style member of traditional society. It would be too naive to think that a member of the old army and an old-style businessman could come out of the mud unsoiled. But his house arrest was certainly associated with the internal conflict in the Yunnan Army and his stand on supporting or betraying the national revolution, for it was only after Fan Shisheng returned from the Northern Expedition that he was released, with Fan Shisheng's help.

While in Guangzhou, my grandfather had made a fortune and served as an officer. So he had written to my grandmother, asking her to come to Guangzhou to share his glory and wealth. My grandmother went to meet her husband in Guangzhou 500 kilometers away, taking her youngest daughter—my mother—with her.

Mother said that at that time she was only four or five years old. She went with her mother from Yunnan to Vietnam, where they boarded a ship to Hong Kong, from whence they traveled to Guangzhou. Because she was so young at the time, Mother could not remember anything about Hong Kong except the tall buildings and narrow, dark streets.

After arriving in Guangzhou, my grandparents and my mother lived together until my grandfather's house arrest in 1925. After Fan Shisheng bailed my grandfather out, the family of three returned to Kunming, the capital of Yunnan, via Hong Kong and Vietnam.

After having been a longtime member of the National Revolutionary Army, my grandfather not only was framed and had to leave the army but also lost nearly all his property in Guangzhou. After returning to his hometown, he never left again, remaining there to engage in the industrial and commercial activities of Kunming and Xuanwei, Yunnan.

As the saying goes, misfortunes never come singly. Just after he had surmounted the difficulties in Guangzhou, he was faced with another danger in Yunnan. As mentioned earlier, when Pu Zaiting founded the Dayouheng Company he used the machines that had belonged to the former Xuanhe Company and promised to return all the capital to the original shareholders when he succeeded. But later, because he had been expanding production and needed more capital, or because he was suffering setbacks, or because he did not want to return the money after all, he did not pay all the money back to the original shareholders. So they joined forces and sued Pu Zaiting. The Yunnan authorities handled the case and sentenced Pu Zaiting to imprisonment. At this critical moment, the Pu family bribed the local officials and let an old housekeeper serve the prison term for him.

In the mid-1930s, after the anti-Japanese war broke out, the situation deteriorated. Especially after the Japanese army invaded Southeast Asia and occupied the Chinese trading ports of Shanghai and Guangzhou, exports of the

Yunnan canned ham stopped completely and domestic sales also dropped considerably. The tin needed for producing cans had all been imported. With the sea route cut off, there was no tin with which to make cans. The noted Yunnan canned ham industry declined and has never been restored to its former glory.

The Pu family was large. Pu Zaiting was one of four sons, and he himself had three sons and four daughters. His eldest son remained an essential assistant to him in running the Dayouheng Company. His second son studied in the Whompoa Military Academy, and after graduation he participated in the Northern Expedition and served as a company commander in the National Revolutionary Army. Pu Zaiting was ambitious. In a society where warlords held sway, he was keenly aware that if one wanted to succeed, he should have his own men in the army, so he had great expectations for his son's army career. However, his son failed to live up to his expectations. He was not used to the military life and left the army to return home, where he joined his father in doing business in Yunnan.

Pu Zaiting sent his third son to Japan for study in 1927. There, under the influence of progressive ideas, he participated in the patriotic activities of some progressive organizations. After returning home, he joined the Communist Party of China in 1928, upon the recommendation of the noted scholar Zheng Yili. Zheng was also a native of Yunnan. The third son did something for the Party's underground work. Later, he quit the Party, returned home, and became a small boss in his father's business.

Three of the four daughters—the eldest married young—that is, Pu Daiying, Pu Shiying, and Pu Qiongying, left home to pursue their studies in the north, joined the Communist Party of China, and became career revolutionaries.

On the eve of liberation, Pu Zaiting was advanced in age and long retired. He had moved from Kunming back to Xuanwei to spend his remaining years. The Pu family property was managed by his three sons separately. At that time, the family's canned ham business was continuing as usual but made little profit. The money earned was not enough to support a big family of some twenty. Fortunately, the family still had some land and a small coal mine, and the income just barely supported the family.

Pu Zaiting died in 1950 at the age of eighty. At that time, the Chinese People's Liberation Army had already swept across southern China. Mother went with the massive Liu-Deng army to advance into southwestern China and arrived in Chongqing, Sichuan Province. Upon hearing that Pu Zaiting was critically ill, Mother took my second elder sister back to Xuanwei and saw her father before he died.

After liberation, the Xuanwei ham industry was taken over by the government. During the decades of ups and downs, the industry had expanded into a fairly large-scale, county-level enterprise. The products sold well at home and abroad and stayed in short supply. At present, efforts are under way to further expand production and sales. Such an expanded industry is what Pu Zaiting dreamed of and worked for all his life but failed to realize.

The life of Pu Zaiting is a vivid history of personal pursuit, a history of a pioneer of Chinese national industry. He had his ambitions, lofty ideals, foresight, and management methods. He was inclined toward the revolution and opposed to feudal monarchy, but at the same time, he was an old-fashioned man with a strong feudal character. He was courageous in developing an industry but constrained by the social conditions of that semifeudal and semicolonial society. On the one hand, he wanted to develop industry, and on the other, he wished to work his way into the army. In these arenas, however, he was suppressed by warlord and bureaucratic forces.

Pu Zaiting's life was filled with successes and failures. Despite his lifelong struggle, he failed to realize his lofty ideals, just like tens of thousands of other Chinese national bourgeoisie (although he could not be put on a par with the fine representatives of the national bourgeoisie and outstanding industrialists in the coastal areas and in the key cities).

In the 5,000 years of the history of civilization, China remained a feudal society for more than 2,000 years. After Qin Shi Huang unified China, the country was under highly centralized and autocratic feudal rule. Despite the changes in the dynasties and the passage of time, feudalism, the feudal system, feudal ideas, and feudal influence adamantly clung to the bulky nation of China.

By the mid-1800s, when bourgeois ideas began to emerge in China, international imperialism had broken in peremptorily and flagrantly. Therefore, the Chinese national bourgeoisie and national capitalism were oppressed and restricted by the two powerful forces of feudalism and imperialism from the very beginning of their development, which remained difficult and tortuous.

Although the Revolution of 1911 broke out under the leadership of the bourgeoisie and the government of the Republic of China had been established, the government gradually lost its revolutionary spirit and became a decadent feudal regime. After that, though national capitalism developed somewhat, its destiny was small, weak, and distorted, just like a bean sprout growing from the seams of rocks.

In the thirty-eight years under the rule of the bourgeoisie, China remained a very feudal society with no industrial system. China was still a backward agricultural country with extremely low productivity. It remained a semifeudal and semicolonial country.

On the eve of the founding of New China in 1949, foreign capitalist forces dominated the old China's economic arteries in many areas, including banking, real estate, foreign trade, industry and mining, and transportation. The national monopoly capitalism of China, that is, the bureaucratic capitalism represented by Chiang Kai-shek and others, controlled about 80 percent of national industrial capital.[1] Genuine national capital existed only in such fields as industry, handcraft industry, commerce, and banking, with total net assets at something like the current figure of two billion yuan. Furthermore, the national bourgeoisie had neither political authority nor a solid foundation for development. They and the feudal landlord class had to depend on each other. They had to rely on the forces of imperialism and bureaucratic capital for support. They did not and could not possibly form a powerful force.

The life of Pu Zaiting, marked by successes, setbacks, and decline, was representative of a fairly large number of the pioneers of national capitalism. They could shine for some time but were suppressed or swallowed up by imperialism, feudalism, and bureaucratic capitalism in the end.

Pu Zaiting was merely an insignificant representative of the national bourgeoisie. Like some other more famous national capitalists, he was never able to straighten his back or to have room for free development. He had to submit to the humiliating position of living under another's roof.

This was the tragedy of Pu Zaiting, and the tragedy of Chinese national capitalism as well.

FROM PU QIONGYING TO ZHUO LIN

When I was a pigtailed little girl who was beginning to understand things, I knew that the two people who loved me most in the world were my parents.

We are closer to Mother because Father was so busy with his work that he spent little time looking after us. My elder brother and sisters were born in wartime, and they had to be entrusted to the care of peasant families because my parents could not take them along with the army. Of the five children, only my younger brother and I were born after liberation and were raised by Mother herself. So she perhaps took more care of the two of us. People often said that Mother spoiled me, so that I have become too squeamish. Indeed, Mother spoiled me a bit, but others do not know that it was I who was beat the most in my family for disobedience. So from an early age, I think what people said about me was wrong.

In our family, Father is certainly the core. But Mother is the center because we children have grown up alongside her. Father was busy, so it was left to Mother to educate the children. Apart from taking care of our daily life, my intellectual mother paid particular attention to ensuring that we acquire some scientific knowledge. Every weekend, when my elder brother and sisters returned home from school, the whole family would sit around the dinner table after supper to listen to Mother teach in various fields, such as nuclear fission and chain reactions. Although we children could not understand anything, we had to sit there. While listening, my brothers and sisters often cut in and sometimes generated a debate. Thus, the dinner table became the "free forum" of our family, and this practice has continued until today, though we now have more children to join in.

Mother not only took care of our daily needs but also had a big influence on our mental development and our life choices. Mother was a graduate, in physics, of Beijing University. My elder brother, my second elder sister, and my younger brother all chose physics and studied at Beijing University. This example alone proves Mother's tremendous influence on us all.

I love Mother from the bottom of my heart. So when I started writing this book, I thought I would do whatever I could to tell Mother's stories. Indeed, I am very eager to write this chapter about Mother.

Hers is a simple and ordinary story. Mother's experiences were not like those of Father, which were magnificent and moving. But they were experiences just as filled with twists and turns, providing just as much food for thought. For her time, her experiences were also quite typical.

Her name was Pu Qiongying, and she came into this world in April 1916. Her father was the industrialist Pu Zaiting, in Xuanwei County, Yunnan Province. She was the seventh and youngest child in her family, which I have described in the preceding chapter, so I will not say more than is needed here.

Although her family was not a noble or scholarly family, it was a family of a famous country gentry. Her father was known as the "King of Ham" of Yunnan. She had three elder brothers and three elder sisters. As she was the youngest in the family, she naturally enjoyed more favor. From her childhood, her parents loved her as a pearl in their hands.

Pu Qiongying looks like her father a bit, and she is very healthy, with a face as red as an apple under the sunshine. Her brows are black and thick, like two arches under her forehead, and below them are a pair of big eyes with double-fold eyelids and long eyelashes. When smiling, she seems unrestrained and looks very lovable.

From the day she was born, she had enough food and clothing, with nothing to worry about. Her parents took care of her, and her elder brothers and sisters protected and accompanied her. In this easygoing environment, she developed an open mind and a lively character, with no thought of gain or loss. The only drawback was that, as the favorite, she was a little spoiled. Her second sister, Pu Daiying, resented this, and sometimes she would give her little sister a lesson.

When they were old enough for school, she and her sisters began to learn to read and write from an old-fashioned private school teacher. They learned to recite *The Three-Character Classic*, *The Book of Family Names*, *The Four Books*, *The Five Classics*, and *The Code of Conduct for Women*. It was rather strange that the teacher taught only how to recite books but not to write, so they read "books without characters."

When she was a little older, her father moved the family to Kunming, the capital of the province, for the sake of his business. There three girls of the family—Pu Daiying, Pu Shiying, and Pu Qiongying—all went to primary school. After graduation, they all passed the entrance examination of the Kunming Girls' School and received their secondary school education there. Their life was peaceful, without worries. But it was not a pool of still water. There were some waves that affected their young souls.

Yunnan is in southwestern China and far from the hinterland. It was inhabited by the primitive people in ancient times. Even up to modern times, its culture was still backward and production was extremely underdeveloped. Although they were influenced by the new ideas and the bourgeois democratic revolution, on the whole the people there were far more backward and conservative than those in the coastal areas. The feudal forces were more stubborn and powerful there. Although Pu Zaiting participated in the national revolution, his family remained a typical old-fashioned feudal family.

In the Pu family, more than one member participated in the national revolution. But when they returned home and lived in the traditional environment, they quickly lost their enthusiasm for the revolution and became ordinary, lackluster persons. The second son of the family became a small boss of a small coal pit after returning from the Northern Expedition. The third son left the Communist Party and returned to become a person of no achievements. He and his wife were addicted to drugs. After Pu Zaiting returned home, his business declined and was never restored to its previous glory.

The Pu family is big. Up to the generation of Pu Qiongying, there were thirteen girls, including her cousins. Pu Qiongying was the thirteenth, the youngest. Above her were numerous aunts.

The women of the Pu family lived in the closed cage of feudalism. The women in old-fashioned families were not independent and had no freedom in marriage. Although they could enjoy all the best of life, they were, at best, the appendages of their fathers, husbands, and even sons. Some were maltreated by their stepmothers and even became disabled. Others whose husbands took concubines were very depressed. Some were married out and were so bullied by their husbands' families that they could tolerate no more and had to commit suicide.

Pu Qiongying and her two sisters saw with their own eyes what was happening around them and felt very indignant about the unequal treatment given to women in their world. From the teachers and their elders, they learned the codes of conduct for women, the three obediences (to father before marriage, to husband after marriage, and to son after the death of husband), and the four virtues (morality, proper speech, modest manner, and diligent work). In daily life, they learned the tragedy of lives of the older women in the Pu family, and they realized that their destiny would be exactly like that of their elders. In their young souls emerged an awareness of revolt. But how? They did not know.

It was only the tide of the times that could clear the oppressive air of that society.

One of the brothers of the Pu family returned from Japan, bringing with him a lot of books and pamphlets about the revolution and communism. When the three sisters of the family got them, they felt very curious. They could hardly understand what the books said about the truth of the revolution. But after reading them, they felt they had bathed in a fresh spring breeze. From then on, they were enlightened by revolutionary ideas.

After going to school in Kunming, they saw many new things. In the middle school, a woman music teacher often taught students the truth about the revolution and communism, disseminating the basic truth that the tiller should have land. These lessons made a deep impression on the three sisters of the Pu family. One day the teacher was suddenly arrested. On the way to the execution ground, the woman teacher, with her hands and feet in cuffs, sang songs and shouted slogans about communism. Her dignity and heroism shocked all their young hearts. From then on, an image of Communists was engraved deep in the hearts of the three sisters. Compared with the destinies of the women with feudal ideas in their hometown, the destiny of that Communist woman teacher seemed so much loftier and more glorious!

The sisters of the Pu family gradually became clear in their ideas about pursuing freedom, women's liberation, the liberation of the individual, and the truth of the revolution.

In 1931 a national games was to be held in Beiping, and athletes for the games were selected by all provinces. Pu Qiongying was selected to represent Yunnan at the games in the girls' sixty-meter run.

The sports delegation set out from Yunnan. But no sooner had they arrived in Hong Kong than the September 18th Incident broke out. The Japanese imperialists launched their large-scale invasion of China and rapidly occupied the three provinces in northeastern China. In the face of the national calamity, the national games were canceled. The Yunnan sports delegation had to return.

It was exciting, of course, to participate in sports games. But the hidden purpose of Pu Qiongying's participation was to take the opportunity to step out of the family and Yunnan and go to Beiping for study. Unexpectedly, she had to return home immediately after she arrived in Hong Kong. She felt discontented and unresigned to her fate.

At that time, Pu Qiongying was fifteen, and she had her own views. She was determined not to return to Yunnan. She wrote to her elder brother to ask for permission to study in Beiping, making it clear that she was determined not to return to Yunnan. Her firm determination made her family yield at last.

Pu Qiongying was so excited! She went to Shanghai by ship and found Zheng Yili. Zheng was a schoolmate of her brother in Japan, and it was on his recommendation that her brother joined the Party. Zheng also had business relations with the Pu family. From then on, Zheng bore her travel expenses from Shanghai to Beiping and her monthly living expenses in school.

With the help of Zheng Yili, Pu Qiongying arrived in Beiping. There she first asked a cousin for help. Very soon, she moved into a dormitory run by the Young Women's Christian Association. As her educational level in Yunnan was not equal to what was expected of her in Beiping, she had to study for a few months to catch up, and then she was admitted into the Beiping No. 1 Girls' Middle School.

This girls' school was a famous school in Beiping and renowned for its style, lively ideas, and the outstanding performances of its students. Pu Qiongying adapted to the new environment very soon and lived a very happy life.

She was born bright and lively and took her new lessons in stride. She got to know a lot of fellow provincials and made new friends. Many well-known personages were her schoolmates, including the noted film star Zhang Ruifang, Chen Yun's wife Yu Ruomu, and Hu Qiaomu's wife Gu Yu. She even performed on the same stage in school with Zhang Ruifang! Zhang played the role of a girl servant, and Pu Qiongying played a young lady.

After school she often went for an outing with a few friends from Yunnan, drank tea in teahouses, and went to theaters. They lived a free and unrestrained life. The oppressive air in Yunnan was gone.

In Beiping she came to love Peking opera. One of her countryfolk even invited a Peking opera singer to teach them. It was a pity that Pu Qiongying was unable to master the tone of the opera. As she was the youngest among the students and had a good memory, she could recite all the famous stanzas of the well-known Peking operas and could still call them to mind decades later.

During the one year she spent at the middle school, Pu Qiongying contracted pneumonia. She was hospitalized in Nanjing. She took the opportunity to pay a visit to Shanghai. All this travel was arranged by Zheng Yili.

The family gave Pu Qiongying fifty silver dollars a month to cover her living expenses. It was a good sum at the time! With one silver dollar, she could buy a sack of flour. She was young and liked to play. Although she had a lot of money, she spent all of her fifty silver dollars every month. She had never learned to be frugal, and she never would. (After liberation, the supply system of military communism was changed to the wage system, so Mother did not know how to use the money for a big family of eight.)

Life in Beiping should have been perfect, as Pu Qiongying had shaken off the bondage of her feudal family and was happily studying, with no worries about food or clothing. But when the nation faced calamity and the situation was volatile, all young students felt that the nation was at peril. Pu Qiongying was no exception.

After the fall of northeastern China, many students there fled to Beiping. They had lost their homes and kinsfolk. The indignant song "My Home Is on the Bank of Northeastern China's Songhua River" could be heard everywhere in the ancient capital. The national calamity spread shock on the campuses and penetrated young hearts.

The voices against the Japanese invasion and for national salvation became louder and louder and found a response from people from all strata and all circles. The people began to donate money to the anti-Japanese army and to boycott goods made in Japan as part of their protests against Japanese aggression. The demand for stopping the civil war and fighting the sacred war of resistance to Japan had become the voice of the overwhelming majority of the Chinese people.

In 1935 the Japanese invaders put constant pressure on the Kuomintang government in Nanjing and, by taking advantage of some agreements, stretched the tentacles of aggression into the five provinces of northern China and stepped up their efforts to create the so-called self-rule of northern China.

In the face of the obviously aggressive goals of the Japanese invaders and the critical possibility that the country was going to be reduced to a colony, the students in Beiping could no longer keep silent. On December 9, 1935, thousands of students in Beiping took to the streets. They shouted with uncontrollable indignation the slogan "We are not to be slaves of a foreign nation." The nineteen-year-old Pu Qiongying and her schoolmates participated in the protest. They joined the ranks of demonstrators, holding high anti-Japanese placards, shouting patriotic slogans, and denouncing the atrocities of aggression perpetrated by Japanese imperialists and the traitorous acts of the Kuomintang government in Nanjing. Shoulder to shoulder and hand in hand, the patriotic students demanded democracy and freedom and denounced Japanese efforts to enslave their country.

The Beiping authorities were panic-stricken and sent troops to suppress the demonstrators. The army and police used water cannons and batons against the patriotic youths. The demonstrators were forced to scatter, more than thirty students were arrested, and several hundred were wounded. But their protest demonstration gave rise to an even greater tide of mass movements for resistance to Japan and for national salvation.

On December 16, Pu Qiongying and her schoolmates left the school campus again to participate in an even larger protest demonstration. But the army and police had enforced a blockade on the city, and they were unable to come close to the demonstrating students. They climbed up the city walls and shouted and cheered in support of the demonstrating students.

Encouraged by the two student demonstrations on December 9 and 16, many other Beiping students increased their awareness of struggle and their ideological level. Many of them marched onto the battlefield against Japanese aggression and started their revolutionary careers.

The December 9th students' movement in Beiping brought about a qualitative change in Pu Qiongying's mind that helped elevate her ideas about the struggle against feudal bondage and for freedom in marriage and the liberation of the individual to a higher political and ideological level, thus preparing her for taking the revolutionary road later.

In 1936 she graduated from the middle school and, with her outstanding examination results, was admitted into Beijing University, where she majored in physics. Why did a girl of this era want to study physics at Beijing University?

Beijing University was a famous institution of higher learning in China and the birthplace of the May 4th and December 9th movements. It was a place where new ideas and new culture were encouraged and where famous scholars and people with innovative ideas gathered. Many progressive youths held the ideal of saving the nation by going into industry and the sciences. What better place than Beijing University to pursue this dream?

Pu Qiongying was clever and studied hard. It meant a lot to her when she was admitted into a nationally well-known university. She was the first person from Yunnan Province to work her way into the most famous university in Beiping. It must have been fate. Decades later one of her sons and two of her daughters would follow in their mother's footsteps and study physics at Beijing University!

After Pu Qiongying entered the university, she found that it was an entirely new world, an environment even more attractive than her middle school. The university was full of both academic and political atmosphere. Active on-campus was an organization known as the Vanguard of Resistance to Japan and for National Liberation. Under the influence of new ideas, Pu Qiongying participated in the activities of this organization. But her objective was still to study hard and prepare to serve the country after graduation by applying her knowledge of science, technology, and industry.

That year her two other sisters, Pu Daiying and Pu Shiying, also got the family's permission to study in Beiping at last. The three sisters were very excited to be reunited. However, the drastically deteriorating political situation shattered their wonderful aspirations and those of all Chinese people. On July 7, 1937, the invading Japanese troops launched an attack at the Marco Polo Bridge near Beiping. In late July the jackbooted Japanese troops drove straight into the ancient city of Beiping. Beiping was soon under the bayonet rule of the Japanese troops!

Beiping was in a state of terror and chaos. People began to stream out of the city—they could no longer live there! The whole city was in panic. People were depressed because they were losing their country and their homes. Crying together in a dark room, the three Pu sisters hugged each other.

The eldest sister, Pu Daiying, had always had an iron will and was brave enough to revolt. When the Japanese aggressors were pressing toward Beiping, she immediately joined the Vanguard for National Liberation, an organization under the Communist Party of China, choosing firmly the revolutionary road, and then went to Yan'an, the sacred heart of the Chinese revolution that Chinese youths were yearning for at that time.

Pu Shiying and Pu Qiongying were hidden by townsfolk who then helped them flee from Beiping under the disguise of ordinary people. They had to go through a checkpoint manned by Japanese soldiers with dazzling bayonets in their hands and fierce looks on their faces. Pu Qiongying will never forget the terrifying scene of Japanese soldiers, with loaded rifles, searching for students and progressives.

Where did they go after fleeing from Beiping? What was the future of young students? There was only one way out: to go to Yan'an, to join the Eighth Route Army, to join the revolutionary ranks.

Pu Shiying found it very difficult to walk at the time because of her congenital heart disease. Pu Qiongying tried to persuade her sister: "You'd better not go to Yan'an!" "I will crawl to Yan'an," she replied. "If I have to die, I would rather die in Yan'an!" There was no straight route to Yan'an. They had to go to Tianjin first, where they went to Qingdao by ship and from there to Ji'nan by train. In Ji'nan they had a hard time buying train tickets among the crowds of refugees. In the end they arrived in Xi'an and found the office of the Eighth Route Army there.

Through examination and on the recommendation of the Vanguard for National Liberation, the two Pu sisters were admitted into the Northern Shaanxi College in Yan'an. They were extremely excited and overjoyed. They had found their destination at last, after making a journey of more than 1,000 kilometers and going through many hardships.

With a group of other students, they reached Yan'an on foot. Carrying only a small bag, they walked day and night. On the way, they saw no Japanese invaders and found everything entirely new. Pu Shiying lay down often because of her heart trouble. Their fellow students hired a donkey to carry her along. After seven days and nights, they arrived in Yan'an.

Northern Shaanxi, Yan'an, the Eighth Route Army, revolution, resistance to the Japanese invaders—it was an entirely new world, with entirely new faces. Compared with the helter-skelter of Beiping, where people had lost their homes and their country, Yan'an was a place where even the sky and the moon seemed brighter. They had arrived at the sacred heart of the revolution at last!

That was November 1937. Winter had not come yet, and the autumn sun still shone warmly. A pagoda stood loftily on top of a hill. The Yanhe River was flowing along the loess bends. Row upon row of cave dwellings lined the loess hills. On the road, shepherds were singing folk songs while driving their herds along. On the hillsides and at the foot of the hills, revolutionary army men and women in the uniforms of the Eighth Route Army were strolling and chatting.

There the two Pu sisters felt refreshed and reinvigorated. What pleased them even more was that their elder sister, Pu Daiying, was also there. She had already graduated from the Anti-Japanese Military and Political College and joined the Communist Party of China. She had also gotten married—her husband was Yue Shaohua, a noted workers' movement activist and a veteran Red Army man.

The three sisters were very excited to have been reunited. Reviewing the events that had depressed them in Beiping and now excited them as they faced an entirely new life, they felt that they had a lot to tell each other.

Pu Shiying and Pu Qiongying entered the Northern Shaanxi College. The college had been set up in Yan'an in September 1937 for training cadres. It enrolled progressive youths from all over the country. Courses offered included Marxist philosophy, political economy, and mass movement. The course of the women's movement was taught by Cai Chang, the noted leader of the women's movement.

After three to four months, the two sisters graduated. Pu Qiongying was assigned to work in the library of the college, while Pu Shiying, because she was not in good health, was assigned to do a temporary job at a small shop attached to the college. In early 1938 both became members of the Communist Party of China.

After serving as the leader of the 12th Group of the college for some time,

Pu Qiongying was sent to study in a special training class run by the Security Division of the government of the Shaanxi-Gansu-Ningxia Special Region. The leadership wanted her to meet all the requirements for doing secret work behind enemy lines, perhaps because she was lively, intelligent, and eager to study. She was sent to the special training class to receive training so that she could do resistance work in the Japanese-occupied area, that is, behind enemy lines.

As required by the work, she changed her name at that time, to Zhuo Lin.

How time flies! Two years had passed. The progressive youths coming to Yan'an from all parts of the country had become accustomed to the hard life there, a life full of vitality and militancy. Soon they had thrown themselves into the revolutionary cause and the sacred cause of resistance to Japan. These revolutionary youths and intellectuals renewed the vigor of the Communist Party and the Eighth Route Army and expanded significantly the ranks of the revolution and of the struggle against the Japanese invasion.

Pu Qiongying had also grown, from a young, lively college student into a fighter for communism and a revolutionary worker who vowed to adhere to her chosen course and to give all her life to the revolutionary cause.

The road from Pu Qiongying to Zhuo Lin was not perhaps as legendary and exciting as those of many other revolutionaries. But this ordinary road was the one traversed by tens of millions of progressive youths in their efforts to seek light, to join the struggle against Japan, and to join the revolutionary ranks. This road was not as glorious as the road of the Jinggang Mountains; nor was it so magnificent as the road of the 12,500-kilometer Long March. However, in that era, the road to Yan'an was a bright road leading to truth and revolution. It was marked by the strength and warm blood of youth and the powerful spirit of millions of people who were longing for progress.

Toward the end of the summer of 1939, Zhuo Lin was introduced to a man named Deng Xiaoping. Perhaps because she was not worldly or because she was a bit muddle-headed, she only knew that this man was a veteran Red Army man and a commander on the front of the anti-Japanese war. She was not clear about his job or his responsibilities. In fact, she had had a scanty experience of life. Even if you had been able to tell her about life's ins and outs, she might not have had a thorough understanding of what you were saying. But the revolutionary ideals they held in common brought them together.

One evening in early September 1939, a dinner party was given in front of the cave dwelling of Mao Zedong at Yangjialing in Yan'an. All the senior leaders of the Central Committee in Yan'an were present. Mao Zedong and his wife Jiang Qing, Liu Shaoqi, Zhang Wentian and his wife Liu Ying, Bo Gu, Li Fuchun and his wife Cai Chang, and others were all there. It was, in fact, a small celebration of the marriage of two couples—Deng Xiaoping and Zhuo Lin, Kong Yuan and Xu Ming.

Kong Yuan was an old Party member who had joined the revolutionary ranks in 1924. He was then the deputy head of the Special Committee of the Central Committee. After liberation, he served as the director-general of the General Administration of Customs, as vice minister of foreign trade, and as director of the Investigation Department of the Central Committee. He was a famous revolutionary activist. His wife, Xu Ming, is a talented and capable woman cadre. After liberation, she served as a deputy secretary-general of the State Council of the People's Republic of China. The two of them were both activists in Yan'an and noted for their good character and good relations with others.

There was no extravagant food on the dinner table, which was made of several pieces of wood and set outside the cave dwelling. On the table was cooked millet, a common food in Yan'an. Although those sitting around the table were noted figures in Yan'an, they wore the Eighth Route Army uniforms of handwoven cloth, with patches on the knees and elbows, and cloth shoes.

A photo was taken of the two couples. Owing to someone's poor photographic skills, the photo looks a bit blurred. It shows the four of them standing shoulder to shoulder, not in wedding attire but in the handwoven army uniforms, reflecting their simple living and lofty revolutionary ideals, their broad smiles reflecting their happiness.

This was indeed a special gathering and a unique marriage celebration. The guests were all mainstays of the People's Republic of China, which was founded later. These great people were the brave battle-tested soldiers and comrades-in-arms who held each other in brotherly affection. Although the dinner was simple, the guests were joyful and chatted to their hearts' content.

In the jubilant atmosphere, the veteran revolutionaries became young at heart once again and, just like ordinary people, tried to make fun of the newlyweds. Kong Yuan was forced to drink too much, and the very night of the wedding got a good dressing-down by his bride. Deng Xiaoping was more fortunate—although he drank to every toast, he did not get drunk.

Afterward, Liu Ying told Zhang Wentian, "Xiaoping can drink a lot of liquor!" Smiling, Zhang Wentian said, "There was a trick to it!" It turned out that Li Fuchun and Deng Fa prepared a bottle of drinking water to pass off as liquor, preventing their old friend Deng Xiaoping from getting drunken.

The light breeze caressed their faces. Late in the evening, the moon looked as cold as water. All of Yan'an was quiet as the bugle call resounded in the distant hills.

The dinner party guests were merry and lively with drinking. In front of the cave dwelling at Yangjialing in Yan'an, these veterans of the Communist Party of China and the Eighth Route Army celebrated the weddings of their old comrades-in-arms sincerely and simply, giving a send-off party for those who were about to go to the front.

But two people were missing from the dinner table, Zhou Enlai and Deng Yingchao. They were then in the Soviet Union, where Zhou Enlai was receiving treatment for an arm injury; he had sustained it falling from his horse when it shied at the sight of Jiang Qing's horse galloping by. Otherwise, he would have been present and would have drunk to the happiness of his close comrades-in-arms.

A few days later, as the autumn morning sun shone over the yellow soils of Yan'an, Zhuo Lin and her new husband set off for the front in the Taihang Mountains. Deng Xiaoping was thirty-five, while Zhuo Lin was twenty-three.

CHAPTER 50

IN THE TAIHANG MOUNTAINS

Over the East the red sun shines,
To the heart's content the God of Liberty is singing;
Look!
The thousands of mountains and gullies;
They are like iron walls;
The anti-Japanese flames are burning on the Taihang Mountains,
Raging sky-high.
Listen!
Mothers are urging their sons to fight the Japanese invaders.
Wives are sending their spouses to the battlefields.
We are fighting on the Taihang Mountains;
We are fighting on the Taihang Mountains;
The mountains are tall and the forests are thick,
The soldiers are strong and horses are sturdy.
Wherever the enemy attacks,
We will let him die then and there!

This is the magnificent anti-Japanese song sung by the people living in the Taihang Mountains. It is sung to this day by patriotic Chinese people.

The Taihang Mountains extend more than 700 kilometers from north to south, at elevations ranging from 1,500 to 2,000 meters. The range divides northern China into two parts, forming a natural screen. To its west are the hilly areas of Shanxi Province, and to its east are the boundless, fertile northern China plains in Hebei and Henan provinces.

It was there that the general headquarters of the Eighth Route Army commanded by Zhu De and Peng Dehuai was located, and from there that the war of resistance fought by the Eighth Route Army was directed.

It was there that the 129th Division, led by Liu Bocheng and Deng Xiaoping, fought doggedly and heroically against the Japanese invaders, under the command of the Central Committee, Mao Zedong, and the general headquarters of the Eighth Route Army. The Taihang Mountains were like a strong backbone of the national war against Japan.

The end of the 1930s found China in a crisis-ridden and extremely complex situation. After capturing Wuhan and Guangzhou in China, the Japanese troops had driven right into the heartland of northern and central China, controlling Hebei, Shanxi, Shandong, and Jiangsu and parts of Henan, Anhui, Hubei, Jiangxi, Zhejiang, and Guangdong, as well as the whole of the Hainan Island. Although the invading Japanese troops occupied vast territories of northeastern, northern, central, eastern, and southern China, they had overextended the battlelines. Their weaknesses—lack of manpower and materials—gradually were coming to light. By 1939 the Eighth Route Army and the New

Fourth Army had established scores of anti-Japanese base areas on the battlefields behind enemy lines in northern and central China. They were fighting on all fronts of the enemy's rear area, forming a jagged, interlocking pattern of warfare against the Japanese troops. They had launched attacks on their own initiative, pinned down and wore down a large number of Japanese troops, and forced them to halt their aggression. The anti-Japanese war had thus entered a phase of strategic stalemate.

While the Japanese troops were frenziedly invading China, fascist Germany was also intensifying its expansion in Europe. The whole world situation was very tense, portending the approach of a violent storm. Leaders of the major Western powers were shuttling around to discuss how to check Hitler's expansionist ambitions.

The changes in the situation in the West were bound to affect the war pattern in the East. Intrigues and compromises in the international arena inevitably had an impact on the rise or decline of the various forces in China.

On September 1, 1939, the German fascists invaded Poland, which was then under the protection of Britain and France. On September 3, Britain and France declared war against Germany. World War II thus broke out in Europe.

To concentrate their forces against the German and Italian fascists, Britain, the United States, and France tried their best to avoid direct conflicts with Japan in the East, adopted the policy of appeasement toward Japan, and energetically plotted the "oriental Munich."[1] As the large-scale frontal attacks by the Japanese stopped, however, the policies of Britain, the United States, and other countries had a serious impact on the situation of resistance to Japan in China.

When it changed its policy of "not making the national government its adversary," Japan adopted the tactic—supplemented by military attacks—of luring the Kuomintang into surrender politically by encouraging the Kuomintang to oppose the Communist Party of China, thus realizing its goal of "using Chinese to subdue Chinese." At the same time, the Japanese troops established the principle of holding on to the occupied areas and employed their main forces to attack the Eighth Route Army and the New Fourth Army, concentrating on northern China.

In such a complex situation, the dangers of compromise, capitulation, split, and retrogression began to emerge in the Chinese War of Resistance Against Japan.

In December 1938 the vice president of the Kuomintang, Wang Jingwei, sneaked into Hanoi and openly threw himself into the arms of the Japanese invaders, betraying his country. Thus he turned from a pro-Japanese person into a shameless traitor and collaborator. Then he openly established a puppet regime in Nanjing in opposition to the Kuomintang government in Chongqing.

Under the influence of many forces behind the scenes, the pro-British and pro-American faction of the Kuomintang, represented by Chiang Kai-shek, gradually took a position of passive resistance to Japan but active opposition to the Communist Party. In January 1939, at the Fifth Plenary Session of the Fifth Central Committee, the Kuomintang established its guideline of "dissolving, guarding against, restricting, and opposing the Communists." Later it issued secret orders to its troops to attack the Eighth Route Army and the New Fourth Army. In December of the same year, the Kuomintang troops under Hu Zongnan flagrantly launched a large-scale attack on the Shaanxi-Gansu-Ningxia Border Region where the Central Committee of the Communist Party was located, and the troops under Yan Xishan attacked the New Army, led by the Communist Party, throughout Shanxi Province. This was the first anti-Communist onslaught.

In light of the actions taken by the diehards of the Kuomintang, the Central Committee put forward the political proposition of "persisting in resistance and opposing capitulation; persisting in unity and opposing split; and persisting in progress and opposing retrogression." It called on people throughout the country to fight the anti-Communist diehards of the Kuomintang.

Meanwhile, the Central Committee had issued the directive that "the position of our Party and army on the local armed conflicts is the clear-cut principle of self-defense." "We will not attack unless we are attacked; if we are attacked, we will certainly counterattack!"

The Communist Party was thus facing a two-pronged attack: the large-scale mopping-up operations of the Japanese aggressors and the armed provocations of the diehards of the Kuomintang. That was the dangerous war situation that Deng Xiaoping faced when he returned to the front in the Taihang Mountains after attending the meeting of the Political Bureau in Yan'an.

Father and his party returned to the Taihang Mountains in September 1939. Mother stayed at the general headquarters of the Eighth Route Army and led the women's training class. Father then hurried back to Tongyu Village, where the headquarters of the 129th Division was located. No sooner had Father arrived at the headquarters than he was busy with his work.

In early October Father made a report at the meeting attended by cadres of the 129th Division, relaying the gist of the meeting of the Political Bureau of the Central Committee and assigning jobs. In his report, Father spoke clearly about the characteristics of the current situation, pointing out that the tendency of capitulation was the biggest danger, while the anti-Communist propaganda was a step in preparation for capitulation. He analyzed in detail the roots of the situation from the perspectives of not only the Chinese and Japanese sides but in the international and domestic contexts as well. He went on to describe the characteristics of the stalemate in the anti-Japanese war and the favorable and unfavorable impacts of the changes in international and domestic conditions on the Chinese War of Resistance Against Japan. He stressed in his report that the Party and the whole nation were faced with three major tasks: to mobilize all the patriotic forces to fight against capitulation and strive for the final victory; to build up the Communist Party of China and prepare for any contingencies; and to continue to consolidate and expand the anti-Japanese united front and persist in cooperation between the Kuomintang and the Communist Party and in the resistance to Japanese invaders. In the end, he pointed out the importance of persisting in resistance to Japan behind enemy lines in northern China and analyzed the favorable conditions for doing so with the support of the broad masses of the people. He then handed out detailed assignments in the political, military, organizational, disciplinary, and economic fields.

The detailed minutes of this report were kept by Zhang Nansheng, an old Red Army solider who was then the director of the Organization Department of the 129th Division. There is a moving story about those minutes.

Zhang Nansheng was a native of Liancheng, Fujian Province. He joined the revolutionary ranks in his youth. He had a habit of keeping a diary in his army life. He never stopped keeping it, even long after his Red Army days. He died in Beijing in 1989. His wife, Lin Renli, known as Da Lin, was then more than seventy years old and seriously ill. She found all the speeches delivered by Liu Bocheng and Deng Xiaoping, for whom she had high respect, in the scores of diaries kept by her husband for a decade and copied them. It took her several years to make them into a book, cataloged, paginated, and indexed.

One spring day in May 1992, Aunt Da Lin and Aunt Zhu Lin, wife of Huang Zhen, who was formerly the deputy director of the Political Department of the 129th Division and a famous diplomat, came to see Mother. Zhuo Lin, Zhu Lin, and Da Lin were the three famous "Lins" in the Taihang Mountains. They were all about the same age, all gray-haired, and almost all the same in height and build. When the three old Lins gathered together, they were as cheerful as in the Taihang Mountains!

Aunt Da Lin took out of her bag the book of speeches delivered by Political Commissar Deng and gave it to Mother. When I handled the thick book of more than 70,000 characters, copied by hand, my eyes became wet. I was moved by Zhang Nansheng's persistent efforts to keep his diary, by Aunt Da Lin's great and conscientious effort in copying them, and by the revolutionary friendship and respect shown by the old comrades-in-arms of the 129th Division to their commanders Liu Bocheng and Deng Xiaoping. Aunt Da Lin had already given the speeches by Liu Bocheng that she had copied to Aunt Wang Ronghua, Marshal Liu Bocheng's widow.

These recorded speeches of Deng Xiaoping are indeed very precious. They have filled in the gaps in the historical materials on the ten years of the anti-Japanese war and the War of Liberation. I have invited some experts to sort out the original records of the speeches, hoping that they will be added to the *Selected Works of Deng Xiaoping*.

Now let's turn back to the Taihang Mountains. Father and Commander Liu Bocheng began to implement the guidelines of the meeting of the Political Bureau, preparing to repel the attacks mounted by the Japanese and puppet troops and the provocations perpetrated by the diehards of the Kuomintang. With the support of the people in the vast Japanese-occupied areas, who were unreconciled to being the slaves of a foreign power, the units of the 129th Division attacked the aggressors, dealt continuous blows to the enemy in protracted operations, and won big victories.

In the three months from late August to early December 1939, the 129th Division fought more than 200 small and big battles, killing or wounding over 2,800 Japanese and puppet troops and shooting down one enemy aircraft. As a result, the roads and railways occupied by the Japanese army were destroyed, its transportation was disrupted at times, and it was thrown into a more passive position. In December the general headquarters of the Eighth Route Army ordered the 129th Division to launch the Handan-Changzhi Campaign by taking advantage of the Japanese army's reduced strength after its relief operation.

On December 8, the campaign started. By the end of the campaign, on December 26, the division had fought dozens of fierce battles, wiping out over seven hundred Japanese and puppet troops and recapturing twenty-three strongholds. More important, the division recovered the two cities of Licheng and Shexian in the Taihang Mountains.

New Year's Day 1940 came. At the small mountain village of Tongyu, Liaoxian, in the Taihang Mountains, all members of the detachments directly under the headquarters of the 129th Division gathered at the drill ground to celebrate the new year. Liu Bocheng and Deng Xiaoping spoke at the gathering. In a joyful atmosphere, Liu Bocheng and Deng Xiaoping dined with all the members of the detachments and then made more speeches.

Political Commissar Deng outlined the work of the 129th Division for the coming year: to do a good job in the political, military, health, supply, and other fields, and to heighten the sense of political responsibility. He pointed out that in 1940 struggle would be more ruthless than ever. He also said that that day was New Year's Day, everything should look fresh, and all the old shortcomings should be overcome.[2]

Indeed, the first year of the new decade should have been the starting point for ever greater achievements in the war of total resistance. However, the Kuomintang slipped back into its old ways and seized the opportunity of the Japanese army's halt in its large-scale attack to attack the Communist Party.

The situation was graver for the Communist Party and the Eighth Route Army. In January 1940 Chiang Kai-shek issued an order that the Eighth Route Army withdraw to the areas east of the Baigui-Jincheng Railway and north of the Handan-Changzhi Highway. Then the Kuomintang troops pressed forward and attacked it, and the people in the base areas, from several directions. To smash the attack mounted by Chiang Kai-shek and Yan Xishan, the 129th Division decided to take advantage of the internal conflicts among the Kuomintang troops to attack first the most anti-Communist KMT general, Sun Chu, so as to consolidate the Taiyue area. Under the command of the famous Gen. Chen

Geng, commander of the 386th Brigade, the troops of the 129th Division resolutely gave a telling blow to Chiang's troops, consolidated the Taiyue Base Area, and recovered parts of the Tainan area. At the same time, the effectives of the Eighth Route Army increased.

Yan Xishan was a local tyrant who was in frequent conflict with the orthodox Chiang Kai-shek. This time, he failed in his plot to attack the Communist Party; instead, his position was weakened, allowing Chiang Kai-shek the chance to encroach on his territory. He was indeed the one who went for wool and came back shorn. In late February Yan Xishan reached an agreement of conciliation with the representative sent by Mao Zedong. This was the first round of friction between the Kuomintang and the Communist Party in 1940.

Another Kuomintang unit, led by Zhu Huaibing, joined forces with the troops under Lu Zhonglin and others and launched an attack on the anti-Japanese base area in the northern part of the Taihang Mountains toward the end of 1939. They killed the anti-Japanese people and destroyed the anti-Japanese democratic governments. Liu Bocheng went personally to western Hebei to persuade Zhu Huaibing, Lu Zhonglin, and others to follow the right path but was spurned by them. In the face of the large-scale armed attacks mounted by the anti-Communist diehard troops, the Eighth Route Army was forced to counterattack. In January 1940 the Eighth Route Army wiped out most of over 8,000 diehard troops. The heavy losses forced Zhu and Lu to lead their troops to retreat southward in February. This was another round of friction between the Kuomintang and the Communist Party.

In early February 1940 the Kuomintang troops led by Shi Yousan launched a large-scale attack on the Eighth Route Army in southern Hebei and colluded with the Japanese troops in their mopping-up operations. Mao Zedong ordered the 129th Division to annihilate them resolutely and completely. The 129th Division organized the campaign against the diehard troops in southern Hebei. Between February 9 and 16, after fierce fighting, the main forces under Shi Yousan suffered heavy casualties and retreated in panic under the cover of the Japanese army. The campaign came to a triumphant end, with the troops of Shi Yousan driven out of southern Hebei. This was the third round of friction between the Kuomintang and the Communist Party.

Mao Zedong said, "In the period of the anti-Japanese united front, struggle is the means to the end of unity, while unity is the end of the struggle."

To thoroughly smash the anti-Communist offensives fought by the diehard troops, the general headquarters of the Eighth Route Army decided to launch the Weidong Campaign and the Cixian-Wu'an-Shexian-Linxian Campaign. On February 22, the Weidong Campaign started.

By April 8, after continuous valiant fighting, 17 regiments of the 129th Division had wiped out over 6,000 Kuomintang diehard troops led by Shi Yousan and others, driving them to the edges of the Hebei-Shandong-Henan area and reversing the serious situation in southern Hebei. The Southern Hebei Base

Area and the Hebei-Shandong-Henan Anti-Japanese Base Area were preserved. Among those who directed and participated in the campaign were such important generals as Cheng Zihua, Song Renqiong, Yang Dezhi, Li Jukui, Chen Zaidao, and Liu Zhijian.

On March 5, the Cixian-Wu'an-Shexian-Linxian Campaign started. Political Commissar Deng Xiaoping of the 129th Division went personally to the front to direct the campaign. At the time, 22,000 diehard troops of the Kuomintang were stationed in the areas of Cixian, Wu'an, Shexian, and Linxian, including the two divisions over 8,000 strong led by Zhu Huaibing. Political Commissar Deng said, "Zhu Huaibing is the vanguard that will attack us. In light of the attitude of the diehard troops at present, our strategy is to concentrate our main forces to annihilate the troops under Zhu Huaibing." By making use of the opportune moment and this correct assessment—to take advantage of the internal conflicts among the diehard troops to attack the most reactionary and to win over the middle—and adopting the tactics of envelopment and penetration at lightning speed, 13 regiments of the 129th Division wiped out more than 10,000 Kuomintang diehard troops in only five days, and the forces under Zhu Huaibing were almost completely annihilated. The campaign was a great success, just as Mao Zedong said: "If we have to fight, we shall win." Among those who participated in the campaign were such senior generals as Li Da, Gui Gansheng, Zhou Xihan, and Wang Shusheng.

The major victory in the Cixian-Wu'an-Shexian-Linxian Campaign marked a decisive victory for the 129th Division and the people of the Shanxi-Hebei-Shandong-Henan Base Area in their fight against the first anti-Communist onslaught launched by the Kuomintang diehards.

At that time, Mao Zedong immediately issued a directive to "stop before going too far" and decided to make major concessions to the Kuomintang, thus easing the tense situation. The ultimate purpose of doing so was to consolidate the anti-Japanese national united front, which was hard to form, and to unite with all the forces that could be united with so as to make concerted efforts to resist the common enemy of the Chinese nation—the Japanese aggressors.

In the struggle against the anti-Communist diehard troops, the Shanxi-Hebei-Shandong-Henan Base Area, led by Liu Bocheng and Deng Xiaoping, was further consolidated. The armed forces of the whole area had expanded to 110,000, and the military and political quality of the army was considerably improved as well. The 129th Division controlled a total of seventy-one counties in all of southern Hebei, northern Taihang, and northern Taiyue, with a population of about eight million.

On April 11, 1940, the Taihang Military and Administrative Committee was established to unify the leadership over the areas of Taihang, Taiyue, and southern Hebei. Deng Xiaoping served as the secretary. On the same day, the Northern Bureau of the Central Committee called a meeting attended by high-ranking cadres from the areas of Taihang, Taiyue, and southern Hebei. Secretary of the Northern Bureau Yang Shangkun made a report on the situation and the tactics of the united front. The meeting reviewed the experience of resistance to Japan behind enemy lines in northern China over the previous three years and put forward the three major principles—building the Party, building the army, and building up political power—and the task of actively attacking the Japanese troops' policy of building "prisoners' cages." Not long after the ceasefire with the Kuomintang anti-Communist diehards, the 129th Division, led by Liu Bocheng and Deng Xiaoping, immediately threw itself into the battles against the Japanese invaders.

To destroy the anti-Japanese base areas, the Japanese invaders encircled and cut them off by using railways, highways, and pillboxes supplemented by blockade ditches and walls. Gen. Liu Bocheng named this practice the policy of building "prisoners' cages." To destroy the enemy's artery of traffic and to frustrate its "prisoners' cages" policy, Liu and Deng called on all the people and the army of the area to be "oriented to the communication line," arranged large-scale operations to destroy railways and highways, and conducted sabotage operations against the Beiping-Hankou, Baigui-Jincheng, and Dezhou-Shijiazhuang railway lines.

Subsequently, Liu and Deng launched and personally directed the Baigui-Jincheng Campaign in May. In one day and two nights, their troops completely destroyed fifty kilometers of the Baigui-Jincheng Railway and fifty bridges, blasted an enemy train, and killed or wounded more than 350 enemy troops. In the next three months, the 129th Division conducted more than forty sabotage operations.

While conducting the sabotage operations, the 129th Division punctured resolutely the arrogance of the puppet troops and launched counterattacks against several mopping-up operations by the Japanese army. In these battles, the heroic people and army in the Taihang Mountains dealt an effective blow to the policy of "prisoners' cages."

Between the summer and autumn of 1940 the international situation deteriorated. Fascist Germany and Italy won alarming victories in Europe. In September Japan formed a military alliance with them and stepped up its aggression against and control over China. Under Japan's strong political and military pressure, the Chinese Kuomintang wavered all the more. The danger of compromise was unprecedentedly serious. Mao Zedong pointed out, "We should be fully aware of the possibilities of the most difficult, the most dangerous, and the darkest days."[3]

To crush the all-around attack mounted by the Japanese troops on the Chinese army in northern China, deal a blow to its policy of "prisoners' cages," and remove the danger of capitulation by the Kuomintang, in August 1940 the Eighth Route Army, under the command of Commander-in-Chief Zhu De and Deputy Commander-in-Chief Peng Dehuai, started a large-scale offensive campaign against the enemy-occupied communication lines and strongholds in northern China, a campaign known as the Hundred-Regiment Campaign.

The Eighth Route Army employed 105 regiments over 200,000 strong in the campaign. During the two operational phases and the counterattacks against the Japanese retaliatory mopping-up operations, it fought 1,824 battles in three and a half months, killing or wounding more than 20,000 Japanese troops and more than 5,000 puppet troops, destroying 474 kilometers of railways, more than 1,500 kilometers of highways, and over 260 bridges or tunnels. The Eighth Route Army suffered over 17,000 casualties.

The Hundred-Regiment Campaign was the largest and longest strategic offensive campaign conducted by the Chinese army in northern China in the anti-Japanese war. The Chinese army and people in northern China, numbering in the hundreds of thousands, fought bloody battles against the strong Japanese forces. The large-scale campaign shocked the land of China.

Fighting shoulder to shoulder with other friendly units in the campaign, the 129th Division won brilliant victories in the sabotage operations and counter–mopping-up campaigns. It destroyed more than 240 kilometers of railways, more than 500 kilometers of highways, and fought 529 big or small battles. It recaptured 9 county towns and killed or wounded over 7,500 Japanese and puppet troops.

In those three months, Liu Bocheng and Deng Xiaoping personally commanded the battles on the front. They sometimes climbed the Taihang Mountains and sometimes marched dozens of kilometers. One day an enemy aircraft threw a bomb just in front of the cave they lived in. Liu and Deng went out to have a look and set out again, braving the heavy firepower of the enemy. As soon as the danger was over, they camped and called meetings at which Political Commissar Deng made reports on the situation and disseminated policies.

It was November. In Songjiazhuang Village, "the golden millet on the fields withered because of wind and sunshine; the corn was flattened onto the fields; the soybeans were lying in the fields, ungathered. Not a single house could be seen in many villages, and even the tiny temples housing the village god were destroyed. The enemy burnt all and killed all in the base areas. The miserable condition was beyond description. Political Commissar Deng immediately mobilized his men to help the local people put their houses in order and gather in crops. Everyone must go and must work hard."[4]

The autumn and winter of 1940 passed quickly. People felt time fleeing because they were busy fighting and launching frequent counterattacks against the mopping-up operations of the enemy.

That year there were several other events worth mentioning. One was making a film.

On February 23, in Tongyu Village, Yang Guoyu and

> Division Commander Liu, Political Commissar Deng, and Chief-of-Staff Li Da went out of the village to pose for a film. Most of us had never done this before, and

> some had not even seen the shooting of a film. The cameraman, whom someone said had been sent by the Soviet Union, came from Yan'an. He ordered us about, sometimes to sit down, now to lie down, and then to write on paper. We really felt fed up, and everyone was standing there with a stern expression. The cameraman asked us to look natural. The more he demanded, the more unnatural we looked. The shooting had to stop. Division Commander Liu told the cameraman, Xiao Xu, "Natural, but we cannot be natural." Political Commissar Deng and others laughed, and this time they looked rather natural.[5]

It was perhaps the first time Father had posed for a film. The film is nowhere to be found now. Ordinarily, Father does not like to pose for photos, let alone for a film. So it was indeed very hard to make him look natural.

Another event was getting a home.

In September Mother was transferred from the general headquarters of the Eighth Route Army to the headquarters of the 129th Division and worked in the secretarial section. From then on, she and Father began to live, march, and fight battles together. Although they were sometimes separated because of war, Father finally had a home, even if it was a home at the front, with no fixed place. On December 4, 1940, the headquarters of the 129th Division moved to Chi'an Village, Shexian County, in the Taihang Mountains in Shanxi Province.

From then on, for five years, this unknown small village in the Taihang Mountains became the location of the headquarters of the 129th Division and the heart and capital of the Shanxi-Hebei-Shandong-Henan Base Area.

Liu Bocheng and Deng Xiaoping settled down in the courtyard of a small temple in this village. From there, the two directed more arduous struggles by the people of the Taihang Mountains against the enemy and spent more difficult days with them.

CHAPTER 51

DIFFICULT YEARS

The war of the Chinese nation against the Japanese aggressors had already lasted more than three years, from the July 7th Incident of 1937 to the end of 1940.

The intricate political situation in China made the war of resistance to aggression extremely complex, and the great changes in the international situation made the future of China's war of resistance more uncertain. However, neither the brutal aggression of the Japanese aggressors nor the base conduct of the Kuomintang diehards—passively fighting Japan but actively opposing the Communist Party—could stem the determination of the Communist-led army to wage a war of resistance to aggression.

After heroic, arduous, and valiant fighting, the Communist-led army not only consolidated the situation in the war of resistance behind enemy lines in northern China but also developed the situation in central and southern China. In two years, these poorly equipped and not very strong armed forces resisted some 60 percent of the Japanese troops and all the puppet troops, crushed nearly 100 mopping-up operations in which 1,000 to 50,000 enemy troops participated each time, fought over 10,000 battles, and annihilated large numbers of Japanese and puppet troops.

With the anti-Japanese support of the broad masses who loved their country, hometown, and families, and while fighting wars on two fronts, the armed forces led by the Communist Party consolidated and expanded anti-Japanese democratic base areas. By the end of 1940 its troops had increased to half a million and the base areas had a population of 100 million. Anti-Japanese democratic governments were established in the base areas in northern and eastern China, and the policies of rent and interest reduction were eventually implemented in rural areas. The implementation of these policies received the wholehearted support of the people, poor peasants in particular, and their support further enhanced the strength of the army in resisting Japan behind enemy lines.

Nevertheless, the national forces for resisting aggression in the land of China were still not strong enough, and the basic situation of the enemy being strong and China weak remained unchanged.

On the battlefield in northern China, the Japanese army continued to focus on "suppressing the Communists." The Japanese army had built up its forces, increasing its troops to 300,000 and the puppet troops to 100,000 in northern China. It planned to conduct continuous and more ruthless mopping-up and "nibbling" operations against the anti-Japanese base areas and to launch the "campaign to tighten public security" in an attempt to eliminate the anti-Japanese forces led by the Communist Party.

On the battlefield in northern China, the strength of the Kuomintang troops was about half a million, more than the Japanese troops. However, the Kuomintang diehards did not join hands with the Communist Party to resist Japan. Instead, they regarded the Communist Party as a serious hidden enemy and incessantly conducted military attacks and imposed economic blockades on the anti-Japanese democratic base areas. Some brazen people went so far as to propose "to save the nation by a devious path" in an attempt to find a pretext for their defection to the enemy and betrayal of their country. Thus, within a short time, more than 30,000 Kuomintang troops openly defected to the enemy under this impudent pretext and then wantonly launched attacks on the anti-Japanese democratic base areas of the Communist Party.

Japan invaded China. These Chinese did not fight the Japanese but other Chinese. What generally acknowledged truth was there in the world?

In 1941, under the pincer attacks of the Japanese troops and the Kuomintang diehards, the war of resistance waged by the Communist Party behind enemy lines in northern China entered the most difficult period of the eight-year war. Immediately after New Year's Day 1941, the Communist-led New Fourth Army, which had persisted in the war of resistance in southern Anhui, moved to southern Jiangsu in accordance with Chiang Kai-shek's order to move to the north of the Yangtze River. To the great surprise of the Communist Party, Kuomintang and Chiang Kai-shek had already issued secret orders to deploy troops to encircle the New Fourth Army in three directions and to launch attacks in an attempt to annihilate it. Both sides fought a fierce battle for one week. By January 14, over 6,000 anti-Japanese officers and men of the New Fourth Army had been killed by the Kuomintang diehards. The famous anti-Japanese general Ye Ting was detained by the Kuomintang as well. This was the Southern Anhui Incident, which shocked the whole country. It was the most horrifying incident in the second anti-Communist onslaught launched by the Kuomintang.

This was an extremely painful page in the history of the anti-Japanese war waged by the Chinese nation. Chiang Kai-shek was never softhearted about killing Communists.

In February 1941 the Japanese army issued its plan to suppress Communists. Then it started the six-month spring and summer mopping-up operations against the anti-Japanese base areas on the plains of western Shandong, the Hebei-Shandong-Henan border area, and eastern and central Hebei—by stages and by areas—in an attempt to capture the main forces and leading organs of the Chinese army.

In the Shanxi-Hebei-Shandong-Henan area, the Japanese army started its operations in January. Between January 10 and 15, 5,000 Japanese troops conducted mopping-up operations in the areas of Yushe, Liaoxian, Heshun, and Xiyang. Between January 15 and February 6, over 7,000 Japanese troops conducted mopping-up operations in the area of western Shandong. Between January 24 and February 4, over 4,000 Japanese troops conducted mopping-up operations in the Taihang Mountains. On March 3, over 1,000 Japanese troops conducted mopping-up operations in the area southeast of Puyang.

On March 21, the Japanese troops stationed in Xiyang invaded the Taihang Mountains. On March 29, the Japanese army launched the first "campaign to tighten public security" in northern China. On April 3, over 1,400 Japanese and puppet troops conducted mopping-up operations in the areas south of Nangong, east of Guangzong, west of Wucheng, and north of the Xingtai-Jinan Highway. Between April 10 and 20, over 10,000 Japanese and puppet troops with 100 trucks and tanks conducted devastating mopping-up operations in the Shaqu Base Area of the Hebei-Shandong-Henan border area.

Between May 7 and 25, some 50,000 troops of six Japanese divisions launched attacks on the Kuomintang troops stationed in the Zhongtiao Mountains and occupied the area. On May 27, over 2,000 Japanese troops invaded the areas of Shouzhang and Fanxian. On May 29, over 1,000 Japanese troops conducted mopping-up operations in the Taiyue area. In May the Japanese army constructed the second blockade line west of the Beiping-Hankou Railway. On June 18, over 5,000 Japanese and puppet troops "cleaned up" the Taixi area. On June 19, over 1,000 Japanese troops conducted mopping-up operations in the Taihang Mountains. On June 28, 2,000 Japanese troops conducted mopping-up operations in the southern Hebei area.

During the spring and summer mopping-up operations, the Japanese troops repeatedly "cleaned up" the anti-Japanese democratic base areas in northern China and carried out a policy of unparalleled savagery: "Burn all, kill all, loot all." In the over 140 villages in the Shaqu area alone, more than 3,400 people were slaughtered, almost all the 50,000 jujube trees on which the villagers depended for their existence were felled, and countless village houses were burned.

To launch the "campaign to tighten public security," the Japanese army divided the areas of northern China by grade into different "security areas," made great efforts to construct railways and highways, dug blockade ditches, and built blockade walls. The Japanese army dug 500-kilometer blockade ditches north of the Beiping-Hankou Railway in an attempt to cut off the ties between the Beiyue and Taihang Mountain base areas on the one hand and the central and southern Hebei plains base areas on the other so as to make the mountain base areas run out of supplies. On the plains, the Japanese army constructed a watchtower every one and a half kilometers and a stronghold every two and a half kilometers and divided the plains base areas into small grid-type areas under strict blockade.

Because of the mopping-up operations, blockades, and nibbling operations by the Japanese army, the anti-Japanese democratic base areas diminished. In view of the difficult situation, the Central Committee of the Communist Party called for the launching of a countercampaign against the nibbling operations by combining political offensive with military attack.

On April 28, Political Commissar Deng Xiaoping published an article entitled "Oppose Insensitiveness and Reverse the Serious Situation in the Taihang Mountains." He pointed out the need to overcome right-deviationist sentiments and oppose insensitiveness and panic; to unite as one to face difficulties squarely, turn in the direction of the enemy and the communication line, and wage tenacious struggles against the enemy; and to reverse the current serious situation with a strong will, a militant spirit, and hard work.

In late May the 129th Division issued orders calling for improving and strengthening guerrilla groups, conducting guerrilla warfare and consolidating the base areas, and repeatedly conducting a warfare of attack and destruction against the enemy troops.

While China was fighting the bitter war of resistance, rapid changes were taking place in the world situation.

On June 22, fascist Germany suddenly launched large-scale blitzkrieg attacks on the Soviet Union on the 2,000-kilometer front. The war between the Soviet Union and Germany subsequently broke out. When international fascist aggression was most furious, the Japanese troops in China were all the more arrogant. In June the Japanese army formulated the policy of persisting in setting up the "Greater East Asia Co-prosperity Sphere." While increasing the strength of the Kwantung Army to 700,000, it sent troops to occupy the southern part of French Indochina.

To maintain the stability of the rear area of the southward advancing troops and to establish a secure "depot base for the Greater East Asia War," the Japanese army stepped up its "public security" operations in the Chinese territory it occupied and started the frenzied and brutal autumn and winter mopping-up operations. The focus of its mopping-up operation was shifted this time from the plains of northern China to mountain areas. Japanese bayonets and guns were aiming directly at the hinterland of the Communist-led anti-Japanese base areas in northern China.

In the second half of 1941 the invading Japanese troops conducted frenzied mopping-up operations against the Communist-led anti-Japanese democratic base areas in northern China for a long time and on a larger scale by using the brutal operational methods of "iron-ring encirclement" and "village combing." On August 12, over 40,000 Japanese troops conducted large-scale iron-ring encirclement–style mopping-up operations in the Shanxi-Chahar-Hebei border area. On September 22, in the Yuenan area, over 20,000 Japanese troops conducted large-scale mopping-up operations.

On October 6, over 30,000 Japanese troops conducted large-scale iron-ring encirclement–style mopping-up operations in the Yuebei area. On October 17, over 1,500 Japanese troops conducted mopping-up operations in the southern Hebei area. On October 25, over 3,000 Japanese and puppet troops conducted mopping-up operations in the Hebei-Shandong-Henan area. On October 31, over 7,000 Japanese troops conducted surprise-attack mopping-up operations in the Taihang Mountains in an attempt to capture and wipe out the general headquarters of the Eighth Route Army and the headquarters of the 129th Division and to make a night raid on the place where the headquarters of the 129th Division was located.

On November 1, the Japanese army in northern China launched its third "public security" campaign. On November 25, over 4,000 Japanese troops conducted mopping-up operations in southern Hebei. On December 9, over 6,000 Japanese troops conducted mopping-up operations in the areas of Nangong and Weixian in southern Hebei. On December 26, over 3,000 Japanese troops "cleaned up" the southern Hebei area.

Confronted with these many intense operations of the Japanese troops, the 129th Division, while leading the masses in strengthening the defenses, clearing the fields, and persisting in the guerrilla war on interior lines, organized the main forces and local armed forces to launch large- and small-scale attacks, sabotage activities, and counterattacks against the enemy. Under the unprecedentedly fierce and brutal attacks by the enemy, the 129th Division preserved its main forces, maintained the anti-Japanese base areas, and made several counterattacks against the invasions by the Kuomintang diehard troops under Yan Xishan from another direction.

The year 1941 was one in which the grimmest struggle of resistance was waged behind enemy lines in northern China. That year, the war was fought intensely and relentlessly, and the troops moved frequently. In the meantime, life for the anti-Japanese army and people was extremely hard because of both the enemy's tight blockade and natural disasters that occurred in the base areas.

In September of that year, my eldest sister, Deng Lin, was born at Chi'an. As the war situation was pressing and the troops were fighting from place to place, Mother reluctantly entrusted her first child to the care of a local family in Licheng County seven days after she was born. After handing over the child, Mother immediately left and followed the troops and never looked back.

In 1941 our army and people behind enemy lines in northern China crushed sixty-nine mopping-up operations launched by over one thousand enemy troops each time, nine large-scale mopping-up operations by anywhere from ten thousand to seventy thousand enemy troops each time, and three "public security" campaigns. They also shattered the Japanese nibbling and cutting-apart operations and blockade. However, with the powerful offensives of the Japanese army, the area of the anti-Japanese base areas and the strength of the Eighth Route Army were reduced, and economic conditions became extremely difficult. Even under these extraordinary hardships, the anti-Japanese army and people were still prepared to greet the year 1942 with high revolutionary morale and a heroic spirit devoted to defending their hometowns and their country.

A stunning change took place in the international war situation on the eve of 1942. On December 7, 1941, the arrogant Japanese militarists launched a surprise attack on the U.S. naval base in the Pacific—Pearl Harbor, in Hawaii. The Pacific War broke out!

On December 9, 1941, the Kuomintang government of China declared war on Japan at last (and on Germany and Italy as well), more than six full years after Japan had invaded China.

After the Pacific War broke out, the CPC Central Committee's analysis of the situation was that the outbreak of the Pacific War was undoubtedly favorable to China's resistance and that Japan's forces of aggression against China could not but be weakened because Japan had antagonized over twenty countries. But to support the Pacific War, Japan might be more anxious than ever to plunder China's resources and consolidate the occupied territory and would be bound to intensify its mopping-up operations and economic blockade against the anti-Japanese base areas behind enemy lines more brutally. The anti-Japanese army and people could be confident that they could persist in protracted resistance and win victory, but at the same time, they needed a full understanding of the increasingly serious difficulties. The Central Committee was in favor of continuing the general principle of persisting in protracted guerrilla war and preparing for future counteroffensives. The whole Party and the whole army should "endure with dogged will and pull through the most difficult struggle in the next two years."[1]

It occurred in the last few days of 1941. After the enemy's mopping-up operations, the headquarters of the 129th Division returned to Chi'an, Shexian County. It was chilly, the earth was frozen, and the wind was bitingly cold in the Taihang Mountains.

Liu and Deng had been busy for a whole year. No battle was fought at the end of the year, so they dined with the staff of the headquarters. "A dozen bowls were put on the table, including square fat meat and Sichuan-style vegetables. Deng's wife, Comrade Zhuo Lin, and all staff were present at the dinner," Yang Guoyu wrote in his diary. He and all the others ate their fill.

It was December 31, the last day of 1941. According to Yang Guoyu,"The headquarters and the Political Department gave a dinner party. Prominent figures from all walks of life and their wives came to Chi'an to pay a formal visit to Liu and Deng, and they had dinner together. All the dishes were of Sichuan style and were plentiful."[2]

The year 1942 came. It was January 1 at Chi'an. "This New Year's Day is not celebrated as well as the last one. Last year we sang songs, paid group New Year calls, and killed pigs to have a dinner party. This year we have mutton cooked with porridge as well as carrots and sweet potatoes. It is not at all bad."[3]

New Year's Day 1942 was not as well spent materially as New Year's Day

1941, but it passed just as busily. On January 3, the 129th Division issued the 1942 outline of military work. On January 7, Division Commander Liu made a report on "better troops and simpler administration." Improving the troops and simplifying administration were important measures that all anti-Japanese base areas were directed by the Central Committee to take, in the face of the increasingly intense mopping-up and nibbling operations by the enemy troops and the increasingly difficult economic conditions.

Mao Zedong said, "Now it is imperative for us to do a little changing and make ourselves smaller but sturdier, and then we shall be invincible."[4] "Better troops" meant reducing the main forces and leading organs and strengthening the companies. Part of the main forces became local forces, local armed forces and people's militias were strengthened, and training and consolidation were intensified to improve combat effectiveness. "Simpler administration" meant strengthening organizations, simplifying structures and reducing staff, strengthening grass-roots units, improving efficiency, economizing on human and material resources, and opposing bureaucratism.

The implementation of these measures helped settle the conflicts between the overstaffed organization and the growing inadequacy of the war-devastated social economy and lightened the burdens of the people. They enabled all anti-Japanese base areas to surmount extreme difficulties.

The 129th Division began to carry out the policy of better troops and simpler administration. Political Commissar Deng Xiaoping reminded the officers and men of the 129th Division that they were the people's army, and as such, they should be particularly concerned about the welfare of the people, whose lives had been extremely hard because of years of continuous war, plunder by the Japanese bandits, and both natural and man-made disasters. Strictly carrying out the policy of better troops and simpler administration would reduce the burdens of the people and only then could they give the army greater support in defeating the Japanese aggressors.[5] On January 15, Liu and Deng issued the order that the policy of better troops and simpler administration was to be implemented. Military orders had to be obeyed!

> Since Political Commissar Deng Xiaoping has taken the lead, other leaders, especially leading organs, have immediately taken action. So the personnel of the leading organs have been organized to go down separately to military subareas and brigades to conduct thorough mobilization. They start off today or tomorrow. Before they depart, Comrade Xiaoping has set four rules. First, organizations are to be adjusted, structure simplified, personnel and horses reduced, and combat companies strengthened according to set percentages. Second, a number of fairly capable local cadres are to be transferred to local armed forces committees for the purpose of strengthening local armed forces and conducting guerrilla war. Third, arrangements are to be made for old, weak, and disabled soldiers to learn some production skills on a work-study program. Although there are only four rules, they are

> important: they have a bearing on whether the war of resistance can be sustained and on the life of the army and the people.[6]

In mid-January Political Commissar Deng went to Qiyuan Village on the western bank of the Qingzhang River to make an important speech to the cultural workers in the Taihang Mountains in which he discussed the importance of the policy of better troops and simpler administration in the light of the current situation.

On January 25, he led a group of the most capable personnel to the 6th Subarea of the Taihang Military Area, in the areas of Wu'an and Shahe, to provide specific guidance on the work of implementing the new policy.

The policy was implemented three times on a large scale in the 129th Division and the Shanxi-Hebei-Shandong-Henan border area. As Liu and Deng attached importance to this work and did it personally, it was done successfully. The combat effectiveness of the army was enhanced, the grass-roots leadership was strengthened, human and financial resources were preserved more economically, and the burdens of the people were lightened. Meanwhile, the staff of the leading organs was reduced, work efficiency was improved, and bureaucratism was overcome. All this made it possible to meet the needs of the new and more difficult war situation.

Mao Zedong praised the work done in the Shanxi-Hebei-Shandong-Henan border area. He said, "The leading comrades of the Shanxi-Hebei-Shandong-Henan border area have really taken this work in hand, setting an example of 'better troops and simpler administration.'"[7]

Father consistently attached great importance to this work, and even after the founding of the People's Republic, he stressed and promoted more than once the work of simplifying administration. In May 1992, after he had retired, he still paid close attention to this question as he followed the structural reforms of the State Council. He said, "The work of simplifying administration remains all along a problem for us."

For a whole year, from the beginning of the new year, Liu and Deng had sole responsibility in the 129th Division for the work of implementing the policy of improving the troops and simplifying administration. Also from the beginning of the new year, Liu and Deng were responsible for the work of promoting self-help through production.

On January 13, 1942, the general headquarters of the Eighth Route Army ordered the units under the command of the Taihang and Taiyue Military Areas to increase production to overcome economic difficulties. This was the most difficult period for the people in the anti-Japanese base areas. In the previous year, the enemy had been brutally conducting large-scale mopping-up operations in the base areas, nibbling incessantly at the base areas, and plundering the people and resources there. During the "public security" campaign and the mopping-up operations, the enemy's policy of "burn all, kill all, loot all" extended even to poisoning the water in the wells and cooking pots and savagely and cruelly imposing a blockade on the army and the people in the base areas.

There were enemies both on the ground and in the heavens. That year natural disasters, including flood, drought, insect infestations, and hail occurred one after another, making the situation worse.

The burning, killing, and looting by the invading Japanese troops wherever they went aroused the great indignation of the people in the base areas and also inspired their ardent love for the Eighth Route Army. In the Taihang Mountains, some army cadres had only five slices of radish, with no salt, for a meal and did not even have enough crude millet to eat. It made the hearts of the people ache to see this, and they gave the cadres their sole stores in small baskets: persimmon skins and steamed maize buns. These were the best offerings the people had; they had nothing more.

Father once told me that when Liu Shaoqi stopped in the Taihang area in September 1942 on his way back to Yan'an from central China, Liu and Deng invited him to dinner at the headquarters of the 129th Division at Chi'an and the only dish was dried mutton. Father said, "That was the best food at that time, and we had not eaten meat for a long time!"

Rong Zihe, then the vice-chairman of the Shanxi-Hebei-Shandong-Henan Border Area Government, recalled:

> During the most difficult time, the grain ration of cadres was reduced from one and a half *jin* [one jin equals half a kilogram] of millet to seven *liang* [sixteen liang to a jin, according to the old system] per day, and my weight was reduced from 125 to 100 jin [from 138 to 110 pounds]. Comrade Xiaoping had an appointment with Comrade Yang Xiufeng, me, and Comrade Li Yiqing of the Border Area Government to talk about work. I and Li Yiqing dozed off from lack of energy. On seeing this, Comrade Xiaoping was touched and told Comrade Yang Xiufeng to his face that the monthly allowance of the department-level cadres of the Border Area Government should be increased to ten dollars.[8]

Even the increase of the allowance to ten dollars, however, could not solve the basic problem! What to do then?

In Yan'an, as early as 1940, the 359th Brigade of the Eighth Route Army, led by Brigade Commander Wang Zhen, the "bearded Wang," marched to Nanniwan to carry out the policy of stationing troops to engage in farming by reclaiming wasteland and increasing production.

In the Taihang Mountains, Liu and Deng called on the army and people in all base areas in the Shanxi-Hebei-Shandong-Henan border area to launch a great production campaign, and Political Commissar Deng gave the mobilization speech himself. The anti-Japanese morale of the army and the people in the base areas was very high then. After Deng delivered his speech, all the base areas responded one after another. Even the cadres of the headquarters of the 129th

Division vied with each other to apply for membership on the wasteland-reclaiming teams to strike back at the enemy's policy of "burn all, kill all, loot all."

Liu's wife Wang Ronghua and Deng's wife Zhuo Lin were in charge of the applications of women comrades, and both of them led women comrades into the mountains to reclaim wasteland together with their male comrades. The harvest reaped by the division headquarters was fairly good that year. Large turnips weighed six jin each. Seeing these turnips, Political Commissar Deng said happily, "They can be called king turnips!"[9]

The Red Army and Eighth Route Army men were originally the sons of Chinese peasants. When the gun-shouldering armymen shouldered hoes, the resulting increase in production improved the life of the army and lightened the burden on the people. Can you find any other army similar to this army in the world?

What should win even greater admiration is that to this day the Chinese People's Liberation Army continues to hold a gun in one hand and a hoe in the other and has engaged in production while defending the motherland in the course of the realization of China's four modernizations and in the modern life of the 1990s. Defending the motherland is the sacred duty of Chinese armymen, and developing production is the traditional way to cherish the people by lightening their burden. This is one of the reasons the Chinese people and the Chinese People's Liberation Army have maintained very close relations, like fish and water. No other army in any country of the world has had such special relations with those it defends.

In the Taihang Mountains in 1942, thanks to the effective and timely implementation of these two major measures—the policy of better troops and simpler administration and the production campaign—the army and the people in the Shanxi-Hebei-Shandong-Henan border area resolutely defeated countless of the Japanese mopping-up operations.

Between early February and early March the Japanese army conducted spring mopping-up operations, dispatching over 12,000 troops to the Taihang area and over 7,000 troops to the Taiyue area. The Japanese army adopted brutal operational methods—such as "seizure and surprise attack," "iron-ring encirclement," "vertical and horizontal mopping-up," "passing through many places to pick out and throw away," "lightning tip-and-run raid," and "marching at night and raiding at dawn"—as it attacked Matian, Liaoxian County, where the general headquarters of the Eighth Route Army was located. During the mopping-up operations, which lasted thirty days, the Japanese invaders burned houses, slaughtered civilians, raped women, looted materials, and perpetrated all kinds of atrocities.

After the spring mopping-up operations, the Japanese army built the third blockade line west of the Beiping-Hankou Railway. In May the Japanese troops launched their large-scale mopping-up operations in the Taihang and Taiyue areas. During that thirty-eight-day mopping-up operation, the Japanese army divided the battle into three stages and employed more troops. Over 25,000 troops were employed at one time against the Taihang area alone. The Japanese army used the vicious warfare method of "concentrating forces to pass through many places for mopping up" to make a long-range raid on the general headquarters of the Eighth Route Army and the leading organs of the 129th Division. Once again, it employed the intense, cruel, and barbarous methods of "iron-ring encirclement," "passing through many places for cleaning up," and "picking out, throwing away, and cleaning up" to kill the Chinese troops and people everywhere, to loot people's properties wantonly, and to destroy the base areas.

While conducting these attacks in the Taihang and Taiyue areas, the Japanese army carried out especially frequent mopping-up and village-combing operations in southern Hebei and nibbling and mopping-up operations in the Hebei-Shandong-Henan area.

In the summer and autumn of 1942 the Japanese army launched the fourth and fifth "public security" campaigns—they were even more cruel and arrogant than the previous campaigns. The Japanese North China Front Army issued the order: "All persons in the enemy area, men and women, old and young, should be killed, all houses should be burned, all grains and fodders should also be burnt if they cannot be moved, all pots and bowls should be broken, and all wells should be filled or poisoned."[10] They went so far as to create shocking "depopulated zones" in northern China!

As autumn came, the mopping-up operations started again. On September 27, over 10,000 Japanese and puppet troops conducted large-scale mopping-up and cleanup operations in the Hebei-Shandong-Henan area. On October 20, over 16,000 Japanese troops conducted mopping-up operations in the Taihang and Taiyue areas simultaneously.

The mopping-up operations in the anti-Japanese base areas behind enemy lines in northern China in 1942 were conducted by the Japanese army for an unusually long time and in an unprecedentedly cruel and ferocious way. Their bizarre warfare methods had more and more queer names that cannot be looked up in the classic *Master Sun's Art of War*. The Chinese have never practiced such unusual warfare methods, let alone methods of such cruel and ferociously murderous character.

The Eighth Route Army neither panicked nor retreated. Liu and Deng said, "The enemy is fierce in appearance; actually he is outwardly strong but inwardly weak, just like the clay idol fording the river."[11]

Thanks to the implementation of the policy of better troops and simpler administration, the Chinese army became smaller but more mobile. Therefore, it boldly used the strategy of advancing as the enemy advanced—that is, attacking the enemy rear area when the enemy attacked a Chinese base area—and attacked the enemy resolutely during the bitter one-year counterattack. It adopted flexible and diverse guerrilla tactics to attack the Japanese army, launched a series of political offensives against the enemy, and crushed the enemy's repeated mopping-up operations with a system of armed forces that combined the main forces, local forces, and the people's militia.

Of course, some alarmingly dangerous situations occurred during the cruel battles of the counterattack campaign. For instance, during the mopping-up operations in June, an enemy unit suddenly approached and encircled the headquarters of the 129th Division. Liu Bocheng calmly and unhurriedly directed the operation. By taking advantage of his knowledge of the terrain and the enemy's situation, he led his men out of the enemy ring of encirclement and then broke out of the encirclement, reaching safety at last. Among those who followed him to safety were his wife Wang Ronghua, Deng's wife Zhuo Lin, Li Da's wife, Huang Zheng's wife Zhu Lin, and others.

In March of that year of pressing battles, Political Commissar Deng Xiaoping went to the Taiyue area. There he personally directed the 385th Brigade and other units in their successful Fuyi campaign of counterattack and self-defense against the invading diehard troops under Yan Xishan on April 15 and 16. Subsequently, he crossed the Baigui-Jincheng Railway through a secret communication line and headed for the newly established Zhongtiao Mountain Base Area to inspect the work there.

Fighting side by side, Liu and Deng had become closer and more inseparable. When Deng went to the Zhongtiao Mountains, Liu stayed in the Taihang Mountains. After they said good-bye to each other, "Division Commander Liu had no intimate comrade-in-arms to consult on the work, and he could only discuss matters with Li Da and Cai Shufan."[12] Deng crossed the Baigui-Jincheng Railway in the enemy-occupied area at the end of March, and Liu sat in the office of the Branch of Operations, waiting for the telegram. When the telegram came, he read it word for word, and only after learning that Deng was safe did he feel enough at ease to go to sleep.[13]

After Liu and Deng were separated, the telegrams issued by the headquarters of the 129th Division continued to be signed jointly by "Liu and Deng." Yang Guoyu wrote, "As they took so serious an attitude toward their work, they united closely as one and set an example themselves, so all the 'instructions,' 'commands,' or 'orders' issued in their names jointly were carried out firmly by their troops. How couldn't they be deeply respected! The enemy is afraid of Liu Bocheng and of Deng Xiaoping as well. The enemy distributed the photographs of Deng Xiaoping to the troops."

The end of 1942 approached. This very difficult year was about to be over. And the people of the Taihang Mountains were still standing as firm as rock!

On December 16, the 129th Division congratulated Division Commander Liu Bocheng on his fiftieth birthday. A congratulation meeting was held at Chi'an. The Central Committee sent him a congratulatory telegram, and Peng

Dehuai wrote a congratulatory poem. Telegrams and letters of congratulation came from many places, and the representatives of the people of Shexian County also offered birthday presents. Political Commissar Deng Xiaoping, Chief-of-Staff Li Da, and the Political Department's Director Cai Shufan and Deputy Director Huang Zheng were present at the meeting.

Deng Xiaoping delivered a long, ebullient, and affectionate speech. He said that to deeply love one's country, people, and Party is a quality that a good Communist must have. Comrade Bocheng not only had this quality but had also devoted all his energy to his country, his people, and his Party. He had made unique contributions to the revolutionary cause. Deng said, on the occasion of Comrade Bocheng's fiftieth birthday, "I wish him good health, and the victory of the cause with our joint efforts."[14]

To the Eighth Route Army, which persisted in the war of resistance behind enemy lines in northern China, 1941 and 1942 were the two most difficult years of that eight-year struggle. The cruel war and extraordinary hardships of those two years remain fresh in the memories of the people of the Taihang Mountains as well. Nevertheless, they held out! The people of the Taihang Mountains endured this most difficult period. When darkness was over, there was light ahead.

The people of the Taihang Mountains realized that the road leading to light was not smooth and the struggle would continue. However, the great road leading to victory was beginning to present itself.

TOWARD REHABILITATION AND DEVELOPMENT

The difficult years of 1941 and 1942 passed at last, and the arm of the clock of history pointed to 1943.

The year 1943 was a crucial year both in the international antifascist war, which was advancing toward victory, and in the Chinese anti-Japanese war, in which China was beginning to overcome difficulties and to enter a stage of rehabilitation and development.

Justice is justice, and it will ultimately triumph over evil. Everything starts to fall after it has reached its zenith. When the arrogant aggression of international fascists had reached its zenith, their doomsday was coming.

The spring of 1943 brought good tidings to people all over the world. The Soviet Red Army won the great victory in the Stalingrad Campaign and began its strategic counteroffensive against the German fascists, forcing the German army to turn to strategic defense. On May 13, the last group of German and Italian fascist troops surrendered to the Allied forces on the battlefield of North Africa. On July 10, the Allied forces of Britain and the United States landed on Sicily, pressing toward Italy proper. On September 3, Italy surrendered to the Allies, and the European fascist front was utterly disintegrated.

On the Pacific battlefield, the U.S. army occupied Guadalcanal Island in February and turned to strategic counteroffensives. Japan shifted in panic to strategic defense.

The defeat on the Pacific battlefield and the protracted war of attrition in China aggravated Japan's domestic contradictions. In 1943 the Japanese cabinet was reshuffled twice. The political situation was unstable, and anxiety was widespread among the Japanese people. Japan's economy was on the verge of bankruptcy. Antiwar sentiments began to spread among the people and among Japanese soldiers at the front.

Facing unfavorable situations both at home and abroad, the Japanese invaders were eager to end their war against China as soon as possible so that they could employ more troops on the Pacific battlefield to check the counteroffensives by the U.S. army. The primary tasks of the Japanese troops in China were ensuring the safety of the occupied areas, the important resource development areas, the key cities, and the main communication lines, redeploying its forces, and maintaining a troop strength of 600,000.

The morale of the people of the worldwide antifascist front was high. But what was the Chinese Kuomintang planning? What did Chiang Kai-shek think of these developments of 1943?

Chiang Kai-shek was Chiang Kai-shek. He could not and would never change his ultimate purpose: to annihilate the Communist Party. He adopted the policy of resisting the invading Japanese troops and, at the same time, "watching others fight." In 1943 the Kuomintang army conducted only a limited number of defensive operations on the frontal battlefield in China, and the India-Burma Expeditionary Army in northern Burma won some counteroffensive operations. Chiang Kai-shek reserved his men and guns to attack the Communist Party.

To the Communist Party of China, which had been fighting a bitter war against Japanese invaders for more than five years, the encouraging victory in the international antifascist war undoubtedly increased their confidence in the war of resistance. However, Chinese Communists had always made a realistic, dialectical, sober, and objective analysis of the situation. Only such an analysis could enable them to win the victory.

In January 1943 Mao Zedong pointed out that it would not be too long before Hitler totally collapsed. When he did, there would be a turn for the better in the situation in China. He encouraged the army and the people to take heart from the situation and to persist in the struggle. But he thought it would take another two years to achieve victory, and we must try every means to tide over these two difficult years. The Communist Party of China made a four-point proposal to the Kuomintang government: (1) to strengthen the military operations, (2) to enhance unity, (3) to introduce political reform, and (4) to develop production. The ultimate aim was to vanquish Japanese imperialism!

In the Shanxi-Hebei-Shandong-Henan area, through the struggle waged since 1942, the situation of great reduction in the base areas had changed a great deal. But because of the enemy's mopping-up operations and its policy of "burn all, kill all, loot all," production in the base areas had dropped drastically and economic conditions were unprecedentedly difficult. To the army and people in the Shanxi-Hebei-Shandong-Henan area, the overall situation was favorable, but the difficulties remained tremendous.

On January 25, 1943, the Taihang Sub-bureau of the Central Committee called a meeting attended by high-ranking cadres at Wencun, Shexian County. All the military and government chiefs from the Taihang, Taiyue, Hebei-Shandong-Henan, and southern Hebei areas, as well as all their subordinate anti-Japanese democratic governments in the base areas, attended the meeting. Secretary of the Taihang Sub-bureau Deng Xiaoping delivered a report in which he reviewed the struggle against the enemy over the last five years and defined the guiding principles for the future struggle.

Deng Xiaoping pointed out that the current struggle called for "all skills" and that in the future the struggle would be more intricate and acute. The army and the people needed to stay keenly aware of their own weakness and the enemy's strength. The principle was to weaken the enemy, preserve our own forces, and accumulate strength for a counteroffensive. People were the mother of everything and the source of all strength in fighting the enemy. The outcome of the struggle would be determined by the people. It was important to follow the correct policy and to expand the anti-Japanese national united front to unite all anti-Japanese people from all strata to fight the enemy. Base areas still needed to be built up, and it was still imperative to use the tactic of advancing when the enemy advanced. As guerrilla warfare was carried out, mobile war had to be waged whenever conditions permitted. Deng concluded by pointing out that the masses had been mobilized to reduce rents and interest rates and to develop production so as to establish a self-sufficient economy.[1]

The Wencun meeting, by defining the basic principles for the future work, suggested that the Shanxi-Hebei-Shandong-Henan area had entered a period of rehabilitation and redevelopment. Besides setting the orientation of future work in the base areas and in the anti-Japanese democratic governments, the meeting also created important conditions politically, ideologically, and organizationally for future victories in the struggle against the enemy.

The traditional CPC approach is to size up a situation, formulate policies, unify the thinking and action, and unite as one to lead the people to win victory in the struggle. The gist of this approach is an effort to seek truth from facts, to formulate correct policies, and to achieve unity both in thinking and action. This approach has its distinct characteristics, as compared with other political parties in China, and they are characteristics that the Kuomintang did not have.

The Communist Party of China is a political organization of ideology and action. It has a greater deterrent power than did the Kuomintang, which seemed to be characterized by internal harmony but was actually at odds, with each member seeking his own gains and sphere of influence. Chiang Kai-shek understood these differences. His utmost regret was that he failed to strangle the Communist Party of China in its cradle!

Making a correct assessment of a situation guarantees the formulation of correct policies, and correct policies, in turn, guarantee victory. Thanks to clear thinking and correct policies and measures, the army and the people in the Shanxi-Hebei-Shandong-Henan area, under the leadership of Liu Bocheng and Deng Xiaoping, overcame one difficulty after another and won victory after victory.

In the first half of 1943 the Japanese troops conducted the spring and summer mopping-up operations against the anti-Japanese base areas. In early May more than 20,000 Japanese and puppet troops launched a mopping-up operation against the Taihang base area. The Japanese army employed the combing formation to press forward in an attempt to encircle and annihilate the general headquarters of the Eighth Route Army and the main forces of the 129th Division at a narrow stretch of land on both banks of the Qingzhang River between Liaoxian and Shexian counties. Since the Eighth Route army had assessed the situation correctly, made a timely movement, adopted the principle of advancing when the enemy advanced, and organized effective counter–mopping-up operations, the Japanese mopping-up operations had ended by late May and over 25,000 enemy troops had been wiped out.

In May the Kuomintang troops, led by Pang Bingxun and Sun Dianying, in northern Henan surrendered to the Japanese army and colluded with the enemy to attack the anti-Japanese base area. The 129th Division launched the southern Weihe Campaign and the southern Linxian Campaign. It successfully used the tactics of raid, frontal attack in force, bold penetration, breaking up, and encirclement and annihilated over 12,000 Japanese and puppet troops in the two campaigns. Meanwhile, it established vast base areas in southern Weihe and northern Henan.

Victory in war should belong to the brave, the wise, and the upholders of justice.

In the second half of the year the Japanese army conducted once again autumn and winter mopping-up operations against base areas behind enemy lines in northern China. On September 21, more than 30,000 Japanese and puppet troops conducted mopping-up operations against the Hebei-Shandong-Henan Base Area. On October 12, more than 15,000 Japanese and puppet troops conducted another mopping-up operation against this area. By mid-November, the mopping-up operations had ended. Our troops had transferred to other places in advance and launched attacks at the right times. They fought more than 300 small and big battles, annihilated over 4,000 enemy troops, and recovered and established some base areas.

On October 1, more than 20,000 Japanese and puppet troops conducted devastating mopping-up operations against the Taiyue Base Area under air cover by employing the new "iron-rolling" method of warfare. The mopping-up operations ended at the end of November. Defending Chinese troops fought more than 700 battles and killed or wounded over 3,500 enemy troops.

While launching counter–mopping-up operations, the Eighth Route Army actively conducted guerrilla warfare with the support of the broad masses of the people, going deep into enemy-occupied areas and dealing powerful blows to the enemy troops.

The struggle against the Kuomintang diehards continued as well. In March 1943 Chiang Kai-shek published his book *The Destiny of China,* in which he preached vigorously "one doctrine [the Three People's Principles], one political party [the Kuomintang], and one leader [Chiang Kai-shek]."

The villain's scheme and motives for whipping up such opinions were obvious to anyone. Chiang Kai-shek indeed had a bit of foresight. He feared that once the anti-Japanese war ended victoriously, he would not be able to deal with the Communist Party. What Chiang Kai-shek hated and feared all his life was the Communist Party.

While whipping up opinion, Chiang Kai-shek issued secret orders to all Kuomintang troops to encircle the anti-Japanese base areas and launch attacks on them in central and eastern China. Meanwhile, his troops were prepared to launch a large-scale attack on the Shaanxi-Gansu-Ningxia Anti-Japanese Base Area in an attempt to start the third anti-Communist onslaught.

To foil the anti-Communist plot of the Kuomintang diehards and avoid a civil war, the Communist Party of China, while countering the attacks by the diehard troops and preparing for large-scale attacks, made strong appeals to the whole country and exposed the war plots of the Kuomintang. Chiang Kai-shek's base act of fighting the compatriots without eliminating foreign aggressors was strongly denounced by the public in China and was even openly opposed by Britain, the United States, and other countries. Under this strong pressure, the third anti-Communist onslaught, which had been expected to spread to northwestern, northern, and central China, fell through in July 1943.

It was true that "the tree may prefer calm, but the wind will not subside"! In spite of the people's will, the struggle between the CPC and the Kuomintang went on before, during, and even after the anti-Japanese war. The Kuomintang understood that they did not have the support of the people in their anti-Communist efforts. So did the Communists.

In 1943 all the anti-Japanese base areas behind enemy lines and the Shanxi-Hebei-Shandong-Henan Base Area, as well as Liu Bocheng and Deng Xiaoping, had an important task: to develop production and the economy and to resist natural disasters and provide disaster relief.

In 1943 there were many natural disasters. The most serious drought in fifty years occurred between 1942 and 1943 and lasted from the autumn of 1942, through the winter, and then into the spring and summer of the next year. The harvest that year was only 20 to 30 percent of the previous year's yield in some parts of the Shanxi-Hebei-Shandong-Henan area. In a few villages, not even a grain was gathered. There were over one and a half million disaster victims.

In the summer and autumn of 1943 there was an unprecedentedly serious plague of locusts that lasted until 1944 and affected most of the border area, reducing the harvest in the base areas to almost nil. In August and September of the same year, a rainstorm struck the area and washed away a large expanse of arable land on the banks of the Zuozhang and Qingzhang rivers, and the dykes of the Weihe, Yunhe, and Fuhe rivers in southern Hebei and the Hebei-Shandong-Henan area were breached at many points. Quite a number of counties and districts, covering 3,000 to 4,000 villages, were inundated.

> The enemy took advantage of the natural disasters to spread rumors. Some cadres and people in the disaster-stricken areas appeared to be pessimistic and disappointed. Prices fluctuated. Food prices skyrocketed, while the price of clothing and furniture plummeted. Society was in a turmoil, and people were uneasy. In the face of this situation, the leading organs of the Party, the government, and the army—such

> as the Northern Bureau of the Central Committee, the Taihang Sub-bureau, the front headquarters of the Eighth Route Army, the headquarters of the 129th Division, and the Border Region Government—all issued directives on providing disaster relief.[2]

The situation was indeed very difficult. The base areas were already short of food and clothing, but disaster victims kept flowing into them. The cadres in the base areas had a daily ration of one jin of grain but, in response to the directives, had to save two liang for the disaster victims.

To allay their hunger, the officers and men of the Eighth Route Army collected wild vegetables and tree leaves to cook together with some grain, although there was not much grain in the small kitchens of the high-ranking cadres. In the headquarters of the 129th Division, people saw only wild vegetables and porridge.

Political Commissar Deng Xiaoping pointed out that natural disasters in successive years in the Taihang area had reduced harvests there, but the enemy's repeated mopping-up operations had been exceptionally destructive as well. Furthermore, the small population of the Taihang area shouldered a heavy burden. The population of one and a half million could support only 30,000 anti-Japanese troops. But the number of troops and other military personnel in residence far exceeded that figure. The Taihang people had to support not only the troops and local cadres but also the leading organs of the Party, the government, and the army, including the 129th Division, the Taihang Sub-bureau, and the government of the Shanxi-Hebei-Shandong-Henan Border Region. Because other base areas in northern China also had their difficulties and the enemy had imposed blockades on them, it was hard for these bases to support the Taihang area. Deng Xiaoping advocated relying on the personnel of the Party, the government, and the army and the people of the Taihang Mountains to develop production as the way to overcome the difficulties. As for wild vegetables, Deng Xiaoping observed that people in the Taihang area had been eating melons and vegetables in place of grain anyway over the past few years![3]

What Father said was true. At that time, the overwhelming majority of the people in the Taihang area ate wild vegetables to make up for the grain shortage. In some areas where the disasters were the most serious, local people could still collect elm fruits and leaves and cook them with coarse sorghum flour. Later, even the leaves of locust trees, willows, and poplars became very precious.

Father said:

> Last winter and this spring, one-fifth of the Taihang Base Area suffered drought, and there were a large number of drought victims flowing in from the enemy-occupied areas. This is the most difficult moment in recent years. . . . For disaster relief, we mainly relied on production, apart from some mutual assistance. . . . First of all, we must recognize that the development of production is the base for economic construction and that the development of agriculture and handicraft industry is the focus of production.[4]

In his report to Mao Zedong and Peng Dehuai, Father said: "The resources in the Taihang area are almost depleted. It is imperative to pay attention to developing production in the future and to saving. We must encourage the people to store a quarter of the harvests, and the army and the government should also attach importance to storing grain and materials."[5] It was imperative to practice economy!

As soon as Liu and Deng issued the order, the whole base area went into action. The millet supply to the army was cut from one and a half jin to one jin per capita for the main forces, and from one jin to thirteen liang for the personnel of the leading organs. Monthly allowances of between one and a half and five dollars were issued to all personnel, ranging from soldiers to division-level cadres. Administrative expenses and expenses for nonstaple foods were all stopped and had to be raised by all units from production. As grain was in inadequate supply, wild vegetables were to be collected to make up for the shortage. Liu and Deng themselves set an example for the whole army by economizing on food.

All the soldiers and cadres were in a state of semistarvation. But the army was well disciplined, and it did not encroach on the interests of the local people to the slightest degree. In the autumn of 1943 ripe persimmons hung from the branches all over the Taihang Mountains. They were bright red and very alluring. But not a single soldier of the Eighth Route Army plucked a single persimmon.

When winter came, the army overcame a lot of difficulties and acquired some handwoven cloth and cotton. As there was no time for mass-production of winter clothes, the cloth and cotton were issued to all units and all personnel were encouraged to make their own cotton-padded clothes. They used plant ash and tree roots to dye the cloth and asked the local people to help cut out the clothes. The armymen learned to sew with their big hands, which were more accustomed to holding guns.

Like all the others, Liu and Deng also wore the gray cotton-padded clothes made of handwoven cloth. Once the Supply Division made a suit of fine gray cotton-padded clothes for each of them, but Liu and Deng insisted that they be returned, criticizing the personnel and saying that presenting them with fine clothes did not serve them well but only divorced them from the masses.

Liu and Deng had the same food and clothes as everyone else in the army. So did all the other cadres. All shared weal and woe. No wonder the 129th Division and the army and people of the Shanxi-Hebei-Shandong-Henan Base Area were united as one.

It only remained to develop production. Deng Xiaoping presided over a special meeting of the Taihang Sub-bureau of the Central Committee to discuss the question of economic construction in the Taihang area. He delivered a report at the meeting on production in the base areas, entitled "Make Great Efforts to Develop Production and Withstand the Difficulties to Greet the Victory." He said: "We must strengthen leadership over production, and in the future we must make production the central task in all aspects of work in the base areas."

In its disaster relief efforts, the government of the border region certainly focused on distributing relief grain and funds to victims, but it also stressed self-help through production. The government helped peasants to work out their self-help plans through production and to overcome their superstitious ideas and pessimism, encouraging them to overcome difficulties. Where conditions permitted, the government also helped peasants organize producers' cooperatives.

Thus, there was a great upsurge in relief-related production throughout the Taihang, Taiyue, and Hebei-Shandong-Henan areas, in southern Hebei, and in the Shanxi-Hebei-Shandong-Henan area. Father wrote an article entitled "Economic Construction in the Taihang Mountains" that was published in *Liberation Daily* in Yan'an. The article gave a vivid and detailed account of the course of relief through production. He wrote:

> Agricultural production runs through the whole year and is also very seasonal work. Strictly speaking, there is no slack season in farming. Ploughing, seed selection, seed sowing, seedling selection, weeding, summer harvest, autumn harvest, as well as timely accumulation and application of manure . . . we have done a great deal in the spring ploughing, autumn sowing, summer harvesting, and autumn harvesting. We have aroused the people's enthusiasm for production, opposed laziness, organized and exchanged labor, improved seeds, met the needs for livestock and farm implements, mobilized children to collect manure, called on women to participate in production, mediated the relations between landlords and hired labor—and between the employers and the employees—and mobilized the people to plant trees, dig ditches and wells, and make windmills. . . . It is exactly because we have paid attention to organizing and leading the people in production that the people have overcome many difficulties. The slogan of "increasing production, improving life, and preparing for counteroffensive" has resounded in every corner of the Taihang Mountains. We have won victory after victory on the front of production.[6]

Liu and Deng directed agricultural production and also took the lead in participating in the work themselves. In early August there was a rain after a long dry spell. Political Commissar Deng directed all institutions and schools in the border region to mobilize all personnel to help the peasants to replant. He went personally to organize and direct the staff of the leading organs to help peasants rush in the crops.

In the office of Political Commissar Deng, there was a handmade spinning wheel. He learned how to spin thread. His wife, Zhuo Lin, and other women comrades also helped with the fieldwork, and after returning home, they spun thread and made clothes for the soldiers. Mother learned the good skill of knitting in the Taihang Mountains. After liberation, she knitted all our family's woolen sweaters and trousers.

Led by Liu and Deng, the army units in the Taihang area alone planted 100,000 *mu* (607,000 acres) of land in 1943, including over 80,000 mu of reclaimed wasteland, producing a total income of over fifteen million dollars. The cigarettes, cloth, and towels they produced exceeded their demand.

While developing production, the Southern Hebei Bank issued southern Hebei banknotes in October to stabilize the currency and prices. The notes began to be circulated throughout the Shanxi-Hebei-Shandong-Henan area. The monetary issue made up for the revenue deficit and effectively supported production.

That year, with the help of Deng Xiaoping, the Consultative Council of the Shanxi-Hebei-Shandong-Henan Border Region adopted and promulgated "unified progressive taxation." This tax system took into consideration the interests of all strata. It laid the foundation for the local revenue and also aroused the enthusiasm of people of all strata for production.

In August of that year a serious plague of locusts struck the Taihang area. It was a devastating plague. The mountains and gullies were covered with locusts. When they migrated, the sky turned dark. The locusts seriously affected 46 percent of the Taihang area, an area of 1,500 square kilometers, and more than 600,000 mu of crops. Not a grain was gathered in from 270,000 mu.

What was to be done? Some people proposed killing the locusts by mixing pesticide with sugar. But sugar was a rare and expensive luxury. Political Commissar Deng said, "Kill them by hand." Thus, all the armymen and people in the area began to kill locusts by hand and with whatever instruments they could find. Leading comrades at all levels participated. Locusts were not only killed but eaten. Some people were suggesting that eating locusts could satisfy hunger and furthermore, that locusts contained high amounts of protein. While developing production and combating locusts, during the counter–mopping-up campaign the Eighth Route Army strove to protect the summer and autumn harvests so as to ensure the successful fulfillment of the task of collecting grain. Father wrote with emotion:

> It cannot but be a miracle for the Eighth Route Army to have overcome all kinds of difficulties and to have fought closely with the powerful enemy, considering its poor weapons and the fact that it has not received any weapons or money from elsewhere for four years. What is the secret? As everyone knows, we have the guidance of the strategy and tactics devised by Mao Zedong. It is by following these principles that the Eighth Route Army established, safeguarded, and consolidated the various anti-Japanese base areas through numerous battles, pinned down half of the total forces of the Japanese invading army, and lightened the burdens of frontal operations in the great rear area. As everyone knows, in carrying out the fierce struggle against the enemy in the political, cultural, and counterespionage fields, we have given full scope to the enthusiasm of the people in the base areas and enemy-occupied areas, thus boosting the national dignity and confidence of the people. In addition, as everyone knows, we have also waged struggles on the economic front under the

> extremely difficult conditions behind enemy lines and won big victories. It is these victories on the economic front that enabled us to persist in the resistance to Japan for six years and to continue doing so.[7]

What unrivaled proud words! People who did not endure the difficult life of that time can hardly understand the meaning of this passage.

Autumn came. The autumn scenery could be seen everywhere in the Taihang Mountains.

On October 6, the Central Committee decided to combine the Northern Bureau with the Taihang Sub-bureau, and the general headquarters of the Eighth Route Army with the 129th Division. The Party committees of the Taihang, Taiyue, southern Hebei, and Hebei-Shandong-Henan base areas in the Shanxi-Hebei-Shandong-Henan area were placed under the direct leadership of the Northern Bureau, and the 129th Division and the four military areas of Taihang, Taiyue, southern Hebei, and Hebei-Shandong-Henan were put under the direct command of the general headquarters of the Eighth Route Army. Deng Xiaoping succeeded Peng Dehuai as the acting secretary of the Northern Bureau of the Central Committee.

Between August and September, Peng Dehuai and Luo Ruiqing left the Taihang area for Yan'an to study there. On October 9, Division Commander Liu Bocheng also left for Yan'an to study and to attend the Seventh National Congress of the Communist Party of China. Such high-ranking cadres as Cai Shufan, Chen Geng, Bo Yibo, Chen Zaidao, Chen Xilian, and Yang Dezhi also left for Yan'an.

Mother took back my eldest sister from the peasant family she had been living with and asked Chen Shulian, the wife of Cai Shufan, to take her to Yan'an. Life was too hard at the front, and it was impossible to raise a child there. After my eldest sister arrived in Yan'an, she was sent to a nursery, and Chen Shulian cared for her for a very long time. Later, Cai Shufan and Chen Shulian became my elder sister's foster father and mother.

In the Taihang Mountains, Father began to take overall charge of the military and administrative work of the Northern Bureau and the Shanxi-Hebei-Shandong-Henan area. It was a heavy job. Putting such a weight upon his shoulders showed the Central Committee's trust in him and in his political integrity and capability.

For Father himself, it was indeed a heavy responsibility to lead the army and people of the whole area, to conduct operations, to build the base areas, to carry out Party and army building, and to renew the situation. But by then Father was just a little over thirty-nine and had become a leading cadre of considerable experience in the political and military fields. He had the ability and the confidence to perform his new tasks well. In the two years from October 1943 to June 1945, when he went to Yan'an to attend the Seventh National Congress of the Communist Party of China, Deng Xiaoping splendidly fulfilled the tasks assigned by the Central Committee and the Central Military Commission.

On December 6, 1943, the Northern Bureau called a meeting attended by its members and presided over by the acting secretary, Deng Xiaoping, to discuss and decide on the work for 1944. The meeting pointed out that the guiding principle for the work in the whole of northern China in 1944 was to unite the people across the region to overcome all difficulties, persist in resisting Japan in northern China, persist in building and consolidating the anti-Japanese base areas, and accumulate the strength to launch the counteroffensive and greet the victory!

Confidence in China's ability to win the war of resistance to Japan was growing. The dawn of victory was already in sight!

CHAPTER 53

VICTORY IN THE SACRED ANTI-JAPANESE WAR

The dawn of victory began to glimmer on the horizon.

The rapid changes in the world war situation brought the hope of victory to the Chinese people. In 1944 the antifascist war entered a period of large-scale counteroffensive. On the European front, the Soviet army continued to deal the German fascists crushing blows and completely gained the initiative in the war. On the Pacific front, the U.S. army approached the Mariana Islands and the Philippines and began bombing Japan proper. On the South Asian front, the British army and the Chinese Expeditionary Army to India and Burma launched a major counteroffensive against the Japanese troops on the battlefields of India and Burma. On the Chinese front, the Chinese army and people, who had persisted in fighting the war of resistance for nearly seven years, turned their defensive posture around and took the initiative in starting offensive operations behind enemy lines.

A change began to occur in the war situation that was favorable to the Chinese anti-Japanese army and people. However, the arrogant, fascist aggressors had long been blinded by lust for gain and would never lay down their arms of their own accord except at the end of a rope. To reverse the extremely unfavorable war situation, the invading Japanese troops issued, on January 24, 1944, the operational order to open up the communication lines on the Chinese mainland for through traffic.

In the increasingly favorable war situation, the Kuomintang regime's priority was to preserve its strength and prepare for civil war. When the Japanese army launched a large-scale surprise frontal attack, the KMT was not at all prepared. In the face of massive attacks mounted by Japanese troops, large numbers of KMT troops rapidly retreated in defeat and panic.

By October the Japanese invaders had not only opened up the north-south and southwestern communication lines but also occupied Zhengzhou, Xuchang, and Luoyang as well as other major cities in southern China, including Changsha, Hengyang, Guilin, Liuzhou, Nanning, and Longzhou, thereby attaining their goal of joining forces with those Japanese troops that had advanced north from Vietnam.

The Kuomintang army suffered one defeat after another on the frontal battlefield and in only a few months lost thousands of square kilometers of territory and several hundred thousand troops. The Japanese army was indeed powerful, but quite unreasonably, the Kuomintang troops could not withstand even a single blow, making people disheartened.

Recalling that their country had been subjected to bullying, humiliation, and aggression by foreign powers and carved up by them for a century, and seeing that the Kuomintang troops suffered defeat without fighting, all Chinese became grieved and indignant. Some people said that China's anti-Japanese war was a war fought by a weak but vast country against a powerful but small country. Why had China, a once powerful country with an ancient civilization, become a "weak but vast" country?

It was unfortunate that the enemy was so powerful, but it was more unfortunate that China was so weak. Fortunately, Chinese history has not been about the Kuomintang regime alone. Otherwise, how could China have gained hope and a future?

In sharp contrast with the Kuomintang, in 1944 the heroic Eighth Route Army, in the rear of the vast enemy-occupied areas in northern China, started to reverse the difficult situation and took the initiative in launching flexible and energetic offensive operations against the Japanese aggressors.

In preparing to write this book, I drew a detailed chronological table of the major events in the war. I marked with blue all the places where the invading Japanese troops conducted mopping-up operations and attacks, and I marked with red all the places where our army launched sabotage operations and attacks. For the first two years of the anti-Japanese war, every page of the table was full of blue marks and showed a dull color. Beginning in 1944, there were more and more red marks, until the whole page was red in the end.

In April 1944 Mao Zedong said with pride that the Chinese people's present task was to prepare themselves for a still greater responsibility. They had to prepare to drive the Japanese invaders out of China, whatever the circumstances.[1]

The Shanxi-Hebei-Shandong-Henan area established the following principles: to conduct offensive operations, to launch local counteroffensives against the enemy, to reduce the enemy-occupied areas, to expand base areas, to adhere to the principle of advancing when the enemy advances, to wage resolute struggles against the enemy's mopping-up operations, and to protect production, grain, and base areas. In 1944 the army and the people in the Shanxi-Hebei-Shandong-Henan area launched spring and summer offensives, then autumn and winter offensives, and began local counteroffensives. The spring and summer offensives were aimed at recovering the strongholds of Japanese and puppet troops within the base areas, foiling the enemy's attempts to cut off and blockade the base areas, reducing the enemy-occupied areas, and expanding the base areas.

The army and people in the Taihang, Taiyue, Hebei-Shandong-Henan, and southern Hebei base areas courageously began local counteroffensives against the Japanese invading troops. In the spring offensives, the operations in the Taihang, Hebei-Shandong-Henan, southern Hebei, and Taiyue military areas won one victory after another, and many county towns and strongholds were recovered. At the same time, the troops of these military areas smashed several Japanese mopping-up operations and repulsed the attacks launched by the diehard troops under Yan Xishan in the Fushan area in the Taiyue Military Area.

When summer came, to consolidate its rear area in northern China, the Japanese army repeatedly harassed the anti-Japanese base areas behind enemy lines in an attempt to plunder grain and other provisions and halt the Chinese army's offensives. Beginning in May, the army and people in the Shanxi-Hebei-Shandong-Henan area conducted summer offensive operations and waged extensive struggles to protect the summer harvests.

In the Hebei-Shandong-Henan area, the Chinese army continued to fight battles, thereby creating a favorable military situation in the northeastern part of the area and in southwestern Shandong and expanding the base areas. In the Taihang Area, the Communist-led troops stepped up their efforts to pin down and attack the enemy in the strongholds in the towns. As a result, they established base areas in Xinxiang and Huixian, Henan Province, destroyed most of the enemy's third blockade line west of the Beiping-Hankou Railway, and moved forward more than ten kilometers along this railway.

In the Taiyue Area, the Communist-led troops wiped out over 800 enemy troops, recovered 2,600 square kilometers of land, and controlled a section of the ferry along the Yellow River. Meanwhile, in the Zhongtiaoshan area of the southwestern part of the Taiyue area, six governments were established in these counties, whose population was 70,000.

While fighting fierce battles with the Japanese army, the Communist-led troops in the Shanxi-Hebei-Shandong-Henan area continued to repulse the savage onslaughts launched by the troops under Yan Xishan against the Fushan area in Shanxi Province and by the diehard troops in northern Jiangsu against the eastern area of Weishan Lake.

The Communist-led army's spring and summer offensives made it impossible for the Japanese troops to fight too many battles simultaneously, and they soon found they were short of men. So, between July and August, the Japanese army moved some units back to northern China. From September to December over 50,000 Japanese troops conducted fourteen cruel and ferocious local mopping-up operations against the Shanxi-Hebei-Shandong-Henan area and applied the inhumane policy of "burn all, kill all, loot all."

The army and people in the Shanxi-Hebei-Shandong-Henan area courageously smashed the enemy's mopping-up operations and dealt heavy blows to the enemy. In the meantime, knowing that the Japanese rear was weakly defended, they took the initiative to launch autumn and winter offensives. Along the Zhengding-Taiyuan, Baigui-Jincheng, Datong-Puzhou, and Tianjin-Pukou railway lines, they repeatedly launched attacks, dealt heavy blows to enemy troops, and destroyed the major communication lines.

To consolidate and expand base areas, Deng Xiaoping, who was in charge of the work of the Northern Bureau and of the front headquarters of the Eighth Route Army, made arrangements himself and applied the strategy of advancing when the enemy advances, in accordance with the directives of the Central Committee. He dispatched troops twice to cross the Yellow River to the south and establish the Western Henan Anti-Japanese Base Area, comprising twenty counties with a population of more than three million. The Communist-led army succeeded in bringing the vast area of eastern Henan under control.

Under the direct leadership of the Northern Bureau and the general headquarters of the Eighth Route Army, by the end of 1944 the army and people in the Shanxi-Hebei-Shandong-Henan area had fought bravely and launched offensives to wipe out over 76,000 enemy troops, recapture eleven county towns, liberate a population of over five million, and recover more than 60,000 square kilometers of territory from the aggressors, thus turning the war situation in China's favor. The army and people in the anti-Japanese democratic base areas did not get bottled up or starve to death. Instead, turning defense into offense and passivity into initiative, they advanced toward a greater victory.

When Father became the acting secretary of the Northern Bureau, he assumed heavier responsibilities. He took charge of the work of the Northern Bureau and of the general headquarters of the Eighth Route Army, as well as the overall work of the Party, the government, and the army of the Shanxi-Hebei-Shandong-Henan area. As he was responsible for handling military affairs, he resolutely and effectively carried out the strategic plan made by the Central Committee and Chairman Mao Zedong. He not only made strict demands on the leading cadres in all areas and army units under his command but also elicited their initiative and their ability to boldly use it in directing operations. All the cadres under him said that Political Commissar Deng was very strict and serious, but they were able to display their talents working under him. Father attached great importance to military operations. Before conducting operations, he often took personal charge of the arrangements, talking with the chief military and political commanders and making plans in a very detailed, careful, and comprehensive way.

As he took charge of the work of the Northern Bureau, he often called meetings to study the situation and to discuss, propagate, and carry out the principles and policies formulated by the Central Committee. He paid particular attention to analysis. In many of his speeches, he first sized up the international fight against fascism, then the anti-Japanese war in China, and finally, the position the Chinese army and people were in, their advantages and difficulties. Acting on the directives of the Central Committee, he then assigned the tasks to be accomplished in their area. As to his style of work, he resolutely carried out the directives of the Central Committee, to the letter. But when he undertook the tasks, he knew what to do and bore in mind the overall situation and the strategy and tactics. He never fought a battle he was unprepared to fight, nor did he ever do something he had not planned.

As he took charge of building base areas, he paid great attention to establishing political power and doing economic work. No foreign assistance was available for the base area–building effort, which was fraught with many difficulties. Since the war was ruthless, one natural disaster followed another, and the people led a hard life, the Communist-led troops had to provide war supplies while fighting battles, use their own hands to develop production, and combine farming with fighting. Only by self-reliance and self-sufficiency was it possible for the people's army to overcome its economic difficulties, lighten the burden on the people in the base areas, and ensure its enduring combat effectiveness to achieve the final victory.

As he took charge of political work, Deng Xiaoping devoted great efforts to Party consolidation and the rectification movement, in accordance with the directives of the Central Committee. The rectification movement was a political education movement within the Party launched by the Central Committee in 1941. As early as 1935, the CPC had ended the rule of dogmatism and the "left"-adventurism in the military field at the Zunyi Meeting. After six years, Party organizations had expanded considerably, the armed forces had swelled enormously, and the revolutionary cause had greatly developed. However, as the Party was much involved in military affairs, there were many unhealthy tendencies and styles of work within the Party ideologically, politically, and organizationally. The rectification movement conducted in the early 1940s was designed to fight subjectivism, sectarianism, and stereotyped Party writing and to educate the entire Party membership in Marxism. The movement was of great significance because it helped the Party achieve unity of thinking and improve its quality. It was precisely because of this movement that the Communist Party of China was able to adapt itself to the situation, wage the struggle, and achieve a brilliant victory. Only when people had gained a clear understanding of the situation could they know what to do and display greater vitality and combat effectiveness. This movement helped the army and people in the whole area improve politically, ideologically, and in their work style, thus laying a good foundation for resisting the powerful enemy, overcoming difficulties, and achieving victory.

When Peng Dehuai, Liu Bocheng, and large numbers of leading Party, government, and army cadres went to study in Yan'an, Father, along with other comrades-in-arms, led the Northern Bureau, the general headquarters of the Eighth Route Army, the 129th Division, and the Shanxi-Hebei-Shandong-Henan area to fulfill successfully the political, military, and production tasks assigned by the Central Committee. In the face of difficulties and hardships, and in the hope of victory, he led the army and people of the area to fight doggedly.

I once asked Father, "You were then alone at the front. It was not easy to lead, was it?" Father replied with a smile, "I didn't do many things, but only one thing, that is, to endure hardships."

During the conversation, I was accompanying Father at Beidaihe, sitting in the courtyard surrounded by many trees, with a sea breeze and fragrant flowers. The sea wind was blowing, and great waves were surging. Father, who was almost eighty-eight years old, was sitting silently in a cane chair. After giving that simple reply, he fixed his eyes on the green trees, lost in thought. I sat beside him without a word. I did not want to disturb him. Perhaps he was thinking of those years of hardships and of the Taihang Mountains.

It was already midwinter in the Taihang Mountains. The cold wind was piercing, and dripping water froze. Even in severe winter, however, the message of spring could be heard in the Taihang Mountains. The layers of ice and snow covering the mountains began thawing in the hearts of the people.

The Taihang Mountains finally saw the year 1945. It was a year in which victory was won in the antifascist world war and in which the Chinese people, who had fought doggedly for nearly eight years, achieved the final victory in the anti-Japanese war.

After the protracted war, the overall strategic situation was more unfavorable to the Japanese troops than before. One defeat after another had lowered their morale, and their combat effectiveness had deteriorated considerably.

After launching local counteroffensives and conducting Party consolidation, the rectification movement, and the great production campaign, the Communist-led Eighth Route Army had grown politically, militarily, and economically. It had base areas with a population of over ninety million, over two million militia men, and a regular army 780,000 strong. The primary tasks of the army and people in the liberated areas were to wipe out Japanese and puppet troops, expand the liberated areas, and reduce enemy-occupied areas.

The Northern Bureau of the Central Committee, the general headquarters of the Eighth Route Army, and the headquarters of the 129th Division in the Taihang Mountains issued orders to the army and people in the Shanxi-Hebei-Shandong-Henan area to concentrate a superior force and launch attacks on the weakly defended areas of the enemy forces.

In January 1945, the Communist-led army launched spring offensives. Actually, spring was yet to come, but spring offensives had begun anyway. The people were more enthusiastic about victory than about the advent of spring. The army in the Hebei-Shandong-Henan area launched a campaign in Daming and captured that ancient city. The army in the Taihang area launched a campaign in Daoqing, wiping out over 2,500 enemy troops and recovering more than 2,000 square kilometers of territory with a population of 750,000. The army in the Taiyue area launched a campaign in northern Henan, capturing over forty enemy strongholds and wiping out over 2,800 enemy troops. After one victory, the army in the Hebei-Shandong-Henan area launched a campaign in Nanle, capturing the county town and thirty-two surrounding enemy strongholds and wiping out over 3,400 enemy troops. The Chinese army's many victories forced the Japanese troops to keep drawing back.

When spring was over, summer came. It was hot, and people's hearts were more passionate. The good news of victory, along with high combat fervor, spread over the anti-Japanese battlefronts.

In the wake of the spring offensives, the Eighth Route Army launched summer offensives swiftly and violently all over the vast land of northern China. As with the spring offensives, news of victory kept pouring in from the summer battlefields.

In the spring and summer offensives, the army in the Shanxi-Hebei-Shandong-Henan area fought over 2,300 small or big battles, capturing over 2,800 enemy strongholds, recovering twenty-eight county towns, and wiping out over 37,800 enemy troops. The Taihang and Taiyue areas were joined together, and links were established between the mountain areas and the plains. The base areas expanded considerably in human and material resources. The anti-Japanese army and people were more confident of victory, and the situation in the anti-Japanese war was excellent!

In Europe, German fascists surrendered unconditionally on May 8, 1945. In fighting battles with the Chinese, American, and British forces, Japanese troops put up desperate fights and soon met with failure.

The Seventh National Congress of the CPC was held at Yan'an, northern Shaanxi, from April 23 to June 11, 1945. The congress reviewed history, summed up experience, and set forth tasks and policies in all fields of endeavor.

The congress pointed out that China had to choose between a destiny of light and a destiny of darkness, between an independent, free, democratic, unified, prosperous, and powerful new China and a semicolonial, semifeudal, divided, poor, and weak old China. The congress further pointed out that the Party's task was to strive with all its might for a bright future and against a dark future. The Party's political line was to boldly mobilize the masses and expand the people's forces so that, under Party leadership, they would defeat the Japanese aggressors, liberate the whole people, and build a new, democratic China.

The congress was held in an atmosphere of democracy and unity and was full of vitality, making the participants more confident. Without external interference, a new Central Committee, a new leading nucleus, and the chairman of the Central Committee—Mao Zedong—were elected at the congress. After twenty-four years, the Communist Party had independently established its first-generation leading nucleus.

Father did not attend the Seventh National Congress, for he had been instructed to direct operations at the front. In February 1945, in accordance with the directive of the Central Committee, he led some members of the Northern Bureau, set off from Matian, Liaoxian County, crossed the enemy's blockade line along the Beiping-Hankou Railway, and arrived in the Hebei-Shandong-Henan area in March to make an inspection tour and conduct investigations and studies there. In mid-June, Father was notified by the Central Committee to go to Yan'an to attend the First Plenary Session of the Seventh Central Committee. On June 29, Father left the Hebei-Shandong-Henan area and hurried to Yan'an to attend the session.[2] He was one of the forty-four members of the Central Committee elected at the congress. Father's election to the Central Committee was another important point in his revolutionary career.

In the summer of 1945, hot summer days had just begun, and the sun was scorching. But the developing situation was more intense than the scorching sun!

On July 26, the governments of China, the United States, and Britain published the Potsdam Declaration, demanding that the Japanese government surrender unconditionally at once. On August 8, the Soviet Union declared war on Japan. On August 9, one million Soviet troops launched offensives against the invading Japanese troops in northeastern China from three directions.

On August 6 and 9, the United States dropped two atom bombs on Japanese cities, Hiroshima and Nagasaki, respectively. On August 9, Mao Zedong issued a statement, "The Last Round with the Japanese Invaders," calling on all the Chinese people's anti-Japanese forces to launch nationwide counteroffensives. On August 10, Commander-in-Chief of the Eighth Route Army Zhu De issued an order that all-out counteroffensives begin. On the same day, Liu Bocheng and Deng Xiaoping sent a telegram from Yan'an to all military areas of the Shanxi-Hebei-Shandong-Henan area to arrange for military operations. The army units of all military areas immediately went into action.

On August 15, Japan announced its surrender. On September 2, the Japanese government signed the instrument of surrender. World War II had come to a triumphant conclusion.

The anti-Japanese war that the Chinese people had fought doggedly for eight years had come to a victorious end.

The history of imperialist brutal aggression that had subjected China to humiliation had begun to come to an end.

The eight-year anti-Japanese war was eventually crowned with victory. The Chinese people were incredibly jubilant. The hearts of the Chinese armymen and people were full of triumphant enthusiasm.

Mother told me that when people in Yan'an heard the news of victory, they were overwhelmed with excitement. They were laughing and dancing, beating drums and gongs. They hung and fired off firecrackers. Those who did not have drums or gongs or firecrackers tore cloth from their clothes or pulled out cotton wadding from quilts and burned them. People all over the country were jubilant.

The eight-year anti-Japanese war, in which the Chinese nation had fought bloody battles with the aggressors, was engraved on the memory of those who had experienced it. During the eight years of the war of resistance, the Chinese army and people suffered greatly, with casualties of over eighteen million civilians and nearly four million armymen. But they had achieved a brilliant victory.

In particular, the people's army, led by the Communist Party of China, fought over 125,000 battles on the battlefields behind enemy lines, wiping out over 527,000 Japanese troops and almost two million puppet troops.

One history book sums up the anti-Japanese war objectively and correctly:

> The anti-Japanese war fought by the Chinese people was a wonder in the history of human war and a great undertaking for the Chinese nation. The anti-Japanese war was a national liberation war in which the Chinese people won complete victory for the first time in a century-long struggle against foreign invaders and of vital significance in the history of the Chinese revolution.[3]

The eight-year anti-Japanese war finally came to an end. However, the struggle between justice and injustice and between light and darkness was far from over.

CHAPTER 54

GIVING TIT FOR TAT AND FIGHTING FOR EVERY INCH OF LAND

Although the anti-Japanese war had come to an end, fighting was far from over. Because the KMT and Chiang Kai-shek were resolved to wipe out their sworn enemy—the Communist Party of China—and its armed forces, so as to maintain their feudal and decadent rule, war would continue in the vast land of China.

Since the Communist Party was founded, the ultimate goal of its members had been to overthrow the reactionary rule of the KMT, to seize political power, and to build a rich, strong, prosperous, independent, and democratic new China. The outbreak of civil war was inevitable because of the objective reality.

After the anti-Japanese war was concluded, the Communist-led army had 1.2 million men and 2.2 million militiamen. The liberated areas were scattered over nineteen provinces and regions and covered an area of one million square kilometers, with a population of 100 million.[1] The KMT army had 4.4 million men and controlled most of China's territories and population, but its main forces were stationed in the rear areas of southwestern and northwestern China and far away from eastern and northern China.

The Chinese people, who had long suffered from war, opposed civil war and autocratic rule and wanted peace and democracy. Operating on their own interests, the United States, Britain, and the Soviet Union were opposed to civil war in China as well.

Because Chiang Kai-shek was not fully prepared to launch a civil war immediately and he was feeling the pressure of public opinion, he made peace overtures by sending telegrams to Mao Zedong inviting him to Chongqing to "discuss jointly" "important domestic and international questions." At the same time, Chiang deployed his troops to seize strategic points and took advantage of the KMT's economic takeover after Japan's surrender of properties to take away people's properties by force or trickery. To seize large cities and important communication lines, Chiang Kai-shek used the aircraft and warships provided by the U.S. government to rapidly transport his troops to the north. Meanwhile, he went so far as to plot and collude with the Japanese invaders and even engaged the Japanese war criminals guilty of the most heinous crimes as his advisers so as to prevent the Communist-led Eighth Route Army and the New Fourth Army from accepting the surrender of Japanese troops.

Chiang Kai-shek assumed that no matter how courageous Mao Zedong was, he did not dare to come to Chongqing for peace negotiations. Contrary to Chiang's expectations, Mao Zedong, who had unusual courage, resourcefulness, and boldness of vision, went so far as to fly from Yan'an to Chongqing with Zhou Enlai, Wang Ruofei, and others to negotiate peace and to sit face to face with Chiang Kai-shek.

Of course, the Communist Party and Mao Zedong were not naive. They had a sober estimate of the danger of a civil war and repeatedly exhorted the Party and the whole army to base all their work on the possibility that the KMT would launch a civil war and to resolutely defend themselves against the attack mounted by the KMT on the liberated areas.

First, the Central Committee ordered the army's main forces to rapidly organize regular armies so as to facilitate concentrated action and unified command. Then the senior commanders from various liberated areas who were attending the meeting called by the Central Committee in Yan'an quickly returned to the battlefronts to prepare for operations. On August 25, a U.S. plane took off from Yan'an Airport. Just before takeoff, those who saw the party off discovered that everyone aboard carried a parachute and did not know why the plane had no hatch. Those aboard the plane were more than twenty of the supreme commanders of the Communist-led liberated areas at the front, among them, Liu Bocheng, Chen Yi, Lin Biao, Deng Xiaoping, Bo Yibo, Chen Geng, Chen Xilian, Chen Zaidao, Zhang Jichun, Teng Daiyuan, Yang Dezhi, Xiao Jingguang, Deng Hua, Deng Keming, Song Shilun, and Li Tianyou.

The pilot was an American and could not speak any Chinese, whereas those aboard were all Chinese and could not speak any English. So Yang Shangkun had Huang Hua, an interpreter who could speak English, aboard the plane to be responsible for staying in contact with the pilot of the U.S. plane. This DC-9 U.S. military transport airplane rumbled and rocked as it took off. In hindsight, we can see that it was very dangerous for all the commanders to be traveling together in such a plane. But they would not have taken such a great risk if the situation had not been so urgent.

Later that day, the plane landed at Changning Airport, Licheng County, in the Taihang Mountains. Chief-of-Staff Li Da sent a cavalry platoon to the airport to receive them. Liu Bocheng and Deng Xiaoping did not dare to stay there for long but immediately hurried to Chi'an, where the headquarters of the military area was located. After returning to the Taihang Mountains, Liu Bocheng and Deng Xiaoping were busy right away with the military and administrative work.

In accordance with the decision of the Central Committee, the Shanxi-Hebei-Shandong-Henan Bureau of the Central Committee was established to unify the leadership of the Taihang, Taiyue, Hebei-Shandong-Henan, and southern Hebei liberated areas, and Deng Xiaoping served as secretary of the bureau. In the meantime, the 129th Division was redesignated the Shanxi-Hebei-Shandong-Henan Military Area—covering the Hebei-Shandong-Henan, southern Hebei, Taihang, and Taiyue military areas—with five columns under its command, Liu Bocheng serving as the commander, and Deng Xiaoping as the political commissar. The whole military area had over 80,000 field troops and over 230,000 local troops.

The anti-Japanese war was concluded, but the political situation in China was still complex. While putting up the smoke screen of peace negotiations, the KMT stepped up its efforts to prepare for war and seize territories. Persons of good sense knew immediately that the KMT was only paying lip service to the idea of peace negotiations while in fact it prepared for civil war.

On August 28, Mao Zedong led the CPC delegation to Chongqing to negotiate with the KMT and Chiang Kai-shek. Mao Zedong's courageous and resolute action shook people both at home and abroad. He went to Chongqing, with no regard for his personal safety, to demonstrate to the world that the CPC was sincere in seeking peace. This magnificent feat put Chiang Kai-shek, who was energetically preparing for war, in a defensive position.

As expected, it was hard to conduct negotiations. Also as expected, the negotiations were constantly retarded or obstructed. But people of goodwill did not expect that, while negotiating at the peace table, the KMT would be starting military actions and rapidly assembling its troops to advance on and invade the liberated areas.

In mid-September, while negotiations were under way in Chongqing, KMT assembled thirty-six corps with seventy-three divisions to invade the liberated areas in an attempt to control strategic points in northern and eastern China as quickly as possible, to open the passage to northeastern China, and to compel the CPC to yield during the peace negotiations by exerting strong military pressure.

While striving for peace, Chinese Communists were not so naive as to stop their own military preparations. Mao Zedong had already predicted that Chiang Kai-shek would try to wrest every ounce of power and every ounce of grain from the people. The CPC policy was to give him tit for tat and to fight for every inch of land.

After the KMT troops began invading the liberated areas, civil war was imminent. The Shanxi-Hebei-Shandong-Henan Liberated Area was the central gate to the strategic areas in northern China. To its west were the Taihang, Taiyue, and Zhongtiao Mountains. To its east was a boundless stretch of the plains in Hebei and Shandong. To its south was the rushing Yellow River. To its north was the winding communication line—the Zhengding-Taiyuan Railway.

Liu Bocheng called this land of the ancient states of Yan and Zhao the land of four battles, and his field army the "army of four battles." The land of the ancient states of Yan and Zhao is a land that has been contested by many strategists over the years. There are many moving and sad ballads about the brave fighters of these ancient states.

This strategic area became the main target of attack by the KMT troops. In southeastern Shanxi is the Shangdang area. It is surrounded by the Taihang, Taiyue, and Zhongtiao Mountains. Its center is Changzhi County, which is a stretch of slightly flat area amid high mountain ridges, and it is another place that has been highly contested since ancient times.

In mid-August, 16,000 KMT troops under the command of Yan Xishan invaded the Shangdang area in the heartland of the Taihang Mountains and seized Xiangyuan, Lucheng, Changzhi, Changzi, Huguan, Tunliu, and other counties.

For a while, the KMT troops fiercely invaded the Communist-led liberated areas from all directions, and the crisis of civil war was unprecedentedly serious. Mao Zedong said "The Kuomintang is negotiating with us, on the one hand, and vigorously attacking the liberated areas on the other hand." He also said, "The Shangdang area, rimmed by the Taihang, Taiyue, and Zhongtiao Mountains, is like a tub. This tub contains fish and meat, and Yan Xishan sent thirteen divisions to grab it. Our policy also was set long ago—to give tit for tat, to fight for every inch of land."[2]

The main task of the Shanxi-Hebei-Shandong-Henan Military Area was to smash the attack mounted by the KMT troops along the Beiping-Hankou and Datong-Puzhou railway lines, but now the enemy troops attacking the Shangdang area posed a serious threat to our troops. If the main KMT forces were not rapidly wiped out as they advanced northward, our troops would be attacked front and rear. Based on this judgment, Liu Bocheng and Deng Xiaoping reported to the Central Committee their decision to launch the Shangdang Campaign to resolutely wipe out the invading troops under the command of Yan Xishan.

This was a difficult decision. During the anti-Japanese war, to fight flexibly, the Communist-led army broke up its main forces into parts, which were gradually concentrated for operations until the later stage of the war. At the time, it was making a gradual shift from scattered guerrilla warfare to concentrated mobile warfare. But the army was understrength, and most of the regiments had less than 1,000 men. It was also poorly equipped. The whole military area had only six mountain guns. Only half of the regiments were equipped with as many as two to four mortars and three to four heavy machine guns. Most of the new recruits still used broadswords and spears to fight the enemy. The army was also very short of ammunition, and only several bullets were available for many rifles.

It was under these conditions that the Communist-led troops were determined to fight the well-equipped core troops under the command of Yan Xishan. That decision required determination, courage, and unusual skill in operational command.

The Central Committee approved the strategy formulated by the Shanxi-Hebei-Shandong-Henan Military Area.

Liu Bocheng and Deng Xiaoping immediately conducted combat deployments in the whole military area, concentrating the main forces of the Taihang, Taiyue, and southern Hebei military areas and some local forces totaling 31,000 men. Famous generals participated in the campaign, including Li Da, Chen Xilian, Chen Geng, Xie Fuzhi, Wang Xinting, Chen Zaidao, and Qin Jiwei.

On September 7, Liu and Deng gave orders to launch the Shangdang Campaign. Mao Zedong had just gone to Chongqing for the peace negotiations, and many people were concerned about his safety. Deng Xiaoping said, "The better we conduct the Shangdang Campaign and the more thoroughly we wipe out the enemy, the safer Chairman Mao will be and the greater initiative Chairman Mao will have at the conference table."[3]

On September 10, the campaign formally started. The Communist-led army captured Tunliu, Lucheng, Huguan, Changzi, and Xiangyuan, isolated Changzhi, and seized large quantities of weapons and ammunition. After winning this victory, the Communist-led army launched a converging attack on Changzhi. Liu Bocheng and Deng Xiaoping issued operational orders to capture Changzhi by fighting bravely and winning a speedy victory and to wipe out the enemy's effectives by using the warfare method of besieging a city to annihilate enemy reinforcements. The enemy reinforcements were pinned down. After several days of fierce fighting, their morale was shaken and they attempted to break out of encirclement and advance northward. The Communist-led army quickly pursued and attacked them, intercepted and penetrated them, so that they were utterly routed and almost totally wiped out. The enemy troops in Changzhi found that it would be impossible to wait for the arrival of reinforcements, abandoned the city, and broke out of the encirclement; they were then pursued, attacked, and rounded up. Despite hunger and fatigue, the Communist-led army marched day and night to pursue and attack the enemy troops and finally wiped out their main forces on the bank of the Qinshui River.

On October 12, the Shangdang Campaign concluded triumphantly. During the campaign, the Communist-led troops wiped out eleven enemy divisions and one advance column totaling over 35,000 men, seized large quantities of mountain guns, machine guns, rifles, and pistols, and captured Shi Zebo, commander of the 19th Corps of the troops under the command of Yan Xishan. At the same time, except in some cities, the Communist-led army eliminated all Japanese and puppet troops along the 250-kilometer Xinxiang-Shijiazhuang Railway.

After the Shangdang Campaign ended, Mao Zedong said, "This time we gave tit for tat, fought, and made a very good job of it. In other words, we wiped out all thirteen divisions. Their attacking forces had 38,000 men, and we employed 31,000 men. Of their 38,000 men, 35,000 were destroyed, 2,000 fled, and 1,000 scattered. Such fighting will continue."[4]

After the victory in the anti-Japanese war, the Shangdang Campaign was the first campaign fought by the Communist-led army against the KMT troops and the first large-scale war of annihilation fought by that army. The campaign ended in its great victory.

This campaign enhanced the Party's position in the peace negotiations in Chongqing, increased the confidence of the army and people in the liberated areas that they could defeat the KMT troops, consolidated the rear area of the Communist-led military area, and accelerated the shift from guerrilla forces to regular troops suitable for large-scale mobile warfare. After the Shangdang Campaign ended, the Shanxi-Hebei-Shandong-Henan Field Army organized the troops of the four military areas under its command into the 1st, 2d, 3d, and 4th columns and established the artillery troops.

After directing the Shangdang Campaign at the battlefront, Father and Liu Bocheng returned to Chi'an. An atmosphere of joy and victory pervaded the small village. Father was jubilant for another reason as well—he had a new daughter.

Chi'an Village was the birthplace of my two elder sisters. My eldest sister is named Deng Lin, and her pet name is Lin Er. As I mentioned before, she was born in the difficult year of 1941. After she was born, she was entrusted to the care of a village family. When she was four years of age, she was sent to Yan'an and lived in the nursery school there.

My elder brother is called Deng Pufang, and his pet name is Pan Pan, perhaps because he was plump when he was born! He was born in 1944 in Matian Village, Liaoxian County, in the Taihang Mountains. After he was born, because Mother had no milk to feed him, he was entrusted to the care of a peasant family living on the other bank of the river in Matian Village.

My second elder sister was born in 1945 after the Shangdang Campaign ended. Her name is Deng Nan, but her pet name from early on was Nan Nan. Mother said my elder brother gave her this name. At the time, he was only one year old and could not speak, but when he saw his younger sister, he repeatedly called "Nan Nan, Nan Nan." After she was born, she was also entrusted to the care of a peasant wet nurse.

After my two sisters and brother were born, they were all entrusted to the care of villagers in the Taihang Mountains. The Taihang Mountains mean a great deal to our family. Mother is more than seventy years old now and generally does not go outdoors. But when the people from the old liberated areas in the Taihang Mountains come, she still goes to see them, despite severe cold weather. While I am writing this chapter, Mother is thinking hard about what she can do for the poor liberated areas in the Taihang Mountains and the people there.

On October 10, 1945, the CPC and the KMT signed the "Summary of Conversations between the Representatives of the Kuomintang and the Communist Party of China," also known as the "October 10th Agreement." The two sides agreed to seek long-term cooperation and avoid civil war.

On October 11, Mao Zedong returned to Yan'an. Immediately on his return, Mao Zedong predicted, "The agreements reached are still only on paper. Words on paper are not equivalent to reality." "Our task is to uphold the agreement, to demand that the Kuomintang honor it, and to continue to strive for peace. If they fight, we will wipe them out completely."[5] Indeed, the agreement had been signed, but whether civil war would break out was not dependent on any individual's will or even on any one party's will.

Just as Mao Zedong predicted, before the ink of the October 10th Agreement was dry, the KMT troops intensified their attacks on the liberated areas. With the assistance of the United States, the KMT used planes and warships to transport its troops to Beiping and Tianjin and, at the same time, increased their troops attacking the liberated areas to 800,000. The primary objective of the KMT was to capture Beiping and Tianjin and then northeastern China.

In mid-October, the KMT troops mounted attacks from several directions. The troops under the command of the KMT generals Hu Zongnan, Fu Zuoyi, and Sun Lianzhong fiercely attacked northern China from four directions, and their attacks seemed to be overwhelming.

Of those KMT troops, the 45,000 troops commanded by Sun Lianzhong mounted an attack northward along the Beiping-Hankou Railway from Xinxiang, Henan Province, under the command of their deputy commanders-in-chief, Ma Fawu and Gao Shuxun, in an attempt to launch a pincer attack on the important city of Handan in Hebei Province. Liu Bocheng said vividly, "Chiang Kai-shek has kicked the ball toward the goal of the liberated area."[6] Handan, the southernmost city of Hebei Province, is located on the Beiping-Hankou Railway and is a strategic point in the North China Plains. The capital of the ancient state of Zhao, the city has a history of more than 3,000 years, a period during which many moving and legendary stories and historical events happened there.

The KMT massive troops had come to capture this famous city in northern China. The Military Commission of the Central Committee gave a directive: it was a pressing task of great strategic importance to block and retard the northward advance of the KMT troops. The outcome of the forthcoming campaign along the Beiping-Hankou Railway had a vital bearing on the overall situation.

The Military Commission gave another directive: drawing on the experience gained in the Shangdang Campaign and mobilizing all forces, Liu Bocheng and Deng Xiaoping were to direct personally and carefully organize all operations and strive to win a victory in the second Shangdang campaign.

After receiving the directives, Liu Bocheng and Deng Xiaoping analyzed the situation. Although the enemy troops outnumbered the Communist-led army and were well equipped, they had very conspicuous weaknesses. First, they had just come from far away, were unfamiliar with the place and the people, were far from their rear, and were not good at field operations. Second, there were many factions within the enemy troops, and many conflicts among them. In particular, the troops of the New 8th Corps and the 40th Corps had miscellaneous allegiances and were at odds with the troops directly under the control of Chiang Kai-shek.

On the Communist side, although the field formations were recently organized and poorly equipped and the troops had not had much rest or time for reorganization after continuous fighting, morale was high following the first Shangdang campaign victory, and they enjoyed enormous support from the people in the base area. Liu Bocheng and Deng Xiaoping made the judgment that their troops were absolutely capable of fighting a larger war of annihilation by making use of their advantages.

On October 6, the headquarters of the military area gave orders to launch the campaign along the Beiping-Hankou Railway. It resolved to concentrate the 1st, 2d, and 3d columns, totaling 60,000 men, and mobilize more than 100,000 militiamen to fight continuously for two months and wipe out the invading enemy troops along the Beiping-Hankou Railway. It carefully conducted the military deployment.

On October 20, Liu Bocheng and Deng Xiaoping moved the headquarters from Chi'an Village, Shexian County, to the Fengfeng Coal Mine, which was at the foot of the Taihang Mountains and very close to Handan, so as to direct the military operations at the front along the Beiping-Hankou Railway.

Like the Shangdang Campaign, the campaign along the Beiping-Hankou Railway, also known as the Handan Campaign, was a large-scale war of annihilation. During this campaign, Liu Bocheng and Deng Xiaoping carefully conducted strategic deployment.

The campaign started in mid-October. One column first blocked the northward advance of the enemy troops from Xinxiang, Henan Province, in the area south of Handan. On October 24, after all three enemy corps had crossed the Zhanghe River, the Communist-led army immediately surrounded them and launched continuous attacks on them. The enemy troops hurriedly shrank back and sent an urgent telegram to Chiang Kai-shek to ask for reinforcements. On October 26, the enemy troops from Shijiazhuang and Anyang came as reinforcements, and the Communist-led army concentrated troops to block them. The military situation was extremely grim, and the fighting was very fierce and intense. The Communist-led army stepped up its efforts to attack the surrounded enemy troops, strike at the enemy reinforcements, and split and demoralize the enemy troops. Li Da, chief-of-staff of the military area, went personally to the headquarters of the enemy's New 8th Corps to work on the corps commander, Gao Shuxun, urging him to serve the best interests of the nation by revolting and coming over to the Communist side. On October 28, the Communist-led army mounted a general attack on the surrounded enemy troops, and under a careful plan, Gao Shuxun led more than 10,000 KMT troops to revolt and come over.

After the New 8th Corps revolted, the strength of the enemy troops was suddenly reduced. Gaps appeared in their deployments, and their morale was shaken. On October 31, the enemy troops began breaking out of encirclement and moving southward. According to plan, the Communist-led army had moved in advance to the two sides of the enemy's retreat route and immediately made an assault from several directions on the enemy troops fleeing southward and rounded them up. By November 2, almost all enemy troops had been wiped out in the area north of the Qingzhang River.

During this campaign, not only did the New 8th Corps revolt and come over, but the Communist-led army killed or wounded over 3,000 enemy troops, captured Ma Fawu, deputy commander-in-chief of the enemy's 11th War Zone as well as commander of the 40th Corps, and the 17,000 enemy troops under his command, and seized large quantities of weapons and matériel.

The two major campaigns conducted in the Shangdang area and along the Beiping-Hankou Railway ended triumphantly. Any mention of these two campaigns continues to bring up all sorts of feelings for Father. He has told us on several occasions that "after the victory was achieved in the anti-Japanese war, our field army never stopped fighting for one day. We could only receive training for a week at most, and it was hard for us to have ten days to spare."

He added,

> In fact, if we use the term *counteroffensive* in its true sense, it was during the campaigns in the Shangdang area and along the Beiping-Hankou Railway that we began fighting the enemy head-on. Since we fought the enemy head-on, we forced Chiang Kai-shek to sign the October 10th Agreement. . . . It was dangerous to fight the two battles. Without enough ammunition, a soldier had only several bullets for his rifle. It was hard to storm heavily fortified positions, and at the critical moment we had to rely on charging and hand-to-hand fighting. Since these two campaigns were battles of annihilation, after we won the victory, we had more weapons and men.

Father did not like to talk much and has never told people what he did in the past. But when speaking about these battles, he had more than usual to say. He always said,

> Who was born to be capable of fighting a battle! One has to learn to fight from fighting and from defeat. When I had just come to the 7th Corps of the Red Army, I knew nothing about military affairs. When I served as the secretary-general of the Central Committee, Chen Yi came to report to the Central Committee on the work of the 4th Corps of the Red Army and it was then that I learned a lot of things. This was also a kind of study. Later, as I fought many battles and suffered several defeats, I came to learn how to fight.

When my good friend Shan Shan, the youngest daughter of Marshal Chen Yi, once visited our home, Father said upon seeing her, "I heard a lot of things from your father and afterwards applied them in the 7th Corps of the Red Army!" Father told the truth. He learned to be a strategist step by step.

On November 20, 1989, the old comrades who compiled *The Military History of the Second Field Army* gathered in the Great Hall of the People. (Because Liu and Deng's field army was finally organized into the Second Field Army, its officers and men usually referred to it later as the Second Field Army.) Liu Bocheng had passed away, so it was important that Deng Xiaoping should attend the gathering. Father came to the Great Hall of the People, met with his old comrades-in-arms, and made a speech. He said,

> During wartime, in every stage the Second Field Army accomplished the tasks assigned by the Military Commission of the Central Committee and did so well. This is my appraisal of the Second Field Army. During the War of Liberation, from the beginning to the end, the Second Field Army waged a tit-for-tat struggle against the enemy and bore the brunt of the enemy's attack. In the Shanxi-Hebei-Shandong-Henan Liberated Area, as Comrade Liu Bocheng put it, that liberated area was a gate to the liberated areas in northern China and the enemy was expected to mount an attack from this gate. As we expected, when Chairman Mao went to Chongqing to sign the October 10th Agreement, the enemy launched attacks from two directions. Yan Xishan attacked from one direction, and we fought the Shangdang Campaign. Ma Fawu and Gao Shuxun attacked from the other direction, and we fought the campaign along the Beiping-Hankou Railway.
>
> Looking back further, during the anti-Japanese war, we gave tit for tat to the enemy and were at the gate of the liberated areas. At the time, we had frictions with the KMT troops in several major liberated areas, but mainly in the Shanxi-Hebei-Shandong-Henan Liberated Area, which extended from Shanxi, Hebei, and Shandong to Henan. The area was a gate, and it was the gate the enemy attacked. But we did not have adequate strength to defend the gate. When Yan Xishan's 30,000 troops attacked the Shangdang area, how many troops did we have? We had slightly fewer troops than he did, just over 30,000. Moreover, we did not even have the organization of a regiment, not a truly complete regiment. In terms of strength and organization, our troops could be said to have been an assembly of guerrilla forces with poor equipment and little ammunition.
>
> Furthermore, on the eve of the campaign, we did not have generals to direct operations. At the time, Li Da was at the front, and all the other generals were away. Nor was Chen Zaidao present then. He flew back on the same plane with us to the Taihang Mountains. Among those flying back together were Marshal Liu Bocheng, me, Chen Xilian, and Chen Geng, as well as some leading comrades of the Second Field Army and other field armies. Song Renqiong was then in southern Hebei, and he did not have generals at the front either.
>
> The battle had been fought intensely when we returned to the Taihang Mountains. It was the Americans who helped us, because we took the transport plane of the observation group of the U.S. army stationed in Yan'an to fly back to the Taihang Mountains. As soon as we alighted from the plane, we went to the front. Under these circumstances, it was hard to wipe out the enemy completely. We should say we overfulfilled the task. It was more difficult to fight the campaign along the Beiping-Hankou Railway than to fight the Shangdang Campaign, but we had one advantage, that is, we had replenished ammunition and were slightly better equipped. However, our troops were still just an assembly of guerrilla forces. After fighting the Shangdang Campaign, our troops were extremely fatigued, but they continued to fight the campaign along the Beiping-Hankou Railway. That time we mainly fought our political war successfully, that is, we persuaded Gao Shuxun to revolt and come over. This was a political war.
>
> So we say we gave tit for tat to the enemy during the anti-Japanese war and in

> the struggle to counter friction. Our troops fought the battles against friction in all the liberated areas throughout the country, but mainly in the Shanxi-Hebei-Shandong-Henan Liberated Area. When Chiang Kai-shek mounted attacks, the gate his troops first attacked was this area where the Second Field Army was stationed. Only when our troops started fighting did the pattern of a field army really begin to take shape.[7]

As he spoke, Father, who was more than eighty-five years old, was full of intense emotions. When those white-haired old generals of the Second Field Army heard his remarks, they all felt an upsurge of emotion and memories flashed across their minds.

In November, autumn is in the air in Beijing and cold wind is beginning to blow. But at that time, the Great Hall of the People was warm and brightly lit. Looking back on these brilliant historical episodes, the old soldiers, all of whom were more than seventy years old, were filled with springlike feelings.

After the campaign along the Beiping-Hankou Railway, Father called a plenary meeting of the Shanxi-Hebei-Shandong-Henan Bureau of the CPC on the Fengfeng Coal Mine. At the meeting, participants drew up an overall plan for the work in the whole liberated area and made arrangements for the mass and economic work. After the meeting, in accordance with the directive of the Central Committee, the Shanxi-Hebei-Shandong-Henan Military Area concentrated twenty-five regiments to support northeastern China and, at the same time, further reorganized its troops.

The troops of the military area were organized into six columns and were led by the following:

1st Column: Commander Yang Dezhi and Political Commissar Su Zhenhua

2d Column: Commander Chen Zaidao and Political Commissar Song Renqiong

3d Column: Commander Chen Xilian and Political Commissar Peng Tao

4th Column: Commander Chen Geng and Political Commissar Xie Fuzhi

5th Column: Commander Wang Hongkun and Political Commissar Duan Junyi

6th Column: Commander Yang Yong and Political Commissar Zhang Linzhi

By then, 200 county and municipal governments had been established in the whole Shanxi-Hebei-Shandong-Henan Liberated Area, which had over 100 cities, and the troops in the liberated area had swelled to more than 310,000 men, with better weapons and equipment. Thus, the shift from scattered guerrilla warfare to concentrated mobile warfare had basically been completed.[8]

In mid-November, Liu Bocheng and Deng Xiaoping led the forward headquarters to return to Chi'an, Shexian County. At Chi'an, a mass rally was held to celebrate victory.

In light of the situation in the war of self-defense, the Shanxi-Hebei-Shandong-Henan Bureau of the Central Committee and the Shanxi-Hebei-Shandong-Henan Military Area decided to leave Shexian County and move to the area of Xiabaishu and Longquan, Wu'an County, west of Handan. One day toward the end of December, the headquarters of the field army set out in orderly formation, leaving behind the small village in the Taihang Mountains and the Qingzhang River, which was glistening under the sun.

Liu Bocheng and Deng Xiaoping and their troops had lived in the small village for more than five years. During those five years, they had studied their situation and the enemy's situation, called many meetings, and issued many operational orders. The small village had become the heart and soul of the Communist-led army, which resisted the Japanese invaders. The many ruthless and fierce battles and hard life were closely associated with this small mountain village. That day, they triumphantly left it and marched to the battlefronts to win more brilliant victories.

The Taihang Mountains and the people there reared this people's army and fostered many heroes. Liu Bocheng and Deng Xiaoping, as well as all the officers and men of their army, would never forget the lofty Taihang Mountains and the industrious and honest people there. After Marshal Liu Bocheng passed away, some of his ashes were buried in the Taihang Mountains, thereby fulfilling his wish to lie silently in the Taihang Mountains forever.

I was not born in the Taihang Mountains, but my two elder sisters and one brother were born there. When I was young, I often heard my parents talk about the Taihang Mountains, the mountain, the water, the villagers, and the militant life of extreme hardship there. In my heart, I felt an intimate link with every tree and every blade of grass in the Taihang Mountains.

I have interviewed many old soldiers who once fought in the Taihang Mountains. When they talked about the Taihang Mountains, the mountains and rocks there, the golden persimmons and big red dates, and the militant life during that period, they were very proud and sentimental. Their intense feelings often moved me, making me unusually fascinated by the Taihang Mountains.

In December 1945, Liu Bocheng and Deng Xiaoping led their troops, their morale high, to march eastward to the vast North China Plains.

ON THE EVE OF CIVIL WAR

The year 1945 had gone by.

After the victory in the War of Resistance Against Japan, the Kuomintang and the Communist Party held peace negotiations and at the same time fought battles with each other. The Communist-led army shattered the large-scale attack mounted by the KMT troops on the areas of Shangdang, Handan, and Suiyuan, wiping out over 70,000 of them. By January 1946, the liberated areas had covered an area of almost 2.4 million square kilometers, with a population of 149 million, including 506 cities. When encircled by the invading Japanese troops, the Communist-led armed forces were not wiped out, so they were not afraid of the attacks launched by Chiang Kai-shek's KMT troops. But what was the KMT doing?

The KMT was busy having a trial of strength with the Communist Party. After Japan's surrender, the KMT immediately stepped up its efforts to seize more land and conducted the so-called takeover. How did the KMT conduct the takeover?

In the Jiangsu-Zhejiang-Anhui, Hunan-Hubei-Jiangxi, Guangdong-Guangxi-Fujian, Hebei-Chahar-Rehe, and Shandong-Henan-Shanxi areas, northeastern China, and Taiwan, the KMT took over 2,411 factories run by the Japanese and the puppet regime, worth two billion U.S. dollars, as well as a large quantity of materials, gold, silver, and real estate worth one billion U.S. dollars.

All kinds of high-ranking KMT officials were responsible for the takeover. All KMT officials and government organs vied with each other in plundering gold bars, housing, and cars and dividing up the assets of the Japanese and the puppet regime in all parts of the country. They took advantage of the takeover to lay private claim to all these things. In Beiping alone, four-fifths of the materials taken over did not go to the state treasury. Wu Shaoshu, director-general of the Shanghai Municipal KMT Headquarters, took advantage of his post and power to seize over 1,000 buildings, over 800 cars, and over 10,000 gold bars formerly owned by the Japanese and the puppet regime. Qian Dajun, the mayor of Shanghai, went so far as to steal and sell the materials owned by the Japanese and the puppet regime, in the amount of 4.2 billion yuan of the KMT paper currency. These KMT officials looted in the name of takeover.[1]

While KMT officials avariciously looted money, to increase revenue and guarantee huge military spending the KMT government did three things: (1) greatly devalued the paper currency issued by the puppet regime in circulation in the enemy-occupied areas and replaced it with the official paper currency issued by the KMT; (2) printed and issued a huge amount of paper currency for emergency use; and (3) increased taxes and levies. As a result, inflation became serious, people's property lost its value, and two-thirds of the industrial enterprises and mines owned by the Japanese and the puppet regime could not be operated. National industrial and commercial enterprises went bankrupt, unemployment increased in the cities, and the rural economy languished.

Just after they won the anti-Japanese war, how happy the Chinese people were! They hoped for peace, tranquillity, and a hopeful life. But after only six months, their hopes had vanished. They anticipated the arrival of the KMT central government with high hopes, but when it arrived, they suffered a greater disaster. They exclaimed, "So many people were overjoyed at the victory, but now they lived in an abyss of misery and in dire poverty. They are in more pain now than before the victory."[2]

In November 1945, over 500 representatives from all walks of life held a rally in Chongqing against the KMT's policy of civil war and against U.S. interference in China's internal affairs. On November 25, over 6,000 university and secondary school students in Kunming held a rally against civil war. The KMT troops surrounded the schools and fired shots to intimidate the students. On November 26, over 30,000 university and secondary school students in Kunming started a general strike. On December 1, the KMT authorities sent large numbers of troops, policemen, and secret agents onto the campus of the Southwest Associated University to beat the teachers and students and even threw hand grenades at them, killing four and wounding several dozen. Between December 2 and 20, 150,000 people in Kunming held public memorial ceremonies for the four martyrs in the mourning hall, expressing their great indignation at the atrocities perpetrated by the KMT regime.

The land was recovered, but the KMT lost the support of the people. This was the fundamental cause for the final collapse of the KMT regime.

The Americans saw what the KMT was doing. A few years later, in a letter to President Harry Truman, U.S. Secretary of State Dean Acheson said that what the KMT civil and military officials were doing in the regions recovered from Japanese had made the KMT rapidly lose the support of the people and its own reputation there.[3] Although Acheson bewailed the KMT's plight, the U.S. government feared that the KMT did not have the military strength to suppress the Communist Party, and that if a civil war broke out in China, the CPC would bring all of China under its control. At the same time, to maintain the compromise it had reached with the Soviet Union on the question of China, the U.S. government sent to China the former chief-of-staff of the U.S. Army, Gen. George Marshall, as the special envoy of the U.S. president.

Marshall's mission was to "mediate" the conflicts between the KMT and the CPC and to continue to give all-out support to the KMT and help the KMT move its troops to northeastern and northern China. The ultimate U.S. aim was to establish, without war, a pro-American—and American-controlled—China under the rule of the KMT and Chiang Kai-shek. This was wishful thinking.

On January 5, 1946, the KMT and the CPC signed a preliminary agreement on the cessation of military conflicts in China. On January 7, the KMT representative Zhang Qun, the CPC representative Zhou Enlai, and the representative of the American government, George Marshall, formed a three-man meeting to discuss ways to handle armed conflicts and related affairs.

On January 10, representatives of the CPC and the KMT signed a truce agreement. On the same day, the Political Consultative Conference was held in Chongqing. The CPC sent a delegation, which included Zhou Enlai, to attend the conference. The participants discussed the establishment of a democratic political system and a national army. At the same time, the three-man meeting held serious consultations inside and outside the conference room.

For about a month, it seemed that the situation was starting to be conducive to ending the civil war and to promoting peace and democracy. But events ran contrary to people's wishes. The KMT and Chiang Kai-shek could not tolerate any genuine democratic reforms. Chiang Kai-shek said, "I am not satisfied with the draft constitution either, but since the matter has already reached this point, there is no way to turn back. For the time being, we have to adopt it and see what we can do in the future."[4]

The CPC had a sober estimate of the situation: it predicted that the road to democracy in China would be long and tortuous. Accordingly, the Central Committee instructed that it was important for the army to defend their positions and that military training, rent reduction, and production should be made the three major tasks for the liberated areas at present.

People's good wishes were one thing, but facts were another. The KMT signed the truce agreement in name, but it prepared for civil war in reality. The United States mediated a truce in name, but it helped the KMT prepare for war in reality.

From September 1945 to June 1946, the United States used military planes and warships to transport fourteen KMT corps of some 540,000 men from the rear area in southwestern and northwestern China to northern, eastern, southern, and northeastern China. The United States also landed 90,000 marines in China; the number of American troops stationed in China came to 113,000 at its peak. The U.S. government did not hesitate to appropriate a huge sum of money to arm forty-five divisions of the KMT army, train 150,000 military personnel, and equip the KMT's air force with 936 aircraft. The U.S. government also extended enormous amounts of economic and military aid to the KMT government. In the first half of 1946 alone, the United States provided the KMT with $51.7 million (in U.S. dollars) worth of military items. The United States also formally organized a military advisory group, consisting of 2,000 members, which in fact was a military organization directly involved in planning and directing the civil war in China.[5]

The U.S. government was determined to achieve its goal. Just after the end of World War II, the United States was at the peak of glory. It seemed that if Americans made a plan, they would attain their goal—as if China's internal affairs could be completely conducted with the American baton. But God does not comply with man's wishes, and events would finally prove the Americans wrong.

What is God's will? It is the people's will. The Chinese people had suffered too much, and they did not want others to determine their destiny anymore. But the U.S. government made a serious mistake, that is, it chose to support a government detested by the Chinese people, a "son of heaven whom it could not prop up," and a future that was doomed to failure.

The truce agreement was signed, but the war never stopped. Between January and June 1946, the KMT troops mounted 4,365 small or big attacks on the various liberated areas, employing a total of 2.7 million men and occupying 40 cities and 2,577 villages and towns in the liberated areas. During the attacks, Chiang Kai-shek and other high-ranking generals frequently flew to the fronts to supervise operations themselves.

In March 1946 the Soviet army withdrew from northeastern China. The KMT army seized Shenyang, a city of strategic importance in northeastern China, and, with five corps, attacked the Communist-led troops in Benxi and Siping to the south of Shenyang. After fighting an intense, monthlong battle, the KMT army gained control over the area south of the Songhua River in Heilongjiang Province.

The battle in northeastern China enabled the KMT army to gain control over vast stretches of territory. It also enabled the United States to achieve its goal of containing the Soviet Union. Consequently, the intensified civil war in northeastern China aggravated the civil war crisis in the whole country. From January 14 to the end of April 1946, the KMT army mounted over 920 big and small attacks on the Shanxi-Hebei-Shandong-Henan Liberated Area—in other words, an average of eight attacks a day. There were four attacks involving 10,000 KMT troops each, forty attacks involving 1,000 KMT troops each, and 110 attacks involving 100 KMT troops each. The army and people in the Shanxi-Hebei-Shandong-Henan Liberated Area did not decline to shoulder the responsibility, waged a tit-for-tat struggle, fought for every inch of land, and, at the same time, fought battles on just grounds, to their advantage and with restraint.

Although the KMT and the CPC had signed the truce agreement and the Americans continued to mediate, the people in the liberated area were clear about what would happen. In the face of frequent attacks by the KMT army, the army and people in the liberated area had to cast away illusions, keep on guard, and be ready at all times to fight any enemy that dared to invade their territory.

This was a relatively quiet period. In December 1945 the headquarters under the command of Liu Bocheng and Deng Xiaoping had moved to Wu'an County, where my parents lived temporarily. Mother took her three children with her. Since the family had been established in 1939, all of its five members were thus reunited for the first time.

After they married, my parents had left Yan'an, and more than five years had passed in a flash. During those five years, Mother had led a very hard life! She was a graduate of a famous university, but when she, a young revolutionary, joined the revolutionary ranks for a short time, she went to the Taihang Mountains and was thrown into a hail of bullets in the anti-Japanese war. She accepted the baptism of war without hesitation or fear. She had been working in the office, but in the Taihang Mountains, there was no rear area. She marched, fought in different places, and took shelter from the enemy's mopping-up operations along with the army units. She first worked in the general headquarters of the Eighth Route Army based in Matian, Liaoxian County, which was a small "rear area" at the front. Not long after that, she packed up her bedding, went to Chi'an, Shexian County, and stayed in the front headquarters of the 129th Division with Father.

When she gave birth to her first child, she could have kept the baby at her side. But at that time, when the enemy was conducting frequent mopping-up operations, it would have been very inconvenient to move about with a child. Mother said, "I didn't want to have some soldiers protect us specially." She reluctantly entrusted Lin Er to the care of a villager, then marched with our army and moved to other places. Not until a year later, when our army passed by the place where her child lived, did Mother have a chance to see her. She went with Chen Shulian, the wife of Cai Shufan, deputy director of the Political Department of the 129th Division. When they entered the house, they found Lin Er looking thin and small, in dirty and worn-out clothes. What a pity! Mother controlled her grief, held back her tears, and bathed Lin Er and made clothes and a quilt for her with the help of Chen Shulian. Our army stayed for three or four days and left. Mother hated to part with her child, but she went away.

When Lin Er was one and a half years old, her wet nurse became pregnant, so Mother took her back. Mother said that when Lin Er came back, she was too weak to whisk the flies off. At the time, the anti-Japanese war was at its most difficult stage and it was indeed hard for Mother to look after a child and take shelter from the enemy mopping-up operations. So Mother had to entrust Lin Er to the care of another villager. The following year, when Cai Shufan went to Yan'an to attend a meeting, Mother asked him to send Lin Er to a nursery school there. When my elder brother, Pang Pang, and my second elder sister, Nan Nan, were born, in 1944 and 1945, respectively, Mother also entrusted them to the care of villagers.

Mother's first three children were born in these war-ridden years and grew up during a ruthless war. After they were born, they did not spend a peaceful day and could not enjoy any of the basic necessities. They were brought up on the milk and millet gruel of the villagers in the Taihang Mountains. They survived the war and all kinds of natural and man-made calamities. When my parents came to Wu'an County, the three children were sent there. Father was very happy, but Mother was very worried.

Why was she worried? Because the eldest child, just returned from Yan'an, neither spoke nor ate. She could grasp an apple but did not know how to eat it. She was very thin and clearly malnourished. Her second child suffered from diarrhea, which prevented her and her husband from sleeping well at night. The third child was only one and a half years old. Mother did not have milk and could not feed her with millet gruel. Mother had to ask a little country girl and an old woman for help, but they gave her no help and could not even make a fire. All this made Mother too worried. However busy she had been with work, however urgent the military situation had been, and however difficult the march had been, Mother had not worried. But this time she was really worried.

This is life. Perhaps the seemingly important things do not put one in difficult positions, but the seemingly less important things put one at a loss as to what to do.

Fortunately, life always goes on. If one has been exposed to just about everything, it is easier to take it more lightly. Things finally went well for our family in Wu'an County. Three children began playing together happily. Their faces became tanned, and they got fat and tall and followed their parents around. Father loved his children very much. It seemed to me that few people on earth loved children from the bottom of their hearts as Father did.

On March 2, 1946, the leading bodies of the Shanxi-Hebei-Shandong-Henan Military Area moved from Wu'an to Handan City.[6] From then on, Handan, the capital of the ancient state of Zhao, was the capital of the Shanxi-Hebei-Shandong-Henan Border Region, the headquarters for the troops under the command of Liu Bocheng and Deng Xiaoping, and the southern gate to the liberated areas in northern China.

My parents took their three children to Handan. Of course, the children knew nothing and played happily every day as usual. But the adults were very happy, too, because this was the first time the leading bodies of the military area had moved to a big city.

Mother was busy looking after her three noisy children, and Father was busy handling military and administrative affairs. After the troops entered the city, Political Commissar Deng instructed the leading bodies and troops not to live in civilian houses but in warehouses. The headquarters was located in a former barracks of the Japanese army.

After the troops entered the city, they began receiving political training. Political Commissar Deng told them to gain a clear understanding of the current situation and to tighten discipline. To acquire that clear understanding, they had to know about the possibility of a large-scale civil war, and about resolute struggle being the important guarantee of success as they strove for peace and democracy. To tighten discipline, they had to carry out the policies and decrees and retain the nature of a people's army. Special stress was placed on the importance of implementing the policies in urban areas.[7]

In May, on the basis of the political training, a large-scale campaign of military training was vigorously launched. All officers and men participated in many kinds of training in preparation for war, raising their military skills to new levels.

While political and military training was being conducted, and in accordance with the instructions of the Central Committee, it was decided that a land reform movement would be launched in the hinterland of the liberated area, on the basis of reduction of land rent and interest, so as to ensure that the tillers had land. The poor peasants in the vast liberated area not only obtained land but enthusiastically supported the people's army in the revolutionary war.

By the end of June 1946 the Shanxi-Hebei-Shandong-Henan Liberated Area had expanded considerably. In the whole area, there were 270,000 troops and 600,000 militiamen, the population had grown to over 30 million, and the number of counties under its jurisdiction increased from some 80 to more than 110.

Within six months, the KMT army had concentrated eleven reorganized divisions and three corps to "nibble at," attack, harass, and invade the Shanxi-Hebei-Shandong-Henan Liberated Area.

The sky over the vast land of China was covered with the dark clouds of war. In mid-June 1946 a meeting attended by high-ranking cadres from the Shanxi-Hebei-Shandong-Henan Liberated Area was held in Handan. Liu Bocheng and Deng Xiaoping once again pointed out the serious danger of an all-out civil war, and the importance of making preparations for it.

The Americans continued to mediate, but it did not matter whether their intentions were serious or spurious—China's civil war was imminent.

CHAPTER 56

THE OUTBREAK OF THE ALL-OUT CIVIL WAR

In late June 1946 the sun blazed over the land of China. People could not endure the intense heat of summer.

Starting with a large-scale encirclement of the Communist-led Central Plains Liberated Area, the KMT and Chiang Kai-shek launched their full-scale attack on the liberated areas. China's large-scale civil war thus broke out.

The KMT had totally disregarded the peace negotiations and mediation efforts. Civil war was no longer a crisis on paper but an extremely grim reality. Determined to achieve success, Chiang Kai-shek had sent troops to mount major offensives against the Communist-led liberated areas.

Chiang Kai-shek had a total strength of 4.3 million troops, enough military equipment—taken over from the Japanese invaders—to arm one million troops, and enormous amounts in aid from the United States. Of Chiang's eighty-eight reorganized divisions, twenty-two were anywhere from half to completely armed with U.S. equipment. Chiang also possessed a large quantity of artillery guns as well as aircraft, warships, and tanks. He controlled war resources far superior to those of his opponent, 76 percent of the territory of China, and 71 percent of the total population. He controlled nearly all the big cities, major national communication lines, and almost all the modern industries.

The Communist Party had a total strength of 1.27 million troops equipped only with infantry weapons and a few guns captured from the Japanese and puppet troops. It controlled an area of 2.3 million square kilometers, with a population of 136 million, and controlled no modern industries by and large.

The ratio of strength between the KMT and the CPC was 3.4 to 1. The superiority of the KMT was apparent but not absolute.

After the Japanese left, Chiang Kai-shek thought he could concentrate on fighting the Communist Party. He had not been able to vent his anger for almost twenty years. Now that he had troops and weapons as well as foreign aid, he wanted to express his great hatred. He vowed to apply the strategic principle of fighting a war of quick decision by employing 80 percent of his regular troops to mount a full-scale nationwide attack on the Communist Party in a bid to wipe out the Communist-led troops in areas south of the Great Wall within three to six months. Then he would turn to solving the question of northeastern China.

Chiang Kai-shek had made up his mind and immediately started large-scale attacks. He did not realize, however, that once he took this step, he was doomed to failure. It was a step toward complete defeat and self-destruction!

The Communist Party and Mao Zedong had long foreseen the intention of the KMT and Chiang Kai-shek to wage war and had long been preparing for it. The Central Committee made it clear to the entire Party that, "although Chiang Kai-shek receives American aid, he has lost the support of the people, the morale of the KMT troops is low, and he has economic difficulties. Although we have no foreign aid, people support us, our army's morale is high, and our economy operates. So we can defeat Chiang Kai-shek. The whole Party should be confident of victory."[1]

Comrade Mao Zedong said,

> We have only millet plus rifles to rely on, but history will finally prove that our millet plus rifles is more powerful than Chiang Kai-shek's aeroplanes plus tanks. Although the Chinese people will face many difficulties and will long suffer hardships from the joint attacks of U.S. imperialism and the Chinese reactionaries, the day will come when these reactionaries are defeated and we are victorious. The reason is simply this: the reactionaries represent reaction, we represent progress.[2]

In June 1946 the large-scale civil war began. The first shot was fired in the Central Plains.

On June 26, Chiang Kai-shek assembled more than twenty divisions to mount large-scale attacks on the Communist-led Central Plains Liberated Area. In accordance with the directives of the Central Committee, the Central Plains Liberation Army, under Li Xiannian and others, resolutely broke the siege there and triumphantly arrived in southern Shaanxi and other places.

At the same time, fifty-eight brigades of KMT troops attacked the Communist-led East China Liberated Area. Over 40,000 troops of the Shandong Field Army, under the command of Chen Yi, and over 30,000 troops of the Central China Field Army, under the command of Su Yu and Tan Zhenlin, met the enemy troops head-on. By October 1946 they had wiped out over 70,000 enemy troops.

The Shanxi-Hebei-Shandong-Henan Liberated Area extended west as far as the Datong-Puzhou Railway, east as far as the Tianjin-Pukou Railway, north to the Zhengding-Taiyuan Railway and Dezhou-Shijiazhuang Railway, and south to the Yellow River and the Longhai Railway. It bordered on the Shaanxi-Gansu-Ningxia, Shanxi-Suiyuan, Shanxi-Chahar-Hebei, and East China liberated areas, and it was adjacent to the Central Plains Liberated Area. Its important geographical location made it the center of all the liberated areas, so it became one of the focuses of the attacks by the KMT troops.

Chiang Kai-shek had assembled over 300,000 troops around the Shanxi-Hebei-Shandong-Henan Liberated Area. The troops under the command of Hu Zongnan, Yan Xishan, Xue Yue, Sun Lianzhong, Liu Zhi, and others were ready to attack, encircle, and suppress the Communist-led troops, to control and open the communication lines, and to divert the Yellow River to its old course in an attempt to separate and inundate the liberated areas.

To gain the initiative in the war and coordinate the operations in eastern China, and with the approval of the Central Committee, Liu Bocheng and Deng Xiaoping decided to concentrate the main forces of their field army and take the initiative to flexibly wipe out enemy troops. They divided the whole field army into two parts: one part, over 40,000 strong, under the command of Liu and Deng fought all the way to eastern Henan, and the other part, over 20,000 strong and under the command of Chen Geng—who, in the chain of command, was directly under the Central Military Commission—fought all the way to southern Shanxi.

On June 28, 1946, the sun had just risen in the east and was shining brightly over the land and the open country. At the Matou Railway Station south of Handan, the Shanxi-Hebei-Shandong-Henan Field Army stood in formation, fully armed. From the platform of a railcar of a small train, Liu Bocheng and Deng Xiaoping were prepared to deliver speeches at this oath-taking rally of the Shanxi-Hebei-Shandong-Henan Field Army.

Political Commissar Deng Xiaoping began:

> Chiang Kai-shek did not comply with the agreements of the Political Consultative Conference and the truce agreement and has openly torn up the truce agreement to mount full-scale attacks on the liberated areas. . . . [We] should make all preparations rapidly to smash Chiang Kai-shek's attacks. . . . Although Chiang Kai-shek receives American aid, he meets with opposition from the people throughout the country because he has launched an antipeople civil war. His army's morale is low, and he has economic difficulties that he is unable to overcome. Although we do not have foreign aid, people support us, our army's morale is high, and our economic foundation is solid. We are convinced that we can defeat Chiang Kai-shek. We should be confident of victory in fighting this battle of counterattack in self-defense.[3]

In the Central Plains, the Beiping-Hankou Railway runs south to north, and the Longhai Railway from east to west. These two railway lines take the shape of a cross and constitute the main artery of traffic through the Central Plains.

Liu Bocheng and Deng Xiaoping first chose the Longhai Railway as a target of attack. In August 1946 the Central Committee approved their plan of operations. On August 10, the Shanxi-Hebei-Shandong-Henan Field Army launched the Longhai Campaign.

On August 10, field army units, coming from all directions, rapidly marched thirty kilometers into the heart of the region and suddenly attacked the enemy troops on the 150-kilometer front along the Kaifeng-Xuzhou section of the Longhai Railway. By August 12, they had captured a dozen cities and railway stations, including Lanfeng, Henan, and Dangshan, Anhui, wiped out over 5,000 enemy troops, and controlled over 100 kilometers of railway lines.

Dangshan was seized by the army unit under the command of the brave generals Yang Yong and Zhang Linzhi. When they were celebrating victory, Political Commissar Deng was wading through water and mud to the battlefront. Since they had won a victory, it was expected that they would be cited. Instead, Yang Yong and Zhang Linzhi were "severely criticized" by Political

Commissar Deng. Although Yang Yong's unit had always been famous for its bravery, it did not observe strict discipline in the battle and damaged the furniture, pots, bowls, and other utensils of the local people.

When Political Commissar Deng arrived, the column immediately called the cadres at and above the regiment level to a meeting. They sat silently on sorghum stalks in the mud and water. Political Commissar Deng spoke earnestly:

> You have fought the Longhai Campaign for four days. At the first stage, you've performed quite well, liberating Dangshan and capturing several thousand enemy troops and many weapons. But I must point out that some of you have violated the discipline in relation to the masses. Many of you have laid down your lives during the battles, but what for? Why do some of you harm the interests of the masses? You should seriously compensate them for their losses.

As he was speaking, enemy aircraft flew overhead. People were worried about Deng's safety, and Yang Yong personally went to an upland to keep a close watch on the aircraft. Deng looked at Yang Yong, shouting, "Yang Yong, what is there to be afraid of? It doesn't matter. Doesn't the aircraft fly here every day?" Deng continued to speak to the assembled company: "Those who violate the discipline in relation to the masses will lose the support of the masses. Without their support, it will be impossible for us to achieve victory!"

Yang Yong and Zhang Linzhi admitted their mistakes then and there and immediately ordered the troops to compensate the masses for their losses and to apologize to them.[4]

In accordance with the Party's style, Father always attached great importance to army discipline and set strict standards for his subordinates. He had emphasized discipline in the previous operations, he did so during the Longhai Campaign, and its importance to him would not diminish in the future as the CPC forces liberated Nanjing, entered Shanghai, marched into southwestern China, and settled the question of Tibet by peaceful means. Today he sets the same requirements for the Chinese People's Liberation Army. His faith has been that "people are the mother of everything and the source of all forces in the struggle against the enemy."[5]

To do all for the people is the basic purpose of the people's army. This purpose enabled the people's army to achieve the final victory. So Father has never been vague on the need to maintain and tighten discipline in relation to the people.

The Longhai Campaign was still under way. Because of the Communist-led army's rapid and sudden attacks, the enemy was forced to rush three divisions that had been pursuing and attacking the Central Plains Field Army to reinforce troops in Kaifeng, and to move a corps attacking Huainan and another two divisions to reinforce the troops in Dangshan and Xuzhou. Liu and Deng directed their troops in the capture of Qixian, Tongxu, Yucheng, and other places and wiped out a unit of the enemy troops. When the enemy reinforcements approached from both east and west, the Liu-Deng troops immediately moved to the area north of the Longhai Railway for rest and reorganization.

On August 22, the Longhai Campaign came to an end. The Communist-led army had wiped out over 16,000 enemy troops, captured five county towns and ten railway stations, and destroyed over 150 kilometers of railway.

With Liu and Deng's elaborate and daring plan of operations, their troops applied the tactics of in-depth penetration and surprise attack, rapidly achieved splendid results on the battlefront, and forced the enemy to pull back troops attacking other liberated areas for reinforcement. This campaign was an example of defeating the enemy by surprise. The KMT army was appalled.

Chiang Kai-shek immediately assembled fourteen reorganized divisions and thirty-two brigades, over 300,000 strong, along Zhengzhou and Xuzhou. He planned to take advantage of the Communist-led army's lack of time for rest and reorganization and to mount with overwhelmingly superior force a converging attack on it in the area of Dingtao and Caoxian north of the Longhai Railway in Shandong Province. The enemy troops were approaching the Liu-Deng troops from six directions.

On August 22, the Central Committee issued a directive to the field army: Fight no battle you are not sure of winning. But if you fight, you must win. When you fight a battle with the enemy's regular troops, concentrate a superior force to destroy them one by one.

In the operations room of the headquarters of the Shanxi-Hebei-Shandong-Henan Field Army, Liu and Deng were studying the enemy's situation. Deng went in front of the military map and said:

> There are a total of three divisions marching north along the Tianjin-Pukou Railway, of which two are Chiang Kai-shek's crack troops. Chiang Kai-shek has five crack troops altogether [the New 1st Corps, the New 6th Corps, the New 5th Corps, the Reorganized 11th Division, and the Reorganized 74th Division]. This time he dispatches two crack troops. The New 5th Corps and the 11th Division are armed totally with American equipment and have a high level of combat effectiveness. It will be hard for us to deal with them. A large number of enemy troops are coming from the west, but their combat effectiveness is poor.
>
> In view of this, I am thinking of two options. One is that our army temporarily evades the enemy's cutting edge to withdraw rapidly our main forces to the north of the Yellow River for rest and reorganization for a short time and later seizes an opportunity to advance south to wipe out the enemy. This option is relatively favorable to us when local conditions are taken into consideration. But the implementation of this option will inevitably put a heavier burden on Chen Yi and Li Xiannian and harm the overall interests. The other option is that we endure hardships with dogged will to fight another battle. In this way, we will bear a heavy burden, but the load of Chen Yi and Li Xiannan will be far more lightened! I think the second option is good.

Liu said with a smile, "I quite agree with you. Chiang Kai-shek applies the tactics used in a restaurant, that is, to give you a dinner first and then give you another dinner before you have finished the first, and forcing you to have it. It is impolite not to reciprocate, and since he gives us another dinner, we will just eat it to the limit of our capacity!" All present burst into laughter.

In Liu and Deng's analysis, the enemy troops coming from the west were relatively weak, so it was appropriate to concentrate their main forces to annihilate those enemy troops rather than the ones coming from the east. Liu and Deng decided to launch the Dingtao Campaign.

After the orders of operation were drafted, Deng said with emotion,

> Our army is well known and is called the massive Liu-Deng army, but actually we have a limited strength of less than 50,000 men plus several mountain guns and mortars and insufficient ammunition. These soldiers of our army are mostly the sons of emancipated peasants, and their quality is good. During the Longhai Campaign, we suffered 5,000 casualties, but we don't have many replacements. If we dispatch these backbone forces to fight the battle, I will be really distressed.[6]

The war was ruthless, but some of those who directed it did not become so.

On September 2, 1946, the campaign began.

The enemy troops from the west strutted to the battlefield prepared by the Liu-Deng army, which had concentrated a strength outnumbering the enemy four to one to surround the enemy's Reorganized 3d Division. The enemy commander Liu Zhi urgently ordered another four divisions to reinforce the 3d Division from two directions. Under the Liu-Deng army's heavy attack, the enemy's 3d Division tried to break out of the encirclement southward. The field army launched attacks on all fronts, however, and wiped out the enemy completely and rapidly. Having no hope of success, the enemy's reinforcements withdrew immediately. The field army quickly moved its forces to wipe out some of the reinforcements as they withdrew.

On September 8, the Dingtao Campaign came to an end. At a cost of 3,500 casualties, the Liu-Deng field army wiped out the enemy's four brigades, numbering some 17,000 men, and captured the division commander of the enemy's Reorganized 3d Division, Lt. Gen. Zhao Xitian.

This campaign, together with the victory won by the Communist-led army in central Jiangsu, helped to reverse the serious situation on the southern front of the liberated areas of the Communist Party. It also resulted in the removal from office of Liu Zhi, the director of the KMT Zhengzhou Pacification Headquarters and one of Chiang Kai-shek's key and trusted generals.

Mao Zedong sent a telegram to Liu and Deng congratulating them on the great victory of annihilating the 3d Division and expressing the hope that they would cite their troops in a dispatch to the whole army.[7]

In capturing Dingtao, the Liu-Deng troops fought very bravely, but after some people claimed credit for themselves, certain army units slackened their discipline. Deng Xiaoping was always well known for his strictness with his army. After the campaign, a meeting of senior army cadres was held. All the participants were beaming with satisfaction, but unexpectedly, Political Commissar Deng said straightforwardly, "Today we don't shake hands before the meeting. There is no need to shake hands to express joy after winning several battles." Those present have still not forgotten this meeting of "no handshaking" and Deng Xiaoping's way of leading the army.

The Communist-led army had won great victories in two campaigns but was extremely battle-fatigued. The civil war had broken out more than two months before. The army needed to take a rest, reorganize, and find replacements, but the enemy prevented it from doing so. Right after the Dingtao Campaign ended, the KMT employed its main forces, the New 5th Corps and the Reorganized 11th Division, to attack Dingtao and Heze. To avoid fighting the enemy's main forces, the Liu-Deng field army withdrew from Heze, and its main forces moved to southwestern Juye, north of Heze, for rest and reorganization.

The enemy troops reached out for a yard after taking an inch; they were determined to prevent the Liu-Deng field army from having its rest and reorganization. In early October they invaded the Juye area. To keep the enemy troops from advancing, Liu and Deng decided to start the Juye Campaign on October 3.

In Juye, the adversary was the entirely American-equipped KMT main force, the New 5th Corps, with strong combat effectiveness. After fighting fierce battles for four days, the field army had killed or wounded 5,000 enemy troops but had also suffered over 4,000 casualties. The battle was at a stalemate. To avoid finding itself in a passive position, the field army stopped its attack.

During this campaign, the Liu-Deng field army encountered a trial of strength with the strong enemy troops for the first time. Although it won a small victory and achieved the goal of preventing the enemy troops from mounting attacks, it had acquired some experience and learned some lessons.

Liu Bocheng, a key strategist of the Communist-led army, was known as the "ever-victorious general." He was also well known for his adeptness at macrostrategy as well as the art of military operations and tactics. But he always approached problems by seeking truth from facts and was good at analyzing advantages and disadvantages as well as strengths and weaknesses. His work style was identical to Deng Xiaoping's.

After conducting a realistic, in-depth analysis of the Juye Campaign, Liu Bocheng and Deng Xiaoping submitted a report to the Central Committee that concluded: "We are prepared to conduct long-distance maneuvers and fight wherever there is an opportunity!"[8]

The enemy did not allow the Liu-Deng field army to have a breathing spell. In mid-October, the enemy's Reorganized 27th Corps, its main forces led by Corps Commander Wang Jingjiu, mounted an attack. This time Liu and Deng decided to avoid engaging the enemy's main forces and to seize an opportunity to attack the troops commanded by Sun Zhen from Zhengzhou. On October 29, the field army launched the Juancheng Campaign in areas northeast of Juye, Shandong Province.

Avoiding the enemy's main forces, the Liu-Deng field army concentrated a strength outnumbering the enemy four to one and suddenly encircled and attacked the troops commanded by Sun Zhen.

On October 31, the campaign ended. The Liu-Deng field army had wiped out over 9,000 enemy troops, captured the enemy's brigade commander, Liu Guangxin, and seized 8 U.S.-made howitzers, 7 mountain guns, 37 mortars, 95 small artillery pieces, and 208 light machine guns. The field army not only wiped out the enemy's effectives and stemmed the momentum of enemy attacks but also seized large quantities of weapons and equipment. The field army's might was redoubled, as if it had enjoyed a good meal.

The Liu-Deng field army had already fought three campaigns in which it launched attacks on its own initiative and flexibly seized opportunities to wipe out ten enemy brigades of over 50,000 men. These victories were won when its weak forces were facing the enemy's strong forces. Although the field army had lost seventeen cities and towns, the enemy forces were scattered and their offensives were beginning to slow down. The field army did not give a thought to seizing or losing a specific city or place; its main objective was to wipe out the enemy's effectives in mobile warfare. It had gained the initiative strategically.

In coordination with the operations conducted by the Liu-Deng field army on the battlefields in Henan and Shandong provinces, the troops under the command of Chen Geng achieved victories in the Wenxi-Xiaxian, Datong-Puzhou, and Linfen-Fushan campaigns in southern Shanxi, wiping out over 50,000 enemy troops in three months.

On other battlefronts, the troops commanded by He Long in the Shanxi-Suiyuan area and those led by Nie Rongzhen in the Shanxi-Chahar-Hebei area wiped out a total of over 38,000 enemy troops.

On October 1, 1946, Comrade Mao Zedong issued "A Three Months' Summary." The Central Committee pointed out that the Communist-led army had annihilated a total of twenty-five brigades of the Kuomintang army over the previous three months. Concentrating a superior force to wipe out the enemy forces one by one was the only correct method of fighting. The committee decreed that this method continue to be used in the future, with the goal of eliminating another twenty-five enemy brigades in the next three months (an average of eight enemy brigades a month). This would be the key to changing the military situation.

After four months of fighting, the strength of the Liu-Deng field army had increased from 270,000 to 310,000 troops, and the number of militiamen from 600,000 to 740,000. The stock of weapons and equipment had improved because of the "gifts" presented by the KMT troops. The Liu-Deng field army had gained more experience in conducting large-scale mobile warfare, and its morale was high. Land reform had been carried out in two-thirds of the areas of the Shanxi-Hebei-Shandong-Henan Liberated Area—with a population of two million—so that the tillers had land and the emancipated peasants were enthusiastically supporting the army's war effort.

In November 1946 Liu and Deng decided that their army should wipe out another six to seven enemy brigades in three or four months in the whole liberated area. Yang Guoyu wrote at the time:

> Liu and Deng are determined to take action only when they have made plans in advance. When they submitted to the Central Military Commission the report on the military situation in Hebei, Shandong, and Henan areas, they put forth the plan of annihilating six to seven enemy brigades in the next three or four months. They would not have written such a telegram unless they were very confident of victory.[9]

On November 4, Deng Xiaoping called on the army units in the liberated area to achieve great victories in several more battles.[10]

In the vast land of Hebei, Henan, and Shandong, the KMT was also assembling its troops in an attempt to advance to Xingtai, seize Handan, and open the Beiping-Hankou Railway for traffic.

On November 18, the Liu-Deng field army launched the Huaxian Campaign in the area east of the Beiping-Hankou Railway. It wiped out 12,000 enemy troops in four days and five nights, attracting some KMT troops to reinforce the enemy in Huaxian from east and west and frustrating the enemy's plan of advancing north to open the Beiping-Hankou Railway for traffic.

The tactics used in this campaign were excellent. Liu Bocheng said humorously, "Our method of fighting is strange. We pay no attention to the enemy's stretching hands, keep clear of them, pass through their small strongholds and hold their waists in our arms at one swoop, pulling out their hearts and hitting their vulnerable points."[11] In the wake of the campaign, the East China Field Army won a great victory near Suqian, wiping out 50,000 enemy troops.

To try once again to carry out the plan of opening the Beiping-Hankou Railway for traffic, nine enemy brigades over 50,000 strong fiercely advanced northward from Huaxian. In accordance with the directives of the Central Committee, the massive Liu-Deng army was determined to advance as the enemy advanced, disregarding the enemy's invasion of the heartland. The main forces attacked the enemy's rear in the area northwest of Xuzhou.

Between December 30, 1946, and January 16, 1947, the Liu-Deng army conducted the Juye-Jinxiang-Yutai Campaign. In this campaign, Liu and Deng displayed their superb leadership skills. Their army marched without rest for over 200 kilometers in twenty days and attacked the enemy's rear by surprise. It wiped out 16,000 enemy troops, seized large quantities of weapons, and recovered nine county towns, forcing the enemy to urgently withdraw for reinforcement and frustrating the enemy plan of opening the Beiping-Hankou Railway for traffic once again.

Liu Bocheng and Deng Xiaoping directed this campaign very well, and all unit commanders also performed very well. In their report to the Central Committee, Liu Bocheng and Deng Xiaoping wrote:

> In accordance with the concept of a mobile campaign, the commanders at all levels must foresee the changes in the situation, make the best use of them, act flexibly and resolutely, and try to fulfill the task of wiping out the enemy. In view of the changes in the campaign, the commanders at all levels generally could, under the guidance of the general strategy, seize opportunities to fight it on their own and achieve great victory. Especially at the last stage of the campaign, when we were going to achieve victory, all columns could make great efforts to conduct convergent operations from all directions, thus achieving such results.
>
> In fighting, the enemy troops suffered one defeat after another, and they were as wily as hares, so it was hard for us to catch them. This required the commanders at all levels to avoid wishful thinking on their own to trust chance and windfalls. Instead, they should, while following the progress of the campaign closely, make use of the factors of our positive actions to create the war situation and move toward the annihilation of enemy troops. That is, we should know how to enable the enemy to show his weaknesses, how to lure him to advance, how to pursue him, and how to outflank him. . . . As it was mobile warfare, we must give play to the flexibility and sense of responsibility of the commanders at the front to seize opportunities to fight battle. The leadership of commanders at the higher level should be provided in advance in the form of instructions (assigning tasks without specifying the means) so that the subordinates act at their discretion according to circumstances.[12]

Achieving victory in a military campaign depends on the wisdom of the commanders. But the wisdom of only a few commanders is far from enough. The superb art of leadership is to give play to the subjective initiative of the commanders at all levels and to encourage the bravery and wisdom of all officers and men. Liu Bocheng and Deng Xiaoping commanded troops this way. Mao Zedong did even better. As a result, the whole army of the Communist Party of China displays collective wisdom and initiative, its units cooperating with and supporting each other and even making self-sacrifices for the overall interest. This spirit provides a major military advantage over any enemy.

The year 1946 went by. Half a year had elapsed since the civil war had broken out. The war between the KMT and the CPC became fiercer. However, the war was not developing as Chiang Kai-shek had carefully planned. Moreover, a strange omen had begun to appear. The year 1947 began. Chiang Kai-shek became impatient. He urgently sent a trusted general, Chen Cheng, to Zhengzhou and Xuzhou to organize a mass campaign in southern Shandong and dispatched over 300,000 troops in fifty-three brigades to attack southern Shandong.

In accordance with the order of the Central Committee to win great victories and attract the enemy's main forces that were attacking southern Shandong, Liu Bocheng and Deng Xiaoping decided to launch a campaign in the Henan-Anhui border area on January 24, 1947, that is, the second Longhai campaign. During the campaign, the Liu-Deng field army conducted mobile operations in the vast areas both south and north of the Longhai Railway. Deng Xiaoping personally went to command the southern group led by Yang Yong in operations. Braving the bombing and strafing of enemy aircraft and heavy artillery fire, he marched and fought with the troops and was a battlefield commander. After Boxian in northern Anhui was liberated, Deng Xiaoping found that the local people were suffering from famine and immediately gave an order to open the granary to help the poor people. The local people were very excited. They said the KMT army looted the people whereas the Communist-led army saved the people.

On February 11, the second Longhai campaign ended in triumph. The massive Liu-Deng army recovered vast areas on both sides of the Longhai Railway and wiped out over 16,000 enemy troops, thus achieving a great victory.

From November 1946 to February 1947, the Liu-Deng field army strode along to conduct mobile operations on the Hebei-Shandong-Henan battlefield, wiped out a total of eight enemy brigades numbering over 44,000 men, recovered twenty-five county towns and abandoned twenty-four, and frustrated the enemy's plan to open the Beiping-Hankou Railway to traffic. It also pinned down the main forces of the enemy troops commanded by Wang Jingjiu and effectively coordinated the Communist-led army's war effort in Shandong and northern Jiangsu.

In the eight months since the all-out civil war had broken out, the urgent war situation, the fierce battles, and the brilliant victories won by the People's Liberation Army were all beyond people's expectations. Chiang Kai-shek had vowed confidently to wipe out the Communist-led troops stationed in areas south of the Great Wall within three to six months. Three months, six months, then eight months had passed, and the Communist-led troops in areas south of the Great Wall were not wiped out. On the contrary, fighting had made them stronger!

By February 1947, the Communist-led troops had wiped out a total of 710,000 KMT troops, while their own total strength swelled to over one and a half million. It accumulated rich experience in conducting large-scale battles of annihilation, increased its supplies of weapons and equipment, and established the artillery corps. The KMT lost 710,000 troops and failed to accomplish its ambitious strategic plan. During this period, the KMT army captured 105 cities in the liberated areas, but each city was seized at an average cost of 7,000 men. Each seizure also put an additional burden on the Kuomintang army to defend it. Therefore, the KMT's strength at the front fell from 117 brigades in October 1946 to 85 brigades.

After eight months of the trial of strength, the overall strength of the Kuomintang troops was still superior to that of the Communist-led army, but the KMT army had lost the capability to mount full-scale attacks on the liberated areas. Mao Zedong said, "We must wipe out another forty to fifty of Chiang Kai-shek's brigades in the next few months. This is the key that will decide everything."[13]

CHAPTER 57

BREAKING THROUGH THE DEFENSE LINE ALONG THE YELLOW RIVER

After eight months of the trial of military strength between the Kuomintang and the Communist Party, the KMT army had lost its ability to mount all-out attacks on the Communist-led troops. The KMT announced in a rage the dissolution of the "three-man" military subcommittee established to mediate peace, and the United States also withdrew from northern Shaanxi the personnel of its liaison group sent to Yan'an. The negotiations between the Kuomintang and the Communist Party broke down completely.

The Kuomintang was keenly aware that in fighting the Communist Party, it could never win victory by its own strength alone and that it had to have a strong supporter, the United States. The Kuomintang did not hesitate to pay any price for U.S. support, including the betrayal of the interests of the Chinese nation and the state.

The Kuomintang signed the following agreements with the United States:

> the Sino-U.S. Commercial Arbitration Agreement, stipulating that if U.S. nationals committed crimes in China, their cases would be submitted to the American "authorities" for arbitration;
>
> the Sino-U.S. Agreement on Joint Service of Military and Police, stipulating that if American armymen caused trouble, their cases would be handled by the American side and the Chinese side would have only the right to hear the case;
>
> the Sino-U.S. Treaty of Friendship, Commerce, and Navigation, stipulating that U.S. nationals would enjoy, anywhere in China, the right to reside and carry on commercial, manufacturing, processing, financial, scientific, educational, religious, and philanthropic activities; that U.S. commodities would be accorded the same treatment as Chinese commodities; that no prohibition or restriction would be imposed by China on the importation from the United States of any article grown, produced, or manufactured in the United States; and that U.S. vessels, including warships in time of "distress," would have the freedom to sail in any of the ports, places, or waters in China that were open to foreign commerce or navigation;
>
> the Sino-U.S. Air Transport Agreement, stipulating that U.S. aircraft would be allowed to fly anywhere within China, and also according the United States the right to make military landings on Chinese territory when necessary.

The Kuomintang gave away all it could.

Now the American armymen stationed in China could act wildly, in defiance of Chinese law, like an occupation army. Everyone knew how the American soldiers behaved. From August 1945 to November 1946 they committed atrocities on at least 3,800 occasions in the five cities of Shanghai, Nanjing, Beiping, Tianjin, and Qingdao alone, killing or wounding more than 3,300 Chinese. From August 1945 to July 1946 American military vehicles caused 1,500 accidents, and U.S. troops raped over 300 Chinese women.[1] These were evil acts that American soldiers did not dare to perpetrate on their own land.

The Chinese people saw these evil acts of the U.S. soldiers and bore them in mind.

On December 24, 1946, American soldiers gang-raped a Peking University student in Beiping. This incident gave vent to the pent-up indignation of the Chinese people and triggered a mass movement to protest the atrocities perpetrated by the U.S. troops. On December 30, more than 5,000 students from Peking University and Qinghua University held a demonstration to protest those atrocities. The angry students raised their arms and shouted: "The American soldiers withdraw from China," and, "Safeguard China's sovereignty and independence!"

Students throughout the country rapidly responded to the anti-American patriotic struggle waged by the students in Beiping. In Tianjin, Shanghai, Nanjing, Kaifeng, Chongqing, Kunming, Wuhan, Guangzhou, Hangzhou, Suzhou, and Taipei, 500,000 students held demonstrations. Many professors, scholars, cultural figures, and businessmen made statements supporting the students' patriotic movement.

The KMT's political and economic policies enabled American industrial and agricultural products to flood China's markets, thus gravely damaging Chinese industry and commerce. As a result, all domestic businesses languished, workers lost their jobs, and the markets were slow.

In 1946 the revenue of the KMT government was just under two trillion yuan (KMT currency), but its military expenditures were six trillion yuan. Worse still, in 1947 its revenue was thirteen trillion yuan, but total expenditures were forty trillion yuan, with the deficit accounting for 67.5 percent of total expenditures.

To make up the huge financial deficit, the KMT recklessly issued paper currency, thus leading to currency depreciation and serious inflation. In July 1947 commodity prices rose 60,000 times, and even 145,000 times toward the end of the year.

At the same time, American goods almost monopolized China's markets. At the end of 1947, in more than twenty major cities, over 27,000 industrial and commercial enterprises went bankrupt, large numbers of workers lost their jobs, and urban residents lived in poverty.

In the countryside, the KMT imposed enormous grain levies and taxes and press-ganged able-bodied men and carriers so that many fields were laid waste and people had to leave their homes. In Henan, Hunan, and Guangdong, uncultivated land totaled fifty-eight million mu. Large quantities of American farm products were dumped into China, thereby jeopardizing agricultural production in China. China had always been an agricultural, if backward, country, so the depression of the rural economy caused famine of a degree unknown for decades, and tens of millions of peasants suffered from it.

In 1947 a violent wave of rice-seizing occurred in more than thirty large Chinese cities, with 3.2 million people participating. Not long after, the same thing happened in more than forty medium-sized and small cities. Famine victims destroyed grain shops and KMT government organs, and in a few places they captured county magistrates. The KMT had not succeeded on the battlefield, and it was completely losing popular support as well.

Chiang Kai-shek turned a deaf ear to the people and ignored their acts of resistance. He thought that his autocratic rule could suppress the people's resistance and that his troops could defeat the Communist-led army.

After Chiang Kai-shek was forced to give up his all-out attack on the Communist Party, the KMT troops switched to the defensive on the battlefields in the Shanxi-Hebei-Shandong-Henan and Shanxi-Chahar-Hebei areas and in northeastern China. Chiang Kai-shek assembled his troops to launch concentrated attacks on the flanks of the Communist-led troops on the southern front, that is, to attack in the east the troops led by Chen Yi in Shandong and in the west the area in northern Shaanxi where the Central Committee was located.

Chiang Kai-shek concentrated ninety-four brigades and planned attacks from both east and west. At the same time, he ordered that the water of the Yellow River be forced to meet at Huayuankou, that is, to flow into its old course, so as to form a 1,000-kilometer defense line along the Yellow River, stretching westward as far as Fenglingdu and eastward as far as Jinan, Shandong.

In March 1947, thirty-four divisions of KMT troops 250,000 strong attacked Yan'an, where the Central Committee and the general headquarters of the CPC troops were located, from several directions, in an arrogant attempt to wipe out the leading CPC organs and Mao Zedong in the area west of the Yellow River. Only four brigades 17,000 strong and three local brigades protected Mao Zedong and the CPC leading organs. To protect the Central Committee, preserve troops, and attract a major part of the KMT troops to the northern Shaanxi battlefield, Mao Zedong decided to abandon Yan'an temporarily. He then led some troops to fight the enemy from place to place on the loess plateau in northern Shaanxi.

Mao Zedong left Yan'an, but he remained in northern Shaanxi. For his own safety and that of the Central Committee, many people tried to persuade him to cross the Yellow River to the eastern side. But he was determined to remain in northern Shaanxi. At the critical moment, when large numbers of enemy troops had invaded and were pursuing him every day, Mao Zedong was not scared but remained calm. He still had full confidence in victory. He said that since the troops controlled the vast border region and had the people's support, and because the terrain was precipitous, he was sure they could pin down the enemy troops and gradually wipe them out.[2]

After the KMT troops occupied Yan'an, Chiang Kai-shek was overjoyed, and he flew to Yan'an and trod on the land under the control of Mao Zedong. One can imagine how excited and proud Chiang Kai-shek was at that time. However, Mao Zedong was still in northern Shaanxi, and he still commanded all the Communist-led troops throughout the country.

In late March Chiang Kai-shek's troops launched an attack on Shandong. The commander-in-chief of the KMT army, Gu Zhutong, personally commanded sixty brigades some 450,000 strong to attack. The military situation there was critical.

The enemy stretched two fists, one to hit northern Shaanxi and the other Shandong. The troops under the command of Liu Bocheng and Deng Xiaoping were in between.

In accordance with the directive of the Central Committee, Liu and Deng decided to launch counterattacks. They planned to fight continuously for two months. By taking advantage of the enemy's shift to strategic defense in the area, they decided to wipe out a large part of the enemy's effectives, recover all lost territories that could be recovered, expand the liberated areas, and cooperate with the friendly troops in northern Shaanxi and Shandong in smashing the enemy's concentrated attacks. They chose two areas to mount counterattacks. One was northern Henan, and the other southern Shanxi.

In northern Henan, more than 95,000 enemy troops, commanded by Wang Zhonglian, Sun Dianying, and Sun Zhen, defended the areas east of the Beiping-Hankou Railway and north of the Yellow River. Liu and Deng resolved to launch counterattacks on northern Henan with the 1st, 2d, 3d, and 6th columns, 100,000 strong.

On March 23, the campaign started. The Liu-Deng army captured Puyang and other towns, forcing the enemy to move troops north for reinforcement. Their army resourcefully evaded the enemy reinforcements, advanced north, and liberated the vast areas north of the Weihe River and along the Beiping-Hankou Railway. Liu and Deng's main forces approached Anyang and mounted a converging attack on Tangyin. When the enemy troops came close, they were lured in deep and then ambushed by massive numbers of prepositioned troops. After the enemy troops were attacked and retreated south, the Liu-Deng army followed up the victory with hot pursuit, annihilated them all along the way, recovered lost towns, and finally captured Tangyin. The counterattack in northern Henan ended on May 25.

For two months the massive Liu-Deng army conducted mobile warfare on the battlefields of the vast areas of northern Henan and achieved brilliant results. It wiped out over 40,000 enemy troops, liberated nine county towns, expanded the liberated areas by 150 kilometers from south to north and 100 kilometers from east to west, and controlled a 150-kilometer section of the Beiping-Hankou Railway.

In southern Shanxi, Chen Geng led 50,000 men of the 4th Column and other units to launch counterattacks. After one and a half months of operations, they had wiped out over 14,000 enemy troops, recovered and liberated twenty-two county towns and a vast area with a population of three million, and controlled a 230-kilometer section of the Datong-Puzhou Railway. These successes completely changed the situation in southern Shanxi, smashed the joint defense system of the troops of Hu Zongnan and Yan Xishan, and posed a serious threat to the flanks and rear of the troops, commanded by Hu Zongnan, that were attacking northern Shaanxi.

While Liu and Deng's troops in the Shanxi-Hebei-Shandong-Henan Liberated Area achieved successes, in Shandong the troops under the command of Chen Yi annihilated over 32,000 enemy troops, including the Reorganized 74th Division under the direct control of Chiang Kai-shek, in the Menglianggu Mountains and killed his star student division commander, Zhang Lingfu.

In the Shanxi-Chahar-Hebei Liberated Area, the troops commanded by Nie Rongzhen launched a campaign along the Zhengding-Taiyuan Railway, wiping out over 35,000 enemy troops. In northeastern China, the troops commanded by Lin Biao and Luo Ronghuan conducted offensive operations for fifty days, wiping out four regular corps and other irregular troops totaling over 80,000 men.

In four months of operations from March to June 1947, the Communist-led army wiped out over 407,000 enemy troops and captured fifty-eight cities. The enemy troops failed in their concentrated attacks on the two battlefields of northern Shaanxi and Shandong, suffered heavy casualties, and thus became unable to extricate themselves from a difficult position. Strategically, the initiative in the war was gradually shifting to the Communist-led troops.

In the year since the civil war had broken out, Liu and Deng's troops had smashed many enemy attacks, launched regional strategic counteroffensives, liberated forty-three county towns, and annihilated thirty brigades of 300,000 men.

While the war was under way, land reform was resolutely carried out in the Shanxi-Hebei-Shandong-Henan Liberated Area, and peasants continued to show great enthusiasm for the Communist-led army's war efforts. A total of 240,000 emancipated peasants volunteered to join the army, and over 100,000 enemy prisoners did the same after education.

The total strength of the Communist-led army in the liberated area increased from 270,000 in 1946 to 420,000 in 1947; of that number, the strength of the field armies had swelled from 80,000 to 280,000. Liu and Deng's troops further expanded, with the newly organized columns as follows:

8th Column: Commander and Political Commissar Wang Xinting

9th Column: Commander Qin Jiwei and Political Commissar Huang Zhen

10th Column: Commander Wang Hongkun and Political Commissar Liu Zhijian

11th Column: Commander Wang Bingzhang and Political Commissar Zhang Linzhi

12th Column: Commander Zhang Caiqian and Political Commissar Liu Jianxun

Northwestern Democratic Joint Army: Commander Kong Congzhou and Political Commissar Wang Feng

Speaking about the operations of the first year, Father recalled later:

> In the first year of the War of Liberation, we fulfilled the targets of annihilating enemy troops set by the Military Commission.
>
> Three months after the war began, Chairman Mao said so long as we could wipe out eight enemy brigades every month, we were sure to win victory. Just as we expected, we overfulfilled the target in the first year by wiping out ninety-seven to ninety-eight enemy brigades. At the time, Chairman Mao said we were certain to win victory in the war. In our area, the Second Field Army overfulfilled the target a little bit and accomplished its tasks successfully. So the Second Field Army contributed its bit to the nationwide victory in the war.

This is what Father recalled. Indeed, Father fought continuously for a year, thought deeply, and made painstaking efforts to win victory.

In the middle of 1947, great and fundamental changes took place in the war situation in China. Total KMT military strength had fallen from 4.3 million to 3.7 million, with its regular troops reduced from 2 million to 1.5 million. Most of the KMT troops were pinned down on the battlefields in northern Shaanxi and Shandong, and only a small number of KMT troops were deployed to defend the central region, thus presenting a dumbbell-shaped war pattern. Chiang Kai-shek had lost the ability to launch large-scale offensives against the Communist-led troops.

Total Communist military strength had grown from 1.27 million to 1.95 million, with its regular troops increased from 610,000 to over 1 million. Its army had frustrated the enemy's concentrated attacks on northern Shaanxi and Shandong and switched to local counteroffensives on the battlefields of the Shanxi-Hebei-Shandong-Henan and Shanxi-Chahar-Hebei areas and northeastern China.

The outcome of war is often unexpected and can be very dramatic as well. In the game of *weiqi* (go), the player who moves the black piece first and is aggressive does not necessarily win the game.

Mao Zedong continued to lead some troops to fight Chiang Kai-shek's invading troops in the mountains in northern Shaanxi. He and the Central Committee were still in danger, but he began to plan a strategic counteroffensive. He and the Central Committee worked out the following strategic plan:

- The Liu-Deng field army would cross the Yellow River to the south, advance to the Central Plains, and operate on exterior lines.
- The field army commanded by Chen Yi and Su Yu would destroy Chiang Kai-shek's troops commanded by Gu Zhutong in coordination with the Liu-Deng troops.
- The troops commanded by Chen Geng in southern Shanxi would cooperate with the troops in northern Shaanxi in destroying the enemy troops commanded by Hu Zongnan.
- After the Liu-Deng troops crossed the Yellow River, they were to maneuver in the Central Plains in areas south of the Yellow River and north of the Yangtze River.
- The overall strategic intention was to immediately shift the main forces of the Chinese People's Liberation Army to a strategic offensive, select as the main target of assault the Central Plains, where the enemy was weak, and make a breakthrough in the center before the enemy's concentrated attacks were completely defeated and overall Communist strength exceeded that of the enemy.

When Communist military strength was still inferior to the enemy's and the leading organs of the Central Committee were still undergoing a siege, only a man as bold and great as Mao Zedong could have decided so resolutely to launch a strategic offensive.

Mao Zedong and the Central Committee began to direct their troops to start the counteroffensive.

The source of the Yellow River is in the Kunlun Mountains in western China, and the river winds through the land and empties into the Huanghai Sea. The 5,464-kilometer river divides the land of China into southern and northern parts. The mighty waves of the Yellow River have always rolled on and have molded the 5,000-year civilization of the Chinese nation. The Yellow River is the mother river of the Chinese nation. Strategically, the Yellow River has also been a natural protective screen, and Chiang Kai-shek arrogantly claimed that it was as useful as 400,000 massive troops. Liu Bocheng and Deng Xiaoping were to lead their troops to cross the Yellow River. They selected the 150-kilometer section from Zhangqiu Town to Linpuji in western Shandong as the area for crossing.

On June 3, the Central Committee ordered Liu and Deng to break through the defense line along the Yellow River by the end of June. The massive Liu-Deng army immediately began to make active preparations. First, the troops were educated on the launching of this large-scale counteroffensive and a campaign of military training was conducted. Second, the army lost no time in building and repairing vessels so as to have 300 large and small vessels ready before crossing the river. Efforts were also stepped up to train sailors and boatmen. Third, Liu and Deng issued the operational orders for the campaign, as well as the orders of tactical guidance.

On the eve of the crossing, Liu and Deng were very busy. Deng went on an inspection tour of the work in the army units and told his troops that the moment of revolutionary upsurge, when the Chinese people would overthrow the reactionary KMT rule, had approached, and that the time was ripe for the Communist-led army to shift to strategic offensive. Doing so would not allow the enemy a breathing spell before overall Communist strength surpassed the enemy's and the concentrated enemy attacks were completely defeated. He said that the war should be carried into the areas controlled by Chiang Kai-shek and that the enemy should not be allowed to smash our pots and pans. He said the Shanxi-Hebei-Shandong-Henan Liberated Area was like a shoulder pole carrying the two battlefields of northern Shaanxi and Shandong. He called for resolute implementation of the strategic guideline formulated by the Central Committee and Chairman Mao and for pushing the war into the KMT areas so as to pull the enemy out of northern Shaanxi and Shandong. The heavier the load might be after pushing the war into the enemy-occupied areas, the more favorable the overall situation would be.[3]

In June 1947 the battlefields on both sides of the Yellow River were ready. Liu Bocheng and Deng Xiaoping ordered their troops to formally launch the river-crossing campaign on the night of June 30. That date came at last.

That night the south wind was blowing and a bright moon was in the sky. The mighty Yellow River flowed swiftly eastward. There were no human voices or horses' neighs, and only the boundless reeds along the banks of the river rustled in the wind. On the eve of the battle, the banks of the Yellow River were unusually tranquil. But in the vast land north of the Yellow River, people were already brimming over with excitement.

The mighty Liu-Deng army was marching rapidly to the ferries. As the army passed, the people who lived in the area north of the Yellow River did not sleep well. The Liu-Deng troops had marched for hundreds of kilometers, and people in every village and town, men and women, old and young, lined the roads to greet them. The villages and towns presented their own troops, who were marching to the battlefront, with boiled water, shoes, and solid food. They pulled carts, drove draught animals, and carried stretchers to follow our troops so as to support them in the river-crossing. It seemed as if the people of the vast land north of the Yellow River were celebrating a festival. The emancipated peasants wanted to send their own army across the Yellow River.

At midnight, the Liu-Deng artillery forces opened fire. The thunderlike booming shook the night sky, and suddenly the other bank of the Yellow River became a sea of fire.

Several hundred wooden boats that had been concealed in the reeds were rushed into the river. Only then did the enemy troops on the other bank of the river realize that they were faced with imminent disaster. They hurriedly returned fire.

Some people said this unexpected attack of massive numbers of troops was like "divine troops coming down from heaven." If they were divine troops, the enemy certainly could not block their advance.

Overnight over 120,000 men of the four main columns of the Liu-Deng army broke through the natural barrier, crossed the Yellow River, and marched on the land south of the river.

The natural defense line that Chiang Kai-shek had claimed to be equal to "400,000 massive troops" was broken through by the Liu-Deng army at one fell swoop. The KMT army's "Yellow River strategy" had instantly come to nothing!

The crossing of the Yellow River by the Liu-Deng army was the prelude to the strategic counteroffensive launched by the Chinese People's Liberation Army.

CHAPTER 58

ADVANCING 500 KILOMETERS TO THE DABIE MOUNTAINS

The crossing of the formidable Yellow River by the massive Liu-Deng army shocked people throughout the country. John Stuart, the U.S. Ambassador to China, exclaimed:

> This is unbelievable! It is just like the breakthrough of the French Maginot Line! You [he was addressing the Kuomintang] spend thirty million silver dollars of American military aid each month, and your troops are equipped with first-class American equipment. Yet your defense line that you boast of being strong enough to fend off an army 400,000 strong was broken without firing a single shot. Your military capabilities are deteriorating with each passing day![1]

Chiang Kai-shek was enraged. He flew to Zhengzhou to hold a military meeting. His high-ranking generals, including Gu Zhutong, Bai Chongxi, Li Zhi, Sun Lianzhong, Wang Jingjiu, Wang Zhonglian, Hu Lian, Qiu Qingquan, Sun Yuanliang, and Li Mi, attended the meeting.

Chiang Kai-shek ordered Wang Jingjiu to hold firm ground at Juncheng, Heze, and Dingtao with fourteen brigades at all costs. The others were ordered to launch counterattacks on the Liu-Deng troops from all directions in an attempt to drive them back to the Yellow River and annihilate them there.

To fight with one's back to a river is the worse military scenario. But since there is no way to retreat, one can be incredibly daring. Deng said, "We are not cowards. We have only one choice between life and death. We must fight for the people and force the enemy to jump into the Yellow River." Liu added, "We should lose no opportunity and engage the enemy right now."[2]

Speed, bravery, determination, and resourcefulness have been precious assets in war since ancient times. Liu Bocheng was a great military strategist, and Deng Xiaoping a great statesman, so the close cooperation between the two was very productive. When they made up their minds to launch a campaign, they were sure to win.

Liu and Deng decided to launch a campaign in southwestern Shandong. Their troops maneuvered swiftly and resolutely launched the attack.

On July 8, the 1st Column seized Yuncheng. On July 10, the 2d Column recaptured Caoxian, the 6th Column captured Dingtao, and the 3d Column marched to the area southeast of Yuncheng. Thus, within ten days, the Liu-Deng troops opened up a vast battlefield south of the Yellow River and avoided fighting with their backs to the river.

After suffering a series of defeats, the main enemy forces under Wang Jingjiu were exposed and isolated in a single-line battle array from Juye to Jinxiang to the south of our troops. The enemy now thought that the Liu-Deng troops would either turn back to take Heze or proceed to attack Jining. They never expected that the Liu-Deng troops would suddenly turn and strike the long, isolated line of troops commanded by Wang Jingjiu. By July 13, three enemy divisions had been cut apart and rapidly encircled.

To prevent the enemy from putting up a desperate fight, Liu and Deng decided to attack the enemy troops at Liuyingji from three sides instead of four, thus forcing them to flee eastward. On July 14, the enemy troops that fled eastward were annihilated, from prepared positions. By then, most of Wang Jingjiu's troops had been wiped out. He had only one and a half brigades left, which were encircled by our troops at Yangshanji.

On July 19, Chiang Kai-shek flew to Kaifeng again and gave strict orders to Wang Jingjiu to rescue, under the cover of aircraft and tanks, the troops encircled at Yangshanji. Although the enemy troops at Yangshanji were only one and a half brigades, they were Wang's crack units, and the Liu-Deng troops had not been able to defeat them after mounting repeated attacks.

The Central Military Commission sent a telegram to Liu and Deng, instructing them to wipe out the enemy at Yangshanji if they could do so. If not, they were to immediately rest for about ten days and mount no attacks on areas along the Longhai Railway, east of the new Yellow River, or along the Beiping-Hankou Railway, and they were to advance swiftly to the Dabie Mountains in half a month without retaining their rear area.

Clearly, the Central Committee did not want to delay the southward advance of its troops. Liu and Deng thought about this. Deng said that the units attacking Yangshanji should not withdraw. Liu said jokingly that since Chiang Kai-shek had sent them rich meat, they should not put down the chopsticks! Thus, Liu and Deng decided to go ahead with the attack.

On July 27, the Liu-Deng troops launched a general offensive against the enemy troops at Yangshanji and completely wiped out the 66th Division after a full day of fierce fighting.

Gen. Chen Zaidao, who was then commander of the 2d Column, once told me with emotion: "The battle at Yangshanji was the toughest one I had ever fought. More soldiers were killed in that battle than in any other battles!"

The campaign in southwestern Shandong ended after twenty-eight days of continuous fighting. Liu and Deng employed a strength of fifteen brigades and annihilated four reorganized enemy divisions that consisted of nine and a half brigades, totaling over 60,000 troops. A lot of military matériel and 872 guns of various types were captured. Moreover, seven reorganized enemy divisions consisting of seventeen and a half brigades were diverted to reinforce enemy troops in southwestern Shandong.

Like a sword drawn from its scabbard, the Liu-Deng troops totally frustrated the enemy's strategic deployment. Chiang Kai-shek's personal command had twice proved to be of no avail.

While Mao Zedong and the organs of the Central Committee were playing hide-and-seek with the troops under Hu Zongnan on the mountain ridges and in the gullies of northern Shaanxi, the Central Committee had already drawn up a more general plan for strategic counteroffensive:

- The Liu-Deng troops were to march to the Dabie Mountains and conduct a strategic deployment in the Hubei-Henan-Anhui border area north of the Yangtze River.
- The troops under Chen (Geng) and Xie (Fuzhi) were to cross the Yellow River in southern Shanxi and conduct a strategic deployment in the Henan-Shaanxi-Hubei border area.
- The troops under Chen (Yi) and Su (Yu) were to conduct a strategic deployment in the Henan-Anhui-Jiangsu border area.
- The task of these troops was to advance to the Central Plains, support and cooperate with each other in a triangle formation, flexibly wipe out the enemy, and establish a new liberated area on the Central Plains.

Extending from the north-south Grand Canal in the east to the Funiu Mountains and the Hanshui River in the west, facing the Yangtze River in the south and the Yellow River in the north, the Central Plains covers the five provinces of Jiangsu, Anhui, Henan, Hubei, and Shaanxi. It is an area of strategic importance between the Yangtze River and the Yellow River in eastern China. In the south of the Central Plains are the key cities of Nanjing and Wuhan, which were under the rule of the Kuomintang. Further down is the hinterland of south China.

If the whole of China is likened to a rooster, the Central Plains area is its heart and lungs. Since ancient times, the Central Plains have been hotly contested in wars. Those who could take the area controlled at least half of China or could even pose a threat to southern China and the whole country. The CPC aimed to advance on and liberate the Central Plains and from there march to liberate the whole of China.

The Central Committee instructed Liu and Deng to make up their minds, abandon the rear area, march swiftly to the Dabie Mountains, seize the several dozen counties around the Dabie Mountains, mobilize the people there to establish base areas, and conduct mobile warfare by luring enemy troops to attack them. Meanwhile, the Central Committee issued the order that, in support of the Liu-Deng troops, the Chen-Xie troops were to leave western Henan and conduct mobile warfare against the troops under Hu Zongnan. The Northwest Field Army was to move north to attack Yulin, thus drawing the main forces of Hu Zongnan to the north. The Chen-Su troops were to pull Gu Zhutong's forces to eastern Shandong on interior lines and contain the forces under Qiu Qingquan to the north of the Longhai Railway on exterior lines, so as to ensure that the Liu-Deng troops penetrated deep into the strategic area of the enemy.

All troops were involved in the overall strategic plan of the Central Committee, and all were to act on its orders. It was like playing a chess game, with every chessman being put in position under a unified command. It was a comprehensive, active, and flexible game.

The main part of the Dabie Mountains is situated in Anhui, covering Hubei and Henan provinces. From northwest to southeast, they separate the northern China plains in the north from the Jianghan Plains in the south. Their terrain is complex, with narrow paths winding through towering peaks, steep valleys, and dense forests. Father described the importance of the Dabie Mountains:

> The position of the Central Plains is of great strategic importance. It is like the gate opposite the enemy, and the Dabie Mountains are right at the gate.
>
> The situation in the Central Plains rests on two big mountain ranges; one is the Dabie Mountains, and the other is the Funiu Mountains. The enemy is more concerned about the Dabie Mountains, for they are more important than the Funiu Mountains. To secure the Central Plains, it is essential to control the Dabie Mountains.
>
> The Dabie Mountains are a very good forward base, close to the Yangtze River, with easy access to Nanjing and Shanghai in the east and Hankou in the southwest. Therefore, it is an important springboard for crossing the Yangtze River. . . . The enemy wants to control the Dabie Mountains desperately, and we are determined to do so, too.[3]

Liu and Deng were determined to lead their troops to the Dabie Mountains. However, it was by no means easy to march to the Dabie Mountains and gain a foothold there. As a matter of fact, the Communist-led troops had previously penetrated the mountains on several occasions and then had to withdraw.

Both the Central Committee and Mao Zedong were keenly aware of the importance and difficulties of this expedition. In its instructions to Liu and Deng, the committee outlined three possibilities for the march to the Dabie Mountains: (1) the troops would not gain a foothold after fighting and would have to withdraw; (2) the troops would not gain a foothold after fighting and would then persist in struggle in the peripheral areas; and (3) the troops would gain a firm foothold after overcoming difficulties. The committee told Liu and Deng that while preparing for the worst, they should overcome all difficulties resolutely and bravely and strive for the best prospect. The committee also gave them special instructions to give the troops a rest until mid-August and then embark on the march.

The Liu-Deng field army had been fighting continuously for a month in southwestern Shandong. The troops had not had time to rest, and new recruits needed to be trained. All columns had just left the battlefield, and the mobilization and specific preparations had yet to be made for the strategic march to the Dabie Mountains. The troops had only enough funds to cover half a month's expenditures. The ammunition and military uniforms had to be shipped from northeastern China and Handan, respectively. These troops were going to have difficulties mounting a large-scale military operation and badly needed a breathing spell, no matter how brief it might be.

However, Chiang Kai-shek did not allow them to take it, nor did the fast-changing war situation.

It had been raining intermittently since the troops crossed the Yellow River, and the rain turned into a downpour, swelling the Yellow River. The Yellow River began to rise. Standing on the dike of the Yellow River, one could watch the waves surging forward from as far away as the eyes could see. The weather was extremely bad, and so were the motives of one man.

Chiang Kai-shek had decided to breach the dike of the Yellow River at Kaifeng, hoping the ensuing flood would drive the Liu-Deng troops back to areas north of the Yellow River.

He had resorted to dike-breaching warfare before. All those who had gone through the anti-Japanese war remembered what happened in June 1938, when Chiang Kai-shek ordered his troops to explode the dike of the Yellow River at Huayuankou, Henan Province, to prevent the advance of the Japanese troops. The breaching of the dike failed, however, to thwart the advance of the Japanese army. Instead, it caused the Yellow River to change its course to the north, flooding 54,000 square kilometers of land in forty-four counties in Henan, Anhui, and Jiangsu provinces. Over 890,000 people lost their lives and twelve and a half million people were miserably displaced. A vast stretch of barren flooded land appeared on the Central Plains.

Almost nine years had passed, but people's memories were still haunted by the breaching of the dike in 1938.

The Yellow River had become a weapon of war, to be moved at will by Chiang Kai-shek. And now he was plotting the same scheme against the Liu-Deng troops. If the dyke were breached, disaster would befall the 100,000-strong army and several million people on their side of the river. In the operations room of the headquarters of the field army, Liu Bocheng said that he was very anxious. The situation was indeed very critical. More than forty years later, Father told us: "That was the tensest moment in my life. When I learned that the Yellow River dike would be breached again, I could hear my own heart beating fast!"

While the troops were resting and preparing for the mid-August march, Liu and Deng received a "triple A" top-secret telegram from Mao Zedong himself. Father told us:

> The telegram had just one sentence ["The situation in northern Shaanxi is rather serious."] Only I and Uncle Liu read the telegram, which was burned at once. We ourselves were truly in great difficulty. Still, without any hesitation, we immediately sent a telegram in reply back to the Central Committee, telling it that we would start the march after ten days. Ten days were too short to prepare for a 500-kilometer march. Yet we made it in less than ten days.

Father seldom reveals his emotions, yet when he said, "We just defied whatever difficulties faced us and went into action," his voice trembled.

The crossing of the Yellow River by the Liu-Deng troops had two aims: to launch the strategic counteroffensive, and to lure and wipe out the enemy. More important, their military actions were to ease the pressure on northern Shaanxi, the Central Committee, and Mao Zedong.

Neither the swelling of the Yellow River nor Chiang's threat to breach its dike could scare Liu and Deng. They had thought about fighting several more battles and wiping out more enemy troops there. However, since the Central Committee was in a difficult position, they set off on the march ahead of schedule, despite difficulties.

On August 6, Liu and Deng gave the preparatory command and made up their minds to end the troops' rest ahead of schedule and to immediately start the strategic march to the Dabie Mountains. Upon receiving Liu and Deng's telegram, the Central Committee responded in two telegrams, on August 9 and 10: Liu and Deng's "decision is completely correct," and "if there is no time to report for approval, all emergencies are to be handled at your discretion."

Liu and Deng instructed the troops to march bravely forward, to give up the rear area, and to accomplish resolutely and courageously this glorious yet arduous strategic task. On August 7, 1947, Liu and Deng began moving their troops southward from the Yuncheng area in southwestern Shandong. The great 500-kilometer strategic march to the Dabie Mountains thus began.

In southwestern Shandong, Chiang Kai-shek himself went to Kaifeng to assume direct command and assemble a powerful force of thirty brigades from five corps.

To make a breakthrough, Liu Bocheng played a game of deception with the enemy. Some troops made demonstrations along the Yellow River, creating the impression that they intended to cross the river to its northern bank. Some other troops moved west to destroy the Beiping-Hankou Railway so as to cut off the enemy communication line, and still others advanced west toward Xinyang, pretending to march toward the Tongbai Mountains.

While the enemy was confused and failed to make a correct judgment, the main forces under Liu and Deng, before the pursuing enemy could encircle them, suddenly left the enemy far behind, dividing themselves into the central, left, and right columns, and started the march toward the Dabie Mountains.

The southwestern Shandong area is 500 kilometers away from the Dabie

Mountains in Anhui. On the way, there were natural barriers, such as the flooded area, the Shahe River, the Ruhe River, and the Huaihe River, to say nothing of the several hundred thousand enemy troops who were either pursuing the Liu-Deng army or trying to block its way. One needed supreme courage and perseverance to surmount those formidable obstacles.

On August 11, the main forces under Liu and Deng crossed the Longhai Railway and advanced swiftly southward. Several days later, the troops reached the 20-kilometer-wide area that had been flooded by the Yellow River. It was covered with water and knee-to waist-deep mud. There was no road, nor were there signs of human habitation. The towns and villages marked on the military map could no longer be found. All one could see were scattered demolished houses and bare tree trunks here and there. The scene was desolate and miserable.

That was the area flooded by Chiang Kai-shek. Scorched under the August sun, the troops waded through the muddy land with great difficulty. Every step forward required strenuous effort. As the trucks could not move over the muddy land, all the heavy weapons and matériel either had to be pulled by cows and pushed by soldiers from behind or had to be disassembled and carried on shoulders. All the howitzers and field guns as well as the towing vehicles, which were too heavy to be pulled or carried, had to be demolished.

Seeing the destruction of these heavy weapons, which had been captured one by one from the enemy at the cost of blood, many officers and men could not hold back their tears. But they knew very well it had to be done to ensure the smooth advance of the troops.

With the help of sticks, and defying enemy air attack, on August 17 Liu and Deng waded through the deep mud in the flooded area together with all the officers and men. The main forces of the field army passed through the area the same day. Inspired by the exemplary actions of their leaders, the whole field army quickly crossed the flooded area on August 18.

Leaving the flooded area behind them, the main forces of the Liu-Deng troops advanced toward the Shahe River. Braving enemy artillery bombardment and air attack, they built pontoon bridges over the river and over which all the troops marched to the other side.

It was not until then that Chiang Kai-shek finally saw the strategic intention of the southern advance of the massive Liu-Deng troops. However, it was too late to try to organize large-scale blocking operations against their advance.

It was said that General Wedemeyer of the U.S. military advisory group complained bitterly to Chiang Kai-shek on August 24 before leaving China: "What have I seen in China this month? The Communist forces broke without firing a single shot the 'Maginot Line of China' that was supposed to fend off the attack of an army 400,000 strong. They fought for twenty-eight consecutive days and wiped out nine and half KMT brigades. You claim that they are fleeing westward, whereas in fact they are making a southern advance. Then you say they have disappeared. Yet it turns out that they are launching a counteroffensive!" If General Wedemeyer was exasperated, so was Chiang Kai-shek, who retreated to the Lushan Mountain to recuperate.

After crossing the Shahe River, the Liu-Deng troops were able to rest for one day at last. Then Political Commissar Deng urged the whole army: "March to the Dabie Mountains, and victory will be ours!"

To accelerate the pace of advance, the troops buried or destroyed some bulky weapons and vehicles.

On August 23, the 1st and 2d columns crossed the Ruhe River, and the 3d Column reached the Huaihe River. By the time the headquarters of the field army and the 6th Column arrived at the bank of the Ruhe River, the enemy had occupied the ferry on the southern bank. They were blocked by heavy enemy firepower in front; behind them, three enemy reorganized divisions pursuing the field army were less than thirty kilometers away; and above them, enemy aircraft were unrelentingly conducting air raids. Repeated attempts to cross the river failed.

At this critical juncture, Liu and Deng came to the front of the Ruhe River shoulder to shoulder. "Our chiefs Liu and Deng have come!" Their presence greatly boosted the morale of the troops. Deng made a brief yet forceful call: "We must cross the river at whatever cost!" Liu raised his normally low voice and said firmly: "When two adversaries encounter each other on a narrow path, only the daring one will prevail. Charge! Let's fight our way through with our blood!"

Their presence and determination, and their order, had a powerful effect on the officers and men, whose courage was plucked up considerably. The 6th Column started crossing the river at dawn. Braving the fierce enemy firepower, the soldiers charged through the pontoons amid the fallen bodies of their comrades and crushed the enemy defense line on the other side of the Ruhe River. After the crossing, people saw Liu and Deng marching on horseback with the swiftly advancing troops. Calm and composed with full confidence, they from time to time exchanged notes and stopped to observe the terrain. They not only commanded but constantly stayed with their troops.

Upon arriving at Pengdian, the assembly point, Political Commissar Deng told the troops, "We still need to cross the Huaihe River, another formidable natural barrier, before reaching the Dabie Mountains." After a march of more than two days, the troops occupied the county town of Xixian, on the northern bank of the Huaihe River, and several ferries.

The Huaihe River is the largest river system between the Yellow River and the Yangtze River in eastern China. It flows from west to east through southern

Henan and the whole of Anhui, and this area is crisscrossed with its tributaries, forming a natural barrier to Liu and Deng's advancing troops.

As most of the officers and men of the Liu-Deng troops were from northern China, they had little experience fighting in the river network areas in the south. One could not keep count of the rivers there were to cross, or how many times one got tumbled in the mud. No wonder some said humorously, "The landscape is so beautiful, and numerous heroes just stumble and fall." Despite stumbles and falls, however, the Huaihe River had to be crossed.

The current of the Huaihe River was swift in the rainy season of August. Moreover, a downpour in the upper reaches had caused the river to swell by the time the Liu-Deng troops arrived there. The pursuing enemy was only fifteen kilometers behind. If the troops could not cross the river in two days, they would have to fight the enemy against the river.

Yet there were neither bridges nor boats, and the river was wide. How could the more than 100,000 troops get across?

Liu and Deng held an emergency meeting in the headquarters that night. Deng proposed that Liu Bocheng first direct the troops to cross "while I and Li Da stay behind to intercept the pursuing enemy." Liu said decisively, "What the political commissar proposes is what we will do. We act at once." Liu reached the riverside.

A cadre reported, "We cannot wade through the Huaihe River." Was it true that it was impossible to wade through the river? Liu Bocheng got on a bamboo raft to fathom carefully the depth of the river with a bamboo pole in one hand and a lantern in the other. Soon he sent back a message that the water was not very deep and that the troops could actually wade across the river!

At daybreak, the water began to subside. It was indeed an opportunity provided by heaven! All the troops started crossing at once. Liu Bocheng was already on the southern bank of the river, and he was standing on a hilltop watching the spectacular crossing scene with a smile. Later he said, "A careless style would have led to disaster!" By August 27, the whole Liu-Deng field army had crossed the Huaihe River.

Interestingly enough, the flood from the upper reaches caused the water level to rise rapidly just after the troops had crossed the river. Separated by the vast expanse of water, the pursuing enemy could do nothing but watch the Liu-Deng troops marching out of their sight.

This experience is unforgettable to Father. He once talked with us about crossing the Huaihe River and told us:

> The crucial part of the march was the crossing of the flooded area and the Huaihe River. When Uncle Liu went to fathom the depth of the Huaihe River, he found that the water was up to his neck, and one could just wade one's way through. What an opportunity! We had barely finished crossing when the water rose. We

> truly had good luck. No one knew until then that one could wade his way across the Huaihe River. Yet we found out that this could be done. Many things are unbelievable. But it seemed that heaven was on our side, too!

Was it a mere coincidence or a miracle? Actually, it was indeed help from heaven. In accordance with the major principles of heaven, it was right that a helping hand be given to the Liu-Deng troops. In the words of a Chinese poem: "Were Heaven sentient, she too would pass from youth to age." It seems as if sometimes heaven is indeed sentient.

Having crossed the Huaihe River, the last natural barrier, the Liu-Deng field army entered the Dabie Mountains, successfully accomplishing the strategic task of marching 500 kilometers.[4]

The Communist-led troops thus gained the upper hand in the contest on the Central Plains with the Kuomintang forces.

CHAPTER 59

CHASING DEER ON THE CENTRAL PLAINS

Whether chasing deer on the Central Plains or vying for power in China, whoever wins the battle becomes the monarch and whoever loses becomes a bandit.

Over the past 2,000 years or more, emperors, kings, generals, prime ministers, and heroes have persistently, sometimes courageously, contended for supremacy. Chasing deer is a metaphor for contention for political power in China. It refers not to hunting but to the rise of heroes and their contention for political power.

When the Qin Empire, which was established by China's first emperor, Qin Shihuang, was about to collapse around 200 B.C., the heroes in all parts of China rose and contended for power. It was recorded in the history books that "when the Qin Empire was about to lose its deer, all people chased it."[1] That meant that, when all the heroes rose, at whose hands the deer would die depended on many factors.

In terms of territory, it has been true from ancient times to the present that whoever controls the Central Plains can control the whole of China. Therefore, whoever wants to seize power over all of China must chase deer on the Central Plains.

In 1947 the Liu-Deng troops advanced southward for 500 kilometers, thus carrying the war into the Central Plains, which were under the control of the KMT. Then the curtain of the battle, between the Communist Party and the KMT, of chasing deer on the Central Plains was raised.

The Liu-Deng troops reached the Central Plains, but it was not easy to gain a foothold there. Liu and Deng led their troops to the Dabie Mountains, but it was not easy to gain a foothold there either. The rolling, towering Dabie Mountains, with their green forests and wild grasses, could be a natural fort favorable to defense and a dangerous place where a strong army might be annihilated.

Liu and Deng knew very well the difficulties and importance of "chasing deer" on the Central Plains. On August 27, as soon as Deng Xiaoping arrived in the Dabie Mountains, he issued instructions in the name of the Central Plains Bureau of the CPC as follows:

1. All personnel are asked to wholeheartedly carry out their duty to establish and consolidate the Dabie Mountain Base Area and, in cooperation with friendly formations, control the whole of the Central Plains.
2. It is necessary to undergo a process of arduous struggle to fulfill this historical task. Only when we annihilate large numbers of enemy troops and fully mobilize the masses can we gain a firm foothold. Therefore, we should never be conceited, arrogant, or impatient. Instead, we should earnestly fulfill each specific task, with one mind.

3. It is imperative to make it clear to the whole army that we are fully confident of success.

4. It is imperative to make it clear to the masses in the whole area that we, the army of the people of the Hubei-Henan-Anhui area, have come back, and our slogan is to live or die with the people in the Hubei-Henan-Anhui area and to liberate the Central Plains.

5. Militarily, we do not intend to fight a large-scale battle but to capture cities and towns and eliminate bandits and diehards in the first month. We should seek to fight some small battles, to get familiar with the terrain, to get used to the life here, and to learn to conduct mountain warfare, thus creating conditions for fighting large-scale battles of annihilation. But we should also eliminate the strength of more than ten enemy brigades in six months.

6. We should fully mobilize the masses to conduct guerrilla war. The army should strictly observe discipline and pay close attention to the armymen's bearing and discipline.

Father always gave instructions so clearly, simply, and firmly. After the founding of the Liu-Deng army, numerous instructions, orders, and especially reports and telegrams to the Central Committee were written by Father himself. He had no secretary. In those war years, when the military situation was very urgent, time was the key to the success of the war effort and the life of the army. Doing important jobs himself, indulging in no bad habits or red tape, was Father's consistent work style. Many old generals of the Liu-Deng field army emotionally recalled this characteristic of my father.

As soon as the Liu-Deng troops entered the Dabie Mountains, all units were deployed rapidly to predetermined places. The 3d Column advanced to western Anhui, and the 2d and 1st columns were deployed in southeastern Henan. The 6th Column captured a dozen county towns along the line from Henan to eastern Hubei and advanced southward to an area over 100 kilometers east of Wuhan. The nine brigades were thus deployed in the area at the northern foot of the Dabie Mountains and assumed a defensive posture.

Chiang Kai-shek employed twenty-three brigades to pursue closely the Liu-Deng troops after crossing the Huaihe River. Among the commanders dispatched by Chiang Kai-shek were Xia Wei, Zhang Zhen, and Cheng Qian. Defense Minister Bai Chongxi assumed command himself. Their troops launched attacks from the east, the north, and the west in an attempt to engage the main forces of the Liu-Deng army. They intended to try driving away the Liu-Deng troops that had not gained a firm foothold or wiping them out in the Dabie Mountains.

In accordance with the directive of the Central Committee, the Liu-Deng troops evaded the main forces of the Guangxi army commanded by Bai Chongxi and attacked the weak Yunnan army first.

In early September Liu and Deng employed the 1st and 2d columns and part of the 6th Column to fight the first battle with the enemy in the Dabie Mountains. The operations were conducted in areas north of Shangcheng, Henan Province, at the northern foot of the Dabie Mountains. Although this battle accomplished the purpose of luring other enemy troops into the area to serve as reinforcements, the Liu-Deng army failed to attain the goal of effectively annihilating the enemy because the troops from northern China were not familiar with operations in mountains and paddy fields.

On September 19, Liu and Deng concentrated the troops that had joined in the previous battle and wiped out an enemy regiment in areas west of Shangcheng, thereby continuing to lure other enemy troops as reinforcements. This was the second battle fought by the Liu-Deng army after marching into the Dabie Mountains. On September 25, the Liu-Deng troops fought the third battle near Guangshan, northwest of Shangcheng, in which they repulsed the attack mounted by enemy reinforcements. These three operations lured all enemy mobile forces to the northern foot of the Dabie Mountains, thus ensuring the disposition of our 3d and 6th columns in the areas of eastern Hubei and western Anhui.

By the end of September, after fierce battles, the Liu-Deng troops had liberated twenty-three county towns in the Hubei-Henan-Anhui area, annihilated more than 6,000 enemy troops, and established democratic governments in seventeen counties. Liu and Deng had created this situation in one month and established a rear area in the mountain area. But one month had definitely not been long enough to enable the Liu-Deng troops to gain a foothold in the Dabie Mountains.

Many difficulties beyond imagination lay ahead of Liu and Deng. First, although this area was once a base area of the Red Army, the Kuomintang had suppressed the local masses most cruelly after the Red Army left. After the Liu-Deng troops arrived, reactionary forces threatened the masses and cut off the sources of grain for the troops. As a result, they had nothing to eat and could not find guides. They often felt hungry and fatigued and lost their way.

Second, the Dabie Mountains are different from the Taihang Mountains. The Dabies are high, paths are narrow, and the terrain is complex. As the Liu-Deng troops came to the south from the north, and to the mountains from the plains, they could not get used to the food or understand the local dialects, they were unfamiliar with the terrain, and they were not accustomed to wearing straw sandals. Furthermore, they often had night marches. They found it very difficult to adapt themselves to these conditions.

Third, our troops were confronted with a massive number of enemy troops—twenty-three brigades' worth. They were extremely fatigued after continuous operations and thus showing some lack of confidence. Later Father told us:

> We chased deer on the Central Plains and had enemy troops all around us!
>
> China is divided into north and south, with the Huaihe River as the boundary line. The area south of the Huaihe River is called the south. South of the Huaihe River, people grow rice and walk in the mountains. These are the habits of the people of the south. We underestimated the situation. We only knew that the northerners would have the problem of adapting themselves to the circumstances [in the south]. We did not expect that we, the southerners, would also find it difficult to adapt ourselves to the circumstances in the south after staying in the north for a long time. The load was on the shoulder of the Second Field Army, and that was the most difficult time for it throughout the War of Liberation.
>
> The success of the struggle in the Dabie Mountains did not hinge on how many enemy troops were annihilated but on whether we could gain a firm foothold there. This required correct decisions on the concentration and dispersion of forces. The key to success was to evade the enemy main forces without fighting bitter battles.
>
> At that time, our troops were impatient and making strong demands to fight. I called a meeting attended by a dozen commanders. Some column commanders were on horseback for over fifty kilometers to attend the meeting. I persuaded the cadres not to fight tough battles. After they went back to their units, the troops were in a good mood. The situation had changed three months later.

The meeting Father referred to was a meeting attended by column commanders and held at Baiqueyuan, Guangshan County, on September 27, 1947. The meeting pointed out the need to strengthen the fighting will, to oppose right-deviationist sentiments, and to take seriously the need to put an end to violations of the law and discipline. This meeting was very timely and essential to reviewing the month of work accomplished after the field army had marched into the Dabie Mountains and to planning future work.

To gain a firm foothold, it was imperative to mobilize the masses. To mobilize the masses, it was necessary to tighten discipline. Father was always very strict about maintaining discipline in relation to the people. He said, "Poor army discipline is the beginning of political crisis." Immediately after their troops entered the Dabie Mountains, Liu and Deng issued a strict order: whoever shoots at the people, robs their property, or rapes a woman would be shot to death!

Political Commissar Deng once discovered that an armyman was walking on a street in Huanggang County with a bundle of colored cloth and a bundle of noodles hanging from his bayonet. Clearly, he did not get these things in a proper way, and the matter was checked into immediately. Deng found out that the armyman was the deputy commander of a security company and had performed meritorious service before. After weighing the advantages and disadvantages, Deng decided at last to enforce strict discipline.

The execution of a deputy company commander who violated discipline was welcomed by large numbers of local peddlers and the masses. People lost no time in quickly spreading the news to all parts of the Dabie Mountains: the Red Army had really come back!

Liu Bocheng once said that kindness was detrimental to leading troops. If kindness was detrimental to leading troops in wartime, kindness was detrimental to leading troops in disciplinary matters as well.

While he was strict with the troops, Father attached the greatest importance to the ideological work of the army. He went personally to the army units and made speeches to cadres in grass-roots units. One day in early October, Political Commissar Deng, in a gray uniform and with a leather belt around his waist, spoke to the cadres at and above company level from the 2d Column on a hillside lawn. He said,

> Now there are some comrades who don't dare to talk about difficulties to their troops. Even if you do not talk, there are still difficulties objectively. We should not be afraid of them. On the contrary, we should face them squarely and explain to the troops realistically. Only by so doing can we be fully prepared mentally for them and find ways to overcome them actively and on our own initiative.
>
> Since we are far away from our rear area, how can we have no difficulties in the enemy-occupied areas? We move around here, luring hundreds of thousands of enemy troops behind us. We cannot replenish ammunition, grain, bedding, or clothing. Our soldiers are not acclimatized. Many of them are ill and have contracted malaria. The wounded and the sick cannot receive good medical treatment. The mass base and material supply here are far inferior to those in the liberated areas. All these have caused us extremely great difficulties. On the face of it, we have difficulties, but we should not be afraid of them. In making revolution, we have to cope with difficulties inevitably and should have the endurance to overcome difficulties.
>
> Chairman Mao said: It is a victory if you advance to the Dabie Mountains! Why? Because we have penetrated into the heart of the enemy and hit home. We have lured large numbers of enemy troops, and the pressure on us has increased. We are far away from the rear area, and we have more difficulties. But our friendly troops on other battlefields have a lighter burden and thus can win battles.
>
> Our march to the Dabie Mountains is just like playing basketball. Seeing that we come to the Dabie Mountains to shoot and are about to score points, Chiang Kai-shek has moved his forwards and guards to follow us. Thus, he attends to the south and loses sight of the north. He does not allow us to shoot in the south without hesitating to employ hundreds of thousands of his troops to pester us. But his backboard in the north is without guards, and so our friendly troops can shoot and score points in the north. We have enormous difficulties in the Dabie Mountains. We are "gnawing bones," but our friendly troops have started "eating meat" on other battlefields! The more enemy troops we lure and the harder bones we gnaw, the more enemies our friendly troops can annihilate on other major battlefields and the greater victory they can win. The victory they win can, in turn, support us and lessen the pressure on us.
>
> We have difficulties, and Chiang Kai-shek has difficulties, too. Ours are local and temporary difficulties, the ones in our advance and victory, whereas theirs are

> difficulties of overall importance, the insurmountable difficulties in their gradual move toward destruction because they are encircled ring by ring by the people in both the liberated areas and the Chiang-controlled areas. Although we have some difficulties now and we have to lose several jin of weight and pay some price, this doesn't matter much. For the victory of the revolution throughout the country, it is both worthwhile and very glorious.[2]

The conviction of Liu and Deng increased the confidence of all the units and the morale of all the officers and men.

The autumn was not yet over in October in the Dabie Mountains, but there was a nip in the air both in the morning and in the evening. As the Liu-Deng troops were far away from the rear area and short of supply, the officers and men were still wearing unlined clothes. As the cold winter approached, the problem of winter clothes became very conspicuous. No rear area, no supply. What to do, then? Liu and Deng ordered all troops to make winter clothes themselves.

To make winter clothes, the troops had to find cloth first. As the cloth purchased was in diverse colors, the troops had to use indigenous methods to dye it. They dyed white cloth gray with pot and plant ash, to be used as the outside of the clothes, and bright-colored cotton print was used as the inside.

They also had the problem of getting cotton to pad the clothes with. Most of the cotton purchased was unginned and still had cotton seed in it. The troops had to process it themselves. As there were no cotton fluffers, the soldiers employed indigenous methods again. They used tree branches to thrash the cotton and tore and peeled it with their hands.

Now cloth and cotton were available. People were needed to make the clothes. This caused enormous difficulties to the armymen. Since ancient times, males have been used to holding hoes, pens, and guns, but not needles! There was also the problem of cutting the cloth out. It got colder and colder every day, so the troops were forced to make their own cotton-padded clothes. The clumsy hands that were used to holding guns had to take up tiny needles. Women villagers were asked to help cut the cloth out. Commander Liu Bocheng, setting an example, put on his spectacles and took a needle in his hand to stitch his own clothes. He said patiently, "There is a knack to stitching clothes. Hook-shaped needles are used to stitch pockets in a well-distributed and close way, buttonholes are made in reverse direction, and a china bowl can be turned upside down to size a neckband."

The soldiers made their own cotton-padded clothes, and so did Liu and Deng. Liu said, "We will put on our uniforms, no matter how great the difficulties are. We will by no means view raggedness as an honor!" Deng said, "Our army has a great advantage. So long as we get to work ourselves, there will be no insurmountable difficulty!"[3]

When the winter clothes were made, all army personnel put on their new cotton-padded clothes. At first glance, these clothes were uniforms by and large. But on careful examination, one could see that they were made crudely. The stitches were not neat, and the cutting out was not very exact. The clothes were shapeless and twisted, and the cotton was not neatly padded. Nevertheless, these unsightly clothes were made by the soldiers with their own hands, and they felt especially warm. Furthermore, Liu and Deng wore the same winter clothes!

In early October, when the Liu-Deng troops had just adapted themselves to the circumstances in the Dabie Mountains, Chiang Kai-shek concentrated the seven reorganized divisions at the northern foot of the Dabie Mountains to encircle the Communist-led troops in Guangshan and Xinxian.

The Liu-Deng army moved flexibly to wipe out the enemy. By October 27, it had annihilated a total of more than 12,000 enemy troops and captured large quantities of military matériel, thus winning its first major victory since marching into the Dabie Mountains.

By November of the same year, the Liu-Deng troops had annihilated a total of more than 30,000 enemy troops, liberated twenty-four county towns, and established governments in thirty-three counties. The Liu-Deng army thus completed its strategic deployment in the Dabie Mountains.

While the Liu-Deng army marched to the Dabie Mountains, the troops led by Chen Geng and Xie Fuzhi advanced to western Henan. They maneuvered for three months, wiping out a total of over 50,000 enemy troops and liberating a dozen county towns. While completing their strategic deployment, they lured eight enemy brigades, thus cooperating effectively with the operations conducted by Liu and Deng in the west of the Dabie Mountains.

In cooperation with the Liu-Deng and Chen-Xie troops, the massive army led by Chen Yi and Su Yu advanced to the Henan-Anhui-Jiangsu border area. In their rapid three-month advance, they annihilated over 20,000 enemy troops, lured fifteen enemy brigades, including the eight brigades that were supposed to be used to attack the Dabie Mountains, frustrated the enemy's military deployments, and expanded the liberated areas north of the Liu-Deng troops.

Thus, in accordance with the direction of the Central Committee and Mao Zedong, the three groups led by Liu and Deng, Chen and Xie, and Chen and Su were in place. They assumed a favorable tripod-shaped strategic posture on the Central Plains, with the Yellow River in the north, the Huaihe River in the center, the Yangtze River in the south, and the Hanshui River in the west, thereby turning the enemy's important rear area for attacking our liberated areas into an advance base for winning the victory across the country.

In the first round of deer-chasing on the Central Plains, the Communist Party had already completed its deployments.

On October 10, 1947, in light of the changing and developing war situation, the Communists made the proud call, in "The Declaration of the Chinese

People's Liberation Army": "Overthrow Chiang Kai-shek! Liberate the whole of China!"

The Liu-Deng troops advanced to the Central Plains and, after overcoming numerous difficulties and dangers, gained a firm foothold in the Dabie Mountains, thus bringing about the best of the three prospects spelled out by Mao Zedong. However, the enemy situation was as serious as ever. In late November Chiang Kai-shek set up a Defense Ministry Jiujiang headquarters with Defense Minister Bai Chongxi serving as the director. This headquarters exercised a unified control over the military and administrative power of the five provinces of Henan, Anhui, Jiangxi, Hunan, and Hubei on the Central Plains. It was designed to contend for the Central Plains with the Communist Party with so-called total warfare and to guarantee the safety of its major artery, the Yangtze River. After receiving the order, Bai Chongxi first concentrated fifteen reorganized divisions and three brigades to start the all-out converging attack on the Dabie Mountains with the support of fighters and bombers on November 27.

In view of the new enemy situation, Mao Zedong issued directives to the three army groups on the Central Plains: the consolidation of the Dabie Mountains was the key to the eventual establishment of the Central Plains Liberated Area, the consolidation of which could have a considerable impact on the development of the war. Therefore, the three army groups at the southern front were to fight in long-term and close cooperation. The main forces under Liu and Deng were to hold fast to the Dabie Mountains. The troops commanded by Chen and Su and by Chen and Xie were to conduct large-scale sabotage operations along the Beiping-Hankou and Longhai railways and maneuver to annihilate the enemy, so as to lure the enemy troops that were attacking the Dabie Mountains, until the enemy's converging attack on the Dabie Mountains was smashed.

Bai Chongxi's thirty-three brigades started their converging attack and pounded on the Communist-led forces in the Dabie Mountains mercilessly. In the face of the increasingly grim enemy situation, Liu and Deng held that there was little room for maneuver in the Dabie Mountains, the food supply was difficult, and maneuvering operations of great width by large formations were too difficult. Accordingly, it was not correct to concentrate too many troops in the Dabie Mountains.

They decided to adopt the principle of "evading battle." The main forces were to stay in the Dabie Mountains and conduct small operations and guerrilla warfare on interior lines so as to pin down the enemy. Meanwhile, the headquarters, with some troops, was to jump out of the ring of encirclement, turn to exterior lines, and make strategic deployments in the areas of Tongbai and Jianghan west of the Dabie Mountains.

At nightfall on December 10, 1947, Liu Bocheng, Deng Xiaoping, and Li Xiannian, the new deputy commander of the field army who had just arrived some ten days before, studied the operational plans at Wangjiawan, a small village over 100 kilometers north of Hankou. Deng Xiaoping told Liu Bocheng, "I am younger than you. So I will stay in the Dabie Mountains to conduct operations, and you will go to Huaixi to direct the overall operations." Liu Bocheng replied, "The security regiment is to be left to you, and I will take just a platoon with me."[4] Liu and Deng parted that night.

With some units each, one of them held fast to the Dabie Mountains on interior lines under the converging attack mounted by massive numbers of enemy troops, and the other operated on exterior lines. Although Liu and Deng parted, their actions were not separated. There were two headquarters, one inside and one outside. But all telegrams and instructions continued to be signed by Liu and Deng jointly during this time of separation. Even when they were not in the same place, Liu and Deng were an integrated whole.

With only a few capable staff members, Father led a forward command post to direct the 2d, 3d, and 6th columns, which had stayed in the Dabie Mountains to conduct the arduous operations against the strong enemy.

The year 1948 came. Father and his comrades-in-arms spent New Year's Eve in the Jinzhai area at the northern foot of the Dabie Mountains. Father in gray cotton-padded clothes looked thinner. In the pine branch light that flickered in the cold wind, Father listened to the radio broadcast of the voices of the Central Committee and Chairman Mao and the report on the local work simultaneously. That was how he greeted the first dawn of 1948.

In early 1948, by relying on their overwhelmingly superior strength, the enemy troops employed close formations to press on, block, and attack the Communist troops in the Dabie Mountains from south to north and quickly occupied the heartland. They carried out the frenzied policy of "burn all, kill all, and loot all" everywhere, destroyed the newly established democratic governments, arrested and killed local cadres, looted the properties of the people, press-ganged able-bodied men, and even cruelly created a no-man's-land. The situation in the Dabie Mountains was unprecedentedly serious.

The command post led by Father was staffed with less than 1,000 men. The tactic he established was to break the whole into parts: the main forces were to advance outward when the enemy moved either outward or inward, thus luring the enemy to exterior lines. Thus, small units were used to pin down large enemy units, and large units were used to wipe out small enemy units. Father recalled vividly later:

> We three, I, Li Xiannian, and Li Da, led the forward command post with several hundred staff to stay in the Dabie Mountains. Our principle was to evade battle and to gain a firm foothold. All our actions were aimed at gaining a firm foothold. At that time, the 6th Column was assigned the most arduous task. It moved from east to west one day and from west to east the next. It moved to and fro many many times. It shuttled around in that hilly land to lure and mislead the enemy, moving

> from west to east one moment and from east to west the next. Other units did not move very much. They were dispersed in an appropriate way and avoided encounters with the enemy. That was the way we acted for two months.

Father described the critical situation in the Dabie Mountains lightly, with just these few words. But what a heavy load had been on his shoulders, and what a heavy burden in his heart, for those two months!

The military situation was dangerous and urgent, but military operations were just a part of the struggle in the Dabie Mountains. After political power was established in the Hubei-Henan-Anhui Liberated Area, the first major task was to carry out land reform. The Central Committee promulgated the Outline Land Law of China in October 1947, and then a vigorous movement for land reform unfolded in all liberated areas under the leadership of the Communist Party. The Hubei-Henan-Anhui Liberated Area was no exception. The movement for land reform to expropriate local tyrants and distribute land and movable property was carried out vigorously in all parts of the area. However, the situation in the Dabie Mountains was different from that in other liberated areas. There the enemy situation was serious and complex, and the masses had many doubts and misgivings. Moreover, the Communist-led troops had withdrawn from this area four times before. If they withdrew once again, could the ordinary people bear it? During the land reform movement, cadres in some areas suffered from "leftist" impetuousness and made mistakes in policy and tactics. These practices were not only harmful to the mobilization of the masses but alienated them and even infringed on their interests.

Father discovered these problems in time. On January 14, 1948, Mao Zedong sent a telegram to Deng, asking about various problems in the liberated area. Taking the opportunity, Father sent two telegrams, on January 15 and 22, 1948, giving a detailed account of the situation in the Dabie Mountains. He indicated that land reform could be carried out in consolidated zones, but in guerrilla zones equal distribution of land should not be made now. On February 8, Father sent another telegram to the Central Committee, reemphasizing his view that land reform should be carried out by zone.

Mao Zedong attached great importance to Deng's telegrams and personally sent a reply with the following comments: "The experience gained in the Dabie Mountains described by Xiaoping is extremely valuable. I hope that all other localities and armies apply it."[5]

In his report to Mao Zedong dated March 8, Father proposed that land reform be stopped in the Dabie Mountains and that land rents and interests be reduced. This report of Father's was written during the march.

That day, spring had just come, and it was still cold. The army units led by Father had just camped at daybreak for a rest. Under the fernleaf hedge bamboo, the ground was damp. Father had an oil lamp lit and, leaning on his saddle, was thinking intently as he quickly wrote the report at dawn.[6]

Land reform was a matter of prime importance in building liberated areas. Wherever he went, Father carried it out. He proposed different policy measures in the light of different conditions but kept a firm grasp on the basic work.

In May, Father sent another telegram to Mao Zedong, reporting on the situation and stating his views. In the same month, Father called a meeting of the Central Plains Bureau at which it was decided to change from a policy of land reform to one of reducing land rents and interests.

On May 24, Mao Zedong sent a telegram to Deng Xiaoping, making it clear that the policy of reducing land rents and interests was to be pursued in the new liberated areas. On June 6, Father issued a directive on the question of land reform in the name of the Central Plains Bureau. The process of changing the basic policy of the land reform work in the new liberated areas of the Central Plains thus came to an end.

Although this question about the policy arose from a specific question on how to carry out land reform in rural areas, its resolution had a major bearing on whether the new liberated areas could be consolidated and developed. Proper handling of the question was extremely important to the continued expansion of the Communist-led liberated areas until the entire mainland of China was liberated.

It was an extremely difficult period when Father led his units to stay in the Dabie Mountains. After the converging attack by the enemy started, all Communist-led columns were able to jump in time out of the ring of encirclement. They endured hunger and cold. They slept in the open air in the wilderness and forests and marched in muddy fields in rain in spite of their fatigue. And all this in the coldest months of the year! Even young men could not stand the cold weather. Like other officers and men, Father, Li Xiannian, and other commanders wore the thin cotton-padded clothes they had made themselves. They walked through muddy water during the march, and when camped, they lit pine branches to study the enemy situation and their work. It was cold outdoors, but it was even gloomier and colder indoors. The bodyguards tried to burn some rice straw so as to warm the hands of the commanders. Father said, "We need not warm ourselves by fire. As everybody can endure the cold, what are we afraid of! We should know that even a leaf of grass of the masses does not come easily!" Father knew that the Dabie Mountains were poor and the people of the Dabie Mountains were even poorer, so he could not bear to use a leaf of grass or a piece of wood there.

During this period, Father signed and issued many instructions in the name of Liu and Deng and wrote several reports on the work in the Dabie Mountains to the Central Committee.

In early February, the Spring Festival came, and the troops had a rest for several days. At a happy festival time, people naturally thought of having better food. Some soldiers drew off the water of a pond in a bid to catch fish. When they jumped joyfully on seeing the several hundred jin of fresh fish they had caught, Political Commissar Deng came down from a hillside. Looking at the joyous scene of jumping soldiers and jumping fish, Father thought of the difficult time they had had and went up to the soldiers. He first praised the optimism these young soldiers were maintaining under harsh conditions but then criticized them seriously: "The water of the pond was reserved by local people for fighting drought. You drained the pond to get all the fish. You seek your current interest by damaging the interest of the masses." Drained water could not be gathered back up. The army compensated the masses for their losses afterwards.[7]

Political Commissar Deng was always serious, but he was affable to his subordinates and showed them that he cared for them. Indeed, he was very concerned about the survival and lives of his troops. But he knew very well that only when discipline was tight could the army enjoy the support of the people, and that only with the support of the people could the army overcome difficulties to win victory.

While Father overcame all inconceivable difficulties in leading the main forces to hold fast to the Dabie Mountains, the 11th and 12th columns of the Liu-Deng army, which advanced southward later on, launched attacks on the Tongbai and Jianghan areas west of the Dabie Mountains and established liberated areas there; the 1st Column, which had followed Liu Bocheng to exterior lines, advanced to and opened up the Huaixi area west of the Dabie Mountains, thus turning the Huaihe River into an inland river of the Central Plains Liberated Area, and successfully joined forces with the troops led by Chen and Su, and Chen and Xie, respectively, along the Beiping-Hankou Railway; and the Chen-Su field army and the Chen-Xie troops energetically conducted sabotage operations against the Beiping-Hankou and Longhai railways and annihilated more than 20,000 enemy troops. The operations conducted by the Communist-led army on the Central Plains and other battlefields forced the enemy to move thirteen brigades out of the Dabie Mountains, thereby forcefully smashing the enemy's converging attack on the Dabie Mountains.

Through the four-month period of close cooperation, on both interior and exterior lines, in the operations of the troops commanded by Liu and Deng, Chen and Su, and Chen and Xie, the Communist-led army annihilated more than 195,000 enemy troops, liberated nearly 100 county towns, established the new Central Plains Liberated Area amid the Yangtze, Huaihe, and Hanshui rivers, and lured 90 of more than 160 enemy brigades at the southern front to the battlefield on the Central Plains, thereby winning a major victory of strategic importance.

The strategic task of the Liu-Deng army to advance 500 kilometers to the Dabie Mountains had been fulfilled. Father said in recalling this heroic undertaking:

> When the War of Liberation began, Chiang Kai-shek attempted to carry the war to the liberated areas so as to exhaust the human, material, and financial resources of the liberated areas. Chairman Mao decided to carry the war to the Chiang-controlled areas, ordering the Second Field Army to leave the Dabie Mountains and the Third Field Army to advance southward. With this move, we pushed the battleline into the Chiang-controlled areas without consuming the human, material, and financial resources of the liberated areas anymore, and we won the victory in this way.
>
> Strategically, it was a great victory to advance 500 kilometers from north to south up to the Yangtze River at one time. To advance 500 kilometers to the Dabie Mountains was a brilliant strategic move. Such a decision could not have been made without great strategic thinking. This strategic thinking was formulated by Chairman Mao. The strategic thinking of Chairman Mao merits our careful study!

In early 1948, the Liu-Deng army reported to the Military Commission and Chairman Mao that it had accomplished the great and arduous strategic task assigned to it. Father said,

> The main reasons for the victory of the struggle in the Dabie Mountains were that we had made relatively accurate judgments on several questions and handled them in a relatively correct way. Our casualties were not heavy, and the efforts we exerted were not very enormous. But we accomplished the strategic task, extending the battleline from the Yellow River to the Yangtze River. So we can say that the Second Field Army shouldered heavy responsibilities during the strategic counteroffensive. We overcame difficulties of all sorts and fulfilled our task.

In February 1948, in accordance with the directive of the Central Committee, Father led the forward command post in the Dabie Mountains to cross the Huaihe River to the north and leave the mountains. On February 24, the forward and rear command posts met in the Linquan area in Anhui Province. Liu and Deng rejoined forces.

Liu and Deng had led their troops to the Dabie Mountains and gained a firm foothold there. Now they were marching out of the Dabie Mountains and leading their troops to conduct larger operations in a greater space and on wider battlefields.

Although the outcome of chasing deer on the Central Plains was not yet decided, the Communist side was already lit up with the radiance of victory.

CHAPTER 60

BEFORE THE DECISIVE BATTLES

Since the War of Liberation began, the Liu-Deng field army had fought a good many battles—and since the anti-Japanese war, a countless number of battles.

Liu Bocheng and Deng Xiaoping were directing more and more battles, and the scale of their military operations was becoming larger and larger, but they continued to have a surprisingly small headquarters. They required that the headquarters be staffed with only a few competent and efficient people.

When the Liu-Deng field army fought in the areas north of the Yellow River, it had a total of over 100,000 men. In its headquarters, there were several division-level units responsible for operations, confidential work, intelligence, communication, and military administration. Each division had a staff of ten to twenty. The radio station and the teams in charge of security, communication, and motor vehicles also had only a few people. The Liu-Deng field army headquarters had no permanent office and no secretariat, and Liu Bocheng and Deng Xiaoping did not even have their own secretaries. Every day the two commanders worked in the Operations Section; here they directed military operations and handled other affairs.

During wartime, Liu and Deng were very busy handling important military and administrative affairs day and night, with highly strained nerves. Their staff's tasks must also have been very pressing and difficult.

By September 1946 the Operations Division still had only one chief and the Operations Section had only two staff officers for operations. In the spring of 1947 the staff swelled to four. After the Liu-Deng field army went to the Dabie Mountains, Li Xiannian served as its deputy commander. He brought four staff officers for operations with him. Soon afterwards, the staff of the Operations Section swelled to nine. By then, Operations Section staff had more than doubled, and indeed the present was better than the past.

In May 1948 Chen Yi served as the deputy commander of the Central Plains Field Army. He also brought a staff officer with him to work in the Operations Section bringing the number of staff officers in the Operations Section to ten.

In the second half of 1948 the army under the command of Chen Geng was reassigned back, and one army of the East China Field Army was also assigned to the Liu-Deng field army. With these two armies plus its original six field columns, one corps, and seven military areas, the Liu-Deng field army totaled several hundred thousand men plus new liberated areas with a population of forty-five million. At the time, the scale of operations became increasingly larger, and the tasks of the Operations Section multiplied.

Zhang Shenghua, chief of the Operations Section, asked Chief-of-Staff Li Da twice to increase the number of staff officers. One day, after handling affairs in the Operations Section, Political Commissar Deng told Zhang Shenghua,

> I heard you wanted to get more staff officers for the Operations Section, but I think your section has enough staff officers. There are ten people, including the chief and deputy chief of the section. If ten people work hard, you have great strength! At the Hebei-Shandong-Henan and northern Henan battlefronts, there were only three to four people working in your section. At the time, you had few people, but you had to do a lot of work and bore heavy responsibilities. You were forced to work conscientiously and diligently, unite as one, exert your efforts, and handle affairs with one heart and one mind. Numbers are not so important in armies as capability. Now you should solve the question of the shortage of personnel by improving your quality, methods of work, and efficiency. Your personnel cannot be increased any more.[1]

With only ten staff officers, the Operations Section was indeed very small. With this small operational chain of command, how did Liu Bocheng and Deng Xiaoping command their massive army to conduct guerrilla, mobile, and large-formation warfare?

Zhang Shenghua worked beside Liu and Deng every day, and he knew best how they directed operations. He said Liu and Deng took personal charge of directing operations, solved major problems, and did not put off what was to be done today until tomorrow. All telegrams and all kinds of information were sent to the Operations Section to be classified by the staff officers. When Liu and Deng came, they read them and made decisions so as to ensure the timeliness and efficiency of the work. He said Li Da, chief-of-staff of the field army, acted for Liu and Deng. He could correctly understand their intentions, take the initiative to undertake much of the concrete work of directing operations, and help to arouse the enthusiasm of every staff officer in the Operations Section.

They drafted telegrams before the chiefs came, thus significantly improving work efficiency. He said Liu, Deng, and other chiefs persisted in doing important work themselves, such as preparing documents and writing and delivering speeches. Liu Bocheng and Deng Xiaoping had one thing in common—they did not use texts while making speeches. But even when they made long speeches of several hours in length, the speeches were still very interesting and well thought out, as though they had a draft text of the speech worked out in their minds.

After every campaign, the Liu-Deng field army had to prepare a summary and report, which were often written by Liu Bocheng and Li Da personally. Deng Xiaoping took personal charge of preparing the preliminary report on the annihilation of the KMT troops led by Huang Wei in the Huai-Hai Campaign and wrote quite a few other documents and telegrams; those submitted to the Central Committee alone numbered several score every year. All staff officers of the Operations Section said that Political Commissar Deng was the best and fastest at writing documents and telegrams and had an accurate memory for the relevant data.

Zhang Shenghua said that Liu and Deng were very strict about the work carried out at headquarters. Deng often said, "Your work is to carry out our operational plan and determination. The success or failure of your work has a great bearing on the action of our troops and on the outcome of campaigns and battles." Deng earnestly exhorted the staff officers of the Operations Section: "You must never be negligent of your duties."

Liu Bocheng and Deng Xiaoping had three main requirements of their staff officers. First, they had to do their work rapidly, that is, they had to get a prompt and accurate understanding of the enemy's situation and their own army's situation and make reports to their superiors, as Liu Bocheng said, "knowing the situation like the palm of one's hand." They had to handle other affairs rapidly as well. "Time is life, and time is victory"—this maxim finds its expression most obviously, realistically, and profoundly in wartime.

Second, staff officers had to report the military situation accurately and write their reports without making mistakes. "The discrepancy of a single hair will lead to an error of 500 kilometers." On operations questions, an error in one question, one sentence, or even one character could cause serious consequences. So the staff officers of the Operations Section had to constantly improve on the accuracy of their work.

Third, staff officers had to be professionally competent and familiarize themselves with the military situation, the enemy's situation, their own army's situation, the terrain, the weather, and other relevant matters. Deng Xiaoping set an example for knowing the enemy's situation, and he knew it better and more accurately than did the staff officers. Often he asked them unexpected questions. If they could not give answers, he would say with a smile, "How is it that you young men do not remember things any better than we old people!" Liu Bocheng and Deng Xiaoping also required their staff officers to learn how to write documents and draw maps and to handle both important and minor affairs.

Liu Bocheng and Deng Xiaoping directly handled affairs on the spot, and their staff officers worked efficiently. These were the basic characteristics of Liu and Deng's operational command. In the Liu-Deng headquarters, there was no red tape. Important and minor affairs were handled, and operational orders issued, immediately. Strategic decisions were made on time, and orders for directing campaigns were issued on the spot.

Zhang Shenghua wrote, with deep feeling, that the three chiefs, Liu Bocheng, Deng Xiaoping, and Li Da, were on very good terms, intimate with one another and cooperating closely with one another. During the campaigns, Chief-of-Staff Li Da consistently and quietly immersed himself in hard work and handled many concrete matters. When Li Da could not solve problems, Political Commissar Deng personally talked to the column commanders to direct operations. Only when the army encountered difficulties in fighting the campaigns did Commander Liu Bocheng attend to matters personally. All column commanders were very familiar with the operational command style of Liu, Deng, and Li.

About Liu and Deng in particular, Zhang Shenghua wrote:

> The two chiefs, Liu and Deng, trusted and supported each other and united as one. This was most educational. Deng often said, "Commander Liu is old and weak, so the staff of the headquarters should pay particular attention to his health, and if something happens, they should consult more with me and the chief-of-staff. Liu is our strategist, and only on important matters can we ask him to make policy decisions." And Liu often said, "Political Commissar Deng is our good political commissar. He is well versed in both polite letters and military science. All of us should respect him and obey his orders."[2]

Not only Zhang Shenghua but also all members of the Liu-Deng field army knew that Liu Bocheng and Deng Xiaoping were close to each other and learned from their example. Those who said the Second Field Army was the most united were right. The unity of the Second Field Army stemmed from the unity of Liu Bocheng and Deng Xiaoping. Unity was strength—an invincible strength.

In June 1947 the Communist-led People's Liberation Army shifted to the strategic offensive. Six months later, the People's Liberation Army had marched south, along three routes, to gain a foothold in the vast areas between the Yangtze River, the Huaihe River, and the Hanshui River and, at the battlefronts in northern, northeastern, northwestern, and eastern China, had mounted attacks and counteroffensives and wiped out large numbers of enemy troops. A change, a qualitative change, then took place in the war situation in China.

Toward the end of 1947 Mao Zedong pointed out with pride that in northern Shaanxi,

> the Chinese people's revolutionary war has now reached a turning point.
>
> It is the turning point from growth to extinction for Chiang Kai-shek's twenty-year counterrevolutionary rule. It is the turning point from growth to extinction for imperialist rule in China, now over 100 years old. This is a momentous event.
>
> Having occurred, it will certainly culminate in victory throughout the country.[3]

In early 1948, on the southern battlefronts in the Central Plains, the balance of military strength between the KMT and the CPC was as follows:

The three key KMT generals—Defense Minister Bai Chongxi, Commander-in-Chief of the Army Gu Zhutong, and Director of Xi'an Pacification Headquarters Hu Zongnan—commanded three reorganized corps, thirty-seven reorganized divisions, and eighty-six brigades, totaling 660,000 men, accounting for one-third of total KMT strength throughout the country. Under them were many important commanders, such as Hu Lian, Qiu Qingquan, Zhang Zhen, Sun Yuanliang, Fei Huichang, and Zhang Gan.

The three commanders Bai Chongxi, Gu Zhutong, and Hu Zongnan had no ordinary backgrounds, not to mention the 660,000 troops. Bai Chongxi belonged to the old Guangxi warlord clique. He was sly, full of stratagems, and known as "small Zhuge Liang." Gu Zhutong was a graduate of Whampoa Military Academy and had held the important posts of general chief-of-staff and commander-in-chief of the army. He followed Chiang Kai-shek for decades and was one of Chiang's trusted generals for "suppressing Communists." Hu Zongnan was a graduate of the first class from Whampoa Military Academy and a star student of President Chiang. He was not only a general of Chiang Kai-shek's troops that were fighting the Communists but also Chiang's fellow townsman. It was said that the second daughter of the Kong family who was dearly loved and regarded as its daughter by the Song family took a fancy to him, but he was unwilling to marry her. Otherwise, he would long have been a son-in-law of the Song family.

From this deployment of leaders we can see the great importance attached by Chiang Kai-shek to the battlefronts on the Central Plains.

The three major CPC combat groups—the main forces of the Liu-Deng field army, the Chen (Geng)-Xie (Fuzhi) army, and the Chen (Shiju)-Tang (Liang) army of the East China Field Army—had fifty brigades totaling 350,000 men. Under the unified command of Liu Bocheng and Deng Xiaoping, they were prepared to fight medium-scale battles of annihilation by maneuvering in the areas between the Huaihe and Hanshui rivers and the Longhai and Tianjin-Pukou railways. (The troops commanded by Su Yu of the East China Field Army undertook other military tasks.) The general strategy of these military deployments was to continue to wipe out large numbers of enemy troops and smash the KMT's defense system on the Central Plains.

To strengthen the leadership of the Central Plains region, the Central Committee decided:

1. to strengthen the Central Plains Bureau, with Deng Xiaoping serving as the first secretary and Chen Yi as the second secretary;

2. to establish the Central Plains Military Area (with seven military subareas under it), with Liu Bocheng serving as the commander, Chen Yi as the first deputy commander, Li Xiannian as the second deputy commander, and Deng Xiaoping as the political commissar; and

3. to rename the Liu-Deng Shanxi-Hebei-Shandong-Henan Field Army the Central Plains Field Army, with the 1st, 2d, 3d, 4th, 6th, 9th, and 11th columns under it, as well as the Chen-Tang army of the East China Field Army.

In the Central Plains region, the KMT and CPC forces had deployed their troops in battle formation, and fighting was under way. The troops commanded by Mao Zedong began to take military action step by step.

In February 1948 Liu and Deng's Central Plains Field Army left the Dabie Mountains and entered the areas north of the Huaihe River for rest and reorganization. In March the Chen-Xie and Chen-Tang armies mounted an attack, seized the large city of Luoyang in Henan Province, and, at the same time, covered the rest and reorganization of the Liu-Deng forces.

On May 2, Chen Geng commanded the 2d and 4th columns of the Central Plains Field Army and the 10th Column of the East China Field Army to launch the Wanxi Campaign, wiping out over 21,000 enemy troops and liberating nine county towns. The army commanded by Su Yu had finished its rest and reorganization, so the Central Committee ordered it to fight at the battlefront on the Central Plains.

On May 22, to pin down the enemy troops commanded by Zhang Zhen and to ensure the southward advance of the army commanded by Su Yu, the Central Plains Field Army launched the Wandong Campaign, with the forces in the east and the west commanded by Chen Xilian and Chen Geng, respectively. The campaign lasted less than ten days and resulted in the annihilation of over 10,000 enemy troops and the capture of three major generals under Zhang Zhen.

On the Central Plains, Chiang's troops had three maneuver groups, namely, the units in Nanyang commanded by Zhang Zhen, the units in Zhumadian commanded by Hu Lian, and the units in southwestern Shandong commanded by Qiu Qingquan. The East China Field Army entered southwestern Shandong and planned to attack the 5th Corps under Qiu Qingquan. On June 17, the Communist-led army launched the battle in eastern Henan. The East China Field Army attacked the troops under Qiu Qingquan, and the Central Plains Field Army blocked the troops under Hu Lian and Zhang Zhen, which advanced northward to reinforce Qiu's troops. On July 6, the East China Field Army wiped out more than 90,000 enemy troops commanded by Qiu Qingquan. The Central Plains Field Army fought three battles to block the advance of the enemy reinforcements and wiped out 7,000 troops.

After the campaign, the East China Field Army moved to other areas for rest and reorganization. The Central Plains Field Army decided to launch the Xiangfan Campaign on July 2. The CPC armies were fighting battles from north to south, near the Hanshui River.

On July 16, the Xiangfan Campaign concluded. The Central Plains Field Army had wiped out 21,000 enemy troops, liberated Xiangyang, Fancheng, and other cities and towns, controlled the middle sector of the Hanshui River, and captured Kang Ze, a key chief of the KMT special agents.

Between January and July 1948, on the battlefields on the Central Plains, the Central Plains Field Army and the East China Field Army coordinated their operations, wiping out over 200,000 enemy troops and liberating Luoyang, Kaifeng, Xiangfan, and other important cities and towns. By then the KMT defense system north of the Huaihe and Hanshui rivers had been smashed completely.

The Central Plains Liberated Area had a population of thirty million, and its human, material, and financial resources had increased considerably. Father once told us with pride,

> After the Second Field Army left the Dabie Mountains, its strength was weakened. The Second Field Army had been poorly equipped, and when it crossed the areas inundated by the Yellow River, it lost the heavy weapons it had recently seized from the enemy. The size of the Second Field Army was small, and furthermore, it was divided into two groups—one commanded by Liu and Deng and the other by Chen Geng. The main forces of the four columns were weakened, with three of the four columns having only two brigades each. But our morale was very high at all times. Before the Huai-Hai Campaign, we fought some small battles successfully. We didn't miss the opportunity and fought battles that needed to be fought. At the time, our three armies assumed battle postures on the Central Plains, and our troops achieved victory in northeastern China and gained a firm foothold in northwestern China, giving great encouragement to the people throughout the country. The situation as a whole was good.

As Father said, by the end of the second year of the War of Liberation, that is, between June and July 1948, the situation as a whole was very good.

The Communist-led troops wiped out 53,000 enemy troops in four months in northwestern China and 25,000 enemy troops in four months in northern Jiangsu. In northern China, they annihilated 140,000 enemy troops in several campaigns, and in northeastern China they annihilated over 156,000 enemy troops in 90 days and captured Siping, Jilin, Yingkou, and other strategic points.

In the second year of the War of Liberation, the Communist-led troops wiped out 94 brigades of the KMT regular troops totaling over one and a half million men, killed or captured 174 enemy generals, emancipated a population of 37 million, recovered and liberated a territory of 155,600 square kilometers, and seized 164 cities and towns.

The Communist-led troops controlled an area of 2.35 million square kilometers, 579 cities, and a population of 168 million. The People's Liberation Army had swelled to 2.8 million men.

The KMT lost 1.52 million troops, and its regular army was pinned down on the battlefields in northeastern, northwestern, and northern China, the Central Plains, and eastern China. Most of its troops were only capable of defending strategic points and vital communication lines and had lost their ability to conduct mobile operations.

In the areas controlled by Chiang Kai-shek, the KMT government found itself in an unprecedentedly serious financial and economic crisis. The high financial deficits continued to soar, inflation was very serious—like fierce floods and savage beasts—currency constantly depreciated, commodity prices rose every day, and even the mint was unable to print the paper currency needed on any given day. Since Chiang Kai-shek was suffering both military defeats and economic difficulties, he was in a bad mood and in a bad temper. He sighed,

"The KMT troops were constrained everywhere and suffered one defeat after another."[4]

Just like the KMT and the CPC, the American government was devising political strategies as well. The United States was very clear about what was happening in China. In his letter to U.S. Secretary of State George Marshall, the American Ambassador to China, Leighton Stuart, said that the situation had deteriorated to a point close to collapse.[5] The United States did not want to see the CPC achieve victory, so it continued to support the KMT in the civil war, as it had in the past. However, the United States was very dissatisfied with Chiang Kai-shek's incompetence, so it gave its secret support to Li Zongren, of the former Guangxi clique, to force Chiang to resign from the post of president. The United States attempted to find another agent in China so as to reverse the depressing situation.

After World War II, the United States exerted great influence in the world. Accordingly, Americans believed that they could do anything they wanted. In fact, the Americans did not know that in China at that time, nothing they did could reverse the situation, whether they provided weapons and funds or plotted to replace one leader with another. Whether those who went against the wishes of the people and the will of heaven were Chinese or foreign, and no matter how they racked their brains in scheming, they were only going to have more white hair and wrinkles after their inevitable failure.

In July 1948 the War of Liberation entered its third year. At the time, the ratio of military strength between the KMT and the CPC had changed from 3.4 to 1 in 1946 to 1.3 to 1. The third year would be the most crucial year of the war, for both the KMT and the CPC.

Both the KMT and the CPC held meetings to study the war situation and strategic issues. The KMT meeting was a "meeting of military review." The CPC meeting was a meeting of its Political Bureau.

The meeting called by Chiang Kai-shek was held in a tall building in Nanjing. The meeting of the CPC Political Bureau called by Mao Zedong was held in a farmhouse in the small village of Xibaipo, Jianping County, Hebei Province, at the foot of the Taihang Mountains.[6]

Mao Zedong and the leading organs of the Central Committee crossed the Yellow River toward the end of March 1948 and came to Xibaipo between April and May. The village suddenly became the "capital" of the CPC. The Central Committee held the enlarged meeting of the Political Bureau from September 8 to 13, 1948. It was attended by the members and alternate members of the Political Bureau and by leading Party and army cadres from the various liberated areas.

In late July Father set off from Baofeng County, Henan Province, where the Central Plains Field Army was stationed, to attend the meeting in Xibaipo in his capacity as first secretary of the Central Plains Bureau. Father said good-bye to Liu Bocheng, took a U.S.-made military jeep captured from the enemy troops, and drove in the direction of Luoyang. He brought with him two gifts, glossy ganoderma and rock crystal, to the comrades of the Central Committee. These two things were not rare, but they had been captured from the enemy, so they were worth giving to the comrades as souvenirs.

While passing through Linru, Father did not stop to have breakfast. While passing by the Longmen Grottoes, he did not visit the well-known gems of ancient art. At the Yishui River, he stood on the bank of the slowly flowing river, watching a truck pull his jeep across the river. While passing by the Guangyu Tomb, he did not stop to pay homage. While passing by the burial place of Liu Xiu, Emperor Wudi of the Han Dynasty, he did not stop to pay a visit.

During his trip of several days, wherever he went he talked with local people about the revolutionary work. When he came to the areas of operations, he could not travel in the daytime because of enemy aircraft, so he traveled at night. When he came to a part of the main road that had been destroyed, he did not make a detour but insisted on taking the rough road to keep going straight. During this period, he did not dare to stay anywhere for even a short time. He was anxious, because he knew he was going to attend a very important meeting at which China's destiny was to be determined.

The meeting at Xibaipo was indeed an extremely important meeting of the CPC. It set the strategic task for the whole Party: to build the People's Liberation Army to five million strong and to wipe out some 500 brigades or divisions of the KMT regular troops in approximately five years, starting from July 1946, thereby overthrowing the reactionary rule of Chiang Kai-shek and the KMT.

The meeting pointed out the war had entered its third year, a critical year if victory was to be achieved within five years. Militarily, the People's Liberation Army had to continue to mount attacks on exterior lines, carry the war to the areas controlled by the KMT, and prepare to fight several decisive campaigns. The army had to focus on the Central Plains but keep operating in the areas north of the Yangtze River and in northern and northeastern China.

The meeting called on the army to dare to fight unprecedentedly large-scale battles, to dare to fight the powerful enemy formations, and to dare to attack the strongly fortified large cities controlled by the enemy so as to achieve nationwide victory. Politically, the meeting declared that the establishment of the national central government had to be placed on the agenda, and that the Political Consultative Conference was going to be held in 1949 so as to establish the people's country and central government.

Mao Zedong's policy decisions were always made resolutely and in good time. His judgment was always foresighted and accurate, and he had unusual courage and resourcefulness.

In the most difficult days of the anti-Japanese war, he predicted that the Chinese were sure to win the war. When the KMT troops were overwhelmingly superior to the Communist-led army, he was confident of victory. When the army he led was far weaker than the enemy troops, he ordered it to launch the strategic counteroffensive. When his troops were still no stronger than the enemy, he resolved to fight the great strategic decisive battles.

The great, decisive strategic battles between the CPC and the KMT thus began in the land of China.

CHAPTER 61

THE GREAT DECISIVE BATTLES

The autumn of 1948 came.

In this ancient Oriental country with an area of almost ten million square kilometers, it was no longer swelteringly hot. The sunshine was becoming brighter and more brilliant. The vast skies were clear blue with rolling clouds. The rivers turbulently flowed east. The autumn brought the autumn wind.

The advent of autumn gave different people different feelings. Sentimentalists believed that autumn represented desolation, the withering cold coming afterward, and an extreme, unavoidable anxiety. Anxiety was autumn in some people's hearts, and anxiety and autumn were two eternal subjects chanted by the sentimental poets.

But the optimists had an entirely different feeling about autumn. The sky they saw in the autumn was clear blue and boundless. The land they walked was full of ripening and bore an abundant harvest of all crops. Their autumn was flavored with ripening, plenty, mellowness, romance, fruit, and victory. In their eyes the autumn was golden.

The autumn of 1948 would be soul-stirring and unforgettable throughout the lives of both those who succeeded then and those who failed.

The CPC resolved to fight large-scale battles of annihilation in the autumn and winter of 1948. The Central Committee issued a directive: The East China Field Army was to seize Jinan in the autumn and capture Xuzhou in the winter. The Central Plains Field Army was to fight a decisive battle with the KMT troops commanded by Liu Zhi in the Central Plains. The Northeast Field Army was to wipe out the troops commanded by Wei Lihuang in northeastern China. In northern China, the Communist-led army was to capture Taiyuan and finish off the troops commanded by Fu Zuoyi in Beiping. According to the operational plan for the third year, the People's Liberation Army was to wipe out 115 divisions of KMT regular troops and 80 percent of its strength north of the Yangtze River.

From September 1948 to January 1949, the CPC troops successfully mounted the large-scale autumn and winter offensives on the battlefields in northeastern, eastern, northern, and northwestern China and on the Central Plains. These were the three great and decisive strategic campaigns—the Liaoxi-Shenyang Campaign, the Huai-Hai Campaign, and the Beiping-Tianjin Campaign.

The first of the three great campaigns was the Liaoxi-Shenyang Campaign launched in the Shenyang region of Liaoning in northeastern China. The northeast looks like the head of the rooster (China's territory takes the shape of a rooster), borders on the Soviet Union and North Korea, and is a region of China where heavy industry is the most advanced. It was also the first region occupied by the Japanese invading troops in 1931.

By the autumn of 1948 over 97 percent of the land in the northeast had come under the control of the CPC, and 86 percent of the population had been emancipated. In the northeast, CPC regular troops numbered about 700,000 men, and local troops had 330,000 men. The Northeast Field Army had twelve columns and thirty-six main force divisions, with large quantities of various artillery pieces and good light equipment. It was the best equipped main force, with the greatest strength, of the Communist-led troops.

The KMT troops under the command of Wei Lihuang, commander of the Northeast "Bandit Suppression" Headquarters, was one of the KMT's main strategic forces. Although this force had 550,000 men, they had been separated by the CPC forces into the three areas of Changchun, Shenyang, and Jinzhou, and all military supplies to Changchun and Shenyang had to be airlifted. In the northeast, the strength of the CPC troops exceeded that of the KMT troops.

Chiang Kai-shek had all along been indecisive on his strategy for the Northeast—to defend or to withdraw. If he was to defend the northeast, he was afraid his troops could not hold on and might be wiped out completely by the Communist-led troops. Although withdrawing might preserve his military strength, the political impact would be negative. Wei Lihuang was worried that if he were to withdraw his troops from the northeast, they were likely to be annihilated completely by the Communist-led troops in the attempt.

After a period of hesitation, Chiang Kai-shek finally resolved to order his troops to tenaciously defend the northeast, ensure the safety of Shenyang, Jinzhou, and Changchun, and seize opportunities to open up the Beiping-Liaoning Railway leading to areas south of the Great Wall.

Mao Zedong made a correct appraisal of the situation in the northeast: because the CPC's military and economic strength was superior to that of the KMT, he resolved to confine all the enemy troops to the northeast and to wipe them out one by one.

Changchun, Shenyang, and Jinzhou stretched from northeast to southwest. Mao Zedong sent a telegram to Lin Biao and Luo Ronghuan of the Northeast Field Army, ordering them to capture Jinzhou first so as to seal off the enemy's exit from the northeast to areas south of the Great Wall.

Lin Biao hesitated to move south along the Beiping-Liaoning Railway to attack Jinzhou but wanted to capture Changchun without having to go far. Enemy reinforcements had arrived at the seaport of Hulu Island near Jinzhou to ensure the safety of the KMT troops' sea passage while they were retreating. With the enemy troops taking such military actions, the Communist-led army could waste no time in cutting off the enemy's retreat route if it was to take the opportunity to wipe out all enemy troops in the northeast.

Under these circumstances, Lin Biao made up his mind to capture Jinzhou first. From October 9 to 15, after fierce and difficult battles attacking the city and blocking enemy reinforcements, the Northeast Field Army eventually captured Jinzhou—the strategic passage from the northeast to areas south of the Great Wall.

When the Communist-led troops attacked Jinzhou, Chiang Kai-shek was very worried and flew to Beiping and then to Shenyang and planned operations himself. After the fall of Jinzhou, Chiang Kai-shek flared up in anger and flew to Shenyang once again. There, he changed his policy decision to tenaciously defend the northeast and strictly ordered the defending troops stationed in Changchun to break out of the encirclement and retreat southward to Shenyang.

By combining the two forces that could have coordinated their efforts and supported each other, Chiang made them take a position from which they could not flee and in fact could only take a beating together.

The defending troops in Changchun under the command of Zheng Dongguo decided not to jump into the jaws of death. Instead, they revolted, and Changchun was liberated peacefully.

After the fall of Jinzhou and Changchun, Chiang Kai-shek was even more worried. He flew to Shenyang for the third time to plan the general retreat. With a stern voice and countenance, he ordered the main force in Shenyang, the army commanded by Liao Yaoxiang, to retreat southward. But how could it retreat?

While retreating southward, Liao Yaoxiang's troops were blocked and then fell under the converging attack of the Communist-led army and were penetrated by it. Both he and his entire army were in confusion. Eventually, his army was annihilated, in an area of 120 square kilometers.

After Liao's army was wiped out, his immediate superior, Wei Lihuang, knew what was coming and immediately fled from Shenyang by plane. On November 2, the CPC forces captured Shenyang. The Liaoxi-Shenyang Campaign concluded.

After fifty-two days of fighting and suffering 69,000 casualties, the Northeast Field Army wiped out the "Bandit Suppression" General Headquarters, four KMT army headquarters, eleven KMT corps headquarters, and thirty-three KMT divisions, totaling more than 472,000 enemy troops. The Communist-led troops had brought the whole of northeastern China under control.

After the campaign, the KMT's total strength fell to 2.9 million men, while that of the CPC troops swelled to 3 million. The military balance had been reversed. Mao Zedong said confidently, "Accordingly, the war will be much shorter than we originally estimated. . . . As we now see it, only another year or so may be needed to overthrow the reactionary Kuomintang government completely."[1]

After the liberation of the northeast, the CPC set its sights on the next target—the Central Plains. The commander-in-chief of the Chinese People's Liberation Army, Zhu De, said, "Since ancient times, whoever has won victory on the Central Plains has achieved the final victory." Now the CPC planned to end the contention on the Central Plains and take control of China.

The second of the three great campaigns, the Huai-Hai Campaign, was about to begin. The battlefields of the Huai-Hai Campaign were located in the plains of the Yellow and Huihe rivers, where Jiangsu, Anhui, Shandong, and Henan provinces met. The terrain was open, with densely populated villages scattered here and there. The Tianjin-Pukou Railway ran from north to south, from Tianjin to Shanghai, and the Longhai Railway ran from east to west, from Xuzhou to Zhengzhou. The center of the battlefield was Xuzhou—the intersection of the two railway arteries. The plains had vast expanses and a crisscross network of roads—good conditions for battlefields of large formations conducting large-scale mobile warfare.

On the Central Plains, Chiang Kai-shek deployed the troops commanded by Liu Zhi, centered in Xuzhou, and the troops commanded by Bai Chongxi, centered in Hankou. In the north, these troops were to control the Longhai and Tianjin-Pukou railways and defend the capital and heart of the KMT government—Nanjing. In the south, they were to control the southern section of the Beiping-Hankou Railway, hold Wuhan and Xinyang, and defend southern China. The troops under Liu Zhi in Xuzhou and the troops under Bai Chongxi in Hankou constituted Chiang Kai-shek's defense system on the Central Plains.

Xuzhou was situated at the intersection of the Longhai and Tianjin-Pukou railway lines and was the location of the KMT "Bandit Suppression" General Headquarters, with Liu Zhi serving as the commander-in-chief. Xuzhou was the most important strategic point of the battlefield on the Central Plains.

The KMT deployed its troops around Xuzhou as follows: in the west, between Xuzhou and Zhengzhou, was the 2d Army, commanded by Qiu Qingquan; in the east was the 13th Army, commanded by Li Mi, and the 7th Army, commanded by Huang Baitao; in the north were troops belonging to the Third Pacification Zone; in the south, between Xuzhou and Bengbu, west of the railway, was the 16th Army, commanded by Sun Yuanliang; and east of the railway was the 6th Army, commanded by Li Yannian.

Hankou was located at the middle section of the Yangtze River and on the artery running between north and south—the Beiping-Hankou Railway. The KMT Central China "Bandit Suppression" General Headquarters was located there, with Bai Chongxi serving as the commander-in-chief. Deployed in the area of Xinyang north of Hankou were the 12th Army, commanded by Huang Wei, the 3d Army, commanded by Zhang Gan, and the Fifth Pacification Zone, commanded by Zhang Zhen. On the Huai-Hai battlefield, Chiang Kai-shek employed twenty-nine corps and seventy divisions and other troops, totaling 700,000 men.

The main forces of the Central Plains Field Army, commanded by Liu and Deng, were deployed in the area of Kaifeng west of Xuzhou, and the main forces of the East China Field Army, commanded by Chen Yi and Su Yu, were deployed in the area of Linyi northeast of Xuzhou. The Central Plains and East China field armies, plus the troops of the East China, Central Plains, and North China military areas, had a total strength of 600,000 men.

Mao Zedong and the Central Committee, after analyzing the strategic situation at the southern front, believed that the time was ripe for fighting the decisive battle and decided to organize and launch the Huai-Hai Campaign.

To unify the command of the operations at the southern front, the Central Committee decided to organize the General Front Committee, consisting of Liu Bocheng, Chen Yi, Deng Xiaoping, Su Yu, and Tan Zhenlin, with Deng Xiaoping serving as the general secretary. The Central Committee decided that the General Front Committee was to command the East China and Central Plains field armies in a unified way, fight the large-scale decisive battle against Chiang Kai-shek's largest strategic forces in the area, centering on Xuzhou, and wipe out enemy forces one by one in areas north of the Huaihe River within three to five months.

Mao Zedong pointed out that the Huai-Hai Campaign was an unprecedentedly large-scale campaign on the southern front. "Victory in this campaign will not only secure the situation north of the Yangtze River but also lay the foundation for the resolution of the situation nationwide."[2]

The Central Committee gave a directive: when possible, meetings of the five members were to be held to discuss important questions, and three of them, Liu Bocheng, Chen Yi, and Deng Xiaoping, were to act as members of the Standing Committee to handle all matters as the occasion required, with Deng Xiaoping serving as the general secretary of the General Front Committee. Indeed, the Central Committee empowered the General Front Committee to handle all matters as the occasion required.

The CPC and the KMT had completed military deployments on the Huai-Hai battlefield, and the largest and most decisive battle on the Central Plains was imminent.

In the first ten days of November 1948, the Huai-Hai Campaign began. In accordance with the directive of the Central Committee, the General Front Committee decided on the operational plan for the first phase. The seven columns of the East China and Central Plains field armies were to isolate, encircle, and annihilate the army commanded by Huang Baitao in the areas east of Xuzhou, while preventing the 13th Army, commanded by Li Mi and west of Huang's army, from reinforcing the latter by moving east.

The Central Plains Field Army was to be divided into two units. One unit, commanded by Deng Xiaoping and Chen Yi, was to conduct operations in Xuzhou and Bengbu and sever the ties between southern Xuzhou and Bengbu along the Tianjin-Pukou Railway. The other, commanded by Liu Bocheng, was to retard the advance from the southwest of the 12th Army commanded by Huang Wei.

The Communist-led army planned to take the first action by cutting off, whatever the cost, the section of the railway between Xuzhou and Bengbu. On November 6, the enemy troops stationed on both sides of the Xuzhou-Bengbu railway began to draw back toward it. On the same night, the CPC forces launched the Huai-Hai Campaign.

The troops led by Chen Yi and Deng Xiaoping mounted an offensive along the section of the railway between Kaifeng and Xuzhou, quickly captured Dangshan 100 kilometers west of Xuzhou, controlled the 300-kilometer railway line from there to Zhengzhou, and pressed on to Xuzhou. The East China Field Army mounted a fierce converging attack on Huang Baitao's army east of Xuzhou.

On seeing the CPC troops advance toward Xuzhou from many directions, Liu Zhi in Xuzhou was convulsed with fear. He immediately ordered the 2d Army, commanded by Qiu Qingquan in the east, and the 13th Army, commanded by Li Mi in the west, to retreat to Xuzhou and the 7th Army, commanded by Huang Baitao, to draw close to Xuzhou as rapidly as possible.

Liu Zhi was a general, but unexpectedly, he could not withstand a single blow. He really did not know what to do and adopted the tactic of holing up his forces. As soon as the operations had started, Liu Zhi was frightened. Because of his timidity, his troops were doomed to defeat. From the moment the campaign began, he was thinking about the general retreat.

Liu Zhi wanted to make a general retreat, but Chen Yi and Deng Xiaoping would not let him simply run away. The General Front Committee instructed the East China Field Army to step up its efforts to intercept Huang Baitao's army in a bid to launch a converging attack on it and annihilate it.

All units of the East China Field Army boldly pursued and intercepted Huang's troops, despite fatigue, hunger, death, and other difficulties, and on November 10, cut off their route of retreat to the west and surrounded the 7th Army in a narrow area called Nianzhuang. Liu Zhi and even Chiang Kai-shek were worried about the encirclement of Huang's 7th Army. Chiang rebuked his subordinates sternly and angrily, saying, "The Xuzhou-Huaihai battle is crucial to the success or failure of our cause and to the survival of the country."

Chiang Kai-shek was also discontented with Liu Zhi because of his incompetence, so he sent his star student, Du Yuming, to Xuzhou to serve as Liu Zhi's deputy and to direct the operations at the front. Afraid that not enough troops were deployed for the campaign, he also increased the KMT strength to 800,000 men on the Huai-Hai battlefield.

Mao Zedong gave instructions to the commanders at the front: first, the East China Field Army was to wipe out the troops led by Huang Baitao and make them unable to move away; second, the Central Plains Field Army was to quickly capture the strategic point between Xuzhou and Bengbu—Suxian. Situated in the section of the Tianjin-Pukou Railway between Xuzhou and Bengbu, Suxian was the communications hub between north and south. Capturing Suxian severed the ties between enemy troops in Xuzhou and those in Bengbu.

The Central Plains Field Army immediately conducted operations in the area of Xuzhou and Bengbu. Its troops wiped out the enemy along the route and, on November 12, accomplished the important strategic task of capturing Suxian. Father often told us, "Suxian was crucial to the campaign. When we captured it, we cut off Xuzhou from the south. In fact, we formed a strategic encirclement of Xuzhou."

There was a shot in a film about this great decisive battle. Standing at the platform bridge of the Suxian Railway Station and watching the rumbling and whistling trains coming and going below their feet, the actors portraying Liu Bocheng, Chen Yi, and Deng Xiaoping expressed what had been their true emotions: after their troops finished capturing Suxian, they were jubilant. It was artistic method to make the three people express their feelings at the same time. But to the General Front Committee, the capture of Suxian was certainly a happy event, so it was not possible for the filmmakers to exaggerate their feelings about the victory too much.

Since the Communist-led army had cut off the enemy's passage from Xuzhou to the south, it could go all out to wipe out enemy troops in the north. On November 11, the East China Field Army launched a general attack on the troops under Huang Baitao. By November 22, after eleven days of fierce and bitter fighting, the East China Field Army had annihilated the 7th Army completely and killed its commander, Huang Baitao.

While capturing Suxian and wiping out the army under Huang Baitao, the Communist-led troops had effectively blocked the two armies coming from Xuzhou to reinforce the army commanded by Huang in the east and the two armies coming from Bengbu to reinforce Xuzhou in the north.

By then, the first phase of the Huai-Hai Campaign had come to an end. The Communist-led army wiped out a total of eighteen divisions of enemy regular troops, cut off the section of railway between Xuzhou and Bengbu, and divided the troops commanded by Liu Zhi into southern and northern parts.

On November 23, the General Front Committee and the headquarters of the Central Plains Field Army moved into a small village in Suxian—Xiaolijia Village. The three members of the Standing Committee of the General Front Committee, Liu Bocheng, Chen Yi, and Deng Xiaoping, gathered together. All three were from Sichuan, and although they had traveled south and north for decades, they had not changed their local accent. Among them, one was several years older than the others, and each had a unique temperament.

Wearing glasses for his nearsightedness, Liu Bocheng was tall, well versed in both polite letters and military science, refined, and good-natured. His conversation often suited both refined and popular tastes. His use of Sichuan proverbs and humorous and vivid metaphors often surprised people and made them laugh. His subordinates wanted to compile a book of the witty remarks made by Commander Liu, but they were afraid his remarks were too figurative and vivid to translate into words. At the time, he was over fifty-four and the eldest of the three.

Chen Yi was of medium height and fat. Because he had a round face, thick double chin, and was potbellied, he was awe-inspiring and looked like a marshal. He had the sense of humor inherent in people of Sichuan. Besides being humorous, Gen. Chen Yi had a sanguine disposition, and he was unrestrained, forthright, and cheerful. Militarily, he could command massive troops. Literarily, he often felt a strong urge to compose poetry and wrote articles quickly. He could talk wittily about everything under the sun and was unforgettable to those who met him. At the time, Chen Yi was forty-seven.

Deng Xiaoping was the shortest and the youngest among them. At the time, he was forty-four. Compared with Liu and Chen, Deng had a uniquely graceful bearing. He was a man of a few words, staid, and astute. When he was serious, others were afraid of him, but in his meticulousness he was very considerate to others. He acted resolutely and had a firm will. When he met with old friends, he talked cheerfully and humorously. He could tell many stories, both modern and ancient, in Sichuan dialect.

It goes without saying that Deng and Liu were on very good terms. Deng and Chen had in common that they had been students in France and maintained close relations. Most coincidentally, the members of the General Front Committee on the battlefield on the Central Plains cooperated very closely.

The headquarters of Liu, Chen, and Deng was located in a big courtyard in Xiaolijia Village, and all of them lived in a small out-of-the-way courtyard at the end of the village. Deng and Chen lived in an outer room, and they let the elder Liu live in an inner room.

At Xiaolijia Village, the General Front Committee was studying what should be the next target of attack in the second phase of the Huai-Hai Campaign. After repeated analysis of the battlefield situation, the General Front Committee sent a telegram to the Central Committee on November 23, proposing to attack the troops led by Huang Wei first.

The reply from the Central Committee came quickly, on the following day, and expressed its approval of their proposal. The Central Committee also instructed, once again, that in case of emergency, Liu, Chen, and Deng were to handle all matters as the occasion required without asking the Central Committee for instructions. The Central Committee and Mao Zedong had full confidence in the General Front Committee of the Huai-Hai Campaign.

The 12th Army under Huang Wei, 120,000 strong, was Chiang Kai-shek's own crack troops, and its 18th Corps was one of the "five main forces" of the KMT troops equipped with American weapons. Huang Wei was one of Chiang Kai-shek's star students, and in his prime he was very arrogant.

Huang Wei's troops were orginally stationed in the Tongbai Mountains. As Huang Baitao's troops fell under the converging attack mounted by the Communist-led army, Chiang Kai-shek ordered Huang Wei to take immediate action to reinforce them. Acting on Chiang's instructions, Huang Wei led his 120,000 troops to advance northward day and night to reinforce the troops in Xuzhou. But when they arrived at Mengcheng, after hurriedly traversing mountain paths for over 150 kilometers and suffering numerous hardships, the army commanded by Huang Baitao had been wiped out completely.

After the army under Huang Baitao was wiped out, the commanders in Xuzhou were convulsed with fear and immediately ordered the two armies commanded by Qiu Qingquan and Li Mi to draw back to Xuzhou from the west and the east, respectively. As soon as Huang Wei arrived on this critical scene, he immediately drew close to Xuzhou as well. If Huang Wei's relatively strong troops had joined the troops in Xuzhou, it would have been difficult for the Communist-led army to wipe out all enemy troops in Xuzhou.

Nevertheless, Liu, Chen, and Deng resolved to fight a decisive battle with Huang Wei's troops. When the Central Plains Field Army left the Dabie Mountains, it had a total of only 150,000 men and was poorly equipped. If it was going to engage Huang Wei's troops, it had to firmly resolve to fight a fierce battle. Deng told the officers and men of the Central Plains Field Army, "So long as the Central Plains Field Army can wipe out the enemy's main forces at the southern front, even at the cost of its own total destruction, other troops of the People's Liberation Army throughout the country can win the nationwide victory. So it is worth the cost!"[3]

The General Front Committee decided to put 120,000 troops into action. The main forces of the Central Plains Field Army were to block the northward advance of Huang Wei's troops, launch a converging attack on them, and wipe them out. The main forces of the East China Field Army were to keep watch on the enemy troops in Xuzhou and prevent them from moving south to reinforce Huang Wei's troops. Some units of both the East China and Central Plains field armies were to be employed to keep under surveillance the two armies commanded by Li Yannian and Sun Yuanliang, stationed in areas east of Huang's troops.

After Huang Wei's troops reached the area of Mengcheng south of Suxian, they continued to move north. But right after they crossed the Huihe River, Huang Wei suddenly realized that they had entered the pocket prearranged by the Communist-led troops. Huang Wei was astute. As soon as he found that the situation was far from good, he turned back to retreat southward. Taking advantage of the enemy's retreat, the Central Plains Field Army launched an attack on all fronts. On November 25, Huang Wei's army was tightly encircled in the area of Shuangduiji south of Suxian.

After Huang Wei's army was encircled, Chiang Kai-shek was panic-stricken. Chiang ordered Huang Wei to break out of the encirclement eastward, but when he tried, his troops were repulsed by the Communist-led army. Then Chiang Kai-shek ordered Huang Wei to tenaciously defend his positions and wait for reinforcements. Acting on Chiang's order, Huang Wei had his troops construct field fortifications around themselves and shift to defense. Chiang Kai-shek immediately ordered the enemy troops in Xuzhou to reinforce Huang, but they were doggedly intercepted by the East China Field Army and failed to advance. Under these circumstances, Chiang ordered Li Yannian and Sun Yunliang to reinforce Huang. But unexpectedly, Li and Sun did not want to fight the Communist-led troops. They hoped to preserve their own forces, so they just retreated southward.

After these failures suffered by Chiang Kai-shek, Huang Wei's army was tightly encircled by the CPC troops. Seeing that the situation was dangerous, Chiang urgently summoned Du Yuming to Nanjing and personally instructed him on the line of action to pursue. Chiang did not give Du Yuming instructions for dealing with the emergency but ordered him to abandon Xuzhou and retreat southward on all fronts.

On December 1, Du Yuming led three armies and other military personnel, totalling 300,000 men, in a hasty withdrawal from Xuzhou. The great retreat involving 300,000 men was extremely chaotic. Everyone was trying to flee for his own life, and vehicles were extremely crowded. This was not a great retreat but an out-and-out rout.

On December 4, the East China Field Army encircled all the troops under Du Yuming at a small village called Chenguanzhuang. Two days later, it completely wiped out the 16th Army under Sun Yuanliang.

Du Yuming failed to save Huang Wei. Instead, his troops were encircled. During this period, the Central Plains Field Army mounted fierce attacks on the army commanded by Huang Wei but made little progress. The General Front Committee decided that the eastern group under Chen Geng, the western group under Chen Xilian, and the southern group under Wang Jinshan should launch an all-out attack on the enemy troops at Shuangduiji on December 6. By December 10, although they had wiped out 50,000 enemy troops, they had not captured the area. The General Front Committee decided to order some units of the East China Field Army to join in the assault on fortified positions. On December 13, the CPC forces mounted a general attack. The fierce fighting lasted until December 15, and the Communist-led army wiped out over 100,000 men of the enemy's 12th Army, centering on Shuangduiji, and captured the commander of the army, Huang Wei.

After Huang Wei's army was wiped out, the second phase of the Huai-Hai Campaign came to an end. This phase was a period in which fighting was the most intense and the General Front Committee was very busy with military affairs. Then chief of the Operations Section, Zhang Shenghua later recalled:

> In fighting the army under Huang Wei, Political Commissar Deng Xiaoping took charge of the specific work of organizing and directing the campaign on his own initiative.
>
> Deng told Liu and Chen, "You two commanders, I am several years younger and healthier than you, so I shall do more work and be on night duty more frequently."
>
> Liu and Chen laughed heartily. Chen said, "We should not only do all that we can and perform our duties but also respect Political Commissar's views. But we must share the right to be on night duty with you!" Liu added, "Since we are old,

> we cannot direct many more such decisive battles. We have reason to work hard and exert our efforts to fulfil our tasks."
>
> Deng said, "Major decisions should be made by you two commanders and by three of us 'cobblers,' but I'll handle the more mundane affairs." Deng told the staff officers of the Operations Section that they should ask Deng for instructions on the more ordinary work, but when they had important affairs, they should report simultaneously to the three chiefs, Liu, Chen, and Deng.

Throughout the operations of the second phase, the military situation was pressing and the fighting was intense. The telephone rang all night, and telegrams and battlefield reports poured in. Deng worked in the operations room all day long and was on duty every day until late at night and even into the small hours. Only when there was no great change in the situation did he leave. Deng communicated most of the decisions made by the General Front Committee to all columns, heard reports made by the Operations Section on the operation at any given time, and often personally talked directly with column commanders by phone. In order to not disturb Liu and Chen with the phone calls he received at night at their dwelling place, Deng had the telephone line extended so that when the phone rang, he could throw on his clothes and go to the courtyard to receive it.[4]

After the task of wiping out the army under Huang Wei during the second phase was accomplished, Deng wrote a summary report to Mao Zedong on January 11, 1949:

> In wiping out the army under Huang Wei, all our troops have made the greatest resolve to carry out the will to wipe out Huang's army at all costs. So throughout the operations, although all columns underwent reorganization at the front three to four times, they did not complain. However, in launching the general attack, all columns of the Central Plains Field Army suffered over 20,000 casualties, so their strength was insufficient. As a result, the enemy was annihilated only after two columns of the East China Field Army were committed. . . . After the battle, all columns felt that the Central Plains Field Army was understrength, and they regretted being unable to wipe out the army under Huang Wei by themselves and thus increasing the burden on the East China Field Army.[5]

Father often told us that it was indeed very difficult to wipe out Huang Wei's army, which had aircraft, guns, and tanks, and that even his defensive works at Shuangduiji were constructed with tanks, armored vehicles, and trucks lining up. Nevertheless, Huang Wei's army may have been one of the crack forces of the KMT troops, but it was eventually wiped out completely.

After the army under Huang Wei was annihilated, the Central Committee instructed the five members of the Huai-Hai General Front Committee to hold a meeting. Because Su Yu and Tan Zhenlin were commanding troops that were about to launch a converging attack on the army under Du Yuming, on the night of December 16, Liu, Chen, and Deng drove in a car over fifty kilometers to Caiwa Village, where the headquarters of the East China Field Army was located. Since the General Front Committee had been established, this was the first time all five members had gathered together.

The meeting of the General Front Committee lasted a whole day. The topic discussed was not the Huai-Hai Campaign and the annihilation of the troops under Du Yuming but how to operate across the Yangtze River. The Huai-Hai Campaign was yet to be finished, but the General Front Committee was already charged with the important task of conducting the Yangtze-Crossing Campaign.

Because of their recent victory, the five members were in high spirits, and the future strategic plan of quick advance encouraged them all the more. In front of the small house where the headquarters of the East China Field Army was located, the five members of the General Front Committee had a picture taken. In the picture, each of the five important generals of the Communist-led troops wears a cotton-padded jacket made of handwoven cloth, is encumbered by too much clothing, and is taking a natural and casual posture. Father is thin and has a sallow face and long beard. Maybe he was too busy to shave. The five people in this picture are all smiling and looking happy.

After the meeting of the General Front Committee was over, on the instruction of the Central Committee, Liu and Chen left for Xibaipo on December 19 to attend the meeting of the Political Bureau of the Central Committee to work out the military plan for 1949.

Later the same day, Deng returned to the headquarters of the General Front Committee at Xiaolijia Village. On December 30, he led the staff of the General Front Committee to Shangqiu via Suxian and Xuzhou and reached Zhangcaiyuan the next day.[6] There Deng and the staff of the General Front Committee celebrated New Year's Day 1949.

The headquarters of the General Front Committee moved to a new place to direct the operations for the third phase of the Huai-Hai Campaign: completely wiping out the army under Du Yuming. Eight corps of the two armies under Du Yuming were tightly encircled in a small, narrow area less than 100 kilometers southwest of Xuzhou, with Chenguanzhuang as the center, stretching five kilometers from south to north and ten kilometers from east to west. In light of the annihilations of the armies under Huang Baitao and Huang Wei, Du Yuming was certainly aware of his own fate.

But because in the north the third campaign of the great decisive battle—the Beiping-Tianjin Campaign—had begun, the Communist-led troops only encircled his troops without attacking them. To prevent the troops under Fu Zuoyi in Beiping from fleeing south, Mao Zedong decided not to wipe out the troops under Du Yuming just then—so as to let Fu Zuoyi harbor illusions and remain in northern China—but to wipe them out later.

From December 1, 1948, when Du Yuming's troops were encircled, to January 10, 1949, when they were wiped out completely, it was truly tough going for Du Yuming. It was midwinter. Beginning on December 20, the weather brought both rain and snow, and the temperature dropped suddenly. Within the ring of encirclement were 200,000 men of the two armies under Qiu Qingquan and Li Mi who did not have sufficient food, clothing, or lodging. Chiang Kai-shek irregularly airlifted inadequate amounts of food to these troops, so when food was dropped, the hungry soldiers desperately scrambled for it and even killed or wounded many others for it. The extremely hungry soldiers ate whatever they could find and even killed the war-horses to appease their hunger. Suffering from the cold, soldiers burned all that could be found and even dug tombs and fortifications to find wood to burn for warmth. In the barracks at Chengguanzhuang, the despairing enemy troops presented a miserable scene.

In his prime, Du Yuming was bright and brave, but in such a plight, who was to blame? To blame Du Yuming himself was unfair, because he acted in compliance with the directive of President Chiang, even though he disagreed. To blame Chiang Kai-shek was futile, because he wanted to win the campaign but could not. Du Yuming was depressed, sitting silently in the shelter all day long and sighing in front of the wall.

On January 6, 1949, the East China Field Army launched the general attack on the troops commanded by Du Yuming. The enemy could not withstand a single blow, and Du Yuming failed in his attempt to break out of the encirclement. On the afternoon of January 10, the Communist-led army wiped out the troops under Du Yuming completely, killed Qiu Qingquan, commander of the 2d Army, and captured Du Yuming, deputy commander-in-chief of the Xuzhou "Bandit Suppression" Headquarters.

The decisive battle on the Central Plains—the Huai-Hai Campaign—came to an end.

Under the command of the General Front Committee, the two main forces of the Central Plains and East China field armies fought for sixty-six days, suffered over 134,000 casualties, and wiped out one "Bandit Suppression" General Headquarters of the KMT troops, five army headquarters, twenty-two corps headquarters, and fifty-six divisions, totaling over 555,000 men.

By then, the KMT crack forces at the southern front were completely wiped out, and the vast areas north of the middle and lower reaches of the Yangtze River were liberated.

The Huai-Hai Campaign was the only one of the three major decisive battles in which the total strength of the Communist-led army was inferior to that of the KMT. An army of 600,000 defeated an army of 800,000. On hearing this news, Joseph Stalin wrote in his notebook: "600,000 men defeated 800,000, a miracle, really a miracle!"[7]

P. F. Udin, who was later sent by Stalin to serve as the Soviet Ambassador to China, said: "The Huai-Hai Campaign was conducted very well. It was a miracle in the history of the Chinese revolutionary war and a miracle rarely seen in the world history of wars as well." Stalin let Udin go to China to study and understand the reasons the Communist-led army won the Huai-Hai Campaign.[8]

Mao Zedong commended the General Front Committee. After liberation, Mao Zedong, still bearing the campaign in mind constantly, once told Liu, Chen, and Deng that they had fought well in the Huai-Hai Campaign. The campaign was like a pot of half-cooked rice, but you ate it mouthful by mouthful.[9]

While the Huai-Hai Campaign was being intensely fought, the CPC troops in northern China launched the third campaign of the great decisive battle—the Beiping-Tianjin Campaign.

After the Liaoxi-Shenyang Campaign ended, about 500,000 men in forty-two KMT divisions were stationed in a narrow zone extending over 500 kilometers from Shanhaiguan in the east to Zhangjiakou in the west. Two-fifths of these troops belonged to Fu Zuoyi, and the rest were Chiang Kai-shek's own troops. Fu Zuoyi garrisoned Beiping. Fu Zuoyi and Chiang Kai-shek had one common foe—the CPC—but they had their own factions and conflicts of interests.

On November 29, 1948, the CPC employed a total strength of one million to launch the Beiping-Tianjin Campaign. According to the plan made by Mao Zedong, during the first phase of the campaign the Communist-led army was first to cut off the enemy's single-line battle array into several parts and to sever the ties between Beiping, Tianjin, and Tanggu. The CPC troops were then to encircle the enemy troops in Zhangjiakou and Xinbao'an at the western front without attacking them so that the enemy troops in Beiping and Tianjin would not flee by sea in the east. The CPC troops were to separate the troops in Beijing and Tianjin and finish cutting them off strategically without encircling them so that they could unhurriedly wipe them out one by one.

After these strategic objectives were achieved, the CPC troops applied the tactical principle of taking enemy positions on both wings and then capturing the ones in the middle. They first captured Xinbao'an and Zhangjiakou on the west wing. Then they kept watch over the enemy in Tanggu on the east wing and concentrated their forces to attack Tianjin.

Tianjin was the second largest city in northern China, with strong fortifications and an enormous number of defending troops. On January 14, 1949, 340,000 men of the five columns of the Northeast Field Army mounted a general attack on the city. After one day of fierce fighting, it wiped out completely the defending enemy troops, captured the garrison commander, Chen Changjie, and liberated Tianjin. After Tianjin was liberated, the 250,000 troops under Fu Zuoyi in Beiping were in a hopeless position and eventually signed a peace agreement with the CPC.

On January 31, 1949, the People's Liberation Army entered Beiping, thus announcing the peaceful liberation of the largest city in northern China, an ancient capital city with a 600-year history and a treasure of Oriental civilization.

On February 3, 1949, the People's Liberation Army held a solemn and grand ceremony on the occasion of its entering the city. This ancient cultural city, which had repeatedly suffered under the iron heel of the Japanese invading troops and had been exploited and oppressed by warlords and bureaucratic capitalists, eventually radiated the vigor of youth. Two million inhabitants of Beiping welcomed the People's Liberation Army with red flags, colored ribbons, cheers, and excited tears. People beat bronze drums, did the yangko dance, lined up for dozens of kilometers to celebrate this event, and greeted an entirely new bright era.

The Beiping-Tianjin Campaign had come to an end. The three campaigns—the Liaoxi-Shenyang, Huai-Hai, and Beiping-Tianjin campaigns—had all come to an end. The great decisive battles between the CPC and the Kuomintang in the areas north of the Yangtze River had come to an end.

These great decisive battles that lasted from September 1948 to January 1949 were unprecedented in the history of Chinese wars, and rare in world military history. In these three campaigns and in other battles fought at the same time, the Communist-led troops wiped out a total of 2.31 million KMT troops and almost all its main forces. The CPC pushed the battle line to the bank of the Yangtze River. The northern part of China—half of its territory—came under the control of the CPC.

Father said, "Mao Zedong's strategic thinking was to lock Chiang Kai-shek's troops into areas north of the Yangtze River and attack them there, giving them no chance to flee. This was great strategic thinking." Mao Zedong's strategic thinking carried the day, and his goals were achieved.

At that time, what pain and hatred Chiang Kai-shek felt in his heart! The pain of losing major campaigns. The unabated hatred he felt for the CPC, which had defeated him.

On January 21, 1949, Chiang Kai-shek announced his "resignation" and went back to his hometown of Xikou, Fenghua County, Zhejiang Province.

CHAPTER 62

FIGHTING IN THE AREAS SOUTH OF THE YANGTZE RIVER

When the three major campaigns came to an end, the Communist Party of China controlled half of China's territory. The liberation of the whole country could be expected soon.

On March 5, 1949, the Seventh Central Committee of the CPC held its Second Plenary Session at Xibaipo. Liu Bocheng and Su Yu asked for leave to be absent from the meeting. On February 28, Deng Xiaoping, Chen Yi, and Tan Zhenlin went together to the simple meeting place in Xibaipo. Mao Zedong made an important report. The main decisions made at the meeting were:

1. The proposal on the convocation of the new Political Consultative Conference and the establishment of a democratic coalition government was approved.
2. The People's Liberation Army was to try to liberate all provinces in central and southern China south of the Yangtze River as well as northwestern China. After crossing the Yangtze River, the army was to march south steadily.
3. The army was to shift the focus of its work to the cities, first capturing the cities and then the countryside.

The meeting also studied the economic problems, the democratic revolution, and other issues.

At the Second Plenary Session, the Central Committee discussed not only military questions but also the problems of how to rebuild the country and transform China from an old agricultural country into an industrial one and from a new-democratic society into a socialist one.

On March 14, the day after the Second Plenary Session concluded, the Central Committee called a meeting to prepare a plan of personnel arrangements for various administrative regions and to make decisions accordingly. Those who attended were not only leading comrades of the Central Committee but also the leading cadres from various administrative regions, including Peng Dehuai from northwestern China, Gao Gang from northeastern China, Nie Rongzhen from northern China, Deng Zihui and Lin Biao from central China, and Chen Yi and Deng Xiaoping from the Central Plains. Deng Xiaoping was the first one to speak at the meeting. Mao Zedong had asked him to advance his proposal on the jurisdiction of the region of eastern China and on personnel arrangements there. Deng Xiaoping had prepared thoroughly because he was keenly aware of the importance of the task assigned him by Mao Zedong of naming persons for particular jobs. Deng Xiaoping took out a name list. As he read it, he made explanations.

The East China Bureau of the Central Committee was to consist of seventeen members, including Deng Xiaoping, Liu Bocheng, and Chen Yi, with Deng serving as the first secretary. The East China Region had jurisdiction over: Shanghai, Nanjing, Hangzhou, Wuhu, Zhenjiang, Wuxi, Suzhou, Wujin, Nantong, Ningbo, and other cities extending across Shandong, Jiangsu, Zhejiang, Anhui, and Jiangxi provinces. The East China Region was to have two million troops. Chen Yi was to serve as the mayor of Shanghai. Liu Bocheng was to serve as the mayor of Nanjing.

Deng also talked about many other personnel arrangements and about ways to collect grain in new liberated areas after the CPC troops crossed the Yangtze River, about raising funds in the cities, and about using currency. In particular, he spoke about the takeover of Shanghai.

Mao Zedong readily agreed to Deng's detailed and comprehensive report. Mao said, "Now personnel arrangements are thus decided upon. If there are changes in future, we can do that." After the meeting, Mao Zedong met separately with Deng Xiaoping, Chen Yi, and some others to discuss questions about the operation of crossing the Yangtze. Father told us that it was at this time that Mao Zedong himself told him to direct the operation. Mao Zedong told Deng Xiaoping to direct an operation himself more than once—for instance, during the Huai-Hai campaign.

After attending the Xibaipo meeting, Father and Chen Yi traveled together back to the front. By then, both of them felt much more relaxed. The Yangtze-Crossing Campaign would be conducted until April, and the troops would have rest and reorganization in the meantime.

Father told me, "After the meeting, I and Uncle Chen took the opportunity to climb Mount Taishan and went to visit the Confucian Temple in Qufu before we returned to the front." Both Father and Chen Yi had a great interest in history and loved a pleasant occupation like sightseeing very much. Their tour was leisurely and carefree, and they had not enjoyed such a pleasure for more than twenty years, so they must have been quite jubilant. On the way, they must have talked cheerfully and humorously.

After the Second Plenary Session of the Central Committee, in accordance with the directive of the Military Commission, the Northwest, Central Plains, East China, and Northeast field armies were redesignated as the First, Second, Third, and Fourth field armies. Peng Dehuai served as commander and political commissar of the First Field Army. Liu Bocheng served as commander, and Deng Xiaoping as political commissar, of the Second Field Army. Chen Yi served as commander and political commissar of the Third Field Army. Lin Biao served as commander, and Luo Ronghuan as political commissar, of the Fourth Field Army. The total strength of the Chinese People's Liberation Army was four million men. The General Front Committee for the Huai-Hai Campaign became the General Front Committee for the Yangtze-Crossing Campaign, with Deng Xiaoping serving as its secretary.

According to the Central Committee's plan, the General Front Committee was to lead the Second and Third field armies to conduct the Yangtze-Crossing Campaign in mid-April. On March 26, the General Front Committee called a meeting attended by senior cadres of the Second and Third field armies in its headquarters near Bengbu, and Deng Xiaoping presided over this meeting to discuss the plan of operations for crossing the Yangtze River.

On March 31, the General Front Committee moved to an area east of Hefei. Here Deng Xiaoping himself wrote "An Outline Plan for Conducting the Campaign in Nanjing, Shanghai, and Hangzhou" and reported it to the Central Committee in a telegram. In the outline plan, Deng stated that the enemy had twenty-four corps, with a total strength of 440,000 men, and the Second and Third field armies had seven armies and twenty-one corps, with a strength of one million men. The People's Liberation Army was overwhelmingly superior to the enemy's.

The plan was to organize the troops for crossing the Yangtze River into east, central, and west army groups and apply the operational tactic of multi-pronged assault on selected targets on a wide front. The objective of the first phase was to cross the Yangtze River and conduct strategic deployment. The objective of the second phase was to separate and encircle the enemy and cut off his retreat routes. The objective of the third phase was to wipe out the encircled enemy troops one by one, thus completing the campaign. The CPC forces were to annihilate the enemy troops concentrated in the area between Shanghai and Anqing, capture southern Jiangsu, southern Anhui, and the whole of Zhejiang Province, seize Nanjing, Shanghai, and Hangzhou, and completely destroy the political and economic centers of the reactionary KMT government.

On April 1, Mao Zedong sent a telegram approving the "Outline Plan for Conducting the Campaign in Nanjing, Shanghai, and Hangzhou." On April 2, Deng Xiaoping and Chen Yi took a train from Bengbu to Hefei. After they arrived, they immediately drove to the headquarters of the General Front Committee located at Yaogang Village.

The Second and Third field armies began to make thorough preparations for the Yangtze-Crossing Campaign. The Communist Party resolved to fight across the Yangtze River and liberate the whole country.

On the southern bank of the Yangtze River, Chiang Kai-shek, though resigned from the presidency in name, in reality was still exercising overall command of the KMT troops. While Chiang Kai-shek had the KMT government in Nanjing send representatives to pretend to hold peace talks with the CPC, the KMT was simultaneously stepping up its efforts to build the defense line along the Yangtze River. Deployed along a 1,800-kilometer stretch from Hukou in Jiangxi to Yichang in Hubei were 115 divisions of KMT troops, totaling 700,000 men. The troops under Tang Enbo defended the battle line in Shanghai, and the troops under Bai Chongxi guarded the battle line in Wuhan, employing more than forty warships for river defense and four air force groups in the theater.

Chiang Kai-shek planned to take advantage of the natural moat of the Yangtze River to stage an all-out desperate fight with the CPC with his remaining courage and greatest strength so as to preserve half of the territory of China and divide the country at the Yangtze River. The Yangtze River flows from the Qinghai-Tibet Plateau in the west to the Yellow Sea in the east. The mighty river winds 5,800 kilometers from west to east through nine provinces and one city, constituting the largest and longest river system in China. The ever-surging Yangtze River has composed the immortal epic of the Chinese nation.

In areas north of the Yangtze River, the People's Liberation Army was busy preparing to cross it. It was making hydrolic surveys, conducting military training, organizing civilian workers, and repairing and building boats. The masses in the areas north of the Yangtze River showed great enthusiasm and tried every means to support the people's army; there were three million temporary civilian workers. Whatever grain, workers, or boats the army needed were all made available.

Under the unified leadership of the General Front Committee, the troops under Su Yu were stationed in the east, those under Tan Zhenlin in the center, and those under Liu Bocheng in the west. The People's Liberation Army was in single-line battle array, ready for the campaign, and just waiting for the order.

On April 21, 1949, Mao Zedong and Zhu De issued the order to "march toward all parts of the country." The Chinese People's Liberation Army was ordered to "advance bravely, wipe out resolutely, thoroughly, and completely all KMT reactionaries who dare to put up resistance in China, emancipate the Chinese people, and safeguard the independence and integrity of China's territorial sovereignty."

At 8:00 P.M. on April 20, the Yangtze-Crossing Campaign began, according to the predetermined plan. Under the unified command of the General Front Committee, the three massive armies in the center, east, and west forced their way across the Yangtze River with the momentum of an avalanche. Tens of thousands of boats started off at the same time and took advantage of the mighty waves to sail to the southern bank. The flares lit the sky like a fireworks display. The guns and cannons boomed like encouraging war drums, making a deafening noise.

Our troops took tens of thousands of wooden boats and advanced courageously like divine troops with an irresistible momentum. Our heroic troops crossed the Yangtze River, which looked like a giant flood dragon.

The Second and Third Field Armies broke through Chiang Kai-shek's "impregnable" defense line along the Yangtze River at one swoop. After they broke through the KMT defense line, Chiang's troops made a hasty general retreat. The Second and Third Field Armies advanced, in-depth, rapidly.

On April 23, the Communist-led army captured Nanjing—the seat of the KMT central government. At the presidential palace in Nanjing, the KMT flag with a white sun against a blue sky quickly fell to the dust, and the dazzlingly beautiful red flag of the Communist Party was slowly hoisted instead. Several days later, the General Front Committee moved into Nanjing. Deng Xiaoping and Chen Yi went into Chiang Kai-shek's presidential palace as the soldiers of the Chinese People's Liberation Army lined up solemnly.

I asked Father, "Did you enter the presidential palace?"

"Yes, I went with Uncle Chen."

"Where was Uncle Liu then?"

"At the time, he was directing operations at the western front."

"Did you sit on President Chiang Kai-shek's presidential throne?"

"Yes, we did," he said with a smile.

Nanjing was liberated. The next step was to liberate Shanghai.

The General Front Committee moved to Danyang County east of Nanjing along the Beiping-Shanghai Railway. Chen Yi went to Danyang first on May 3. Deng did not arrive there from Nanjing until May 6. It was already the still of the night when Deng arrived at Danyang. Chen wanted very much to take Deng out into the street to find a snack. In the dead of night, all the snack bars were closed. Chen and Deng happened to find a person selling won ton. They ate only a hodgepodge of won ton wrapper and stuffing without knowing what it was.

One of them was a commander and the other a political commissar. Both were founders of the People's Republic, but they had not changed their lifelong Sichuan habit of eating at small restaurants. They did this any time, any place, with no regard for the time it took. What a pleasure for them to saunter around a bit. This time, Deng and Chen were snatching only a moment of pleasure, because they were about to make a plan for capturing Shanghai and had to seize every minute and second.

Danyang was not large. With the General Front Committee located in the small city, there were suddenly many people. The streets were crowded, and people were busy. Some wore the uniforms of the various columns, and others wore long gowns or Western-style clothes. People were coming and going, and the city bustled with noise and excitement.[1]

Liu was still in Nanjing, and Su Yu and Tan Zhenlin were in their own army units. So only Chen and Deng were in Danyang. As secretary of the General Front Committee, Deng was naturally very busy just before the campaign. The General Front Committee had many jobs, such as planning military deployments and coming up with operational plans and plans for the troops' entry into the city and preparing cadres—so it lost no time in making all these arrangements. The General Front Committee's workload was quite heavy!

At the time, Shanghai was the largest port city of China—and of Asia—with a population of six million. Its industrial output value and trade accounted for half of the total in the country. Shanghai was Asia's largest financial center as well. Chen Yi said, "Liberating Shanghai is like catching a rat in a chinaware shop. We should catch the rat without smashing the chinaware." This was what people meant when they spoke of "sparing the rat to save the dishes."

But Chiang Kai-shek had deployed eight corps and twenty-five divisions, totaling 200,000 men, under the command of Tang Enbo near Shanghai. On April 26, Chiang Kai-shek hurriedly sailed by warship from Zhejiang to Shanghai to make defense arrangements in Shanghai in an attempt to take advantage of the city's abundant financial resources and more than 4,000 permanent fortifications to defend the city itself and to gain time for transporting matériel and conducting sabotage.

Chiang Kai-shek had personally gone to the northeast battlefield to direct operations. He had also gone to the Huai-Hai battlefield to direct operations. This time he went to Shanghai to defend it. Father once told us, "Wherever Chiang Kai-shek went, the KMT suffered defeat." After Chiang Kai-shek arrived in Shanghai, the KMT was doomed to failure in the battle in Shanghai.

After careful study, Deng and Chen decided:

- to pin down the troops under Tang Enbo so as to prevent them from fleeing by the sea route;
- to conduct operations on the periphery of Shanghai first and cut off the retreat route by sea at Wusongkou, and in case the city proper was attacked, to strive to protect the lives of inhabitants and their property without firing shells or using explosives;
- to strengthen education on discipline among the army units, tighten military discipline, enter the city without disturbing people, sleep on the streets without living in the people's houses, and make a deep impression on the people of Shanghai by enforcing rigorous discipline;
- to prepare for the takeover of the city and assemble over 5,000 cadres to receive training so as to be prepared in taking over the city; and
- to mobilize over 9,000 underground Party members, as well as the people, to protect factories and schools, maintain public order, and oppose the sabotage activities of KMT soldiers and special agents.

Everything was ready, and the liberation of Shanghai was just around the corner.

On May 12, the Communist-led army mounted an attack on the enemy troops on the periphery of Shanghai. On May 22, the Communist-led army drove the enemy main forces to the area on both sides of Wusongkou. On May 23, the CPC forces launched a general attack on the defending troops in Shanghai. On May 27, the campaign in Shanghai came to an end, 150,000 enemy troops having been wiped out. Shanghai, the bright pearl of the East, was reborn.

By then, Deng and Chen, who had been working in the operations room day and night at the headquarters of the General Front Committee in Danyang, could take a rest.

The soldiers of the Chinese People's Liberation Army entered the large city of Shanghai. They were well disciplined and in good order. These soldiers, in their handwoven cloth uniforms, did their proper duty and treated the people of Shanghai politely. To avoid disturbing the people, they slept at night, even when it rained, under the eaves of the tall buildings of Shanghai. The people's army immediately made an impression on the inhabitants of the city.

Order was rapidly restored in Shanghai, but the General Front Committee was still confronted with many complex problems. In Shanghai, it had to break the armed blockade imposed by the enemy and cope with the bombardments of enemy aircraft. More important, it had to restore production and economic activity there as quickly as possible. Deng entrusted the takeover to the military, administrative, financial and economic, and cultural and educational departments.

There was a great deal of work to be done in Shanghai. The General Front Committee had to maintain public order, restore production, streamline administration, and solve the problem of feeding the six million people of Shanghai.

Father and Chen Yi moved to the building of the former KMT Li Zhi Society at Ruijin Road. Both of them visited the grass-roots units during the day, heard reports in the evening, and often worked all night. At dawn, when the sun rose in the east, they lay down to take a nap. They had only one aim in their minds—to build a new Shanghai.

By May 27, the campaign in Nanjing, Shanghai, and Hangzhou had come to an end. During the campaign, the Second and Third Field Armies cooperated closely with each other and liberated Nanjing, Shanghai, and Hangzhou. They marched further south. Some units entered Fujian and liberated northern Fujian, while others advanced into Jiangxi and gained control of the vast areas in central Jiangxi.

Father once told us a story. He said,

> We fought those battles quickly. Because the enemy ran away quickly. Our troops ran by platoons, companies, and regiments to pursue the enemy. Otherwise, they could not catch up with the enemy. Our troops marched in many directions. The army led by Chen Geng marched the greatest distance and captured the whole of Jiangxi Province. During the years of the Red Army, Chiang Kai-shek caught Chen Geng. Taking into consideration that Chen Geng had once saved his life during the Great Revolution, Chiang Kai-shek set him free. When Chiang released Chen Geng, someone in Nanchang told Chen, "You are welcome to come here again." Chen Geng replied, "If I come here again, I shall bring 100,000 troops with me." Indeed, it was

> Chen Geng who commanded the troops that captured Nanchang. Fortunately, at the time we didn't let Chen Geng attack Nanjing. Instead he fought his way southward. Otherwise, Chen Geng wouldn't have kept his promise and fulfilled his hope.

Chen Geng was a most outstanding general whom Liu and Deng appreciated very much. He had unusual courage and insight, was forthright and somewhat naughty, and won the trust of Liu and Deng. During wartime, the Central Committee and Liu and Deng let him take charge of one heavy military task after another. When he spoke of Chen Geng, Father was always very proud of him and praised him.

After Shanghai was liberated, the families of Chen Yi and Deng came to Shanghai, and the two families lived in the building of the Li Zhi Society.

Mother told me that after Father moved from the Taihang Mountains to the Central Plains in 1945, he seldom lived with his family. After Mother left the Taihang Mountains, she worked in the Organization Department of the Shanxi-Hebei-Shandong-Henan Central Bureau. Afterwards, she took the three children to Handan.

To the children, Handan was the first big city they had experienced. Everything was different from the countryside, and everything was new to them. There was a flush toilet in the bathroom of the house. My elder brother was just over three years old then. He had never seen such a thing before and found it strange, so he often went to the bathroom to let water out of the flush toilet for fun.

As the Second Field Army marched forward continuously, Mother and the other members of the families of the army's senior cadres continued to move forward as well. In these families at the time, each mother looked after several children. Liu Bocheng had three children, Li Da two, Cai Shufan two, and Zhang Jichun three. This group of families moved from Handan to Xingtai, where they lived in a church and the five wives prepared meals in turn. Mother said that when it was her turn to cook, no one liked the dishes she served. Mother has still not learned how to cook.

Mother always wanted to apply scientific methods to the raising of her children. Every day she placed a big tub made of iron sheet on the roof of the church and put water in it. First she sunned the water all morning, then she took off the children's clothes, exposing them to the sun for a while, and let them jump into the water, playing and bathing. Mother said this was a sunbath. As a result, all three children were tanned and in good health.

After Zhengzhou was liberated, these families were going to move there. The railway and highway had been destroyed. The mothers and children took a truck, with no tarpaulin, and hurried to Zhengzhou. On the way, they felt quite nervous, not because of the war but because of the children. At dawn every morning they woke the children up. While the children were still dazed with sleep and had not yet opened their eyes, their mothers helped them put on their clothes. As the mothers packed their belongings, they quickly fed several mouthfuls of food to the children. They began to set off in the dark, with only the starlight to guide them. At noon, they took a nap when the truck stopped, and then they set off again at once. As there were so many children on the truck, a urinal had to be prepared for their use. In this way, they traveled without pause for a few days and finally arrived in Zhengzhou.

After they arrived in Zhengzhou, they immediately moved to the countryside in Luoyang. The Party organs and camping sites were always located in the countryside, probably because the Party could command troops there more conveniently. The CPC troops also did not want to disturb city dwellers but did want to maintain close ties with the peasants.

In the countryside in Luoyang, the mothers lived peacefully, but the children began to be naughty. One day three children sat at a table. The two-year-old Nan Nan held a firework in her hand and played with it, then lit it in the candle. The firework spurted spark, and Nan Nan was afraid and stretched out her hand. The spark was hurting Pan Pan's cheek. Afterwards, the adults found that nothing serious had happened. Several days later Pan Pan got a pair of scissors and jabbed Nan Nan's face with them. Fortunately, Nan Nan was not seriously wounded, but her face was scratched. This episode could be counted as a family skirmish. Mother was angered and beat the three children on their buttocks, whether there was reason to or not.

After Nanjing was liberated, the party of mothers and children moved there. Very soon they went to Shanghai with the army.

Chen Yi's family had three children, and the Deng family had three as well. Each family had five members. One day, when everyone was feeling happy, the adults of the two families brought their children together and had an intimate picture taken. In the picture, Uncle Chen is potbellied, sitting there and looking natural and unrestrained. Father is thin and smiling calmly. The two mothers are very young and pretty. The children are small and lovely. These were two happy families.

In Shanghai, Father paid a special visit to Dr. Sun Yat-sen's widow and a close friend of the CPC, Madam Soong Ching Ling. Father also took Mother to find Zhang Xiyuan's ashes and placed them downstairs in the building of the Li Zhi Society where they lived.

One day, Father and the new mayor of Shanghai, Chen Yi, went to attend a grand celebration. They left the offices and, escorted and protected by many bodyguards, attended a meeting in the hall just across the street. Crossing the narrow street took just a few minutes, but in that time a Parker pen in Father's chest pocket, one he had captured from the enemy, was stolen by a Shanghai pickpocket. Even now, Father still takes this theft to heart. Whenever he goes to Shanghai, he talks about this episode and says, "The pickpockets in Shanghai were terrible."

Seven years had passed since 1938, when Father went to the anti-Japanese battlefield, until 1945, when the People's Liberation Army fired the first shot in the battle against the KMT and crossed the Yellow River, advanced to the Dabie Mountains, conducted the Huai-Hai and Yangtze-Crossing campaigns, and liberated Nanjing and Shanghai.

During those years, Father fought battles, despite wind and rain, and went through many hardships but was never ill. Although he was not strong, he was in good health. For the sake of the war and victory, he had to be healthy. Ever since the anti-Japanese war Father had taken a cold bath every day. Whether it was spring, summer, autumn, or winter—even midwinter, when it was the coldest and the ground was frozen—every morning he poured a bucket of cold water over himself from head to foot.

But when he went to Shanghai and the people's army had achieved a decisive victory in the war, Father became ill. He had headaches and was in too much pain to get up. He was also very tired. The Central Committee agreed to give him a one-month sick leave.

One day in September, my parents went with their three children to Beijing. While under medical treatment there, Father reported on the work to the Central Committee and studied the military options for liberating southwestern China. In his leisure time, he took his children to visit the Summer Palace on the western outskirts of Beijing and zestfully went boating on the Kunming Lake, with its continuously autumnal waters. This was Father's first visit to Beijing in his forty-five years.

During Father's first visit to Beijing, two grand events occurred. One was the First Plenary Session of the Chinese People's Political Consultative Conference. The other was the grand ceremony honoring the founding of the People's Republic of China.

On September 21, 1949, the First Session of the Chinese People's Political Consultative Conference was solemnly held in the Huairen Hall in Zhongnanhai. Outstanding representatives from all circles and strata came from all directions and gathered in an atmosphere of joy and victory. Everyone was smiling and showing their enthusiasm for a new national life. At the session Mao Zedong solemnly declared, "The Chinese people, comprising one-quarter of humanity, have now stood up!"

Representatives, voting by a show of hands, adopted the national flag, national anthem, and provisional constitution (the "Common Program") of the People's Republic of China, approved Beijing as the capital, and elected the first Central People's Government—and Mao Zedong as chairman of that government—of the People's Republic of China.

After the meeting, to commemorate the countless martyrs who had laid down their lives in the people's revolutionary war, Mao Zedong led all the representatives in the solemn ceremony of digging a hole to lay the foundation stone for the Monument to the People's Heroes at Tian'anmen Square.

On October 1, 1949, Mao Zedong and his comrades-in-arms mounted the Tian'anmen Rostrum and solemnly declared: "The People's Republic of China has been founded! The Chinese people have now stood up!"

Father, Liu Bocheng, Chen Yi, and other founders of the People's Republic stood shoulder to shoulder at the Tian'anmen Rostrum, watched the bright, five-starred red flag rise slowly under the sunshine, listened attentively to the majestic, encouraging March of the Volunteers, and watched the 300,000 people in the square cheering and overflowing with excitement and the mighty contingents of paraders. They had lofty emotions and aspirations, longed for new undertakings, and were confident about the future of their new country.

The grand ceremony to honor the founding of the People's Republic of China was not a change of dynasties, as emperors and kings had done throughout Chinese history, and not a change of power between warlords. It was a moment to honor the Chinese people for having founded their own country.

On October 1, 1949, China turned an entirely new page in its 5,000-year history!

CHAPTER 63

MARCHING TOWARD SOUTHWESTERN CHINA

After the Yangtze-Crossing Campaign, the Communist-led troops captured Nanjing, Shanghai, Wuhan, Hangzhou, Jiujiang, Nanchang, Anqing, Jinhua, Shangrao, and other cities, as well as the whole of Jiangsu, Anhui, and Zhejiang and parts of Jiangxi, Hubei, and Fujian south of the Yangtze River.

The People's Liberation Army battle line was rapidly extending to the southern and western areas of China. To thoroughly wipe out the KMT troops in China, the people's army had to mount continuous, spirited, all-out attacks on them, launch speedy operations, and vigorously pursue them. The Central Committee made the following plan: The First Field Army was to leave Shaanxi and Gansu and liberate the five provinces in northwestern China. The Second Field Army was to directly enter Guizhou and Sichuan and liberate southwestern China. The Third Field Army was to march south to Fujian and liberate the southeastern coastal areas. The Fourth Field Army was to capture Guangzhou first and then liberate all provinces in central and southern China. The North China Military Area Command was to capture Taiyuan and liberate the whole of northern China.

The massive People's Liberation Army lost no time in setting off.

Liu and Deng were to lead the Second Field Army in its advance to southwestern China. They said goodbye to Chen Yi. Over the course of a year, Liu, Chen, and Deng had fought shoulder to shoulder, and the Second and Third field armies had fought shoulder to shoulder, achieving victory in the Huai-Hai Campaign, crossing the Yangtze River, and liberating Nanjing and Shanghai. Mao Zedong once said that because the Second and Third Field Armies conducted joint operations, their strength was not only multiplied one to two times over but their qualitative improvement was incalculable. Father said,

> From the very beginning, this qualitative change found its expression in the efforts to strengthen the Central Plains Bureau and appoint Chen Yi its second secretary. In particular, before the Huai-Hai Campaign, the General Front Committee was established. The committee comprised five members. Liu, Chen, and I were members of its Standing Committee, and I served as the secretary. Mao Zedong personally told me, "I give you the power to command." This was the task that Chairman Mao personally assigned me.

The General Front Committee accomplished its mission. The Second and Third Field Armies under Liu, Chen, and Deng were now going to operate separately. In accordance with the directive to march into southwestern China, the Central Committee decided to establish the Southwestern Bureau, with Deng Xiaoping serving as the first secretary and Liu Bocheng and He Long as the second and third secretaries, respectively.

The First and Second Field Armies were responsible for liberating southwestern China. The Second Field Army was to move from east to west and the First Field Army from north to south to mount a two-pronged attack on and wipe out the KMT troops in southwestern China.

On October 20, 1949, Liu Bocheng and Deng Xiaoping led the general headquarters of the Second Field Army in marching westward from Nanjing and began conducting operations in Sichuan and Guizhou. The same day, people from all circles in Nanjing gave a warm send-off to the Liu-Deng troops before their expedition into southwestern China. The battle for southwestern China had begun.

On October 23, the general headquarters led by Liu and Deng arrived in Zhengzhou. On October 28, the general headquarters entered Wuhan.

To liberate southwestern China, it was crucial that the people's army capture Sichuan, because Sichuan was the heart of southwestern China. Chiang Kai-shek went into battle once again and went to Chongqing to command the KMT troops personally.

The KMT concentrated on defending Sichuan. The terrain in eastern Sichuan, Hubei, and Guizhou was strategically situated and inaccessible because of poor communications, and over 100,000 KMT troops under Bai Chongxi were deployed there. So Chiang Kai-shek judged that the main force of the CPC troops would not enter Sichuan from the east but would advance into the south from the north. Chiang Kai-shek's judgment was wrong again.

Liu and Deng conducted a demonstration indicating the movement of their massive army from Zhengzhou to the west, but actually they had ordered the 3d Army under Chen Xilian to advance directly from the east into eastern Sichuan and the 5th Army under Yang Yong to make a detour to the south and enter Guizhou so as to cut off the enemy's retreat route to the south. The sudden, multipronged attack mounted by the Liu-Deng troops along a 500-kilometer section of an area from east to west completely disrupted Chiang Kai-shek's defense deployments in southwestern China.

The enemy troops were routed as if the mountains had collapsed, and they had no choice but to flee. The KMT troops fled so rapidly that the people's army could not carry out the Central Committee's principle of marching forward steadily. As the enemy troops ran so quickly, the Liu-Deng army had to pursue them even faster!

The Liu-Deng army braved the rain, marched over muddy roads, climbed high mountains, and passed through jungles. With no time for rest or food, it tried to pursue the enemy, speedily cut off the Sichuan army's retreat route to the south, and directed its attack at Chongqing and Chengdu.

Chongqing was the largest city in southwestern China and the provisional capital of the KMT government during the anti-Japanese war. To capture Chongqing was tantamount to taking off the crown of southwestern China. Not long before, Chiang Kai-shek had been in Chongqing to supervise the operations. This time, he understood what was about to happen and fled by plane with senior KMT military and government officials.

On November 30, the 3d Army under Chen Xilian captured Chongqing easily. After the fall of Chongqing, Chiang Kai-shek ordered his troops to withdraw to the Chengdu area. Liu and Deng resolved to wipe out completely Chiang Kai-shek's last main force—the army group under Hu Zongnan in the Chengdu Basin.

On December 20, the Liu-Deng army completely cut off Hu Zongnan's retreat route and surrounded the Chengdu area from west, south, and east, with Yang Yong heading up a unified command. Hu Zongnan had long been very arrogant in northern Shaanxi, but now he became a cornered beast. The morale of several hundred thousand KMT troops under his command was shaken.

Hu Zongnan abandoned his desperate officers and men and took a plane to flee alone. On December 27, the Liu-Deng troops wiped out the enemy troops in Chengdu completely. Chengdu was liberated.

At the time, the Yunnan army announced its revolt, and the 4th Army under Chen Geng liberated the whole of Yunnan Province. Afterwards, Chen Geng led his 4th Army in wiping out 10,000 enemy troops in the Xichang area in Sichuan, and it annihilated all the remaining regular troops of Chiang Kai-shek in southwestern China.

In southwestern China there was only one area—Tibet—that was yet to be liberated. On January 31, 1950, Tibet's Bainqen Kanpolija, the highest administrative organ under Bainqen, sent a telegram to Mao Zedong and Zhu De, indicating its opposition to the acts of the Lhasa authorities to betray the motherland. Mao Zedong ordered the Liu-Deng troops to march into Tibet.

In October 1950, the Liu-Deng army spent eighteen days conducting a campaign in Changdu—an eastern entrance to Tibet—wiped out more than 5,700 men of the Tibetan army, and opened the passage for marching into Tibet. Under the influence of the CPC's work among people from all circles in Tibet and its sincerity in seeking peace, the local Tibetan authorities decided to answer the call of the Central People's Government for the peaceful liberation of Tibet and sent a delegation of the local Tibetan government, headed by Apei Awangjinmei, to Beijing to negotiate with the Central Government. On May 23, 1951, the two sides signed "The Agreement on Methods of Peaceful Liberation of Tibet between the Central People's Government and the Tibetan Local Government."

In August and September 1951, the Liu-Deng troops began marching into "the roof of the world." They climbed a dozen high snow-capped mountains, strode over many raging billows and torrents, crossed the magnificent primitive forests, and went through boundless grasslands and marshlands. They walked over the 1,000-year-old snow despite the cold weather and thin air and finally entered Lhasa—the capital of Tibet—in October and November.

Tibet had established religious customs and was still in an era characterized by the co-existence of feudal and slave systems. After the People's Liberation Army entered Tibet, its soldiers did not live in people's houses or the monasteries, and they did not borrow their utensils. They preferred to eat in the wind and sleep in the dew and to be exposed to sunshine and rain, putting their rigorous discipline and commitment to protecting and respecting national and religious customs above all else. At the same time, they worked to contact people from all circles and cured the sicknesses of the Tibetan people, who were short of physicians and medicine. Their strict military discipline and way of treating the Tibetan people as equals moved their Tibetan compatriots, who warmly welcomed them.

After the Liu-Deng army liberated Tibet, its first important act was immediately to begin prospecting and designing the Xikang-Tibet Road—the first one in Tibet.

After Tibet was peacefully liberated, the whole of southwestern China was returned to the embrace of the people. The Liu-Deng troops courageously marched more than 2,000 kilometers and liberated southwestern China gloriously and triumphantly. During the campaign, they annihilated the remaining KMT troops, totaling 900,000 men, on the mainland and 900,000 bandits who were entrenched in southwestern China and stopped at nothing in doing evil.

While the Liu-Deng Second Field Army marched into southwestern China, the First Field Army liberated Shaanxi, Gansu, Ningxia, Qinghai, and Xinjiang. The Third Field Army marched into southeastern China and liberated Fujian and most of the islands along the southeastern coastal areas.

The Fourth Field Army liberated Hubei, Hunan, Jiangxi, Guangdong, and Guangxi in central and southern China. The North China Military Area Command liberated all of northern China. China was reunified. There was no longer war on the Chinese land. Chinese Communists had fulfilled their promise: to eventually eliminate war through war.

Over the last several thousand years, Chinese dynasties changed twenty-five times, and there were countless calls to battles. China's territory was always seized by force to make dynastic changes. It was China's historical law that those who won became the monarchs and those who lost became bandits.

History is so merciless, and people shiver, as though cold, to think of what has happened. But history is also merciful, and after a turn of events, it returns the steering wheel to the people, who become masters of their own fate.

The founding of the People's Republic of China was different from any dynastic change made by the feudal emperors over the previous 4,000 years. It was a product of a genuine people's revolution. The founding of the People's Republic showed that China, the Eastern giant that had been sunk in sleep for years, had now come to. A new China would now appear in the East.

For too long, Chinese men had been mocked for wearing long plaits and Chinese women for binding their feet with bandages. For too long, the Chinese people had been mocked as the "sick people of East Asia." For too long, the Chinese people had looked down on themselves and traveled far away across the sea to live in other countries. Fortunately, all such humiliation was gone forever. Chinese Communists and the 450 million Chinese people could now throw out their chests.

There have never been saviors, celestial beings, or emperors. We must rely on ourselves to create happiness for humanity—rely on ourselves completely! We take the path by ourselves! We take this magnificent path with pride! We have just begun to take the path, which is very, very long. The Chinese people must take the path step by step.

We stride forward along the path.

On December 1, 1949, the Second Field Army of the Chinese People's Liberation Army held a ceremony to enter Chongqing—the heart of southwestern China.

People from all trades and circles in Chongqing, men and women, young and old, solemnly lined up along the streets to welcome the Second Field Army.

On December 8, Liu and Deng led the headquarters of the Second Field Army in entering Chongqing.[1] The Central Committee had decided to establish the Southwestern Bureau, with Deng Xiaoping serving as the first secretary. After the liberation of Chongqing, the Central Committee decided to establish the Southwestern Military and Administrative Council, with Liu Bocheng serving as chairman, and the Southwestern Military Area, with He Long serving as commander and Deng Xiaoping as political commissar.

China was divided into six regions: northwest, southwest, east, central-south, north, and northeast. The CPC forces marched from the northwestern plateau to the battlefield in the Shanxi-Hebei-Shandong-Henan area, and after the major battles on the Central Plains, advanced to the battlefront in eastern China and finally marched to southwestern China.

The Liu-Deng troops had marched from the Taihang Mountains to the Dabie Mountains and from there to the Himalayas. This army wiped out 2.3 million KMT troops and more than one million bandits. It performed immortal deeds for the Chinese revolution and for the great and magnificent people's revolutionary war. Now this army had 600,000 crack troops and commanders and was stationed in southwestern China, where the population was sixty million. The Liu-Deng army had become a strong force in the people's army, rendering outstanding meritorious services and gaining great fame and prestige. Many of the commanders of this army became top generals of the Chinese People's Liberation Army and were selected to be key administrative officials of the government of the People's Republic of China.

Deng Xiaoping, political commissar of this army, calmly described this army:

> The Second Field Army didn't seek undeserved reputation but paid attention to unity in its ranks, and it did this throughout the process of operations. So the army units of the Second Field Army coordinated their actions very well. The relations were very harmonious between all columns, army units, men, and basic units. From the beginning of the war, all column commanders directed all specific operations; Liu and I didn't go to the battlefield personally to direct them. This method had an advantage, that is, when we found something improper, we could contact commanders by telephone. We didn't find anything wrong, and we didn't change the orders issued by the leading comrades of each column. This helped enhance confidence between the superiors and the subordinates and the combat effectiveness of the troops and encourage commanders to take the initiative.
>
> On the whole, during wartime the Second Field Army bore a heavy burden and accomplished its tasks, thus living up to the expectations of the Party and the people. Such is the history of the Second Field Army. We suffered many hardships, but we overcame all difficulties.

He made these few plain remarks with no undue pride and without bragging. It was the story of an old man to his descendants.

His words are plain and not surprising, but meaningful. Nevertheless, since his army suffered many hardships, rendered meritorious services, achieved glory, and developed a heroic spirit, one wonders how its story could be told in only a few words.

It was their subordinates whom Liu and Deng commended. And it was Liu and Deng whom their subordinates respected. To write this book, I interviewed many old white-haired generals. During the conversations, they showed deep and sincere feelings, feelings of great reverence and hearty love for their chiefs Liu Bocheng and Deng Xiaoping. Liu and Deng's subordinates were proud of having such chiefs.

While talking with these old soldiers of the Second Field Army, I was often moved by their sincere feelings. Sometimes it seemed as if I were marching and fighting alongside them—not interviewing them—and accomplishing the arduous tasks assigned by Liu and Deng.

Indeed, to everyone, the thirteen years from the time when the Second Field Army marched to the anti-Japanese battlefronts to its liberation of southwestern China was a long period. Moreover, those thirteen years witnessed much bloodshed and many moving and tragic incidents, as well as victory and glory. How could anyone who participated forget these years?

Time passes quickly. Some people say that time will obliterate the signs of human life. But time cannot obliterate the towering monument in the hearts of the people.

CHAPTER 64

THE FIRST SECRETARY OF THE SOUTHWESTERN BUREAU OF THE CENTRAL COMMITTEE

Father was back again in Sichuan. He was back again in Chongqing. He was back again in his hometown.

Destiny arranged his return with this coincidence. Twenty-nine years before, at a pier by the riverside in Chongqing, a sixteen-year-old boy named Deng Xixian left Sichuan on the passenger ship *Jiqing*, traveling down to the ever-flowing Yangtze River and then sailing overseas to start the first journey of his life career. Who could imagine that twenty-nine years later this Deng Xixian, renamed Deng Xiaoping, would be the supreme commander of the massive army that liberated Sichuan?

Destiny drew a circle for Father's life career in those twenty-nine years. On his return to Sichuan, Father was a middle-aged man of forty-five. He was then the highest ranking official of one of the several major administrative regions under the Central Government.

It was in Chongqing that Father finally began to lead a peaceful family life. Since the military situation was not so critical, both Father and Liu Bocheng took their families along with them when the people's army advanced southward. There were two American jeeps, one for the Liu family and the other for the Deng family. The Deng family had five members, two adults and three children. But, in fact, at that time Mother was with child. That would be me.

My elder brother and sisters often lorded it over me, saying that they went through the anti-Japanese war and the War of Liberation. I would answer sarcastically, "I also took part in the War of Liberation in liberating southwestern China, but in mother's abdomen."

After they had bumped along over 2,000 kilometers, the two jeeps reached Sichuan. It was not long before Sichuan was liberated, and New China was founded, that I was born, on January 25, 1950. Then my younger brother, Fei Fei, was born on August 26, 1951. He experienced no war at all, and we designated him a genuine "Liberation Brand."

In Chongqing we lived upstairs in a building that had been an office building of the Kuomintang. Liu Bocheng and his family lived downstairs. Later Liu was transferred to Nanjing to serve as the president of the Advanced Military Academy of the Chinese People's Liberation Army, and He Long's family moved in.

Reticent as Father is, he is actually easy to get along with. He maintained very good personal friendships with almost all of the ten marshals of the Chinese People's Liberation Army. Father had a very deep respect for Commander-in-Chief Zhu De, who, also a native of Sichuan, was unassuming, easy of access, and held in high esteem.

Father worked with Peng Dehuai in the general headquarters of the Eighth Route Army in the Taihang Mountains. Peng had no children and asked my parents if his family could adopt me. Naturally, my parents hated to part with me. For some time, whenever I saw Commander Peng, I immediately hid behind my parents.

Father worked with Liu Bocheng for thirteen years. The two families lived in the same building for a long time, and Mother and Liu's wife were very close friends.

Like Father, Chen Yi, a native of Sichuan, had been a student in France. They commanded together the People's Liberation Army in the Huai-Hai and Yangtze-Crossing Campaigns, as well as in the liberation of Nanjing and Shanghai. Both were fond of cooking. The two families were next-door neighbors for ten years in Beijing and were so close that they often went for walks and outings together and even shared the rare fruits—like the extremely stinking durian—sent by foreign friends.

Nie Rongzheng, another native of Sichuan, was also a student in France. Father called him "old chap." In the early 1950s, when our family had just moved to Beijing, we lived next-door to Uncle Nie's home. My siblings and I often went to his home for candies. Uncle Nie often invited our whole family to his home for Sichuan-style snacks. Without standing on ceremony, Father took seven or eight members of our big family to enjoy food there for "free." Uncle Nie lived a very long life. Toward the end, he was the only person whom my father, aged himself, went out to see. Uncle Nie passed away at the age of ninety.

Father was with Luo Ronghuan during the Long March. Although they worked in different places later, they knew each other very well. They cared about each other and shared the same viewpoints on political questions. Unfortunately, Uncle Luo died too early. After his death, my parents asked me to stay at his home for a week to be with his two daughters, who were about my age.

Before liberation, Father and Ye Jianying had not worked together very much. But after liberation, especially after the Cultural Revolution, they made deeply sincere joint efforts to save the country from disaster. I still remember that, to have Father reinstated for the third time, Uncle Ye secretly sent his youngest son to pick up Father while he was still under house arrest. Seeing each other, they got very excited. After Father called him "old chap," they clasped hands.

Xu Xiangqian once served as deputy commander in the Liu-Deng army. We were neighbors for a while. Xu, a few years older than Father, was very weak and afflicted with prolonged illness. Father had a deep respect for him and was very concerned about his health.

He Long, also known as "The Bearded He," was very straightforward. In Chongqing, our two families shared the same building, one upstairs and the other downstairs. The children of the two families were close in age and always played together. After liberation, Father often brought us to see Uncle He. Whenever we got together, adults were chatting and laughing and children were playing. We were just like members of one family.

It may seem strange that Father was on very good terms with only nine of the ten marshals. The exception was Lin Biao, with whom father never had any contact, mainly because Lin Biao was a most eccentric person and never had any contact with anyone.

After Liu Bocheng was transferred to Nanjing, it was mainly Father and He Long who took charge of the military and administrative work in Sichuan and southwestern China. Because he took charge of military, political, and economic work as well as minority nationality affairs, it is easy to understand that Father was as busy with his work as ever.

The war was over in southwestern China. Father once asked his subordinates, "Has the war come to an end?" He went on to say that the future work in southwestern China would be much more complicated than an ordinary military struggle, and that it would not be accomplished simply by making several "charges."

He set three tasks for southwestern China: the problems of 900,000, 60 million, and 600,000.

The figure 900,000 referred to the more than 900,000 Kuomintang troops who had been captured or had come over to the CPC side during the war. They had to be remolded into members of the people's army or other kinds of workers who would be able to work and engage in production.

The figure 60 million referred to the 60 million people whom the CPC could rely on out of the total population of over 70 million in southwestern China. The people needed to be organized to carry out land reform and undertake production so as to help the economy recover.

The figure 600,000 referred to the 600,000 troops of the people's army in southwestern China. The fighting force needed to be turned into a political work force, its quality improved and its discipline tightened, so as to create and build a new southwestern China.

Father's work style was the same for both government and military work. He explained things clearly and concisely and handled matters resolutely. The meetings he presided over had two special features. First, they were brief. The First Plenary Session of the Southwestern Military and Administrative Council lasted only nine minutes. His meetings had no excessive formality or superfluous words. As soon as participants finished speaking, the meetings concluded.

Second, Father listened to others' opinions before he made decisions. At the meetings, he first asked people from all departments to speak and raise questions. No matter how many spoke or what questions were raised, when all others finished speaking, it was Father's turn to speak. He answered the questions he was responsible for and made some immediate decisions. As to questions that required further study, he instructed people from the relevant departments to research them.

According to those working in the Southwestern Bureau of the Central Committee, they often came to the meetings with many questions, worries, and misgivings but afterwards were clear about the objectives, tasks, and methods to be adopted.

While working in southwestern China, Father made two major decisions. One was to build the Qinghai-Tibet Highway leading from Lhasa to Qinghai, and the other was to build the Chengdu-Chongqing Railway. Father knew very well that although the war was over, building a new China would entail tasks more arduous than fighting a war.

After two years of hard work, public order in southwestern China had been rapidly restored, the economy had begun to recover, and work was proceeding in all fields on the right track, step by step.

My younger brother and I were born in Chongqing, so the two of us are genuine natives of Sichuan.

Actually, my younger brother was born by luck. Before he was born, there were already four children in our family, three girls and one boy. When Mother was with her fifth child, she intended to have an abortion because she was very busy with her work as the schoolmaster of the People's Primary School in Chongqing. But the director of the Department of Health of the Second Field Army told her, "Maybe it is a boy!" Thanks to his words, my younger brother Fei Fei came into this world. As a result, the child whom my parents did not want became the one they loved best.

In Chongqing, my grandma came to join us from our native place. It is interesting to tell the story.

At that time, my second aunt, Deng Xianfu, was studying in a middle school in her hometown of Guang'an and had taken part in the activities organized by the underground Party organization of Sichuan. When Sichuan was to be liberated, comrades of the underground Party organization told her that her elder brother was going to fight his way back to Sichuan. After Father arrived in Chongqing, with the help of the Party organization, my second aunt also came to Chongqing from her hometown to meet her elder brother, whom she had never seen before. Upon her return to Guang'an, my second aunt told my grandma the news. My grandma was overjoyed.

Shortly before Sichuan was liberated, the chaos caused by the war had brought our kinsfolk in Guang'an to the verge of bankruptcy. Grandma knew nothing about politics, but since Father was a Communist, she firmly believed the Communist Party was good. Her daughter took part in the activities of the underground Party. She once took home several Party members of the guerrilla detachment of the Huaying Mountains, and she hid them in her house without saying anything.

My grandma was a widow. She had neither money nor power. She risked her life to help the Communists go into hiding! In doing so, she saved their lives.

On hearing from my second aunt that Father had returned to Sichuan, my grandma was very excited. She locked up her house and left for Chongqing in a small boat of her father's, a boatman on the Jialing River, bringing only a parcel with her.

She left her hometown and abandoned her land and house. Since then, she has lived with us. Although my grandma was not Father's own mother, my parents treat her very well. Mother in particular has been on very intimate terms with her. Grandma took care of the family and looked after the children when Mother went to work. I was just ten months old when Grandma joined us. It was she who brought me up. My younger brother was brought up by her as well. So we two have been especially close to her. Grandma has really borne much of the load of household duties for Mother. Grandma has lived a very long life. She is now over ninety and still lives with us. In short, all the members of our family lived together in Chongqing. The family remained the same for twenty years, until the members of my generation got married and had children.

In Chongqing both my parents were very busy. Mother served as the schoolmaster of the People's Primary School. All the pupils in this school were the sons and daughters of those working in the Second Field Army or in the Southwestern Bureau of the Central Committee. These children who had grown up in the army were all spoiled and disobedient. Mother made an example of her own children, my eldest sister and elder brother, who were, of course, Mother's students. If they failed to observe discipline, Mother severely criticized them. My second elder sister was only five years old then. Mother had her sit in the last row in the classroom every day, whether she understood the lessons or not.

Some pupils did not pay attention in class or even cried. Mother would take them to her office. No matter how they cried or made noise, she just did her own work, paying no attention to them. Finally they would stop crying.

Mother was the schoolmaster, but she taught all the courses—Chinese, mathematics, and even music. I always wondered how she taught music, for she herself always sang out of tune.

The children of many cadres of the Second Field Army were Mother's students, and they are between forty and fifty now. They often recalled how their schoolmaster taught and "criticized" them.

Father was exactly forty-five years old when the People's Republic of China was founded. Forty-five was nearly half a century.

The sixteen-year-old Deng Xixian, with his childish face, started his long life journey when he left Sichuan. He changed from a patriotic young man into a man strongly espousing the ideal of communism. He went abroad to study and became a Communist Party member before he returned to the motherland. He took part in the Great Revolution and experienced the storm of the White terror. He went to the battlefield and fought twenty years of battles against the warlord forces, the Japanese invading troops, and the Kuomintang army.

When he was forty-five years old, his name was already so closely linked as to be inseparable from the history of the Party, the people's army, the People's Republic of China, and the entire history of the Chinese people's revolution. Under the leadership of the Central Committee and Mao Zedong, Deng Xiaoping and other founders of the People's Republic performed meritorious services and made immortal wartime contributions that ushered in a new era for the Chinese nation.

Deng Xiaoping also grew from a young revolutionary into an all-powerful revolutionary and a battlefield commander capable of directing large-scale operations. Father spent twenty of his first forty-five years fighting battles.

The war was over, and the New China was founded. Father was assigned to take charge of the work of southwestern China as a "governor" of the region.

Forty-five years represent a period of time neither too short nor too long. Some people have passed the peak of their life by the time they are forty, while others have just begun to find the starting point of their career when they are near fifty. I believe that the lifelong experience of the overwhelming majority of people may possibly amount to less than one half of what Father experienced in his first forty-five years.

If Father had looked back at his past on his forty-fifth birthday, he would have had many pleasant recollections and no qualms about what he had accomplished. If he looked forward to the future, he could have been very proud of his success, full of confidence, and determined to spare no effort in creating a completely new situation in China.

The age of forty-five marked a milestone in Father's long career. However, it marked another starting point in Father's political life as well. The day he turned forty-five, neither he nor anyone else in the world could have expected how many more difficulties and dangers he would encounter in his future.

The first half of Father's life was brilliant, but not the most brilliant. The first step in the political career of Deng Xiaoping was taken just at this moment. Now he would start marching toward the most brilliant peak of his political career.

CHAPTER 65

AN UNFINISHED STORY

This book is coming to an end.

However, the story of Deng Xiaoping is far from over. The story of the second half of Father's life does not fall within the scope of this book. I hope that in the near future I will be able to present to readers the story of the second half of the brilliant life of my father. Here I will make only a brief account of the rest of his life.

Father was transferred to Beijing in 1952 on the instruction of the Central Committee. For the second time in his life, he left Sichuan with his family. His first departure from Sichuan was the first step in an as yet unknown career. His second departure was a stride toward his increasingly brilliant future.

In 1952, when he went to Beijing, he was appointed vice-premier of the Government Administration Council of the Central People's Government, and later on, while still serving as vice-premier, he also held positions as minister of finance, minister of communications, and director of the Organization Department of the Central Committee.

In 1954 Father was appointed secretary-general of the Central Committee, a member of the Military Commission of the Central Committee, and vice-chairman of the National Defense Council. In 1955 he was elected a member of the Political Bureau of the Central Committee. In 1956, at the Eighth National Congress of the Communist Party of China, he was elected a member of the Standing Committee of the Political Bureau of the Central Committee and general secretary of the Central Committee. Father has been one of the members of the top leadership of the Party and the government in China ever since.

As the general secretary, he took charge of the day-to-day work of the secretariat of the Central Committee and became an important assistant to Chairman Mao Zedong. As the first vice-premier of the State Council, he became an assistant to Premier Zhou Enlai, according to the division of work.

By the early 1960s Father and Liu Shaoqi had been selected by Mao Zedong to be the successors who would take joint charge of the work of practical leadership.

From 1952 to 1966 the political situation in China was relatively stable. Although some policy decisions were not correct, New China, generally speaking, after seventeen years of economic development, had laid a sound economic and material foundation and was playing an important role in international affairs.

In 1966 the history of New China turned to an unfortunate page. The Cultural Revolution, initiated by Mao Zedong himself, broke out. A furious "leftist" political storm swept across the land of China. Father was overthrown as the "No. 2 capitalist roader." Father, our family, and all the Chinese people had to go through this frantic, puzzling, and unfortunate period when politics was misguided and human nature distorted.

In 1971 the situation changed. Lin Biao, the successor designated by Mao Zedong himself, plotted to assassinate Chairman Mao. After his plot was foiled, Lin Biao fled in a plane that crashed and killed him.

In March 1973 Mao Zedong miraculously reinstated Father by restoring him to his former post as vice-premier of the State Council. In January 1975 Mao Zedong assigned Deng Xiaoping to the important posts of vice-chairman of the Central Committee, vice-premier of the State Council, vice-chairman of the Military Commission of the Central Committee, and chief of the general staff of the Chinese People's Liberation Army.

After his reinstatement, Father was faced everywhere with the desolation and devastation caused by the disastrous Cultural Revolution. He had been overthrown himself, but he performed his duties without the slightest hesitation. With the support of Premier Zhou Enlai, he promptly decided to begin to correct the mistakes of the Cultural Revolution by using the power conferred on him by Mao Zedong in compliance with his sense of responsibility for the future of the disaster-ridden country. Father's bold and resolute actions and clear-cut stand met with strong opposition from Mao Zedong's wife, Jiang Qing, and others. Deng Xiaoping and Jiang Qing's "Gang of Four" became two irreconcilable and antagonistic forces in China's political arena.

Mao Zedong acted very wisely for most of his life, but his later years were fraught with lamentable mistakes. By this time, he trusted no one except his kinsfolk and a few protégés. Mao tipped the political balance in favor of the Gang of Four, who pursued a more "leftist" line than Mao did.

The year 1976 was an unforgettable year full of tragedies. On January 8, 1976, Zhou Enlai passed away and was deeply mourned by his comrades. In April Deng Xiaoping was overthrown once again. On September 9, Mao Zedong passed away. On October 6, Jiang Qing and her gang were arrested and brought to trial at last.

In 1977, Deng Xiaoping was reinstated again and restored to all his former posts in the Party, the government, and the army. In his life, he was overthrown three times and reinstated three times. Each time he drew greater attention, and each time he went on to achieve greater successes. This is neither a myth nor a fabrication. This is the true story of Deng Xiaoping.

Deng Xiaoping was already seventy-five years old when he was reinstated the third time. He did not change his tenacious work style, which he had had for decades, his bold way of thinking, or his firm convictions. His conviction was that a new path of development, suited to China, could be blazed by taking a realistic and scientific approach and by making use of a multitude of assets and strengths, both ancient and modern, both Chinese and foreign. His conviction was that the Chinese people could lead a prosperous life and that China could become a strong and prosperous country.

As the ten-year Cultural Revolution ended, a new epoch began in China. Initiated, led, and propelled by Deng Xiaoping, China, from the ruins of the

Cultural Revolution, took the bright road of reform and of opening to the outside world and started a new revolution and a new long march.

Like an architect, Deng Xiaoping drew a completely new blueprint of development for his motherland. By the end of the 1980s the gross national product of 1980 was to be doubled so as to solve first the problem of food and clothing for the 1.1 billion Chinese people. By the end of the century the 1980 GNP was to be quadrupled so that the Chinese people would have moved from simply having adequate food and clothing to being fairly well off. By the middle of the next century, when we celebrate the one hundredth anniversary of the founding of the People's Republic of China, China, with a population of one and a half billion, is to rank with the moderately developed countries in per capita GNP. The Chinese people will become prosperous, and modernization will be basically realized by then.

This is the three-step strategy for China's development devised by Deng Xiaoping.

Deng Xiaoping has held that the Chinese should build socialism with Chinese characteristics. Under his guidance, China has been trying to discover its own path of development and marching forward steadily. More than fifteen years have elapsed since 1978. China has made progress and achievements that are universally acknowledged. There is a popular view that the next century will be the century of the Asia-Pacific countries, and that among them China will attract the greatest attention. China is proud of this expectation.

Some people say that Deng Xiaoping is one of the great men of this century who has attracted the greatest attention in the world. Others say that Deng Xiaoping is the most notable man in the world today. In China, some ordinary people put up portraits of Mao Zedong and Deng Xiaoping together in their homes.

In 1989 Deng Xiaoping resigned and retired. He retired so as to abolish the feudal life tenure system in China and to promote younger people to the leading posts. Even in retirement, he still pays constant attention to the great undertaking of reform in China and its opening up to the outside world. At the advanced age of eighty-eight, he still goes around calling for the further development of China.

He retired, but the cause he initiated keeps on advancing. Present-day China has achieved initial success in its great undertaking and is going to develop rapidly.

How time flies! The autumn and winter were over in a flash, and another spring had come. The Spring Festival of 1993 arrived.

The year 1993 is the year of the rooster in China. "To rise up upon hearing the crow of a rooster" goes a Chinese saying. Like a majestic-looking rooster and a great dragon that is about to fly, China, full of confidence, pride, and enthusiasm and with its chin up and chest out, was ready to advance valiantly toward the twenty-first century in the midst of the deafening noise of firecrackers on the eve of the Spring Festival.

As usual, Father took the whole family to Shanghai to spend the festival. It was indeed a jubilant festival. Everyone felt happy, and everything looked fresh. Outside the house, colored lanterns were hung up high, their bright lights like spluttering fireworks. Inside the house, we felt warm, cozy, and comfortable, as if spring had already come. A dozen members of our family of three generations were joyously immersed in the jubilation of the festival.

Father, nearly eighty-nine years old, sat in the center among us. He looked calm and composed with his white hair glittering under the lamplight. On his face was a serene smile that came from the bottom of his heart.

His smile transcends the range of time and space and is eternal.

CONCLUDING REMARKS

It has been my long-cherished wish to write about my father.

I want to write about my father because I am often at his side, and I think I know him very well. I want to write about him because I hold him in high esteem.

There are many prominent persons in the world. There are many sons and daughters of the prominent persons in the world as well. Many sons and daughters of prominent persons are writing about their parents. Quite a few of them do not praise their parents. I am different. I love my father from the bottom of my heart.

I have wanted to write about my father for a long time, but it is only in recent years that I have decided to do so. I spent three years collecting materials, interviewing, and reading history books. Over those three years, I have devoted all my energies to writing about the first half of my father's life.

Now my father is eighty-nine years old. When New China was founded, he was only forty-five years old, having made it just about halfway through his life journey. My father has gone through so many things in his long life that there are too many stories about him to be written. It is beyond my ability to write about the colorful and legendary life of my father, or even to make a rough sketch of it. My purpose in writing this book has been to tell readers only what I know about my father. Perhaps it will serve as a supplement to books on Chinese history.

I have finished this book, but I have not finished writing all about my father. The most brilliant chapters of my father's life lie ahead. I know more about and have a deeper understanding of the second half of Father's life because that was when I myself grew up and became mature enough to understand him.

I want to present the whole colorful life of my father to readers. Some people may ask why I have not finished that story, why I have interrupted it.

I want to tell you the reason and ask you to excuse me. My energy and ability have prevented me from continuing right away. But I should like to ask those who have read this book and want to read the next one about the second half of my father's life to be a little patient as I take a rest before finishing the story.

I know that this book absolutely cannot depict the whole of my father's long life and career, which have been full of vicissitudes. However, I have done my best. I believe that the next book will be better than this one.

Finally, I hope that when my father finishes reading this book, he will say, "It is not bad."

Father has never praised us. "It is not bad." That would be enough.

CHAPTER 1. THE DAY OF RETIREMENT

1. *People's Daily*, November 10, 1989.
2. Ibid.

CHAPTER 2. AN AFFECTION FOR SICHUAN

1. Human fossils two million years old were discovered in Longping Village, Damiao District, Wushan County, near Chongqing. These are the most ancient human fossils discovered in China so far. See *People's Daily*, November 19, 1988.
2. Zhou Yong, *Chongqing: The Rise of an Inland City* (Chongqing: Chongqing Publishing House, 1989).
3. Annals of Guang'an County, Guang'an County Archives.

CHAPTER 6. THE TRAGIC HISTORY OF A DECLINING NATION

1. "An Open Call to Arms by Righteous Persons of Ideals of All Villages of Guangdong," in *The Struggle Between China and Britain After the Opium War: Collected Materials*, p. 207. See Lin Xianming, Lu Min, Li Tongnian, Ma Hairong, Jin Shaoqing, Wang Yuling, Zhang Zhijun, Chen Mingming, and Cong Jun, eds., *The History of Modern China* (Beijing: Zhonghua Book Company, 1979), p. 49.
2. Marx to Engels, October 8, 1858, in *Selected Correspondences of Marx and Engels*, p. 111.

CHAPTER 7. THE PEOPLE'S HEROIC REVOLT

1. Mao Zedong, "The Chinese Revolution and the Chinese Communist Party."

CHAPTER 11. THE MOVEMENT TO STUDY IN FRANCE ON WORK-STUDY PROGRAMS

1. Mao Zedong, "On the People's Democratic Dictatorship," in *Selected Works of Mao Zedong*, vol. 4.
2. Cai Yuanpei, a native of Shaoxing, Zhejiang Province, was a famous educator and revolutionary democrat in modern China. He studied in Germany in 1907 and as a young man joined the Chinese Revolutionary League led by Dr. Sun Yat-sen. He served as the minister of education for the Provisional Government in Nanjing in 1912. He became president of Beijing University in 1917 and gave much support to the new cultural movement. Later on, he served as the director of the National Research Institute of the National Government. After the September 18th Incident in 1931, he, together with Madame Soong Qing Ling, Lu Xun, and others, organized the China League for Protecting Civil Rights. He died of illness in 1940.
3. Wu Yuzhang, a native of Rongxian County, Sichuan Province, was a renowned educator and a proletarian revolutionary of the older generation of the Communist Party of China. He studied in Japan in 1903 and was an important member of the Chinese Revolutionary League. He joined the Communist Party of China in 1925 and took part in the August 1st Nanchang Uprising in 1927. Before and after liberation, he held various high posts in the Communist Party of China and the government. He and Xu Teli, Xie Juezai, and Dong Biwu were addressed respectfully as the "four venerables" within the Party. He died of illness in Beijing in 1966.

4. Li Shizeng, a native of Gaoyang, Hebei Province, studied in France in his early years. He professed to believe in anarchism and highly praised French civilization.

CHAPTER 13. MAKING A LONG AND DIFFICULT JOURNEY

1. According to news reports at that time, the number of students who set off from Chongqing was eighty-three (see *The National Gazette*, August 8, 1920) but was increased to eighty-four after they boarded the ship (see *The New Daily of Current Affairs*, September 14, 1920).
2. Letter from Feng Xuezong in Cambrai. See "The May 4th Movement of 1919 in Chongqing," in Party History Working Committee of the Chongqing CPC Municipal Committee, ed., *Collection of Research Materials on CPC History in Chongqing* (Chongqing, July 1984), p. 180.
3. Jiang Zemin, "Recollections of Studying in France and Belgium on a Work-Study Program," in CPC History Teaching and Research Group of Qinghua University, ed., *Historical Materials of the Movement to Study in France on a Work-Study Program*, vol. 3 (Beijing: Beijing Publishing House, November 1981), p. 445.

CHAPTER 14. FROM SCHOOL TO FACTORY DESPITE HARDSHIPS TO PURSUE STUDIES

1. *Petit Marseillais*, October 20, 1921.
2. Letter from Feng Xuezong in Cambrai.
3. Jiang Zemin, "Recollections of Studying in France and Belgium on a Work-Study Program."
4. Ibid.
5. *Bayeux Daily*, October 22, 1920.
6. Files of the French Chinese Relief Committee of the Chinese Youths in France, no. 47AS2, National Archives of France.
7. Report of Bayeux Middle School, March 1921.
8. Huang Lizhou, "The Movement in Sichuan to Study in France on a Work-Study Program," in Committee for Research Into Historical Accounts of Past Events of the Sichuan Provincial Committee of the Chinese People's Political Consultative Conference, ed., *Collection of Selected Historical Accounts of Past Events in Sichuan*, vol. 23 (Chengdu: Sichuan People's Publishing House, November 1980), p. 1.
9. Jiang Zemin, "Recollections of Studying in France and Belgium on a Work-Study Program."
10. Huang Lizhou, "The Movement in Sichuan to Study in France on a Work-Study Program."
11. Barman and Denis Voussetois, *Modification of a Youthhood Portrait: Deng Xiaoping's Years in France*.
12. Zhang Hongxiang and Wang Yongxiang, A *Brief History of the Movement to Study in France on a Work-Study Program* (Harbin: Heilongjiang People's Publishing House, April 1982), p. 55.
13. Shu Guang, "The Student-Workers in Schneider & Cie Factory, Creusot, France," in CPC History Teaching and Research Group of Qinghua University, ed., *Histori-*

cal Materials of the Movement to Study in France on a Work-Study Program, vol. 2, pt. 1, p. 256.

14. Employment registration cards from the Personnel Department of the Schneider & Cie Factory, archive no. 62175, Schneider & Cie Factory Archives.
15. Huang Lizhou, "The Movement in Sichuan to Study in France on a Work-Study Program."
16. Contract of apprenticeship for Schneider & Cie, Creusot, archive no. 47AS8, National Archives of France.
17. Huang Lizhou, "The Movement in Sichuan to Study in France on a Work-Study Program"; Shu Guang, "The Student Workers in Schneider & Cie Factory, Creusot, France."
18. Personnel records of Schneider & Cie, and list of Chinese volunteer workers at Schneider & Cie, in Schneider & Cie Factory Archives.

CHAPTER 15. FOR SURVIVAL AND STUDY

1. CPC History Teaching and Research Group of Qinghua University, ed., *Historical Materials of the Movement to Study in France on a Work-Study Program,* vol. 2, pt. 1, p. 335.
2. Ibid., p. 249.
3. Ibid., p. 365.
4. Huang Lizhou, "The Movement in Sichuan to Study in France on a Work-Study Program," p. 26.
5. Li Weihan, *Recollections and Research* (Beijing: CPC Historical Materials Publishing House, April 1986), pt. 1, p. 23.

CHAPTER 16. IN THE HUTCHINSON RUBBER PLANT

1. Chen Zhiling and He Yang, *Biography of Wang Ruofei* (Shanghai: Shanghai People's Publishing House, August 1986), p. 44.
2. List of recruited workers at Chambrelent, archive no. 47AS8, National Archives of France.
3. Luo Han, "A Period of Life of Studying on a Work-Study Program," *Revolutionary Weekly,* December 15, 1928.
4. Ibid.
5. Chambrelent files, archive no. 47AS8, National Archives of France.
6. Huang Lizhou, "The Movement in Sichuan to Study in France on a Work-Study Program," p. 1.
7. Zheng Chaolin, *Memoirs* (restricted material of the Party History Office of the CPC Shanghai Municipal Committee) (Shanghai, May 1983).
8. Register of Foreigners, Chalette.
9. Working cards of the Hutchinson Plant.
10. Manager of the Hutchinson Plant to Chairman of the French-Chinese Relief Committee of Chinese Youth in France, March 18, 1922, archive no. 47AS9, National Archives of France.
11. Zheng Chaolin, *Memoirs.*
12. Working card of Deng Xixian, Files of the Hutchinson Plant.

13. Register of Foreigners, Chalette.
14. Ibid.
15. Working card of Deng Xixian; register of Chinese workers in the Hutchinson Plant.

CHAPTER 17. ESTABLISHING COMMUNIST ORGANIZATIONS IN EUROPE

1. Li Lisan, "Reminiscences of Comrade Zhao Shiyan During His Years of Study in France," in *Collection of Revolutionary Materials About Martyr Zhao Shiyan* (restricted material).
2. "Historical Record of the Life of Zhao Shiyan," in Committee for Research Into Historical Accounts of Past Events of the Sichuan Provincial Committee of the Chinese People's Political Consultative Conference, ed., *Collection of Selected Historical Accounts of Past Events in Sichuan*, p. 193.

CHAPTER 18. THE STARTING POINT OF THE REVOLUTIONARY COURSE

1. Register of Foreigners, Chalette.
2. Shi Yisheng, "Recollections of the Glorious Achievements of the European Branch of the Communist Party of China," in Committee for Research Into the Historical Accounts of Past Events in Tianjin of the Tianjin Municipal Committee of the Chinese People's Political Consultative Conference, ed., *Selected Historical Accounts of Past Events in Tianjin*, vol. 15 (Tianjin: Tianjin People's Publishing House, May 1981), p. 114.
3. Liao Huanxing, "The European General Branch of the Communist Party of China," in Research Office of the Contemporary History of the Chinese Academy of Social Science and the Party History Research Office of the Chinese Revolutionary Museum, eds., *Around the First National Party Congress*, pt. 2 (Beijing: People's Publishing House, August 1980), p. 502.
4. Jiang Zemin, "Reminiscences of Studying in France and Belgium on a Work-Study Program," in Committee of Research Into the Historical Accounts of Past Events in Tianjin of the Tianjin Municipal Committee of the Chinese People's Political Consultative Conference, ed., *Selected Historical Accounts of Past Events in Tianjin*, vol. 15, p. 93.
5. Cai Chang, "On Studying in France on a Work-Study Program and the European Branch of the Socialist Youth League," in Research Office of the Contemporary History of the Chinese Academy of Social Science and the Party History Research Office of the Chinese Revolutionary Museum, eds., *Around the First National Party Congress*, pt. 2, p. 555.
6. Chen Congshan, "Lei Mingyuan Attempted to Sabotage the Movement to Study in Europe on a Work-Study Program," in Committee for Research Into the Historical Accounts of Past Events in Tianjin of the Tianjin Municipal Committee of the Chinese People's Political Consultative Conference, ed., *Collection of Selected Historical Accounts of Past Events in Tianjin*, vol. 15, p. 146.
7. Ibid.

CHAPTER 19. TEMPERING IN THE PARTY

1. "Fu Zhong's Talks About the Movement to Study in France on a Work-Study Program and the General Branch of the Socialist Youth League in Europe," in Research

Office of the Contemporary History of the Chinese Academy of Social Science and the Party History Research Office of the Chinese Revolutionary Museum, eds., *Around the First National Party Congress*, pt. 2, p. 559.

2. Huang Lizhou, "The Movement in Sichuan to Study in France on a Work-Study Program."
3. Circular No. 77 of the Chinese Communist Youth League in Europe, December 29, 1924.
4. Record of June 20, 1925, archive no. F7 13438, National Archives of France.
5. Shi Yisheng, "Recollections of the Glorious Achievements of the European Branch of the Communist Party of China."
6. Archive no. F7 12900, National Archives of France.
7. Circular of the Provisional Executive Committee of the Chinese Communist Youth League in Europe, July 1, 1925.
8. Record of July 2, 1925, archive no. F7 12900, National Archives of France.
9. Ibid.
10. Circular No. 113 of the Chinese Communist Youth League in the European District, August 17, 1925.
11. Circular No. 2 of the Propaganda Department of the Extraordinary Executive Committee of the Chinese Communist Youth League, July 22, 1925.
12. Record of September 9, 1925, archive no. F7 12900, National Archives of France.
13. Shi Yisheng, "Recollections of the Glorious Achievements of the European Branch of the Communist Party of China."
14. Record of October 25, 1925, archive no. F7 12900, National Archives of France.
15. Record of November 16, 1925, archive no. F7 12900, National Archives of France.
16. French Interior Minister to Foreign Minister, November 20, 1925, archive no. F7 12900, National Archives of France.
17. Files of the Renault Automobile Factory.
18. Ibid.
19. Record of January 7, 1926, archive no. F7 13438, National Archives of France.
20. Search Report, January 8, 1926, document no. 202, archive no. F7 13438, National Archives of France.

CHAPTER 20. ADIEU, FRANCE

1. Letter from the Executive Committee of the CPC European Branch to the Local Committee of the CPC in Moscow, May 29, 1925, Central Archives.
2. Yuan Qingyun to Fu Zhong and others, November 18, 1925.
3. Moscow to Fu Zhong and others, December 9, 1925, Central Archives. "C. P." stands for "Communist Party" in English.

CHAPTER 21. IN THE HOMETOWN OF THE OCTOBER REVOLUTION

1. In writing this chapter, reference was made to A. B. Panzov, *How the Soviet Union Trained Marxist Theoretical Cadres for the Chinese Revolution*.

CHAPTER 23. OUT OF BLOODBATH

1. Li Weihan, *Recollections and Research*, p. 158.
2. Mao Zedong, "On Coalition Government," in *Selected Works of Mao Zedong*, vol. 3.

CHAPTER 24. THE TWENTY-FOUR-YEAR-OLD SECRETARY-GENERAL OF THE CENTRAL COMMITTEE

1. Li Weihan, *Recollections and Research,* vol. 1, p. 243.

CHAPTER 28. GOING TO GUANGXI

1. "Recollections of Xu Fengxiang, Former Aide de Camp of the 7th Corps of the Red Army," in *Bulletin of the Study of the Communist Party History in Guangxi: Special Issue in Commemoration of the 60th Anniversary of the Bose and Longzhou Uprisings*, no. 6 (1989): 51 (restricted material).
2. Yuan Renyuan, "From Bose to Hunan and Jiangxi," in Party Historical Material Collection Committee of the CPC Guangxi Autonomous Regional Committee, ed., *The Zuojiang and Youjiang Revolutionary Base Areas*, pt. 3 (Beijing: CPC Historical Materials Publishing House, 1989), p. 621.
3. Zhang Yunyi, "The Bose Uprising and the Founding of the 7th Corps of the Red Army," in Party Historical Material Collection Committee of the CPC Guangxi Autonomous Regional Committee, ed., *The Zuojiang and Youjiang Revolutionary Base Areas*, pt. 2, p. 585.

CHAPTER 29. THE BOSE AND LONGZHOU UPRISINGS

1. Yuan Renyuan, Wei Guoqing, Chen Manyuan, Mo Wenhua, and Wu Xi, "In Commemoration of the Bose Uprising," in Party History Research Committee of the CPC Guangxi Zhuang Autonomous Regional Committee, ed., *Memoirs of the Revolutionary Struggle in Guangxi* (Guangxi: Guangxi People's Publishing House, April 1984), p. 1.
2. Yuan Renyuan, "From Bose to Hunan and Jiangxi," pt. 2, p. 621.
3. "Recollections of Xu Fengxiang, Former Aide de Camp of the 7th Corps of the Red Army."
4. Zhang Yunyi, "The Bose Uprising and the Founding of the 7th Corps of the Red Army."
5. Yuan Renyuan, Wei Guoqing, Chen Manyuan, Mo Wenhua, and Wu Xi, "In Commemoration of the Bose Uprising."
6. He Jiarong, "Recollections of the 8th Corps of the Chinese Workers' and Peasants' Red Army," in Committee for Research Into Historical Accounts of Past Events of the Guangxi Autonomous Regional Committee of the Chinese People's Political Consultative Conference, ed., *Historical Accounts of Past Events in Guangxi*, vol. 10 (Nanning, June 1982), p. 1 (restricted material).
7. Huang Yiping, "Some Policies in the Early Days of the Founding of the 7th Corps of the Red Army," in Party Historical Material Collection Committee of the CPC Guangxi Autonomous Regional Committee, ed., *The Zuojiang and Youjiang Revolutionary Base Areas*, pt. 2, p. 687.
8. He Jiarong, "Recollections of the 8th Corps of the Red Army," Party Historical Material Collection Committee of the CPC Guangxi Autonomous Regional Committee, ed., *The Zuojiang and Youjiang Revolutionary Base Areas*, pt. 2, p. 867.
9. Li Lisan, "Red Longzhou, March 20, 1930," in Party Historical Material Collection Committee of the CPC Guangxi Autonomous Regional Committee, ed., *The Zuojiang and Youjiang Revolutionary Base Areas*, pt. 2, p. 251.

CHAPTER 30. STATE AFFAIRS, FAMILY AFFAIRS, AND PERSONAL GRIEF

1. "A Discussion on the Work Arrangements of the Red Army in Guangxi, January 1930," in Party Historical Material Collection Committee of the CPC Guangxi Autonomous Regional Committee, ed., *The Zuojiang and Youjiang Revolutionary Base Areas*, pt. 1, p. 174.
2. "Directive of the Central Committee of the Communist Party of China to the Front Committee of the 7th Corps in Care of the Guangdong Provincial Committee, March 2, 1930," in Party Historical Material Collection Committee of the CPC Guangxi Autonomous Regional Committee, ed., *The Zuojiang and Youjiang Revolutionary Base Areas*, pt. 1, p. 218.

CHAPTER 31. THE RISE AND FALL OF THE 8TH CORPS OF THE RED ARMY

1. "Report of the 7th Corps of the Red Army (to the Congress of the Soviets), May 1930," in Party Historical Material Collection Committee of the CPC Guangxi Autonomous Regional Committee, ed., *The Zuojiang and Youjiang Revolutionary Base Areas*, pt. 1, p. 290.
2. Zhou Zhi, "Joining Forces with the 7th Corps of the Red Army," in Party History Research Committee of the CPC Guangxi Zhuang Autonomous Regional Committee, ed., *Memoirs of the Revolutionary Struggles in Guangxi*, p. 53.
3. He Jiarong, "Recollections of the 8th Corps of the Chinese Workers' and Peasants' Red Army."

CHAPTER 32. THE RISE OF THE 7TH CORPS OF THE RED ARMY AND THE YOUJIANG RED REVOLUTIONARY BASE AREA

1. Huang Meilun, "Political Commissar Deng's Arrival in Wuzhuan," in Party Historical Material Collection Committee of the CPC Guangxi Autonomous Regional Committee, ed., *The Zuojiang and Youjiang Revolutionary Base Areas*, pt. 2, p. 682.
2. Jiang Maosheng, "Recollections of the Party Organizations and Soldiers' Committees in the 7th Corps of the Red Army," in Party History Research Committee of the CPC Guangxi Zhuang Autonomous Regional Committee, ed., *Recollections on the Revolutionary Struggles in Guangxi*, pt. 2 (Guangxi: Guangxi People's Publishing House, April 1984), p. 34.
3. Ya Meiyuan, "Escorting Political Commissar Deng," in Party Historical Material Collection Committee of the CPC Guangxi Autonomous Regional Committee, ed., *The Zuojiang and Youjiang Revolutionary Base Areas*, pt. 2, p. 795.
4. "A Letter of Instruction from the CPC Central Committee to the Southern Office of Its Military Commission and to the Front Committee of the 7th Corps, June 16, 1930," in Party Historical Material Collection Committee of the CPC Guangxi Autonomous Regional Committee, ed., *The Zuojiang and Youjiang Revolutionary Base Areas*, pt. 1, p. 315.
5. Yuan Renyuan, Wei Guoqing, Chen Manyuan, Mo Wenhua, and Wu Xi, "In Commemoration of the Bose Uprising," p. 1.
6. Mo Wenhua, "The Storms of Bose," pp. 156–57.
7. "A Letter of Instruction from the Central Committee of the Communist Party of China to the Southern Office of Its Military Commission and to the Front Committee of the 7th Corps, June 16, 1930."

CHAPTER 33. THE ORIGIN OF LI LISAN'S "LEFT"-ADVENTURISM

1. Li Weihan, *Recollections and Research,* pt. 1, p. 237.

CHAPTER 34. THE EXPERIENCE OF THE 7TH CORPS

1. Wu Xi, *A Veteran's Recollections of the Flames of War* (Nanning: Guangxi People's Publishing House, October 1988), p. 73.

CHAPTER 35. ETERNAL GLORY TO THE 7TH CORPS

1. Jiang Maosheng, "From the Separation at the Lechang River to the Rendezvous in Yongxin," in Party History Research Committee of the CPC Guangxi Zhuang Autonomous Regional Committee, ed., *Memoirs of the Revolutionary Struggle in Guangxi,* p. 123.
2. Yuan Renyuan, Wei Guoqing, Chen Manyuan, Mo Wenhua, and Wu Xi, "In Commemoration of the Bose Uprising."
3. Qin Guohan, Huang Chao, and Tan Qingrong, "Revolutionary Militant Friendship," in Party History Research Committee of the CPC Guangxi Zhuang Autonomous Regional Committee, ed., *Memoirs of the Revolutionary Struggle in Guangxi,* p. 162.

CHAPTER 36. CHANGES IN THE EARLY 1930s

1. Li Weihan, *Recollections and Research,* pt. 1, p. 323.
2. Chen Haoren, "General Report on the Work of the 7th Corps, March 9, 1931," and Yan Heng, "Report on the 7th Corps, April 4, 1931," both in Party Historical Material Collection Committee of the CPC Guangxi Autonomous Regional Committee, ed., *The Zuojiang and Youjiang Revolutionary Base Areas,* pt. 1, pp. 358, 382.
3. "Letter from the CPC Central Committee to the Front Committee of the 7th Corps, May 14, 1931," in Party Historical Material Collection Committee of the CPC Guangxi Autonomous Regional Committee, ed., *The Zuojiang and Youjiang Revolutionary Base Areas,* pt. 1, p. 412.

CHAPTER 38. THE FIRST SECRETARY OF THE PARTY COMMITTEE OF THE CENTRAL COUNTY OF HUICHANG

1. Zhong Yaqing, "The Period When I Worked Under Comrade Deng Xiaoping in Huichang," in Yang Guoyu et al., eds., *During the Twenty-eight Years: From Division Political Commissar to General Secretary* (Shanghai: Shanghai Literature and Art Publishing House, July 1989), p. 256.
2. Ibid.

CHAPTER 39. THE "DENG, MAO, XIE, AND GU INCIDENT"

1. In writing this chapter, the Office of Material Collection Group of the CPC History of Huichang County, Jiangxi Province, "The Historical Records of the Huichang Key County Committee of the CPC," was consulted.

CHAPTER 40. THE EDITOR-IN-CHIEF OF *RED STAR*

1. Uli Franz, *Deng Xiaoping: Chinese-type Political Legend* (Hong Kong: Zhong Yuan Publishing House, January 1990).
2. Li Weihan, *Recollections and Research,* pt. 1, pp. 341–42.

CHAPTER 41. THE FAILURE OF THE FIFTH COUNTERCAMPAIGN AGAINST ENCIRCLEMENT AND SUPPRESSION

1. "Presidential Instructions on Foreign Policy," in Party Historical Material Editing Committee of the Chinese Kuomintang Central Committee, ed., *Revolutionary Documents*, vol. 72 (Taipei: Taipei Central Cultural Relic Supply Society, March 1970), p. 136.
2. *Collection of Materials on the North China Incident* (Henan: Henan People's Publishing House, 1983), pp. 83–86.
3. Zhang Xianwen, ed., *Outline History of the Republic of China* (Henan: Henan People's Publishing House, 1985), pp. 402–8.
4. Li Jian, ed., *History of the Communist Party of China*, vol. 1 (Beijing: People's Publishing House, June 1990), p. 461.
5. Liu Bocheng et al., *Recollections of the Long March* (Beijing: People's Publishing House, December 1985), pp. 3–4.
6. Li Weihan, *Recollections and Research*, pt. 1, pp. 343–44.
7. Nie Rongzhen, "Breaking Through the Enemy's First, Second, and Third Blockade Lines," in Liu Bocheng et al., *Recollections of the Long March*, p. 72.
8. The commander of the 1st Column was Ye Jianying.
9. Li Weihan, *Recollections and Research*, pt. 1, pp. 344–46. Luo Fu was Zhang Wentian, Deng Yingchao was the wife of Zhou Enlai, Kang Keqing was the wife of Zhu De. Cai Chang was the wife of Li Fuchun. Ah Jin was Jin Weiying, the wife of Li Weihan.

CHAPTER 42. THE PRELUDE TO THE LONG MARCH AND THE ZUNYI MEETING

1. Nie Rongzhen, "Breaking Through the Enemy's First, Second, and Third Blockade Lines," pp. 72–73.
2. Ibid., p. 72.
3. Li Weihan, *Recollections and Research*, pt. 1, p. 347.
4. Nie Rongzhen, "Breaking Through the Enemy's First, Second, and Third Blockade Lines," p. 75.
5. Liu Bocheng, *Recollections of the Long March*, p. 4.
6. Li Weihan, *Recollections and Research*, pt. 1, p. 348.
7. Liu Bocheng et al., *Recollections of the Long March*, p. 1.
8. Zhang Tinggui and Yuan Wei, *Brief History of the Chinese Workers' and Peasants' Red Army* (Beijing: CPC Historical Materials Publishing House, March 1987), p. 118.
9. Li Jian, ed., *History of the Communist Party of China*, vol. 1, p. 384.
10. Ibid., p. 388.

CHAPTER 43. THE RED ARMY BRAVES ALL DIFFICULTIES ON THE LONG MARCH

1. The eighth issue of *Red Star* is not available, so I cannot be sure whether there were seven or eight issues between the publication of the first issue on October 20 after the start of the Long March and the eve of the Zunyi Meeting. The ninth issue was published on February 10, after the Zunyi Meeting, and surely was not edited by Father.

2. Liu Bocheng et al., *Recollections of the Long March*, p. 1.
3. Li Jukui, *Memoirs of Li Jukui* (Beijing: Liberation Army Publishing House, September 1986), pp. 143–44.
4. Series Biographies of Contemporary Chinese Personages Editing Committee, ed., *Biography of Luo Ronghuan* (Beijing: Contemporary China Publishing House, December 1991), pp. 130–32.
5. Ibid.
6. Liu Bocheng et al., *Recollections of the Long March*, p. 1.
7. Zhang Tinggui and Yuan Wei, *Brief History of the Chinese Workers' and Peasants' Red Army*, p. 129.
8. Mao Zedong, "On Tactics Against Japanese Imperialism, December 27, 1935," in *Selected Works of Mao Zedong*, vol. 1.

CHAPTER 45. THE XI'AN INCIDENT

1. Zhang Xianwen, ed., *Outline History of the Republic of China*, p. 440.
2. Chiang Kai-shek, *Soviet Russia in China*, p. 74.
3. Li Jian, ed., *The History of the Communist Party of China*, pp. 449–50.
4. Zhang Tinggui and Yuan Wei, *Brief History of the Chinese Workers' and Peasants' Red Army*, p. 152.

CHAPTER 46. MARCHING TO BATTLE AGAINST THE JAPANESE

1. Soong Ching Ling, "Thoughts on the Movement of Unification of the KMT and the CPC," *Resistance*, no. 12 (September 26, 1937).
2. Li Jian, ed., *The History of the Communist Party of China*, vol. 1, p. 479.
3. In writing this chapter, reference was made to Fu Zhong, *The Initial March to the Anti-Japanese Battlefield*; and Yang Guoyu et al., eds., *During the Twenty-eight Years: From Division Political Commissar to General Secretary*, p. 1.

CHAPTER 47. THE POLITICAL COMMISSAR OF THE 129TH DIVISION

1. Yang Guoyu, *Thirteen Years Under the Command of Liu and Deng* (Chongqing: Chongqing University Publishing House, May 1991), pp. 36–37.
2. Deng Xiaoping, "Mourning for Liu Bocheng," in Shanghai Literature and Art Publishing House, ed., *Memoirs of Liu Bocheng*, vol. 3 (Shanghai: Shanghai Literature and Art Publishing House, July 1987), p. 5.
3. Yang Guoyu, *Thirteen Years Under the Command of Liu and Deng*, p. 39.
4. Zhang Yixiang, "Several Contacts During My Ten Years in the Taihang Mountains," in Yang Guoyu et al., eds., *During the Twenty-eight Years: From Division Political Commissar to General Secretary*, pp. 14–17.
5. Chen Zaidao, *Memoirs of Chen Zaidao* (Beijing: Liberation Army Publishing House, July 1991), pp. 362, 376.
6. Yang Guoyu, *Thirteen Years Under the Command of Liu and Deng*, p. 93.

CHAPTER 48. MY GRANDFATHER PU ZAITING

1. In 1949, there were 3,489 banks in China, of which 2,448 were controlled by bureaucrat capital, which also controlled 40 percent of the spindles in the textile

industry, 60 percent of the looms, 90 percent of iron and steel output, 33 percent of coal output, 67 percent of electric power output, 45 percent of cement output, 90 percent of sugar output, and all petroleum and nonferrous metal output. At the same time, it controlled over 43 percent of shipping tonnage and a dozen large-scale monopoly trade companies. See Zhao Dexing, ed., *Economic History of the People's Republic of China (1949–66)* (Henan: Henan People's Publishing House, March 1989), pp. 18–20.

CHAPTER 50. IN THE TAIHANG MOUNTAINS

1. The "Munich conspiracy" refers to the Agreement of Munich signed by Britain, France, and other nations with fascist Germany and Italy on September 30, selling out the Czech nation in an attempt to seek compromise with the fascists and divert the fascist aggression to the Soviet Union.
2. Yang Guoyu, *Thirteen Years Under the Command of Liu and Deng*, pp. 139–41.
3. Mao Zedong, "On the Estimation of the Situation and Counter-Measures Against the Possible Attack by the Kuomintang," October 25, 1940.
4. Yang Guoyu, *Thirteen Years Under the Command of Liu and Deng*, pp. 181–82.
5. Ibid., p. 153.

CHAPTER 51. DIFFICULT YEARS

1. "Directive of the CPC Central Committee on the Work of the Anti-Japanese Base Areas Behind Enemy Lines After the Outbreak of the Pacific War, December 17, 1941," in Military History Research Department of the Academy of Military Science, ed., *The Military History of the Chinese People's Liberation Army*, vol. 2, *The Period of the War of Resistance Against Japan* (Beijing: Military Science Publishing House, July 1987), pp. 309–10.
2. Yang Guoyu, *Thirteen Years Under the Command of Liu and Deng*, pp. 195–96.
3. Ibid.
4. Mao Zedong, "A Most Important Policy," in *Selected Works of Mao Zedong*, vol. 3.
5. *The Military History of the Second Field Army of the Chinese People's Liberation Army*, vol. 1, *The Period of the War of Resistance Against Japan* (Beijing: Liberation Army Publishing House, February 1991), pp. 221–22.
6. Yang Guoyu, *Thirteen Years Under the Command of Liu and Deng*, p. 197. It was said in this paragraph that four rules were set, but only three were cited in the original text.
7. Mao Zedong, "A Most Important Policy."
8. Chen Dong, "Comrade Deng Xiaoping and the Building Up of the Shanxi-Hebei-Shandong-Henan Border Area," in Yang Guoyu et al., eds., *During the Twenty-eight Years: From Division Political Commissar to General Secretary*, p. 19.
9. Zhang Yixiang, "Several Contacts During My Ten Years in the Taihang Mountains," p. 14.
10. *Draft History of the Taihang Revolutionary Base Area (1937–49)*, p. 118 (restricted material).
11. *The Military History of the Second Field Army of the Chinese People's Liberation Army*, vol. 1, *The Period of the War of Resistance Against Japan*, p. 228.

12. Yang Guoyu, *Thirteen Years Under the Command of Liu and Deng,* p. 207.
13. Yang Guoyu, "The Majestic Mountain," in Yang Guoyu et al., eds., *During the Twenty-eight Years: From Division Political Commissar to General Secretary,* p. 44.
14. *Memoirs of Liu Bocheng,* vol. 3, p. 107.

CHAPTER 52. TOWARD REHABILITATION AND DEVELOPMENT

1. *The Military History of the Second Field Army of the Chinese People's Liberation Army,* vol. 1, *The Period of the War of Resistance Against Japan,* pp. 257–58.
2. Chen Dong, "Comrade Deng Xiaoping and the Building Up of the Shanxi-Hebei-Shandong-Henan Border Area," pp. 28–29.
3. Chen Heqiao, "Comrade Deng Xiaoping in the Northern Bureau," in Yang Guoyu et al., eds., *During the Twenty-eight Years: From Division Political Commissar to General Secretary,* p. 103.
4. Chen Dong, "Comrade Deng Xiaoping and the Building Up of the Shanxi-Hebei-Shandong-Henan Border Area," p. 29.
5. Chen Heqiao, "Comrade Deng Xiaoping in the Northern Bureau," p. 76.
6. CPC Central Document Editing Committee, ed., *Selected Works of Deng Xiaoping (1938–65)* (Beijing: People's Publishing House, May 1989), pp. 81–82.
7. Ibid., p. 78.

CHAPTER 53. VICTORY IN THE SACRED ANTI-JAPANESE WAR

1. Mao Zedong, "Our Study and the Current Situation," in *Selected Works of Mao Zedong,* vol. 3.
2. *The Military History of the Second Field Army of the Chinese People's Liberation Army,* vol. 1, *The Period of the War of Resistance Against Japan,* p. 506.
3. Li Jian, ed., *The History of the Communist Party of China,* vol. 1, p. 659.

CHAPTER 54. GIVING TIT FOR TAT AND FIGHTING FOR EVERY INCH OF LAND

1. Zhang Xianwen, ed., *Outline History of the Republic of China,* p. 631.
2. Mao Zedong, "On the Chongqing Negotiations," in *Selected Works of Mao Zedong,* vol. 4.
3. Lu Jiasheng and Du Zhongde, "Giving Tit for Tat: A Selection of Materials on the Shangdang Campaign," p. 78 (restricted material).
4. Mao Zedong, "On the Chongqing Negotiations."
5. Ibid.
6. *The Military History of the Second Field Army of the Chinese People's Liberation Army,* vol. 2, *The Period of the War of Liberation,* p. 20.
7. These remarks are quoted from the notes taken by the staff present at that time.
8. Party History Office of Shexian County Party Committee, "The 129th Division in Shexian County: A Selection of Materials," p. 245 (restricted material).

CHAPTER 55. ON THE EVE OF CIVIL WAR

1. Li Jian, ed., *The History of the Communist Party of China,* vol. 1, pp. 687–88.
2. "Appeal for the People of Jiangsu and Zhejiang Provinces," editorial, *Da Gong Bao,* October 24, 1945.

3. U.S. Secretary of State Dean Acheson to U.S. President Harry S. Truman, July 30, 1949.
4. Liang Shuming, "Peace Negotiations Between Kuomintang and the Communist Party I Participated In," in *Draft Series of Historical Materials of the Republic of China*, supp. 6.
5. Li Jian, ed., *The History of the Communist Party of China*, vol. 1, pp. 700–701.
6. Chen Feiqin, "He Is Such a Person," in Yang Guoyu et al., eds., *During the Twenty-eight Years: From Division Political Commissar to General Secretary*, p. 135.
7. Ibid., pp. 135–37.

CHAPTER 56. THE OUTBREAK OF THE ALL-OUT CIVIL WAR

1. CPC Central Committee, "Smash Chiang Kai-shek's Offensive by a War of Self-defense," July 20, 1946, Central Archives.
2. "Talk with the American Correspondent Anna Louise Strong, August 6, 1946," in *Selected Works of Mao Zedong*, vol. 4.
3. Zhang Yunxuan, "Several Stories About How Political Commissar Deng Educates the Army Units," in Yang Guoyu et al., eds., *During the Twenty-eight Years: From Division Political Commissar to General Secretary*, vol. 3, p. 291.
4. Duan Junyi and Qiao Mingfu, "A Leader Who Seeks the Truth from Facts and Upholds Principles," in Yang Guoyu et al., eds., *During the Twenty-eight Years: From Division Political Commissar to General Secretary*, pp. 61, 81.
5. Yang Guoyu, *Thirteen Years Under the Command of Liu and Deng*, p. 216.
6. Wang Wenzhen, "The Person Who Dares to Shoulder Heavy Responsibilities and Explore New Ways," in Yang Guoyu et al., eds., *During the Twenty-eight Years: From Division Political Commissar to General Secretary*, p. 212.
7. *The Military History of the Second Field Army of the Chinese People's Liberation Army*, vol. 3, *The Period of the War of Liberation*, p. 58.
8. Ibid., p. 63.
9. Yang Guoyu, *Thirteen Years Under the Command of Liu and Deng*, p. 281.
10. *The Military History of the Second Field Army of the Chinese People's Liberation Army*, vol. 2, *The Period of the War of Liberation*, p. 90.
11. Ibid., p. 92.
12. Chen Feiqin, "He Is Such a Person," p. 140.
13. Mao Zedong, "Greet the New High Tide of the Chinese Revolution," in *Selected Works of Mao Zedong*, vol. 4.

CHAPTER 57. BREAKING THROUGH THE DEFENSE LINE ALONG THE YELLOW RIVER

1. Li Jian, ed., *The History of the Communist Party of China*, vol. 1, p. 730.
2. Military History Research Department of the Academy of Military Science, ed., *The Military History of the Chinese People's Liberation Army*, vol. 3, *The Period of the War of Liberation*, p. 101.
3. Du Yide, "Before and After Crossing the Yellow River for the Counteroffensive," in Yang Guoyu et al., eds., *During the Twenty-eight Years: From Division Political Commissar to General Secretary*, p. 151.

CHAPTER 58. ADVANCING 500 KILOMETERS TO THE DABIE MOUNTAINS

1. Miao Bingshu, "The Breakthrough of the Central Front by Massive Liu-Deng Troops," in Wang Chuanzhong et al., eds., *The Crossing of the Yellow River by Massive Liu-Deng Troops: Selected Materials* (Jinan: Shangdong University Publishing House), p. 187.
2. Ibid.
3. Deng Xiaoping, "The Situation After the Victorious March into the Central Plains and Future Policies and Tactics, April 25, 1948," in CPC Central Document Editing Committee, ed., *Selected Works of Deng Xiaoping (1938–65)*.
4. The following books were consulted in writing this chapter: *The Military History of the Second Field Army of the Chinese People's Liberation Army*, vol. 2, *The Period of the War of Liberation*; Yang Guoyu et al., eds., *During the Twenty-eight Years: From Division Political Commissar to General Secretary*, and its two sequels; Wang Chuanzhong et al., eds., *The Crossing of the Yellow River by the Liu-Deng Massive Troops: Selected Materials*; Luo Ronxun and Zheng Mingxin, eds., *The March Towards the Dabie Mountains* (Henan: Henan People's Publishing House, June 1987); and Li Jian, ed., *The History of the Communist Party of China*.

CHAPTER 59. CHASING DEER ON THE CENTRAL PLAINS

1. *The History of the Han Dynasty: The Biography of Kuai Tong.*
2. Yan Daiju, "Brilliant Exposition and Great Encouragement," in Luo Ronxun and Zheng Mingxin, eds., *The March Toward the Dabie Mountains*, p. 125.
3. Miao Bingshu, *Liu and Deng at the Front of the Central Plains* (Beijing: China Youth Publishing House, December 1987), p. 136.
4. Yang Guoyu, "The Majestic Mountain," p. 44.
5. Mao Zedong, "An Introductory Note to the Approved Transmission of Deng Xiaoping's 'Additional Opinions on the Land Reform Policy in New Liberated Areas,'" Central Archives.
6. Miao Bingshu, *Liu and Deng at the Front of the Central Plains*, pp. 154–55; Zhu Minggan and Yang Liangxin, "A Great Heroic Undertaking," in Luo Rongxun and Zheng Mingxin, eds., *The March Toward the Dabie Mountains*, p. 327.
7. Wei Jinguo, "A Record of the Forward Command Post," in Yang Guoyu et al., eds., *During the Twenty-eight Years: From Division Political Commissar to General Secretary*, p. 177.

CHAPTER 60. BEFORE THE DECISIVE BATTLES

1. Zhang Shenghua, "The Headquarters Led by Liu Bocheng, Deng Xiaoping, and Li Da," in Yang Guoyu et al., eds., *During the Twenty-eight Years: From Division Political Commissar to General Secretary*, p. 131.
2. Ibid.
3. Mao Zedong, "The Present Situation and Our Tasks," in *Selected Works of Mao Zedong*, vol. 4.
4. "Defense Studies," in *Collected Works of President Chiang*, vol. 2 (Taiwan: Taiwan Defense Institute, August 1961), p. 163.
5. Report submitted by Leighton Stuart to George Marshall, August 10, 1948.
6. Xibaipo now belongs to Pingshan County, Hebei Province.

CHAPTER 61. THE GREAT DECISIVE BATTLES

1. Mao Zedong, "The Momentous Change in China's Military Situation," Central Archives.
2. Military History Research Department of the Academy of Military Science, ed., *The Military History of the Chinese People's Liberation Army*, vol. 3, *The Period of the War of Liberation*, p. 272.
3. He Zhengwen, "Under the Leadership of the General Front Committee," in Yang Guoyu et al., eds., *During the Twenty-eight Years: From Division Political Commissar to General Secretary*, p. 260.
4. Zhang Shenghua, "Political Commissar Deng During the Huai-Hai Campaign," in Yang Guoyu et al., eds., *During the Twenty-eight Years: From Division Political Commissar to General Secretary*, p. 155.
5. Chen Feiqin, "November 27, 1948," in Yang Guoyu et al., eds., *During the Twenty-eight Years: From Division Political Commissar to General Secretary*, p. 271.
6. Ibid.
7. Wu Kebin, "Secretary of the General Front Committee in the North and South of the Yangtze River," in Yang Guoyu et al., eds., *During the Twenty-eight Years: From Division Political Commissar to General Secretary*, p. 298.
8. Hu Qicai, "The Origin of the Miracle," in Yang Guoyu et al., eds., *During the Twenty-eight Years: From Division Political Commissar to General Secretary*, p. 179.
9. Hu Qicai, "The Five-Member General Front Committee with Deng Xiaoping as Secretary," in Yang Guoyu et al., eds., *During the Twenty-eight Years: From Division Political Commissar to General Secretary*, p. 242.

CHAPTER 62. FIGHTING IN THE AREAS SOUTH OF THE YANGTZE RIVER

1. For the above contents of this chapter, reference was made to Wu Kebin, "Secretary of the General Front Committee in the North and South of the Yangtze River," p. 298.

CHAPTER 63. MARCHING TOWARD SOUTHWESTERN CHINA

1. *The Military History of the Second Field Army of the Chinese People's Liberation Army*, vol. 2, *The Period of the War of Liberation*, p. 424.

INDEX